SCHAUM'S OUTLINE OF

THEORY AND PROBLEMS

OF

CONTINUUM MECHANICS

•

BY

GEORGE E. MASE, Ph.D.

Professor of Mechanics
Michigan State University

•

SCHAUM'S OUTLINE SERIES
McGraw-Hill
New York San Francisco Washington, D.C. Auckland Bogotá
Caracas Lisbon London Madrid Mexico City Milan
Montreal New Delhi San Juan Singapore
Sydney Tokyo Toronto

ISBN 07-040663-4

19 20 21 22 23 CUS CUS 06

McGraw-Hill

A Division of The ***McGraw-Hill*** *Companies*

Preface

Because of its emphasis on basic concepts and fundamental principles, Continuum Mechanics has an important role in modern engineering and technology. Several undergraduate courses which utilize the continuum concept and its dependent theories in the training of engineers and scientists are well established in today's curricula and their number continues to grow. Graduate programs in Mechanics and associated areas have long recognized the value of a substantial exposure to the subject. This book has been written in an attempt to assist both undergraduate and first year graduate students in understanding the fundamental principles of continuum theory. By including a number of solved problems in each chapter of the book, it is further hoped that the student will be able to develop his skill in solving problems in both continuum theory and its related fields of application.

In the arrangement and development of the subject matter a sufficient degree of continuity is provided so that the book may be suitable as a text for an introductory course in Continuum Mechanics. Otherwise, the book should prove especially useful as a supplementary reference for a number of courses for which continuum methods provide the basic structure. Thus courses in the areas of Strength of Materials, Fluid Mechanics, Elasticity, Plasticity and Viscoelasticity relate closely to the substance of the book and may very well draw upon its contents.

Throughout most of the book the important equations and fundamental relationships are presented in both the indicial or "tensor" notation and the classical symbolic or "vector" notation. This affords the student the opportunity to compare equivalent expressions and to gain some familiarity with each notation. Only Cartesian tensors are employed in the text because it is intended as an introductory volume and since the essence of much of the theory can be achieved in this context.

The work is essentially divided into two parts. The first five chapters deal with the basic continuum theory while the final four chapters cover certain portions of specific areas of application. Following an initial chapter on the mathematics relevant to the study, the theory portion contains additional chapters on the Analysis of Stress, Deformation and Strain, Motion and Flow, and Fundamental Continuum Laws. Applications are treated in the final four chapters on Elasticity, Fluids, Plasticity and Viscoelasticity. At the end of each chapter a collection of solved problems together with several exercises for the student serve to illustrate and reinforce the ideas presented in the text.

The author acknowledges his indebtedness to many persons and wishes to express his gratitude to all for their help. Special thanks are due the following: to my colleagues, Professors W. A. Bradley, L. E. Malvern, D. H. Y. Yen, J. F. Foss and G. LaPalm each of whom read various chapters of the text and made valuable suggestions for improvement; to Professor D. J. Montgomery for his support and assistance in a great many ways; to Dr. Richard Hartung of the Lockheed Research Laboratory, Palo Alto, California, who read the preliminary version of the manuscript and gave numerous helpful suggestions; to Professor M. C. Stippes, University of Illinois, for his invaluable comments and suggestions; to Mrs. Thelma Liszewski for the care and patience she displayed in typing the manuscript; to Mr. Daniel Schaum and Mr. Nicola Monti for their continuing interest and guidance throughout the work. The author also wishes to express thanks to his wife and children for their encouragement during the writing of the book.

Michigan State University
June 1970

GEORGE E. MASE

CONTENTS

Page

Chapter 1 MATHEMATICAL FOUNDATIONS

1.1 Tensors and Continuum Mechanics 1
1.2 General Tensors. Cartesian Tensors. Tensor Rank 1
1.3 Vectors and Scalars 2
1.4 Vector Addition. Multiplication of a Vector by a Scalar 2
1.5 Dot and Cross Products of Vectors 3
1.6 Dyads and Dyadics 4
1.7 Coordinate Systems. Base Vectors. Unit Vector Triads 6
1.8 Linear Vector Functions. Dyadics as Linear Vector Operators 8
1.9 Indicial Notation. Range and Summation Conventions 8
1.10 Summation Convention Used with Symbolic Convention 10
1.11 Coordinate Transformation. General Tensors 11
1.12 The Metric Tensor. Cartesian Tensors 12
1.13 Transformation Laws for Cartesian Tensors. The Kronecker Delta. Orthogonality Conditions 13
1.14 Addition of Cartesian Tensors. Multiplication by a Scalar 15
1.15 Tensor Multiplication 15
1.16 Vector Cross Product. Permutation Symbol. Dual Vectors 16
1.17 Matrices. Matrix Representation of Cartesian Tensors 17
1.18 Symmetry of Dyadics, Matrices and Tensors 19
1.19 Principal Values and Principal Directions of Symmetric Second-Order Tensors 20
1.20 Powers of Second-Order Tensors. Hamilton-Cayley Equation 21
1.21 Tensor Fields. Derivatives of Tensors 22
1.22 Line Integrals. Stokes' Theorem 23
1.23 The Divergence Theorem of Gauss 23

Chapter 2 ANALYSIS OF STRESS

2.1 The Continuum Concept 44
2.2 Homogeneity. Isotropy. Mass-Density 44
2.3 Body Forces. Surface Forces 45
2.4 Cauchy's Stress Principle. The Stress Vector 45
2.5 State of Stress at a Point. Stress Tensor 46
2.6 The Stress Tensor—Stress Vector Relationship 47
2.7 Force and Moment. Equilibrium. Stress Tensor Symmetry 48
2.8 Stress Transformation Laws 49
2.9 Stress Quadric of Cauchy 50
2.10 Principal Stresses. Stress Invariants. Stress Ellipsoid 51
2.11 Maximum and Minimum Shear Stress Values 52
2.12 Mohr's Circles for Stress 54
2.13 Plane Stress 56
2.14 Deviator and Spherical Stress Tensors 57

CONTENTS

			Page
Chapter	3	**DEFORMATION AND STRAIN**	
	3.1	Particles and Points	77
	3.2	Continuum Configuration. Deformation and Flow Concepts	77
	3.3	Position Vector. Displacement Vector	77
	3.4	Lagrangian and Eulerian Descriptions	79
	3.5	Deformation Gradients. Displacement Gradients	80
	3.6	Deformation Tensors. Finite Strain Tensors	81
	3.7	Small Deformation Theory. Infinitesimal Strain Tensors	82
	3.8	Relative Displacements. Linear Rotation Tensor. Rotation Vector	83
	3.9	Interpretation of the Linear Strain Tensors	84
	3.10	Stretch Ratio. Finite Strain Interpretation	86
	3.11	Stretch Tensors. Rotation Tensor	87
	3.12	Transformation Properties of Strain Tensors	88
	3.13	Principal Strains. Strain Invariants. Cubical Dilatation	89
	3.14	Spherical and Deviator Strain Tensors	91
	3.15	Plane Strain. Mohr's Circles for Strain	91
	3.16	Compatibility Equations for Linear Strains	92
Chapter	4	**MOTION AND FLOW**	
	4.1	Motion. Flow. Material Derivative	110
	4.2	Velocity. Acceleration. Instantaneous Velocity Field	111
	4.3	Path Lines. Stream Lines. Steady Motion	112
	4.4	Rate of Deformation. Vorticity. Natural Strain	112
	4.5	Physical Interpretation of Rate of Deformation and Vorticity Tensors	113
	4.6	Material Derivatives of Volume, Area and Line Elements	114
	4.7	Material Derivatives of Volume, Surface and Line Integrals	116
Chapter	5	**FUNDAMENTAL LAWS OF CONTINUUM MECHANICS**	
	5.1	Conservation of Mass. Continuity Equation	126
	5.2	Linear Momentum Principle. Equations of Motion. Equilibrium Equations	127
	5.3	Moment of Momentum (Angular Momentum) Principle	128
	5.4	Conservation of Energy. First Law of Thermodynamics. Energy Equation	128
	5.5	Equations of State. Entropy. Second Law of Thermodynamics	130
	5.6	The Clausius-Duhem Inequality. Dissipation Function	131
	5.7	Constitutive Equations. Thermomechanical and Mechanical Continua	132
Chapter	6	**LINEAR ELASTICITY**	
	6.1	Generalized Hooke's Law. Strain Energy Function	140
	6.2	Isotropy. Anisotropy. Elastic Symmetry	141
	6.3	Isotropic Media. Elastic Constants	142
	6.4	Elastostatic Problems. Elastodynamic Problems	143
	6.5	Theorem of Superposition. Uniqueness of Solutions. St. Venant's Principle	145
	6.6	Two-Dimensional Elasticity. Plane Stress and Plane Strain	145

CONTENTS

			Page
	6.7	Airy's Stress Function	147
	6.8	Two-Dimensional Elastostatic Problems in Polar Coordinates	148
	6.9	Hyperelasticity. Hypoelasticity	149
	6.10	Linear Thermoelasticity	149
Chapter	7	**FLUIDS**	
	7.1	Fluid Pressure. Viscous Stress Tensor. Barotropic Flow	160
	7.2	Constitutive Equations. Stokesian Fluids. Newtonian Fluids	161
	7.3	Basic Equations for Newtonian Fluids. Navier-Stokes-Duhem Equations	162
	7.4	Steady Flow. Hydrostatics. Irrotational Flow	163
	7.5	Perfect Fluids. Bernoulli Equation. Circulation	164
	7.6	Potential Flow. Plane Potential Flow	165
Chapter	8	**PLASTICITY**	
	8.1	Basic Concepts and Definitions	175
	8.2	Idealized Plastic Behavior	176
	8.3	Yield Conditions. Tresca and von Mises Criteria	177
	8.4	Stress Space. The II-Plane. Yield Surfaces	179
	8.5	Post-Yield Behavior. Isotropic and Kinematic Hardening	180
	8.6	Plastic Stress-Strain Equations. Plastic Potential Theory	181
	8.7	Equivalent Stress. Equivalent Plastic Strain Increment	182
	8.8	Plastic Work. Strain Hardening Hypotheses	182
	8.9	Total Deformation Theory	183
	8.10	Elastoplastic Problems	183
	8.11	Elementary Slip Line Theory for Plane Plastic Strain	184
Chapter	9	**VISCOELASTICITY**	
	9.1	Linear Viscoelastic Behavior	196
	9.2	Simple Viscoelastic Models	196
	9.3	Generalized Models. Linear Differential Operator Equation	198
	9.4	Creep and Relaxation	199
	9.5	Creep Function. Relaxation Function. Hereditary Integrals	200
	9.6	Complex Moduli and Compliances	202
	9.7	Three Dimensional Theory	203
	9.8	Viscoelastic Stress Analysis. Correspondence Principle	204
		INDEX	217

Chapter 1

Mathematical Foundations

1.1 TENSORS AND CONTINUUM MECHANICS

Continuum mechanics deals with physical quantities which are independent of any particular coordinate system that may be used to describe them. At the same time, these physical quantities are very often specified most conveniently by referring to an appropriate system of coordinates. Mathematically, such quantities are represented by *tensors*.

As a mathematical entity, a tensor has an existence independent of any coordinate system. Yet it may be specified in a particular coordinate system by a certain set of quantities, known as its *components*. Specifying the components of a tensor in one coordinate system determines the components in any other system. Indeed, the *law of transformation* of the components of a tensor is used here as a means for defining the tensor. Precise statements of the definitions of various kinds of tensors are given at the point of their introduction in the material that follows.

The physical laws of continuum mechanics are expressed by tensor equations. Because tensor transformations are linear and homogeneous, such tensor equations, if they are valid in one coordinate system, are valid in any other coordinate system. This *invariance* of tensor equations under a coordinate transformation is one of the principal reasons for the usefulness of tensor methods in continuum mechanics.

1.2 GENERAL TENSORS. CARTESIAN TENSORS. TENSOR RANK.

In dealing with general coordinate transformations between arbitrary curvilinear coordinate systems, the tensors defined are known as *general tensors*. When attention is restricted to transformations from one homogeneous coordinate system to another, the tensors involved are referred to as *Cartesian tensors*. Since much of the theory of continuum mechanics may be developed in terms of Cartesian tensors, the word "tensor" in this book means "Cartesian tensor" unless specifically stated otherwise.

Tensors may be classified by *rank*, or *order*, according to the particular form of the transformation law they obey. This same classification is also reflected in the number of components a given tensor possesses in an n-dimensional space. Thus in a three-dimensional Euclidean space such as ordinary physical space, the number of components of a tensor is 3^N, where N is the order of the tensor. Accordingly a tensor of *order zero* is specified in any coordinate system in three-dimensional space by *one* component. Tensors of order zero are called *scalars*. Physical quantities having magnitude only are represented by scalars. Tensors of *order one* have *three* coordinate components in physical space and are known as *vectors*. Quantities possessing both magnitude and direction are represented by vectors. *Second-order* tensors correspond to *dyadics*. Several important quantities in continuum mechanics are represented by tensors of rank two. Higher order tensors such as *triadics*, or tensors of order three, and *tetradics*, or tensors of order four are also defined and appear often in the mathematics of continuum mechanics.

1.3 VECTORS AND SCALARS

Certain physical quantities, such as force and velocity, which possess both magnitude and direction, may be represented in a three-dimensional space by *directed line segments* that obey the *parallelogram law of addition.* Such directed line segments are the geometrical representations of first-order tensors and are called *vectors.* Pictorially, a vector is simply an arrow pointing in the appropriate direction and having a length proportional to the magnitude of the vector. *Equal vectors* have the same direction and equal magnitudes. A *unit* vector is a vector of unit length. The *null* or *zero* vector is one having zero length and an unspecified direction. The *negative* of a vector is that vector having the same magnitude but opposite direction.

Those physical quantities, such as mass and energy, which possess magnitude only are represented by tensors of order zero which are called *scalars.*

In the *symbolic,* or *Gibbs* notation, vectors are designated by bold-faced letters such as $\mathbf{a}, \mathbf{b}$, etc. Scalars are denoted by italic letters such as a, b, λ, etc. Unit vectors are further distinguished by a caret placed over the bold-faced letter. In Fig. 1-1, arbitrary vectors $\mathbf{a}$ and $\mathbf{b}$ are shown along with the unit vector $\hat{\mathbf{e}}$ and the pair of equal vectors $\mathbf{c}$ and $\mathbf{d}$.

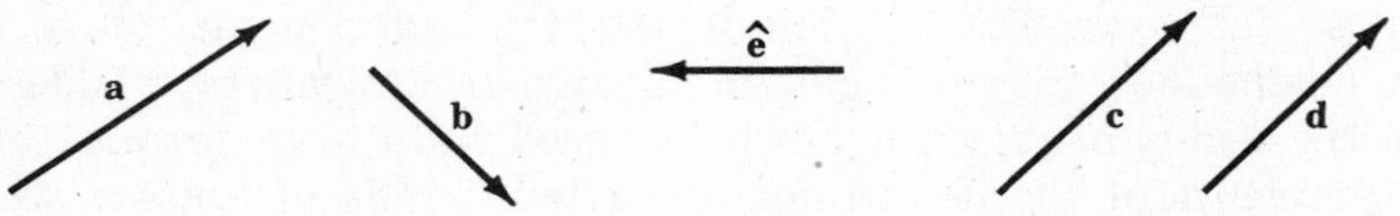

Fig. 1-1

The magnitude of an arbitrary vector $\mathbf{a}$ is written simply as a, or for emphasis it may be denoted by the vector symbol between vertical bars as $|\mathbf{a}|$.

1.4 VECTOR ADDITION. MULTIPLICATION OF A VECTOR BY A SCALAR

Vector addition obeys the *parallelogram law,* which defines the vector sum of two vectors as the diagonal of a parallelogram having the component vectors as adjacent sides. This law for vector addition is equivalent to the *triangle rule* which defines the sum of two vectors as the vector extending from the tail of the first to the head of the second when the summed vectors are adjoined head to tail. The graphical construction for the addition of $\mathbf{a}$ and $\mathbf{b}$ by the parallelogram law is shown in Fig. 1-2(*a*). Algebraically, the addition process is expressed by the vector equation

$$\mathbf{a}+\mathbf{b} = \mathbf{b}+\mathbf{a} = \mathbf{c} \tag{1.1}$$

Vector subtraction is accomplished by addition of the negative vector as shown, for example, in Fig. 1-2(*b*) where the triangle rule is used. Thus

$$\mathbf{a}-\mathbf{b} = -\mathbf{b}+\mathbf{a} = \mathbf{d} \tag{1.2}$$

The operations of vector addition and subtraction are commutative and associative as illustrated in Fig. 1-2(*c*), for which the appropriate equations are

$$(\mathbf{a}+\mathbf{b})+\mathbf{g} = \mathbf{a}+(\mathbf{b}+\mathbf{g}) = \mathbf{h} \tag{1.3}$$

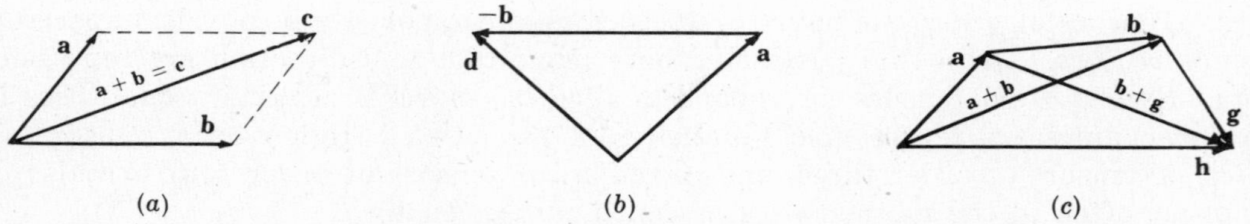

Fig. 1-2

Multiplication of a *vector* by a *scalar* produces in general a new vector having the same direction as the original but a different length. Exceptions are multiplication by zero to produce the null vector, and multiplication by unity which does not change a vector. Multiplication of the vector **b** by the scalar m results in one of the three possible cases shown in Fig. 1-3, depending upon the numerical value of m.

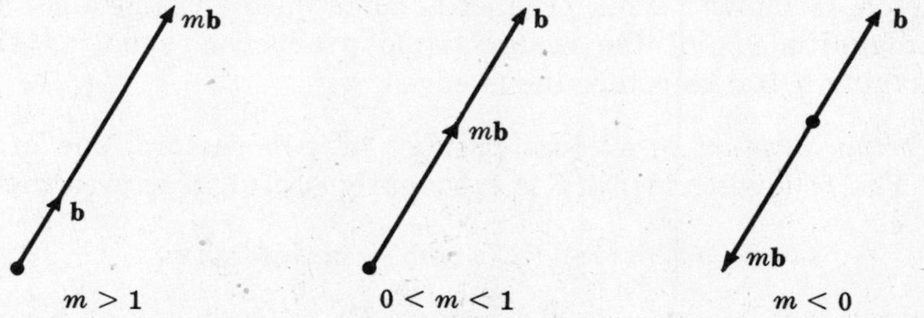

Fig. 1-3

Multiplication of a vector by a scalar is associative and distributive. Thus

$$m(n\mathbf{b}) = (mn)\mathbf{b} = n(m\mathbf{b}) \tag{1.4}$$

$$(m+n)\mathbf{b} = (n+m)\mathbf{b} = m\mathbf{b} + n\mathbf{b} \tag{1.5}$$

$$m(\mathbf{a}+\mathbf{b}) = m(\mathbf{b}+\mathbf{a}) = m\mathbf{a} + m\mathbf{b} \tag{1.6}$$

In the important case of a vector multiplied by the reciprocal of its magnitude, the result is a *unit vector* in the direction of the original vector. This relationship is expressed by the equation

$$\hat{\mathbf{b}} = \mathbf{b}/b \tag{1.7}$$

1.5 DOT AND CROSS PRODUCTS OF VECTORS

The *dot* or *scalar product* of two vectors **a** and **b** is the scalar

$$\lambda = \mathbf{a}\cdot\mathbf{b} = \mathbf{b}\cdot\mathbf{a} = ab\cos\theta \tag{1.8}$$

in which θ is the smaller angle between the two vectors as shown in Fig. 1-4(a). The dot product of **a** with a unit vector $\hat{\mathbf{e}}$ gives the projection of **a** in the direction of $\hat{\mathbf{e}}$.

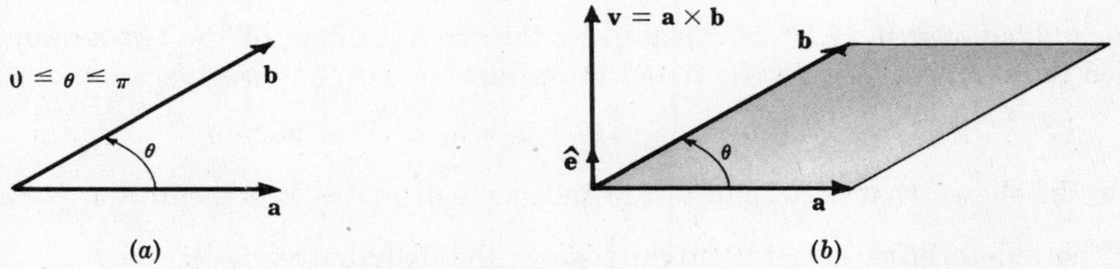

Fig. 1-4

The *cross* or *vector product* of **a** into **b** is the vector **v** given by

$$\mathbf{v} = \mathbf{a}\times\mathbf{b} = -\mathbf{b}\times\mathbf{a} = (ab\sin\theta)\,\hat{\mathbf{e}} \tag{1.9}$$

in which θ is the angle less than 180° between the vectors **a** and **b**, and $\hat{\mathbf{e}}$ is a unit vector perpendicular to their plane such that a right-handed rotation about $\hat{\mathbf{e}}$ through the angle θ carries **a** into **b**. The magnitude of **v** is equal to the area of the parallelogram having **a** and **b** as adjacent sides, shown shaded in Fig. 1-4(b). The cross product is not commutative.

The *scalar triple product* is a dot product of two vectors, one of which is a cross product.

$$\mathbf{a}\cdot(\mathbf{b}\times\mathbf{c}) = (\mathbf{a}\times\mathbf{b})\cdot\mathbf{c} = \mathbf{a}\cdot\mathbf{b}\times\mathbf{c} = \lambda \tag{1.10}$$

As indicated by (*1.10*) the dot and cross operation may be interchanged in this product. Also, since the cross operation must be carried out first, the parentheses are unnecessary and may be deleted as shown. This product is sometimes written $[\mathbf{abc}]$ and called the *box product.* The magnitude λ of the scalar triple product is equal to the volume of the parallelepiped having $\mathbf{a}, \mathbf{b}, \mathbf{c}$ as coterminous edges.

The *vector triple product* is a cross product of two vectors, one of which is itself a cross product. The following identity is frequently useful in expressing the product of $\mathbf{a}$ crossed into $\mathbf{b}\times\mathbf{c}$.

$$\mathbf{a}\times(\mathbf{b}\times\mathbf{c}) = (\mathbf{a}\cdot\mathbf{c})\mathbf{b} - (\mathbf{a}\cdot\mathbf{b})\mathbf{c} = \mathbf{w} \tag{1.11}$$

From (*1.11*), the product vector $\mathbf{w}$ is observed to lie in the plane of $\mathbf{b}$ and $\mathbf{c}$.

1.6 DYADS AND DYADICS

The *indeterminate vector product* of $\mathbf{a}$ and $\mathbf{b}$, defined by writing the vectors in juxtaposition as $\mathbf{ab}$ is called a *dyad.* The indeterminate product is not in general commutative, i.e. $\mathbf{ab} \neq \mathbf{ba}$. The first vector in a dyad is known as the *antecedent,* the second is called the *consequent.* A *dyadic* D corresponds to a tensor of order two and may always be represented as a finite sum of dyads

$$\mathsf{D} = \mathbf{a}_1\mathbf{b}_1 + \mathbf{a}_2\mathbf{b}_2 + \cdots + \mathbf{a}_N\mathbf{b}_N \tag{1.12}$$

which is, however, never unique. In symbolic notation, dyadics are denoted by bold-faced sans-serif letters as above.

If in each dyad of (*1.12*) the antecedents and consequents are interchanged, the resulting dyadic is called the *conjugate dyadic* of D and is written

$$\mathsf{D}_c = \mathbf{b}_1\mathbf{a}_1 + \mathbf{b}_2\mathbf{a}_2 + \cdots + \mathbf{b}_N\mathbf{a}_N \tag{1.13}$$

If each dyad of D in (*1.12*) is replaced by the dot product of the two vectors, the result is a scalar known as the *scalar of the dyadic* D and is written

$$\mathsf{D}_s = \mathbf{a}_1\cdot\mathbf{b}_1 + \mathbf{a}_2\cdot\mathbf{b}_2 + \cdots + \mathbf{a}_N\cdot\mathbf{b}_N \tag{1.14}$$

If each dyad of D in (*1.12*) is replaced by the cross product of the two vectors, the result is called the *vector of the dyadic* D and is written

$$\mathsf{D}_v = \mathbf{a}_1\times\mathbf{b}_1 + \mathbf{a}_2\times\mathbf{b}_2 + \cdots + \mathbf{a}_N\times\mathbf{b}_N \tag{1.15}$$

It can be shown that D_c, D_s and D_v are independent of the representation (*1.12*).

The indeterminate vector product obeys the distributive laws

$$\mathbf{a}(\mathbf{b}+\mathbf{c}) = \mathbf{ab} + \mathbf{ac} \tag{1.16}$$

$$(\mathbf{a}+\mathbf{b})\mathbf{c} = \mathbf{ac} + \mathbf{bc} \tag{1.17}$$

$$(\mathbf{a}+\mathbf{b})(\mathbf{c}+\mathbf{d}) = \mathbf{ac} + \mathbf{ad} + \mathbf{bc} + \mathbf{bd} \tag{1.18}$$

and if λ and μ are any scalars,

$$(\lambda+\mu)\mathbf{ab} = \lambda\mathbf{ab} + \mu\mathbf{ab} \tag{1.19}$$

$$(\lambda\mathbf{a})\mathbf{b} = \mathbf{a}(\lambda\mathbf{b}) = \lambda\mathbf{ab} \tag{1.20}$$

If $\mathbf{v}$ is any vector, the dot products $\mathbf{v}\cdot\mathsf{D}$ and $\mathsf{D}\cdot\mathbf{v}$ are the vectors defined respectively by

$$\mathbf{v}\cdot\mathsf{D} = (\mathbf{v}\cdot\mathbf{a}_1)\mathbf{b}_1 + (\mathbf{v}\cdot\mathbf{a}_2)\mathbf{b}_2 + \cdots + (\mathbf{v}\cdot\mathbf{a}_N)\mathbf{b}_N = \mathbf{u} \tag{1.21}$$

$$\mathsf{D}\cdot\mathbf{v} = \mathbf{a}_1(\mathbf{b}_1\cdot\mathbf{v}) + \mathbf{a}_2(\mathbf{b}_2\cdot\mathbf{v}) + \cdots + \mathbf{a}_N(\mathbf{b}_N\cdot\mathbf{v}) = \mathbf{w} \tag{1.22}$$

In (*1.21*) D is called the *postfactor*, and in (*1.22*) it is called the *prefactor*. Two dyadics D and E are *equal* if and only if for every vector $\mathbf{v}$, either

$$\mathbf{v}\cdot\mathsf{D} = \mathbf{v}\cdot\mathsf{E} \quad \text{or} \quad \mathsf{D}\cdot\mathbf{v} = \mathsf{E}\cdot\mathbf{v} \tag{1.23}$$

The *unit dyadic*, or *idemfactor* I, is the dyadic which can be represented as

$$\mathsf{I} = \hat{\mathbf{e}}_1\hat{\mathbf{e}}_1 + \hat{\mathbf{e}}_2\hat{\mathbf{e}}_2 + \hat{\mathbf{e}}_3\hat{\mathbf{e}}_3 \tag{1.24}$$

where $\hat{\mathbf{e}}_1, \hat{\mathbf{e}}_2, \hat{\mathbf{e}}_3$ constitute any orthonormal basis for three-dimensional Euclidean space (see Section 1.7). The dyadic I is characterized by the property

$$\mathsf{I}\cdot\mathbf{v} = \mathbf{v}\cdot\mathsf{I} = \mathbf{v} \tag{1.25}$$

for all vectors $\mathbf{v}$.

The cross products $\mathbf{v}\times\mathsf{D}$ and $\mathsf{D}\times\mathbf{v}$ are the dyadics defined respectively by

$$\mathbf{v}\times\mathsf{D} = (\mathbf{v}\times\mathbf{a}_1)\mathbf{b}_1 + (\mathbf{v}\times\mathbf{a}_2)\mathbf{b}_2 + \cdots + (\mathbf{v}\times\mathbf{a}_N)\mathbf{b}_N = \mathsf{F} \tag{1.26}$$

$$\mathsf{D}\times\mathbf{v} = \mathbf{a}_1(\mathbf{b}_1\times\mathbf{v}) + \mathbf{a}_2(\mathbf{b}_2\times\mathbf{v}) + \cdots + \mathbf{a}_N(\mathbf{b}_N\times\mathbf{v}) = \mathsf{G} \tag{1.27}$$

The dot product of the dyads $\mathbf{ab}$ and $\mathbf{cd}$ is the dyad defined by

$$\mathbf{ab}\cdot\mathbf{cd} = (\mathbf{b}\cdot\mathbf{c})\mathbf{ad} \tag{1.28}$$

From (*1.28*), the dot product of any two dyadics D and E is the dyadic

$$\begin{aligned}\mathsf{D}\cdot\mathsf{E} &= (\mathbf{a}_1\mathbf{b}_1 + \mathbf{a}_2\mathbf{b}_2 + \cdots + \mathbf{a}_N\mathbf{b}_N)\cdot(\mathbf{c}_1\mathbf{d}_1 + \mathbf{c}_2\mathbf{d}_2 + \cdots + \mathbf{c}_N\mathbf{d}_N)\\ &= (\mathbf{b}_1\cdot\mathbf{c}_1)\mathbf{a}_1\mathbf{d}_1 + (\mathbf{b}_1\cdot\mathbf{c}_2)\mathbf{a}_1\mathbf{d}_2 + \cdots + (\mathbf{b}_N\cdot\mathbf{c}_N)\mathbf{a}_N\mathbf{d}_N = \mathsf{G}\end{aligned} \tag{1.29}$$

The dyadics D and E are said to be *reciprocal* of each other if

$$\mathsf{E}\cdot\mathsf{D} = \mathsf{D}\cdot\mathsf{E} = \mathsf{I} \tag{1.30}$$

For reciprocal dyadics, the notation $\mathsf{E} = \mathsf{D}^{-1}$ and $\mathsf{D} = \mathsf{E}^{-1}$ is often used.

Double dot and cross products are also defined for the dyads $\mathbf{ab}$ and $\mathbf{cd}$ as follows,

$$\mathbf{ab} : \mathbf{cd} = (\mathbf{a}\cdot\mathbf{c})(\mathbf{b}\cdot\mathbf{d}) = \lambda, \quad \text{a scalar} \tag{1.31}$$

$$\mathbf{ab} \overset{\times}{\cdot} \mathbf{cd} = (\mathbf{a}\times\mathbf{c})(\mathbf{b}\cdot\mathbf{d}) = \mathbf{h}, \quad \text{a vector} \tag{1.32}$$

$$\mathbf{ab} \overset{\cdot}{\times} \mathbf{cd} = (\mathbf{a}\cdot\mathbf{c})(\mathbf{b}\times\mathbf{d}) = \mathbf{g}, \quad \text{a vector} \tag{1.33}$$

$$\mathbf{ab} \overset{\times}{\times} \mathbf{cd} = (\mathbf{a}\times\mathbf{c})(\mathbf{b}\times\mathbf{d}) = \mathbf{uw}, \quad \text{a dyad} \tag{1.34}$$

From these definitions, double dot and cross products of dyadics may be readily developed. Also, some authors use the double dot product defined by

$$\mathbf{ab}\cdot\cdot\,\mathbf{cd} = (\mathbf{b}\cdot\mathbf{c})(\mathbf{a}\cdot\mathbf{d}) = \lambda, \quad \text{a scalar} \tag{1.35}$$

A dyadic D is said to be *self-conjugate*, or *symmetric*, if

$$\mathsf{D} = \mathsf{D}_c \tag{1.36}$$

and *anti-self-conjugate*, or *anti-symmetric*, if

$$\mathsf{D} = -\mathsf{D}_c \tag{1.37}$$

Every dyadic may be expressed uniquely as the sum of a symmetric and anti-symmetric dyadic. For the arbitrary dyadic D the decomposition is

$$\mathsf{D} = \tfrac{1}{2}(\mathsf{D} + \mathsf{D}_c) + \tfrac{1}{2}(\mathsf{D} - \mathsf{D}_c) = \mathsf{G} + \mathsf{H} \tag{1.38}$$

for which $$\mathbf{G}_c = \tfrac{1}{2}(\mathbf{D}_c + (\mathbf{D}_c)_c) = \tfrac{1}{2}(\mathbf{D}_c + \mathbf{D}) = \mathbf{G} \quad \text{(symmetric)} \tag{1.39}$$

and $$\mathbf{H}_c = \tfrac{1}{2}(\mathbf{D}_c - (\mathbf{D}_c)_c) = \tfrac{1}{2}(\mathbf{D}_c - \mathbf{D}) = -\mathbf{H} \quad \text{(anti-symmetric)} \tag{1.40}$$

Uniqueness is established by assuming a second decomposition, $\mathbf{D} = \mathbf{G}^* + \mathbf{H}^*$. Then

$$\mathbf{G}^* + \mathbf{H}^* = \mathbf{G} + \mathbf{H} \tag{1.41}$$

and the conjugate of this equation is

$$\mathbf{G}^* - \mathbf{H}^* = \mathbf{G} - \mathbf{H} \tag{1.42}$$

Adding and subtracting (*1.41*) and (*1.42*) in turn yields respectively the desired equalities, $\mathbf{G}^* = \mathbf{G}$ and $\mathbf{H}^* = \mathbf{H}$.

1.7 COORDINATE SYSTEMS. BASE VECTORS. UNIT VECTOR TRIADS

A vector may be defined with respect to a particular coordinate system by specifying the *components* of the vector in that system. The choice of coordinate system is arbitrary, but in certain situations a particular choice may be advantageous. The reference system of coordinate axes provides units for measuring vector magnitudes and assigns directions in space by which the orientation of vectors may be determined.

The well-known *rectangular Cartesian coordinate system* is often represented by the mutually perpendicular axes, *Oxyz* shown in Fig. 1-5. Any vector $\mathbf{v}$ in this system may be expressed as a linear combination of three arbitrary, nonzero, noncoplanar vectors of the system, which are called *base vectors.* For base vectors $\mathbf{a}, \mathbf{b}, \mathbf{c}$ and suitably chosen scalar coefficients λ, μ, ν the vector $\mathbf{v}$ is given by

$$\mathbf{v} = \lambda\mathbf{a} + \mu\mathbf{b} + \nu\mathbf{c} \tag{1.43}$$

Base vectors are by hypothesis linearly independent, i.e. the equation

$$\lambda\mathbf{a} + \mu\mathbf{b} + \nu\mathbf{c} = 0 \tag{1.44}$$

is satisfied only if $\lambda = \mu = \nu = 0$. A set of base vectors in a given coordinate system is said to constitute a *basis* for that system.

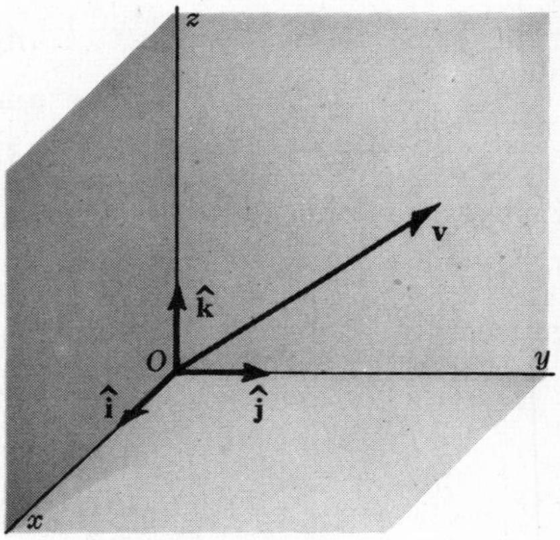

Fig. 1-5

The most frequent choice of base vectors for the rectangular Cartesian system is the set of unit vectors $\hat{\mathbf{i}}, \hat{\mathbf{j}}, \hat{\mathbf{k}}$ along the coordinate axes as shown in Fig. 1-5. These base vectors constitute a right-handed *unit vector triad,* for which

$$\hat{\mathbf{i}} \times \hat{\mathbf{j}} = \hat{\mathbf{k}}, \quad \hat{\mathbf{j}} \times \hat{\mathbf{k}} = \hat{\mathbf{i}}, \quad \hat{\mathbf{k}} \times \hat{\mathbf{i}} = \hat{\mathbf{j}} \tag{1.45}$$

and
$$\hat{\mathbf{i}} \cdot \hat{\mathbf{i}} = \hat{\mathbf{j}} \cdot \hat{\mathbf{j}} = \hat{\mathbf{k}} \cdot \hat{\mathbf{k}} = 1$$
$$\hat{\mathbf{i}} \cdot \hat{\mathbf{j}} = \hat{\mathbf{j}} \cdot \hat{\mathbf{k}} = \hat{\mathbf{k}} \cdot \hat{\mathbf{i}} = 0 \tag{1.46}$$

Such a set of base vectors is often called an *orthonormal basis.*

In terms of the unit triad $\hat{\mathbf{i}}, \hat{\mathbf{j}}, \hat{\mathbf{k}}$, the vector $\mathbf{v}$ shown in Fig. 1-6 below may be expressed by

$$\mathbf{v} = v_x\hat{\mathbf{i}} + v_y\hat{\mathbf{j}} + v_z\hat{\mathbf{k}} \tag{1.47}$$

in which the Cartesian components

$$v_x = \mathbf{v}\cdot\hat{\mathbf{i}} = v\cos\alpha$$
$$v_y = \mathbf{v}\cdot\hat{\mathbf{j}} = v\cos\beta$$
$$v_z = \mathbf{v}\cdot\hat{\mathbf{k}} = v\cos\gamma$$

are the projections of $\mathbf{v}$ onto the coordinate axes. The unit vector in the direction of $\mathbf{v}$ is given according to (*1.7*) by

$$\hat{\mathbf{e}}_v = \mathbf{v}/v = (\cos\alpha)\,\hat{\mathbf{i}} + (\cos\beta)\,\hat{\mathbf{j}} + (\cos\gamma)\,\hat{\mathbf{k}} \qquad (1.48)$$

Since $\mathbf{v}$ is arbitrary, it follows that any unit vector will have the *direction cosines* of that vector as its *Cartesian components.*

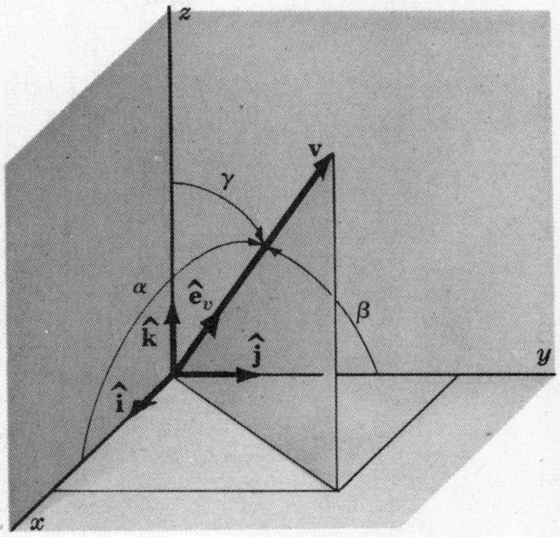

Fig. 1-6

In Cartesian component form the dot product of $\mathbf{a}$ and $\mathbf{b}$ is given by

$$\mathbf{a}\cdot\mathbf{b} = (a_x\hat{\mathbf{i}} + a_y\hat{\mathbf{j}} + a_z\hat{\mathbf{k}})\cdot(b_x\hat{\mathbf{i}} + b_y\hat{\mathbf{j}} + b_z\hat{\mathbf{k}}) = a_xb_x + a_yb_y + a_zb_z \qquad (1.49)$$

For the same two vectors, the cross product $\mathbf{a}\times\mathbf{b}$ is

$$\mathbf{a}\times\mathbf{b} = (a_yb_z - a_zb_y)\,\hat{\mathbf{i}} + (a_zb_x - a_xb_z)\,\hat{\mathbf{j}} + (a_xb_y - a_yb_x)\,\hat{\mathbf{k}} \qquad (1.50)$$

This result is often presented in the determinant form

$$\mathbf{a}\times\mathbf{b} = \begin{vmatrix} \hat{\mathbf{i}} & \hat{\mathbf{j}} & \hat{\mathbf{k}} \\ a_x & a_y & a_z \\ b_x & b_y & b_z \end{vmatrix} \qquad (1.51)$$

in which the elements are treated as ordinary numbers. The triple scalar product may also be represented in component form by the determinant

$$[\mathbf{abc}] = \begin{vmatrix} a_x & a_y & a_z \\ b_x & b_y & b_z \\ c_x & c_y & c_z \end{vmatrix} \qquad (1.52)$$

In Cartesian component form, the dyad $\mathbf{ab}$ is given by

$$\begin{aligned}\mathbf{ab} &= (a_x\hat{\mathbf{i}} + a_y\hat{\mathbf{j}} + a_z\hat{\mathbf{k}})(b_x\hat{\mathbf{i}} + b_y\hat{\mathbf{j}} + b_z\hat{\mathbf{k}}) \\ &= a_xb_x\hat{\mathbf{i}}\hat{\mathbf{i}} + a_xb_y\hat{\mathbf{i}}\hat{\mathbf{j}} + a_xb_z\hat{\mathbf{i}}\hat{\mathbf{k}} \\ &\quad + a_yb_x\hat{\mathbf{j}}\hat{\mathbf{i}} + a_yb_y\hat{\mathbf{j}}\hat{\mathbf{j}} + a_yb_z\hat{\mathbf{j}}\hat{\mathbf{k}} \\ &\quad + a_zb_x\hat{\mathbf{k}}\hat{\mathbf{i}} + a_zb_y\hat{\mathbf{k}}\hat{\mathbf{j}} + a_zb_z\hat{\mathbf{k}}\hat{\mathbf{k}}\end{aligned} \qquad (1.53)$$

Because of the *nine* terms involved, (*1.53*) is known as the *nonion form* of the dyad $\mathbf{ab}$. It is possible to put any dyadic into nonion form. The nonion form of the idemfactor in terms of the unit triad $\hat{\mathbf{i}}, \hat{\mathbf{j}}, \hat{\mathbf{k}}$ is given by

$$\mathbf{I} = \hat{\mathbf{i}}\hat{\mathbf{i}} + \hat{\mathbf{j}}\hat{\mathbf{j}} + \hat{\mathbf{k}}\hat{\mathbf{k}} \qquad (1.54)$$

In addition to the rectangular Cartesian coordinate system already discussed, curvilinear coordinate systems such as the cylindrical (R, θ, z) and spherical (r, θ, ϕ) systems shown in Fig. 1-7 below are also widely used. Unit triads $(\hat{\mathbf{e}}_R, \hat{\mathbf{e}}_\theta, \hat{\mathbf{e}}_z)$ and $(\hat{\mathbf{e}}_r, \hat{\mathbf{e}}_\theta, \hat{\mathbf{e}}_\phi)$ of base vectors illustrated in the figure are associated with these systems. However, the base vectors here do not all have fixed directions and are therefore, in general, functions of position.

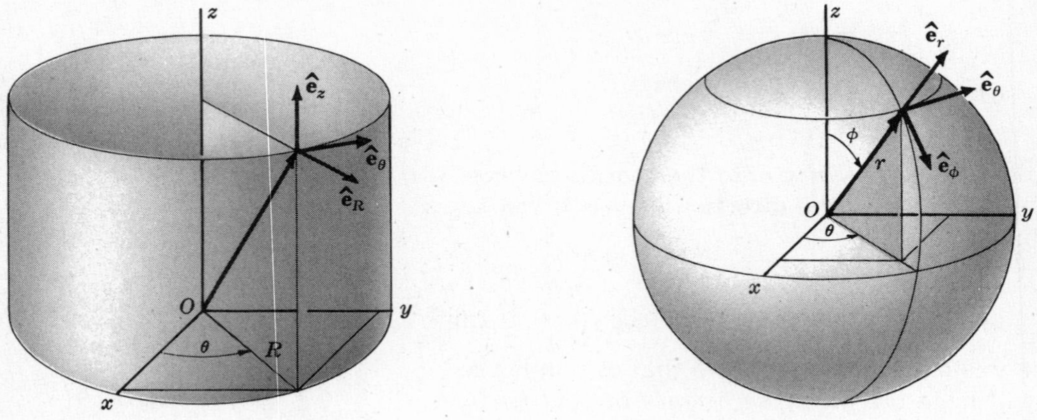

(*a*) Cylindrical (*b*) Spherical

Fig. 1-7

1.8 LINEAR VECTOR FUNCTIONS. DYADICS AS LINEAR VECTOR OPERATORS

A vector **a** is said to be a function of a second vector **b** if **a** is determined whenever **b** is given. This functional relationship is expressed by the equation

$$\mathbf{a} = \mathbf{f}(\mathbf{b}) \tag{1.55}$$

The function **f** is said to be linear when the conditions

$$\mathbf{f}(\mathbf{b}+\mathbf{c}) = \mathbf{f}(\mathbf{b}) + \mathbf{f}(\mathbf{c}) \tag{1.56}$$

$$\mathbf{f}(\lambda\mathbf{b}) = \lambda\mathbf{f}(\mathbf{b}) \tag{1.57}$$

are satisfied for all vectors **b** and **c**, and for any scalar λ.

Writing **b** in Cartesian component form, equation (*1.55*) becomes

$$\mathbf{a} = \mathbf{f}(b_x\hat{\mathbf{i}} + b_y\hat{\mathbf{j}} + b_z\hat{\mathbf{k}}) \tag{1.58}$$

which, if **f** is linear, may be written

$$\mathbf{a} = b_x\mathbf{f}(\hat{\mathbf{i}}) + b_y\mathbf{f}(\hat{\mathbf{j}}) + b_z\mathbf{f}(\hat{\mathbf{k}}) \tag{1.59}$$

In (*1.59*) let $\mathbf{f}(\hat{\mathbf{i}}) = \mathbf{u}$, $\mathbf{f}(\hat{\mathbf{j}}) = \mathbf{v}$, $\mathbf{f}(\hat{\mathbf{k}}) = \mathbf{w}$, so that now

$$\mathbf{a} = \mathbf{u}(\hat{\mathbf{i}}\cdot\mathbf{b}) + \mathbf{v}(\hat{\mathbf{j}}\cdot\mathbf{b}) + \mathbf{w}(\hat{\mathbf{k}}\cdot\mathbf{b}) = (\mathbf{u}\hat{\mathbf{i}} + \mathbf{v}\hat{\mathbf{j}} + \mathbf{w}\hat{\mathbf{k}})\cdot\mathbf{b} \tag{1.60}$$

which is recognized as a dyadic-vector dot product and may be written

$$\mathbf{a} = \mathsf{D}\cdot\mathbf{b} \tag{1.61}$$

where $\mathsf{D} = \mathbf{u}\hat{\mathbf{i}} + \mathbf{v}\hat{\mathbf{j}} + \mathbf{w}\hat{\mathbf{k}}$. This demonstrates that any linear vector function **f** may be expressed as a dyadic-vector product. In (*1.61*) the dyadic D serves as a *linear vector operator* which operates on the *argument* vector **b** to produce the *image* vector **a**.

1.9 INDICIAL NOTATION. RANGE AND SUMMATION CONVENTIONS

The components of a tensor of any order, and indeed the tensor itself, may be represented clearly and concisely by the use of the *indicial notation*. In this notation, letter indices, either subscripts or superscripts, are appended to the *generic* or *kernel* letter representing the tensor quantity of interest. Typical examples illustrating use of indices are the tensor symbols

$$a_i,\quad b^j,\quad T_{ij},\quad F_i^{\ j},\quad \epsilon_{ijk},\quad R^{pq}$$

In the "mixed" form, where both subscripts and superscripts appear, the dot shows that j is the second index.

Under the rules of indicial notation, a letter index may occur either *once* or *twice* in a given term. When an index occurs unrepeated in a term, that index is understood to take on the values $1, 2, \ldots, N$ where N is a specified integer that determines the *range* of the index. Unrepeated indices are known as *free* indices. The tensorial rank of a given term is equal to the number of free indices appearing in that term. Also, correctly written tensor equations have the same letters as free indices in every term.

When an index appears *twice* in a term, that index is understood to take on all the values of its range, and the resulting terms *summed.* In this so-called *summation convention,* repeated indices are often referred to as *dummy indices,* since their replacement by any other letter not appearing as a free index does not change the meaning of the term in which they occur. In general, no index occurs more than twice in a properly written term. If it is absolutely necessary to use some index more than twice to satisfactorily express a certain quantity, the summation convention must be suspended.

The number and location of the free indices reveal directly the exact tensorial character of the quantity expressed in the indicial notation. Tensors of *first order* are denoted by kernel letters bearing *one free index.* Thus the arbitrary vector **a** is represented by a symbol having a single subscript or superscript, i.e. in one or the other of the two forms,

$$a_i, \quad a^i$$

The following terms, having only one free index, are also recognized as first-order tensor quantities:

$$a_{ij}b_j, \quad F_{ikk}, \quad R^{p}_{\cdot qp}, \quad \epsilon_{ijk}u_j v_k$$

Second-order tensors are denoted by symbols having *two* free indices. Thus the arbitrary dyadic **D** will appear in one of the three possible forms

$$D^{ij}, \quad D_i^{\cdot j} \quad \text{or} \quad D^i_{\cdot j}, \quad D_{ij}$$

In the "mixed" form, the dot shows that j is the second index. Second-order tensor quantities may also appear in various forms as, for example,

$$A_{ijip}, \quad B^{ij}_{\cdot\cdot jk}, \quad \delta_{ij}u_k v_k$$

By a logical continuation of the above scheme, *third-order* tensors are expressed by symbols with *three* free indices. Also, a symbol such as λ which has no indices attached, represents a scalar, or tensor of zero order.

In ordinary physical space a *basis* is composed of three, noncoplanar vectors, and so any vector in this space is completely specified by its three components. Therefore the range on the index of a_i, which represents a vector in physical three-space, is $1, 2, 3$. Accordingly the symbol a_i is understood to represent the three components a_1, a_2, a_3. Also, a_i is sometimes interpreted to represent the ith component of the vector or indeed to represent the vector itself. For a range of three on both indices, the symbol A_{ij} represents nine components (of the second-order tensor (dyadic) **A**). The tensor A_{ij} is often presented explicitly by giving the nine components in a square array enclosed by large parentheses as

$$A_{ij} = \begin{pmatrix} A_{11} & A_{12} & A_{13} \\ A_{21} & A_{22} & A_{23} \\ A_{31} & A_{32} & A_{33} \end{pmatrix} \qquad (1.62)$$

In the same way, the components of a first-order tensor (vector) in three-space may be displayed explicitly by a row or column arrangement of the form

$$a_i = (a_1, a_2, a_3) \quad \text{or} \quad a_i = \begin{pmatrix} a_1 \\ a_2 \\ a_3 \end{pmatrix} \tag{1.63}$$

In general, for a range of N, an nth order tensor will have N^n components.

The usefulness of the indicial notation in presenting systems of equations in compact form is illustrated by the following two typical examples. For a range of three on both i and j the indicial equation

$$x_i = c_{ij} z_j \tag{1.64}$$

represents in expanded form the three equations

$$\begin{aligned} x_1 &= c_{11}z_1 + c_{12}z_2 + c_{13}z_3 \\ x_2 &= c_{21}z_1 + c_{22}z_2 + c_{23}z_3 \\ x_3 &= c_{31}z_1 + c_{32}z_2 + c_{33}z_3 \end{aligned} \tag{1.65}$$

For a range of two on i and j, the indicial equation

$$A_{ij} = B_{ip} C_{jq} D_{pq} \tag{1.66}$$

represents, in expanded form, the four equations

$$\begin{aligned} A_{11} &= B_{11}C_{11}D_{11} + B_{11}C_{12}D_{12} + B_{12}C_{11}D_{21} + B_{12}C_{12}D_{22} \\ A_{12} &= B_{11}C_{21}D_{11} + B_{11}C_{22}D_{12} + B_{12}C_{21}D_{21} + B_{12}C_{22}D_{22} \\ A_{21} &= B_{21}C_{11}D_{11} + B_{21}C_{12}D_{12} + B_{22}C_{11}D_{21} + B_{22}C_{12}D_{22} \\ A_{22} &= B_{21}C_{21}D_{11} + B_{21}C_{22}D_{12} + B_{22}C_{21}D_{21} + B_{22}C_{22}D_{22} \end{aligned} \tag{1.67}$$

For a range of three on both i and j, (*1.66*) would represent nine equations, each having nine terms on the right-hand side.

1.10 SUMMATION CONVENTION USED WITH SYMBOLIC NOTATION

The summation convention is very often employed in connection with the representation of vectors and tensors by *indexed base vectors* written in the symbolic notation. Thus if the rectangular Cartesian axes and unit base vectors of Fig. 1-5 are relabeled as shown by Fig. 1-8, the arbitrary vector $\mathbf{v}$ may be written

$$\mathbf{v} = v_1\hat{\mathbf{e}}_1 + v_2\hat{\mathbf{e}}_2 + v_3\hat{\mathbf{e}}_3 \tag{1.68}$$

in which v_1, v_2, v_3 are the rectangular Cartesian components of $\mathbf{v}$. Applying the summation convention to (*1.68*), the equation may be written in the abbreviated form

$$\mathbf{v} = v_i\hat{\mathbf{e}}_i \tag{1.69}$$

where i is a summed index. The notation here is essentially *symbolic*, but with the added feature of the *summation convention*. In such a "combination" style of notation, tensor character is not given by the *free indices rule* as it is in true indicial notation.

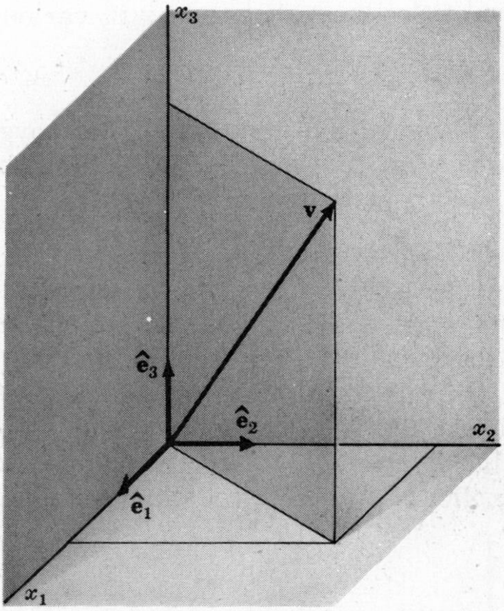

Fig. 1-8

Second-order tensors may also be represented by summation on indexed base vectors. Accordingly the dyad **ab** given in nonion form by (*1.53*) may be written

$$\mathbf{ab} = (a_i\hat{\mathbf{e}}_i)(b_j\hat{\mathbf{e}}_j) = a_ib_j\hat{\mathbf{e}}_i\hat{\mathbf{e}}_j \tag{1.70}$$

It is essential that the sequence of the base vectors be preserved in this expression. In similar fashion, the nonion form of the arbitrary dyadic **D** may be expressed in compact notation by

$$\mathbf{D} = D_{ij}\hat{\mathbf{e}}_i\hat{\mathbf{e}}_j \tag{1.71}$$

1.11 COORDINATE TRANSFORMATIONS. GENERAL TENSORS

Let x^i represent the arbitrary system of coordinates x^1, x^2, x^3 in a three-dimensional Euclidean space, and let θ^i represent any other coordinate system $\theta^1, \theta^2, \theta^3$ in the same space. Here the numerical superscripts are labels and not exponents. Powers of x may be expressed by use of parentheses as in $(x)^2$ or $(x)^3$. The letter superscripts are indices as already noted. The *coordinate transformation equations*

$$\theta^i = \theta^i(x^1, x^2, x^3) \tag{1.72}$$

assign to any point (x^1, x^2, x^3) in the x^i system a new set of coordinates $(\theta^1, \theta^2, \theta^3)$ in the θ^i system. The functions θ^i relating the two sets of variables (coordinates) are assumed to be single-valued, continuous, differentiable functions. The determinant

$$J = \begin{vmatrix} \dfrac{\partial\theta^1}{\partial x^1} & \dfrac{\partial\theta^1}{\partial x^2} & \dfrac{\partial\theta^1}{\partial x^3} \\ \dfrac{\partial\theta^2}{\partial x^1} & \dfrac{\partial\theta^2}{\partial x^2} & \dfrac{\partial\theta^2}{\partial x^3} \\ \dfrac{\partial\theta^3}{\partial x^1} & \dfrac{\partial\theta^3}{\partial x^2} & \dfrac{\partial\theta^3}{\partial x^3} \end{vmatrix} \tag{1.73}$$

or, in compact form,

$$J = \left|\frac{\partial\theta^i}{\partial x^j}\right| \tag{1.74}$$

is called the *Jacobian* of the transformation. If the Jacobian does not vanish, (*1.72*) possesses a unique inverse set of the form

$$x^i = x^i(\theta^1, \theta^2, \theta^3) \tag{1.75}$$

The coordinate systems represented by x^i and θ^i in (*1.72*) and (*1.75*) are completely general and may be any curvilinear or Cartesian systems.

From (*1.72*), the differential vector $d\theta^i$ is given by

$$d\theta^i = \frac{\partial\theta^i}{\partial x^j}\,dx^j \tag{1.76}$$

This equation is a prototype of the equation which defines the class of tensors known as *contravariant vectors*. In general, a set of quantities b^i associated with a point P are said to be the components of a *contravariant tensor of order one* if they transform, under a coordinate transformation, according to the equation

$$b'^i = \frac{\partial\theta^i}{\partial x^j}\,b^j \tag{1.77}$$

where the partial derivatives are evaluated at P. In (*1.77*), b^j are the components of the tensor in the x^j coordinate system, while b'^i are the components in the θ^i system. In general tensor theory, contravariant tensors are recognized by the use of superscripts as indices. It is for this reason that the coordinates are labeled x^i here rather than x_i, but it must be noted that it is only the differentials dx^i, and not the coordinates themselves, which have tensor character.

By a logical extension of the tensor concept expressed in (*1.77*), the definition of *contravariant tensors of order two* requires the tensor components to obey the transformation law

$$B'^{ij} = \frac{\partial\theta^i}{\partial x^r}\frac{\partial\theta^j}{\partial x^s}B^{rs} \tag{1.78}$$

Contravariant tensors of third, fourth and higher orders are defined in a similar manner.

The word *contravariant* is used above to distinguish that class of tensors from the class known as *covariant* tensors. In general tensor theory, covariant tensors are recognized by the use of subscripts as indices. The prototype of the *covariant vector* is the partial derivative of a scalar function of the coordinates. Thus if $\phi = \phi(x^1, x^2, x^3)$ is such a function,

$$\frac{\partial\phi}{\partial\theta^i} = \frac{\partial\phi}{\partial x^j}\frac{\partial x^j}{\partial\theta^i} \tag{1.79}$$

In general, a set of quantities b_i are said to be the components of a *covariant tensor of order one* if they transform according to the equation

$$b'_i = \frac{\partial x^j}{\partial\theta^i}b_j \tag{1.80}$$

In (*1.80*), b'_i are the covariant components in the θ^i system, b_i the components in the x_i system. *Second-order covariant tensors* obey the transformation law

$$B'_{ij} = \frac{\partial x^r}{\partial\theta^i}\frac{\partial x^s}{\partial\theta^j}B_{rs} \tag{1.81}$$

Covariant tensors of higher order and *mixed tensors*, such as

$$T'^{r}_{\cdot sp} = \frac{\partial\theta^r}{\partial x^m}\frac{\partial x^n}{\partial\theta^s}\frac{\partial x^q}{\partial\theta^p}T^{m}_{\cdot nq} \tag{1.82}$$

are defined in the obvious way.

1.12 THE METRIC TENSOR. CARTESIAN TENSORS

Let x^i represent a system of rectangular Cartesian coordinates in a Euclidean three-space, and let θ^i represent any system of rectangular or curvilinear coordinates (e.g. cylindrical or spherical coordinates) in the same space. The vector $\mathbf{x}$ having Cartesian components x^i is called the *position vector* of the arbitrary point $P(x^1, x^2, x^3)$ referred to the rectangular Cartesian axes. The square of the differential element of distance between neighboring points $P(\mathbf{x})$ and $Q(\mathbf{x}+d\mathbf{x})$ is given by

$$(ds)^2 = dx^i\,dx^i \tag{1.83}$$

From the coordinate transformation

$$x^i = x^i(\theta^1, \theta^2, \theta^3) \tag{1.84}$$

relating the systems, the distance differential is

$$dx^i = \frac{\partial x^i}{\partial\theta^p}d\theta^p \tag{1.85}$$

and therefore (*1.83*) becomes

$$(ds)^2 = \frac{\partial x^i}{\partial \theta^p}\frac{\partial x^i}{\partial \theta^q}\,d\theta^p\,d\theta^q = g_{pq}\,d\theta^p\,d\theta^q \qquad (1.86)$$

where the second-order tensor $g_{pq} = (\partial x^i/\partial\theta^p)(\partial x^i/\partial\theta^q)$ is called the *metric tensor,* or *fundamental tensor* of the space. If θ^i represents a rectangular Cartesian system, say the x'^i system, then

$$g_{pq} = \frac{\partial x^i}{\partial x'^p}\frac{\partial x^i}{\partial x'^q} = \delta_{pq} \qquad (1.87)$$

where δ_{pq} is the *Kronecker delta* (see Section 1.13) defined by $\delta_{pq} = 0$ if $p \neq q$ and $\delta_{pq} = 1$ if $p = q$.

Any system of coordinates for which the squared differential element of distance takes the form of (*1.83*) is called a system of *homogeneous coordinates.* Coordinate transformations between systems of homogeneous coordinates are *orthogonal transformations,* and when attention is restricted to such transformations, the tensors so defined are called *Cartesian tensors.* In particular, this is the case for transformation laws between systems of rectangular Cartesian coordinates with a common origin. For Cartesian tensors there is no distinction between contravariant and covariant components and therefore it is customary to use subscripts exclusively in expressions representing Cartesian tensors. As will be shown next, in the transformation laws defining Cartesian tensors, the partial derivatives appearing in general tensor definitions, such as (*1.80*) and (*1.81*), are replaced by constants.

1.13. TRANSFORMATION LAWS FOR CARTESIAN TENSORS. THE KRONECKER DELTA. ORTHOGONALITY CONDITIONS

Let the axes $Ox_1x_2x_3$ and $Ox'_1x'_2x'_3$ represent two rectangular Cartesian coordinate systems with a common origin at an arbitrary point O as shown in Fig. 1-9. The primed system may be imagined to be obtained from the unprimed by a rotation of the axes about the origin, or by a reflection of axes in one of the coordinate planes, or by a combination of these. If the symbol a_{ij} denotes the cosine of the angle between the ith primed and jth unprimed coordinate axes, i.e. $a_{ij} = \cos(x'_i, x_j)$, the relative orientation of the individual axes of each system with respect to the other is conveniently given by the table

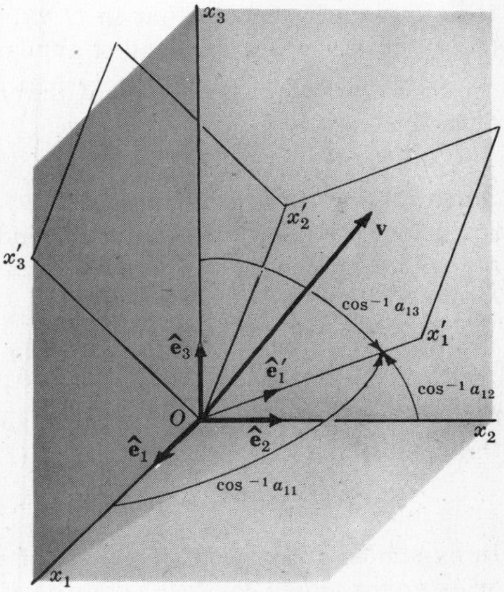

Fig. 1-9

	x_1	x_2	x_3
x'_1	a_{11}	a_{12}	a_{13}
x'_2	a_{21}	a_{22}	a_{23}
x'_3	a_{31}	a_{32}	a_{33}

or alternatively by the transformation tensor

$$\mathbf{A} = \begin{pmatrix} a_{11} & a_{12} & a_{13} \\ a_{21} & a_{22} & a_{23} \\ a_{31} & a_{32} & a_{33} \end{pmatrix}$$

From this definition of a_{ij}, the unit vector $\hat{\mathbf{e}}_1'$ along the x_1' axis is given according to (*1.48*) and the summation convention by

$$\hat{\mathbf{e}}_1' = a_{11}\hat{\mathbf{e}}_1 + a_{12}\hat{\mathbf{e}}_2 + a_{13}\hat{\mathbf{e}}_3 = a_{1j}\hat{\mathbf{e}}_j \tag{1.88}$$

An obvious generalization of this equation gives the arbitrary unit base vector $\hat{\mathbf{e}}_i'$ as

$$\hat{\mathbf{e}}_i' = a_{ij}\hat{\mathbf{e}}_j \tag{1.89}$$

In component form, the arbitrary vector $\mathbf{v}$ shown in Fig. 1-9 may be expressed in the unprimed system by the equation

$$\mathbf{v} = v_j\hat{\mathbf{e}}_j \tag{1.90}$$

and in the primed system by

$$\mathbf{v} = v_i'\hat{\mathbf{e}}_i' \tag{1.91}$$

Replacing $\hat{\mathbf{e}}_i'$ in (*1.91*) by its equivalent form (*1.89*) yields the result

$$\mathbf{v} = v_i'a_{ij}\hat{\mathbf{e}}_j \tag{1.92}$$

Comparing (*1.92*) with (*1.90*) reveals that the vector components in the primed and unprimed systems are related by the equations

$$v_j = a_{ij}v_i' \tag{1.93}$$

The expression (*1.93*) is the *transformation law* for first-order Cartesian tensors, and as such is seen to be a special case of the general form of first-order tensor transformations, expressed by (*1.80*) and (*1.77*). By interchanging the roles of the primed and unprimed base vectors in the above development, the inverse of (*1.93*) is found to be

$$v_i' = a_{ij}v_j \tag{1.94}$$

It is important to note that in (*1.93*) the free index on a_{ij} appears as the second index. In (*1.94*), however, the free index appears as the first index.

By an appropriate choice of dummy indices, (*1.93*) and (*1.94*) may be combined to produce the equation

$$v_j = a_{ij}a_{ik}v_k \tag{1.95}$$

Since the vector $\mathbf{v}$ is arbitrary, (*1.95*) must reduce to the identity $v_j = v_j$. Therefore the coefficient $a_{ij}a_{ik}$, whose value depends upon the subscripts j and k, must equal 1 or 0 according to whether the numerical values of j and k are the same or different. The *Kronecker delta,* defined by

$$\delta_{ij} = \begin{cases} 1 & \text{for } i = j \\ 0 & \text{for } i \neq j \end{cases} \tag{1.96}$$

may be used to represent quantities such as $a_{ij}a_{ik}$. Thus with the help of the Kronecker delta the conditions on the coefficient in (*1.95*) may be written

$$a_{ij}a_{ik} = \delta_{jk} \tag{1.97}$$

In expanded form, (*1.97*) consists of nine equations which are known as the *orthogonality* or *orthonormality conditions* on the direction cosines a_{ij}. Finally, (*1.93*) and (*1.94*) may also be combined to produce $v_i = a_{ij}a_{kj}v_k'$ from which the orthogonality conditions appear in the alternative form

$$a_{ij}a_{kj} = \delta_{ik} \tag{1.98}$$

A linear transformation such as (*1.93*) or (*1.94*), whose coefficients satisfy (*1.97*) or (*1.98*), is said to be an *orthogonal transformation.* Coordinate axes rotations and reflections of the axes in a coordinate plane both lead to orthogonal transformations.

The Kronecker delta is sometimes called the *substitution operator,* since, for example,

$$\delta_{ij}b_j = \delta_{i1}b_1 + \delta_{i2}b_2 + \delta_{i3}b_3 = b_i \tag{1.99}$$

and, likewise,

$$\delta_{ij}F_{ik} = \delta_{1j}F_{1k} + \delta_{2j}F_{2k} + \delta_{3j}F_{3k} = F_{jk} \tag{1.100}$$

It is clear from this property that the Kronecker delta is the indicial counterpart to the symbolic idemfactor **I**, which is given by (*1.54*).

According to the transformation law (*1.94*), the dyad u_iv_j has components in the primed coordinate system given by

$$u'_i v'_j = (a_{ip}u_p)(a_{jq}v_q) = a_{ip}a_{jq}u_pv_q \tag{1.101}$$

In an obvious generalization of (*1.101*), any second-order Cartesian tensor T_{ij} obeys the transformation law

$$T'_{ij} = a_{ip}a_{jq}T_{pq} \tag{1.102}$$

With the help of the orthogonality conditions it is a simple calculation to invert (*1.102*), thereby giving the transformation rule from primed components to unprimed components:

$$T_{ij} = a_{pi}a_{qj}T'_{pq} \tag{1.103}$$

The transformation laws for first and second-order Cartesian tensors generalize for an *N*th order Cartesian tensor to

$$T'_{ijk\ldots} = a_{ip}a_{jq}a_{km}\ldots T_{pqm\ldots} \tag{1.104}$$

1.14 ADDITION OF CARTESIAN TENSORS. MULTIPLICATION BY A SCALAR

Cartesian tensors of the same order may be added (or subtracted) component by component in accordance with the rule

$$A_{ijk\ldots} \pm B_{ijk\ldots} = T_{ijk\ldots} \tag{1.105}$$

The sum is a tensor of the same order as those added. Note that like indices appear in the same sequence in each term.

Multiplication of every component of a tensor by a given scalar produces a new tensor of the same order. For the scalar multiplier λ, typical examples written in both indicial and symbolic notation are

$$b_i = \lambda a_i \quad \text{or} \quad \mathbf{b} = \lambda\mathbf{a} \tag{1.106}$$

$$B_{ij} = \lambda A_{ij} \quad \text{or} \quad \mathbf{B} = \lambda\mathbf{A} \tag{1.107}$$

1.15 TENSOR MULTIPLICATION

The *outer product* of two tensors of arbitrary order is the tensor whose components are formed by multiplying each component of one of the tensors by every component of the other. This process produces a tensor having an order which is the sum of the orders of the factor tensors. Typical examples of outer products are

(*a*) $a_ib_j = T_{ij}$ (*c*) $D_{ij}T_{km} = \Phi_{ijkm}$

(*b*) $v_iF_{jk} = \alpha_{ijk}$ (*d*) $\epsilon_{ijk}v_m = \Theta_{ijkm}$

As indicated by the above examples, outer products are formed by simply setting down the factor tensors in juxtaposition. (Note that a dyad is formed from two vectors by this very procedure.)

Contraction of a tensor with respect to two free indices is the operation of assigning to both indices the same letter subscript, thereby changing these indices to dummy indices. Contraction produces a tensor having an order two less than the original. Typical examples of contraction are the following.

(*a*) Contractions of T_{ij} and $u_i v_j$

$$T_{ii} = T_{11} + T_{22} + T_{33}$$

$$u_i v_i = u_1 v_1 + u_2 v_2 + u_3 v_3$$

(*b*) Contractions of $E_{ij} a_k$

$$E_{ij} a_j = b_i$$

$$E_{ij} a_i = c_j$$

$$E_{ii} a_k = d_k$$

(*c*) Contractions of $E_{ij} F_{km}$

$$E_{ij} F_{im} = G_{jm} \qquad E_{ij} F_{kk} = P_{ij}$$

$$E_{ij} F_{ki} = H_{jk} \qquad E_{ij} F_{jm} = Q_{im}$$

$$E_{ii} F_{km} = K_{km} \qquad E_{ij} F_{kj} = R_{ik}$$

An *inner product* of two tensors is the result of a contraction, involving one index from each tensor, performed on the outer product of the two tensors. Several inner products important to continuum mechanics are listed here for reference, in both the indicial and symbolic notations.

Outer Product	Inner Product	
	Indicial Notation	Symbolic Notation
1. $a_i b_j$	$a_i b_i$	$\mathbf{a} \cdot \mathbf{b}$
2. $a_i E_{jk}$	$a_i E_{ik} = f_k$	$\mathbf{a} \cdot \mathbf{E} = \mathbf{f}$
	$a_i E_{ji} = h_j$	$\mathbf{E} \cdot \mathbf{a} = \mathbf{h}$
3. $E_{ij} F_{km}$	$E_{ij} F_{jm} = G_{im}$	$\mathbf{E} \cdot \mathbf{F} = \mathbf{G}$
4. $E_{ij} E_{km}$	$E_{ij} E_{jm} = B_{im}$	$\mathbf{E} \cdot \mathbf{E} = (\mathbf{E})^2$

Multiple contractions of fourth-order and higher tensors are sometimes useful. Two such examples are

1. $E_{ij} F_{km}$ contracted to $E_{ij} F_{ij}$, or $\mathbf{E} : \mathbf{F}$
2. $E_{ij} E_{km} E_{pq}$ contracted to $E_{ij} E_{jm} E_{mq}$, or $(\mathbf{E})^3$

1.16 VECTOR CROSS PRODUCT. PERMUTATION SYMBOL. DUAL VECTORS

In order to express the cross product $\mathbf{a} \times \mathbf{b}$ in the indicial notation, the third-order tensor ϵ_{ijk}, known as the *permutation symbol* or *alternating tensor*, must be introduced. This useful tensor is defined by

$$\epsilon_{ijk} = \begin{cases} 1 & \text{if the values of } i, j, k \text{ are an } \textit{even permutation} \text{ of } 1,2,3 \text{ (i.e. if they appear in sequence as in the arrangement } 1\,2\,3\,1\,2). \\ -1 & \text{if the values of } i, j, k \text{ are an } \textit{odd permutation} \text{ of } 1,2,3 \text{ (i.e. if they appear in sequence as in the arrangement } 3\,2\,1\,3\,2). \\ 0 & \text{if the values of } i, j, k \text{ are } \textit{not} \text{ a permutation of } 1,2,3 \text{ (i.e. if two or more of the indices have the same value).} \end{cases}$$

From this definition, the cross product $\mathbf{a} \times \mathbf{b} = \mathbf{c}$ is written in indicial notation by

$$\epsilon_{ijk} a_j b_k = c_i \tag{1.108}$$

Using this relationship, the box product $\mathbf{a} \times \mathbf{b} \cdot \mathbf{c} = \lambda$ may be written

$$\lambda = \epsilon_{ijk} a_i b_j c_k \tag{1.109}$$

Since the same box product is given in the form of a determinant by (*1.52*), it is not surprising that the permutation symbol is frequently used to express the value of a 3×3 determinant.

It is worthwhile to note that ϵ_{ijk} obeys the tensor transformation law for third order Cartesian tensors as long as the transformation is a *proper* one ($\det a_{ij} = 1$) such as arises from a rotation of axes. If the transformation is *improper* ($\det a_{ij} = -1$), e.g. a reflection in one of the coordinate planes whereby a right-handed coordinate system is transformed into a left-handed one, a minus sign must be inserted into the transformation law for ϵ_{ijk}. Such tensors are called *pseudo-tensors.*

The *dual vector* of an arbitrary second-order Cartesian tensor T_{ij} is defined by

$$v_i = \epsilon_{ijk} T_{jk} \tag{1.110}$$

which is observed to be the indicial equivalent of $\mathbf{T}_v$, the "vector of the dyadic $\mathbf{T}$", as defined by (*1.15*).

1.17 MATRICES. MATRIX REPRESENTATION OF CARTESIAN TENSORS

A rectangular array of elements, enclosed by square brackets and subject to certain laws of combination, is called a *matrix.* An $M \times N$ matrix is one having M (horizontal) rows and N (vertical) columns of elements. In the symbol A_{ij}, used to represent the typical element of a matrix, the first subscript denotes the row, the second subscript the column occupied by the element. The matrix itself is designated by enclosing the typical element symbol in square brackets, or alternatively, by the *kernel* letter of the matrix in *script.* For example, the $M \times N$ matrix $\mathcal{A}$, or $[A_{ij}]$ is the array given by

$$\mathcal{A} = [A_{ij}] = \begin{bmatrix} A_{11} & A_{12} & \ldots & A_{1N} \\ A_{21} & A_{22} & \ldots & A_{2N} \\ \cdots & \cdots & \cdots & \cdots \\ A_{M1} & A_{M2} & \ldots & A_{MN} \end{bmatrix} \tag{1.111}$$

A matrix for which $M = N$, is called a *square matrix.* A $1 \times N$ matrix, written $[a_{1k}]$, is called a *row matrix.* An $M \times 1$ matrix, written $[a_{k1}]$, is called a *column matrix.* A matrix having only zeros as elements is called the *zero matrix.* A square matrix with zeros everywhere except on the main diagonal (from A_{11} to A_{NN}) is called a *diagonal matrix.* If the nonzero elements of a diagonal matrix are all *unity,* the matrix is called the *unit* or *identity matrix.* The $N \times M$ matrix $\mathcal{A}^T$, formed by interchanging rows and columns of the $M \times N$ matrix $\mathcal{A}$, is called the *transpose matrix* of $\mathcal{A}$.

Matrices having the same number of rows and columns may be *added* (or subtracted) *element* by *element.* Multiplication of the matrix $[A_{ij}]$ by a scalar λ results in the matrix $[\lambda A_{ij}]$. The product of two matrices, $\mathcal{A}\mathcal{B}$, is defined only if the matrices are *conformable,* i.e. if the *prefactor* matrix $\mathcal{A}$ has the same number of columns as the *postfactor* matrix $\mathcal{B}$ has rows. The product of an $M \times P$ matrix multiplied into a $P \times N$ matrix is an $M \times N$ matrix. Matrix multiplication is usually denoted by simply setting down the matrix symbols in juxtaposition as in

$$\mathcal{A}\mathcal{B} = \mathcal{C} \quad \text{or} \quad [A_{ij}][B_{jk}] = [C_{ik}] \tag{1.112}$$

Matrix multiplication is not, in general, commutative: $\mathcal{A}\mathcal{B} \neq \mathcal{B}\mathcal{A}$.

A square matrix $\mathcal{A}$ whose determinant $|A_{ij}|$ is zero is called a *singular matrix*. The *cofactor* of the element A_{ij} of the square matrix $\mathcal{A}$, denoted here by A^*_{ij}, is defined by

$$A^*_{ij} = (-1)^{i+j} M_{ij} \tag{1.113}$$

in which M_{ij} is the *minor* of A_{ij}; i.e. the determinant of the square array remaining after the row and column of A_{ij} are deleted. The *adjoint* matrix of $\mathcal{A}$ is obtained by replacing each element by its cofactor and then interchanging rows and columns. If a square matrix $\mathcal{A} = [A_{ij}]$ is non-singular, it possesses a unique *inverse matrix* $\mathcal{A}^{-1}$ which is defined as the adjoint matrix of $\mathcal{A}$ divided by the determinant of $\mathcal{A}$. Thus

$$\mathcal{A}^{-1} = \frac{[A^*_{ji}]}{|\mathcal{A}|} \tag{1.114}$$

From the inverse matrix definition (*1.114*) it may be shown that

$$\mathcal{A}^{-1}\mathcal{A} = \mathcal{A}\mathcal{A}^{-1} = \mathcal{I} \tag{1.115}$$

where $\mathcal{I}$ is the *identity matrix*, having ones on the principal diagonal and zeros elsewhere, and so named because of the property

$$\mathcal{I}\mathcal{A} = \mathcal{A}\mathcal{I} = \mathcal{A} \tag{1.116}$$

It is clear, of course, that $\mathcal{I}$ is the matrix representation of δ_{ij}, the Kronecker delta, and of $\mathbf{I}$, the unit dyadic. Any matrix $\mathcal{A}$ for which the condition $\mathcal{A}^T = \mathcal{A}^{-1}$ is satisfied is called an *orthogonal matrix*. Accordingly, if $\mathcal{A}$ is orthogonal,

$$\mathcal{A}^T\mathcal{A} = \mathcal{A}\mathcal{A}^T = \mathcal{I} \tag{1.117}$$

As suggested by the fact that any dyadic may be expressed in the nonion form (*1.53*), and, equivalently, since the components of a second-order tensor may be displayed in the square array (*1.62*), it proves extremely useful to represent second-order tensors (dyadics) by square, 3×3 matrices. A first-order tensor (vector) may be represented by either a 1×3 row matrix, or by a 3×1 column matrix. Although every Cartesian tensor of order two or less (dyadics, vectors, scalars) may be represented by a matrix, not every matrix represents a tensor.

If both matrices in the product $\mathcal{A}\mathcal{B} = \mathcal{C}$ are 3×3 matrices representing second-order tensors, the multiplication is equivalent to the inner product expressed in indicial notation by

$$A_{ij}B_{jk} = C_{ik} \tag{1.118}$$

where the range is three. Expansion of (*1.118*) duplicates the "row by column" multiplication of matrices wherein the elements of the ith row of the prefactor matrix are multiplied in turn by the elements of the kth column of the postfactor matrix, and these products summed to give the element in the ith row and kth column of the product matrix. Several such products occur repeatedly in continuum mechanics and are recorded here in the various notations for reference and comparison.

(*a*) *Vector dot product*

$$\mathbf{a}\cdot\mathbf{b} = \mathbf{b}\cdot\mathbf{a} = \lambda \qquad [a_{1j}][b_{j1}] = [\lambda]$$

$$a_i b_i = b_i a_i = \lambda \qquad [a_1, a_2, a_3]\begin{bmatrix} b_1 \\ b_2 \\ b_3 \end{bmatrix} = [a_1 b_1 + a_2 b_2 + a_3 b_3] \tag{1.119}$$

(*b*) *Vector-dyadic dot product*

$$\mathbf{a}\cdot\mathbf{E} = \mathbf{b} \qquad a\mathcal{E} = \mathcal{B}$$

$$a_iE_{ij} = b_j \qquad [a_{1i}][E_{ij}] = [b_{1j}]$$

$$[a_1, a_2, a_3]\begin{bmatrix} E_{11} & E_{12} & E_{13} \\ E_{21} & E_{22} & E_{23} \\ E_{31} & E_{32} & E_{33} \end{bmatrix} = \begin{matrix}[a_1E_{11} + a_2E_{21} + a_3E_{31}, \\ a_1E_{12} + a_2E_{22} + a_3E_{32}, \\ a_1E_{13} + a_2E_{23} + a_3E_{33}]\end{matrix} \tag{1.120}$$

(*c*) *Dyadic-vector dot product*

$$\mathbf{E}\cdot\mathbf{a} = \mathbf{c} \qquad \mathcal{E}a = c$$

$$E_{ij}a_j = c_i \qquad [E_{ij}][a_{j1}] = [c_{i1}]$$

$$\begin{bmatrix} E_{11} & E_{12} & E_{13} \\ E_{21} & E_{22} & E_{23} \\ E_{31} & E_{32} & E_{33} \end{bmatrix}\begin{bmatrix} a_1 \\ a_2 \\ a_3 \end{bmatrix} = \begin{bmatrix} a_1E_{11} + a_2E_{12} + a_3E_{13} \\ a_1E_{21} + a_2E_{22} + a_3E_{23} \\ a_1E_{31} + a_2E_{32} + a_3E_{33} \end{bmatrix} \tag{1.121}$$

1.18 SYMMETRY OF DYADICS, MATRICES AND TENSORS

According to (*1.36*) (or (*1.37*)), a dyadic $\mathbf{D}$ is said to be symmetric (anti-symmetric) if it is equal to (the negative of) its conjugate $\mathbf{D}_c$. Similarly the second-order tensor D_{ij} is *symmetric* if

$$D_{ij} = D_{ji} \tag{1.122}$$

and is *anti-symmetric, or skew-symmetric,* if

$$D_{ij} = -D_{ji} \tag{1.123}$$

Therefore the decomposition of D_{ij} analogous to (*1.38*) is

$$D_{ij} = \tfrac{1}{2}(D_{ij} + D_{ji}) + \tfrac{1}{2}(D_{ij} - D_{ji}) \tag{1.124}$$

or, in an equivalent abbreviated form often employed,

$$D_{ij} = D_{(ij)} + D_{[ij]} \tag{1.125}$$

where parentheses around the indices denote the symmetric part of D_{ij}, and square brackets on the indices denote the anti-symmetric part.

Since the interchange of indices of a second-order tensor is equivalent to the interchange of rows and columns in its matrix representation, a square matrix $\mathcal{A}$ is symmetric if it is equal to its transpose $\mathcal{A}^T$. Consequently a symmetric 3×3 matrix has only six independent components as illustrated by

$$\mathcal{A} = \mathcal{A}^T = \begin{bmatrix} A_{11} & A_{12} & A_{13} \\ A_{12} & A_{22} & A_{23} \\ A_{13} & A_{23} & A_{33} \end{bmatrix} \tag{1.126}$$

An anti-symmetric matrix is one that equals the *negative* of its transpose. Consequently a 3×3 anti-symmetric matrix $\mathcal{B}$ has zeros on the main diagonal, and therefore only three independent components as illustrated by

$$\mathcal{B} = -\mathcal{B}^T = \begin{bmatrix} 0 & B_{12} & B_{13} \\ -B_{12} & 0 & B_{23} \\ -B_{13} & -B_{23} & 0 \end{bmatrix} \tag{1.127}$$

Symmetry properties may be extended to tensors of higher order than two. In general, an arbitrary tensor is said to be symmetric with respect to a pair of indices if the value of the typical component is unchanged by interchanging these two indices. A tensor is anti-symmetric in a pair of indices if an interchange of these indices leads to a change of sign without a change of absolute value in the component. Examples of symmetry properties in higher-order tensors are

(*a*) $R_{ijkm} = R_{ikjm}$ (symmetric in k and j)

(*b*) $\epsilon_{ijk} = -\epsilon_{kji}$ (anti-symmetric in k and i)

(*c*) $G_{ijkm} = G_{jimk}$ (symmetric in i and j; k and m)

(*d*) $\beta_{ijk} = \beta_{ikj} = \beta_{kji} = \beta_{jik}$ (symmetric in all indices)

1.19 PRINCIPAL VALUES AND PRINCIPAL DIRECTIONS OF SYMMETRIC SECOND-ORDER TENSORS

In the following analysis, only symmetric tensors with real components are considered. This simplifies the mathematics somewhat, and since the important tensors of continuum mechanics are usually symmetric there is little sacrifice in this restriction.

For every symmetric tensor T_{ij}, defined at some point in space, there is associated with each direction (specified by the unit normal n_i) at that point, a vector given by the inner product

$$v_i = T_{ij}n_j \tag{1.128}$$

Here T_{ij} may be envisioned as a linear vector operator which produces the vector v_i conjugate to the direction n_i. If the direction is one for which v_i is parallel to n_i, the inner product may be expressed as a scalar multiple of n_i. For this case,

$$T_{ij}n_j = \lambda n_i \tag{1.129}$$

and the direction n_i is called a *principal direction*, or *principal axis* of T_{ij}. With the help of the identity $n_i = \delta_{ij}n_j$, (*1.129*) can be put in the form

$$(T_{ij} - \lambda\delta_{ij})n_j = 0 \tag{1.130}$$

which represents a system of three equations for the four unknowns, n_i and λ, associated with each principal direction. In expanded form, the system to be solved is

$$\begin{aligned} (T_{11} - \lambda)n_1 + T_{12}n_2 + T_{13}n_3 &= 0 \\ T_{21}n_1 + (T_{22} - \lambda)n_2 + T_{23}n_3 &= 0 \\ T_{31}n_1 + T_{32}n_2 + (T_{33} - \lambda)n_3 &= 0 \end{aligned} \tag{1.131}$$

Note first that for every λ, the trivial solution $n_i = 0$ satisfies the equations. The purpose here, however, is to obtain non-trivial solutions. Also, from the homogeneity of the system (*1.131*) it follows that no loss of generality is incurred by restricting attention to solutions for which $n_in_i = 1$, and this condition is imposed from now on.

For (*1.130*) or, equivalently, (*1.131*) to have a non-trivial solution, the determinant of coefficients must be zero, that is,

$$|T_{ij} - \lambda\delta_{ij}| = 0 \tag{1.132}$$

Expansion of this determinant leads to a cubic polynomial in λ, namely,

$$\lambda^3 - \mathrm{I}_T\lambda^2 + \mathrm{II}_T\lambda - \mathrm{III}_T = 0 \tag{1.133}$$

which is known as the *characteristic equation* of T_{ij}, and for which the scalar coefficients,

$$\mathrm{I}_T = T_{ii} = \operatorname{tr} T_{ij} \text{ (trace of } T_{ij}) \tag{1.134}$$

$$\mathrm{II}_T = \tfrac{1}{2}(T_{ii}T_{jj} - T_{ij}T_{ij}) \tag{1.135}$$

$$\mathrm{III}_T = |T_{ij}| = \det T_{ij} \tag{1.136}$$

are called the first, second and third *invariants*, respectively, of T_{ij}. The three roots of the cubic (*1.133*), labeled $\lambda_{(1)}, \lambda_{(2)}, \lambda_{(3)}$, are called the *principal values* of T_{ij}. For a symmetric tensor with real components, the principal values are real; and if these values are distinct, the three principal directions are mutually orthogonal. When referred to principal axes, both the tensor array and its matrix appear in diagonal form. Thus

$$\mathbf{T} = \begin{pmatrix} \lambda_{(1)} & 0 & 0 \\ 0 & \lambda_{(2)} & 0 \\ 0 & 0 & \lambda_{(3)} \end{pmatrix} \quad \text{or} \quad \mathcal{T} = \begin{bmatrix} \lambda_{(1)} & 0 & 0 \\ 0 & \lambda_{(2)} & 0 \\ 0 & 0 & \lambda_{(3)} \end{bmatrix} \tag{1.137}$$

If $\lambda_{(1)} = \lambda_{(2)}$, the tensor has a diagonal form which is independent of the choice of $\lambda_{(1)}$ and $\lambda_{(2)}$ axes, once the principal axis associated with $\lambda_{(3)}$ has been established. If all principal values are equal, any direction is a principal direction. If the principal values are ordered, it is customary to write them as $\lambda_{(\mathrm{I})}, \lambda_{(\mathrm{II})}, \lambda_{(\mathrm{III})}$ and to display the ordering as in $\lambda_{(\mathrm{I})} > \lambda_{(\mathrm{II})} > \lambda_{(\mathrm{III})}$.

For principal axes labeled $Ox_1^*x_2^*x_3^*$, the transformation from $Ox_1x_2x_3$ axes is given by the elements of the table

	x_1	x_2	x_3
x_1^*	$a_{11} = n_1^{(1)}$	$a_{12} = n_2^{(1)}$	$a_{13} = n_3^{(1)}$
x_2^*	$a_{21} = n_1^{(2)}$	$a_{22} = n_2^{(2)}$	$a_{23} = n_3^{(2)}$
x_3^*	$a_{31} = n_1^{(3)}$	$a_{32} = n_2^{(3)}$	$a_{33} = n_3^{(3)}$

in which $n_i^{(j)}$ are the direction cosines of the jth principal direction.

1.20 POWERS OF SECOND-ORDER TENSORS. HAMILTON-CAYLEY EQUATION

By direct matrix multiplication, the square of the tensor T_{ij} is given as the inner product $T_{ik}T_{kj}$; the cube as $T_{ik}T_{km}T_{mj}$; etc. Therefore with T_{ij} written in the diagonal form (*1.137*), the nth power of the tensor is given by

$$(\mathbf{T})^n = \begin{pmatrix} \lambda_{(1)}^n & 0 & 0 \\ 0 & \lambda_{(2)}^n & 0 \\ 0 & 0 & \lambda_{(3)}^n \end{pmatrix} \quad \text{or} \quad \mathcal{T}^n = \begin{bmatrix} \lambda_{(1)}^n & 0 & 0 \\ 0 & \lambda_{(2)}^n & 0 \\ 0 & 0 & \lambda_{(3)}^n \end{bmatrix} \tag{1.138}$$

A comparison of (*1.138*) and (*1.137*) indicates that T_{ij} and all its integer powers have the same principal axes.

Since each of the principal values satisfies (*1.133*), and because of the diagonal matrix form of $\mathcal{T}^n$ given by (*1.138*), the tensor itself will satisfy (*1.133*). Thus

$$\mathcal{T}^3 - \mathrm{I}_T\mathcal{T}^2 + \mathrm{II}_T\mathcal{T} - \mathrm{III}_T\mathcal{I} = 0 \tag{1.139}$$

in which $\mathcal{I}$ is the identity matrix. This equation is called the *Hamilton-Cayley equation*. Matrix multiplication of each term in (*1.139*) by $\mathcal{T}$ produces the equation,

$$\mathcal{T}^4 = \mathrm{I}_T\mathcal{T}^3 - \mathrm{II}_T\mathcal{T}^2 + \mathrm{III}_T\mathcal{T} \tag{1.140}$$

Combining (*1.140*) and (*1.139*) by direct substitution,

$$\mathcal{T}^4 = (\mathrm{I}_\mathrm{T}^2 - \mathrm{II}_\mathrm{T})\mathcal{T}^2 + (\mathrm{III}_\mathrm{T} - \mathrm{I}_\mathrm{T}\mathrm{II}_\mathrm{T})\mathcal{T} + \mathrm{I}_\mathrm{T}\mathrm{III}_\mathrm{T}\mathcal{I} \tag{1.141}$$

Continuation of this procedure yields the positive powers of $\mathcal{T}$ as linear combinations of $\mathcal{T}^2, \mathcal{T}$ and $\mathcal{I}$.

1.21 TENSOR FIELDS. DERIVATIVES OF TENSORS

A *tensor field* assigns a tensor $\mathbf{T}(\mathbf{x}, t)$ to every pair $(\mathbf{x}, t)$ where the position vector $\mathbf{x}$ varies over a particular region of space and t varies over a particular interval of time. The tensor field is said to be continuous (or differentiable) if the components of $\mathbf{T}(\mathbf{x}, t)$ are continuous (or differentiable) functions of $\mathbf{x}$ and t. If the components are functions of $\mathbf{x}$ only, the tensor field is said to be *steady*.

With respect to a rectangular Cartesian coordinate system, for which the position vector of an arbitrary point is

$$\mathbf{x} = x_i\hat{\mathbf{e}}_i \tag{1.142}$$

tensor fields of various orders are represented in indicial and symbolic notation as follows,

(*a*) scalar field: $\phi = \phi(x_i, t) \quad \text{or} \quad \phi = \phi(\mathbf{x}, t)$ (*1.143*)

(*b*) vector field: $v_i = v_i(\mathbf{x}, t) \quad \text{or} \quad \mathbf{v} = \mathbf{v}(\mathbf{x}, t)$ (*1.144*)

(*c*) second-order tensor field:

$$T_{ij} = T_{ij}(\mathbf{x}, t) \quad \text{or} \quad \mathbf{T} = \mathbf{T}(\mathbf{x}, t) \tag{1.145}$$

Coordinate differentiation of tensor components with respect to x_i is expressed by the differential operator $\partial/\partial x_i$, or briefly in indicial form by ∂_i, indicating an operator of tensor rank one. In symbolic notation, the corresponding symbol is the well-known differential vector operator ∇, pronounced *del* and written explicitly

$$\nabla = \hat{\mathbf{e}}_i \frac{\partial}{\partial x_i} = \hat{\mathbf{e}}_i\partial_i \tag{1.146}$$

Frequently, partial differentiation with respect to the variable x_i is represented by the *comma-subscript convention* as illustrated by the following examples.

(*a*) $\dfrac{\partial \phi}{\partial x_i} = \phi_{,i}$ (*d*) $\dfrac{\partial^2 v_i}{\partial x_j\,\partial x_k} = v_{i,jk}$

(*b*) $\dfrac{\partial v_i}{\partial x_i} = v_{i,i}$ (*e*) $\dfrac{\partial T_{ij}}{\partial x_k} = T_{ij,k}$

(*c*) $\dfrac{\partial v_i}{\partial x_j} = v_{i,j}$ (*f*) $\dfrac{\partial^2 T_{ij}}{\partial x_k\,\partial x_m} = T_{ij,km}$

From these examples it is seen that the operator ∂_i produces a tensor of order one higher if i remains a free index ((*a*) and (*c*) above), and a tensor of order one lower if i becomes a dummy index ((*b*) above) in the derivative.

Several important differential operators appear often in continuum mechanics and are given here for reference.

$$\operatorname{grad}\phi = \nabla\phi = \frac{\partial\phi}{\partial x_i}\hat{\mathbf{e}}_i \quad \text{or} \quad \partial_i\phi = \phi_{,i} \tag{1.147}$$

$$\operatorname{div}\mathbf{v} = \nabla\cdot\mathbf{v} \quad \text{or} \quad \partial_i v_i = v_{i,i} \tag{1.148}$$

$$\operatorname{curl}\mathbf{v} = \nabla\times\mathbf{v} \quad \text{or} \quad \epsilon_{ijk}\partial_j v_k = \epsilon_{ijk}v_{k,j} \tag{1.149}$$

$$\nabla^2\phi = \nabla\cdot\nabla\phi \quad \text{or} \quad \partial_{ii}\phi = \phi_{,ii} \tag{1.150}$$

1.22 LINE INTEGRALS. STOKES' THEOREM

In a given region of space the vector function of position, $\mathbf{F} = \mathbf{F}(\mathbf{x})$, is defined at every point of the piecewise smooth curve C shown in Fig. 1-10. If the *differential tangent vector* to the curve at the arbitrary point P is $d\mathbf{x}$, the integral

$$\int_C \mathbf{F} \cdot d\mathbf{x} \quad \equiv \quad \int_{\mathbf{x}_A}^{\mathbf{x}_B} \mathbf{F} \cdot d\mathbf{x} \tag{1.151}$$

taken along the curve from A to B is known as the *line integral* of F along C. In the indicial notation, (*1.151*) becomes

$$\int_C F_i\, dx_i \quad \equiv \quad \int_{(x_i)_A}^{(x_i)_B} F_i\, dx_i \tag{1.152}$$

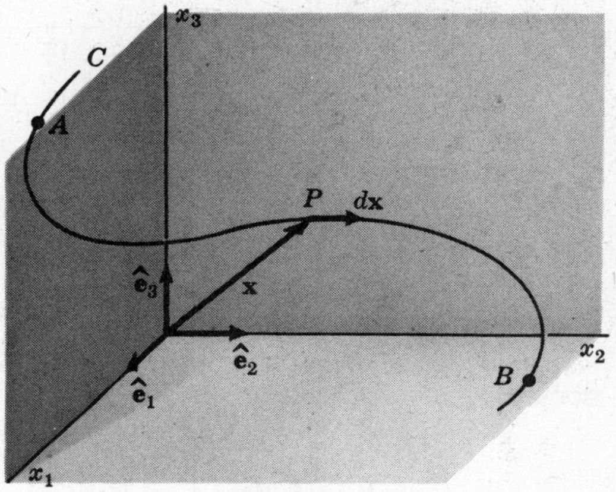

Fig. 1-10

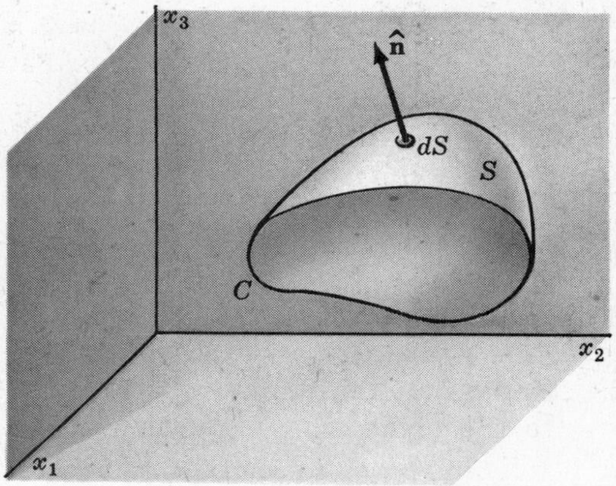

Fig. 1-11

Stokes' theorem says that the line integral of $\mathbf{F}$ taken around a closed reducible curve C, as pictured in Fig. 1-11, may be expressed in terms of an integral over any two-sided surface S which has C as its boundary. Explicitly,

$$\oint_C \mathbf{F} \cdot d\mathbf{x} \quad = \quad \int_S \hat{\mathbf{n}} \cdot (\nabla \times \mathbf{F})\, dS \tag{1.153}$$

in which $\hat{\mathbf{n}}$ is the unit normal on the positive side of S, and dS is the differential element of surface as shown by the figure. In the indicial notation, (*1.153*) is written

$$\oint_C F_i\, dx_i \quad = \quad \int_S n_i \epsilon_{ijk} F_{k,j}\, dS \tag{1.154}$$

1.23 THE DIVERGENCE THEOREM OF GAUSS

The *divergence theorem of Gauss* relates a volume integral to a surface integral. In its traditional form the theorem says that for the vector field $\mathbf{v} = \mathbf{v}(\mathbf{x})$,

$$\int_V \operatorname{div} \mathbf{v}\, dV \quad = \quad \int_S \hat{\mathbf{n}} \cdot \mathbf{v}\, dS \tag{1.155}$$

where $\hat{\mathbf{n}}$ is the outward unit normal to the bounding surface S, of the volume V in which the vector field is defined. In the indicial notation, (*1.155*) is written

$$\int_V v_{i,i}\, dV \quad = \quad \int_S v_i n_i\, dS \tag{1.156}$$

The divergence theorem of Gauss as expressed by (*1.156*) may be generalized to incorporate a tensor field of any order. Thus for the arbitrary tensor field $T_{ijk\ldots}$ the theorem is written

$$\int_V T_{ijk\ldots,p}\, dV \quad = \quad \int_S T_{ijk\ldots} n_p\, dS \tag{1.157}$$

Solved Problems

ALGEBRA OF VECTORS AND DYADICS (Sec. 1.1-1.8)

1.1. Determine in rectangular Cartesian form the unit vector which is (*a*) parallel to the vector $\mathbf{v} = 2\hat{\mathbf{i}} + 3\hat{\mathbf{j}} - 6\hat{\mathbf{k}}$, (*b*) along the line joining points $P(1,0,3)$ and $Q(0,2,1)$.

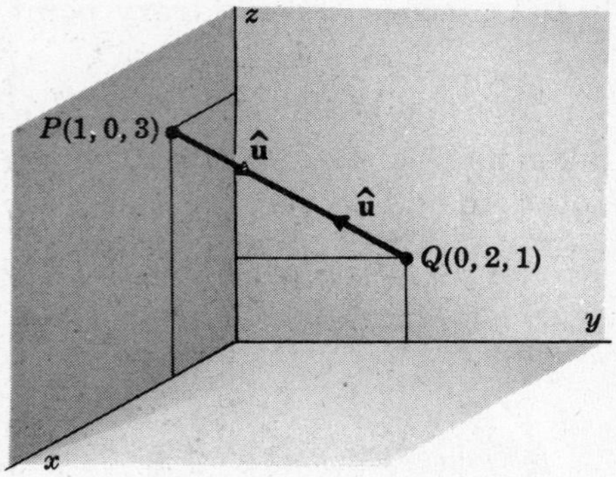

Fig. 1-12

(*a*) $|\mathbf{v}| = v = \sqrt{(2)^2 + (3)^2 + (-6)^2} = 7$

$\hat{\mathbf{v}} = \mathbf{v}/v = (2/7)\hat{\mathbf{i}} + (3/7)\hat{\mathbf{j}} - (6/7)\hat{\mathbf{k}}$

(*b*) The vector extending from P to Q is

$$\mathbf{u} = (0-1)\hat{\mathbf{i}} + (2-0)\hat{\mathbf{j}} + (1-3)\hat{\mathbf{k}} = -\hat{\mathbf{i}} + 2\hat{\mathbf{j}} - 2\hat{\mathbf{k}}$$

$$u = \sqrt{(-1)^2 + (2)^2 + (-2)^2} = 3$$

Thus $\hat{\mathbf{u}} = -(1/3)\hat{\mathbf{i}} + (2/3)\hat{\mathbf{j}} - (2/3)\hat{\mathbf{k}}$ directed from P to Q

or $\hat{\mathbf{u}} = (1/3)\hat{\mathbf{i}} - (2/3)\hat{\mathbf{j}} + (2/3)\hat{\mathbf{k}}$ directed from Q to P

1.2. Prove that the vector $\mathbf{v} = a\hat{\mathbf{i}} + b\hat{\mathbf{j}} + c\hat{\mathbf{k}}$ is normal to the plane whose equation is $ax + by + cz = \lambda$.

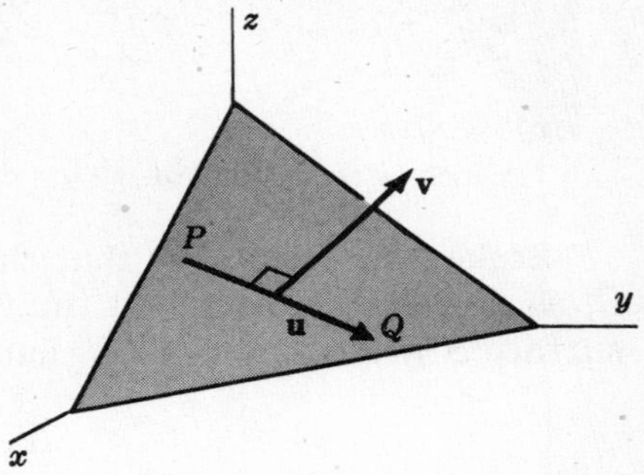

Fig. 1-13

Let $P(x_1, y_1, z_1)$ and $Q(x_2, y_2, z_2)$ be any two points in the plane. Then $ax_1 + by_1 + cz_1 = \lambda$ and $ax_2 + by_2 + cz_2 = \lambda$ and the vector joining these points is $\mathbf{u} = (x_2 - x_1)\hat{\mathbf{i}} + (y_2 - y_1)\hat{\mathbf{j}} + (z_2 - z_1)\hat{\mathbf{k}}$. The projection of $\mathbf{v}$ in the direction of $\mathbf{u}$ is

$$\frac{\mathbf{u}\cdot\mathbf{v}}{u} = \frac{1}{u}[(x_2-x_1)\hat{\mathbf{i}} + (y_2-y_1)\hat{\mathbf{j}} + (z_2-z_1)\hat{\mathbf{k}}]\cdot[a\hat{\mathbf{i}} + b\hat{\mathbf{j}} + c\hat{\mathbf{k}}]$$
$$= \frac{1}{u}(ax_2 + by_2 + cz_2 - ax_1 - by_1 - cz_1) = \frac{\lambda - \lambda}{u} = 0$$

Since $\mathbf{u}$ is any vector in the plane, $\mathbf{v}$ is $\perp$ to the plane.

1.3. If $\mathbf{r} = x\hat{\mathbf{i}} + y\hat{\mathbf{j}} + z\hat{\mathbf{k}}$ is the vector extending from the origin to the arbitrary point $P(x,y,z)$ and $\mathbf{d} = a\hat{\mathbf{i}} + b\hat{\mathbf{j}} + c\hat{\mathbf{k}}$ is a constant vector, show that $(\mathbf{r}-\mathbf{d})\cdot\mathbf{r} = 0$ is the vector equation of a sphere.

Expanding the indicated dot product,

$$(\mathbf{r}-\mathbf{d})\cdot\mathbf{r} = [(x-a)\hat{\mathbf{i}} + (y-b)\hat{\mathbf{j}} + (z-c)\hat{\mathbf{k}}]\cdot[x\hat{\mathbf{i}} + y\hat{\mathbf{j}} + z\hat{\mathbf{k}}]$$
$$= x^2 + y^2 + z^2 - ax - by - cz = 0$$

Adding $d^2/4 = (a^2 + b^2 + c^2)/4$ to each side of this equation gives the desired equation

$$(x - a/2)^2 + (y - b/2)^2 + (z - c/2)^2 = (d/2)^2$$

which is the equation of the sphere centered at $\mathbf{d}/2$ with radius $d/2$.

1.4. Prove that $[\mathbf{a}\cdot\mathbf{b}\times\mathbf{c}]\mathbf{r} = (\mathbf{a}\cdot\mathbf{r})\mathbf{b}\times\mathbf{c} + (\mathbf{b}\cdot\mathbf{r})\mathbf{c}\times\mathbf{a} + (\mathbf{c}\cdot\mathbf{r})\mathbf{a}\times\mathbf{b}$.

Consider the product $\mathbf{a}\times[(\mathbf{b}\times\mathbf{c})\times\mathbf{r}]$. By direct expansion of the cross product in brackets,

$$\mathbf{a}\times[(\mathbf{b}\times\mathbf{c})\times\mathbf{r}] = \mathbf{a}\times[(\mathbf{b}\cdot\mathbf{r})\mathbf{c} - (\mathbf{c}\cdot\mathbf{r})\mathbf{b}] = -(\mathbf{b}\cdot\mathbf{r})\mathbf{c}\times\mathbf{a} - (\mathbf{c}\cdot\mathbf{r})\mathbf{a}\times\mathbf{b}$$

Also, setting $\mathbf{b}\times\mathbf{c} = \mathbf{v}$,

$$\mathbf{a}\times[(\mathbf{b}\times\mathbf{c})\times\mathbf{r}] = \mathbf{a}\times(\mathbf{v}\times\mathbf{r}) = (\mathbf{a}\cdot\mathbf{r})\mathbf{b}\times\mathbf{c} - (\mathbf{a}\cdot\mathbf{b}\times\mathbf{c})\mathbf{r}$$

Thus $-(\mathbf{b}\cdot\mathbf{r})\mathbf{c}\times\mathbf{a} - (\mathbf{c}\cdot\mathbf{r})\mathbf{a}\times\mathbf{b} = (\mathbf{a}\cdot\mathbf{r})\mathbf{b}\times\mathbf{c} - (\mathbf{a}\cdot\mathbf{b}\times\mathbf{c})\mathbf{r}$ and so

$$(\mathbf{a}\cdot\mathbf{b}\times\mathbf{c})\mathbf{r} = (\mathbf{a}\cdot\mathbf{r})\mathbf{b}\times\mathbf{c} + (\mathbf{b}\cdot\mathbf{r})\mathbf{c}\times\mathbf{a} + (\mathbf{c}\cdot\mathbf{r})\mathbf{a}\times\mathbf{b}$$

This identity is useful in specifying the displacement of a rigid body in terms of three arbitrary points of the body.

1.5. Show that if the vectors $\mathbf{a}, \mathbf{b}$ and $\mathbf{c}$ are linearly dependent, $\mathbf{a}\cdot\mathbf{b}\times\mathbf{c} = 0$. Check the linear dependence or independence of the basis

$$\mathbf{u} = 3\hat{\mathbf{i}} + \hat{\mathbf{j}} - 2\hat{\mathbf{k}}$$
$$\mathbf{v} = 4\hat{\mathbf{i}} - \hat{\mathbf{j}} - \hat{\mathbf{k}}$$
$$\mathbf{w} = \hat{\mathbf{i}} - 2\hat{\mathbf{j}} + \hat{\mathbf{k}}$$

The vectors $\mathbf{a}$, $\mathbf{b}$ and $\mathbf{c}$ are linearly dependent if there exist constants λ, μ and ν, not all zero, such that $\lambda\mathbf{a} + \mu\mathbf{b} + \nu\mathbf{c} = 0$. The component scalar equations of this vector equation are

$$\lambda a_x + \mu b_x + \nu c_x = 0$$
$$\lambda a_y + \mu b_y + \nu c_y = 0$$
$$\lambda a_z + \mu b_z + \nu c_z = 0$$

This set has a nonzero solution for λ, μ and ν provided the determinant of coefficients vanishes,

$$\begin{vmatrix} a_x & b_x & c_x \\ a_y & b_y & c_y \\ a_z & b_z & c_z \end{vmatrix} = 0$$

which is equivalent to $\mathbf{a}\cdot\mathbf{b}\times\mathbf{c} = 0$. For the proposed basis $\mathbf{u}, \mathbf{v}, \mathbf{w}$,

$$\begin{vmatrix} 3 & 1 & -2 \\ 4 & -1 & -1 \\ 1 & -2 & 1 \end{vmatrix} = 0$$

Hence the vectors $\mathbf{u}, \mathbf{v}, \mathbf{w}$ are linearly dependent, and indeed $\mathbf{v} = \mathbf{u} + \mathbf{w}$.

1.6. Show that any dyadic of N terms may be reduced to a dyadic of three terms in a form having the base vectors $\hat{\mathbf{e}}_1, \hat{\mathbf{e}}_2, \hat{\mathbf{e}}_3$ as (*a*) antecedents, (*b*) consequents.

Let $\mathsf{D} = \mathbf{a}_1\mathbf{b}_1 + \mathbf{a}_2\mathbf{b}_2 + \cdots + \mathbf{a}_N\mathbf{b}_N = \mathbf{a}_i\mathbf{b}_i$ $(i = 1, 2, \ldots, N)$.

(*a*) In terms of base vectors, $\mathbf{a}_i = a_{1i}\hat{\mathbf{e}}_1 + a_{2i}\hat{\mathbf{e}}_2 + a_{3i}\hat{\mathbf{e}}_3 = a_{ji}\hat{\mathbf{e}}_j$ and so $\mathsf{D} = a_{ji}\hat{\mathbf{e}}_j\mathbf{b}_i = \hat{\mathbf{e}}_j(a_{ji}\mathbf{b}_i) = \hat{\mathbf{e}}_j\mathbf{c}_j$ with $j = 1, 2, 3$.

(*b*) Likewise setting $\mathbf{b}_i = b_{ji}\hat{\mathbf{e}}_j$ it follows that $\mathsf{D} = \mathbf{a}_i b_{ji}\hat{\mathbf{e}}_j = (b_{ji}\mathbf{a}_i)\hat{\mathbf{e}}_j = \mathbf{g}_j\hat{\mathbf{e}}_j$ where $j = 1, 2, 3$.

1.7. For the arbitrary dyadic D and vector $\mathbf{v}$, show that $\mathsf{D}\cdot\mathbf{v} = \mathbf{v}\cdot\mathsf{D}_c$.

Let $\mathsf{D} = \mathbf{a}_1\mathbf{b}_1 + \mathbf{a}_2\mathbf{b}_2 + \cdots + \mathbf{a}_N\mathbf{b}_N$. Then

$$\mathsf{D}\cdot\mathbf{v} = \mathbf{a}_1(\mathbf{b}_1\cdot\mathbf{v}) + \mathbf{a}_2(\mathbf{b}_2\cdot\mathbf{v}) + \cdots + \mathbf{a}_N(\mathbf{b}_N\cdot\mathbf{v})$$
$$= (\mathbf{v}\cdot\mathbf{b}_1)\mathbf{a}_1 + (\mathbf{v}\cdot\mathbf{b}_2)\mathbf{a}_2 + \cdots + (\mathbf{v}\cdot\mathbf{b}_N)\mathbf{a}_N = \mathbf{v}\cdot\mathsf{D}_c$$

1.8. Prove that $(\mathsf{D}_c\cdot\mathsf{D})_c = \mathsf{D}_c\cdot\mathsf{D}$.

From (*1.71*), $\mathsf{D} = D_{ij}\hat{\mathbf{e}}_i\hat{\mathbf{e}}_j$ and $\mathsf{D}_c = D_{ji}\hat{\mathbf{e}}_i\hat{\mathbf{e}}_j$. Therefore

$$\mathsf{D}_c\cdot\mathsf{D} = D_{ji}\hat{\mathbf{e}}_i\hat{\mathbf{e}}_j\cdot D_{pq}\hat{\mathbf{e}}_p\hat{\mathbf{e}}_q = D_{ji}D_{pq}(\hat{\mathbf{e}}_j\cdot\hat{\mathbf{e}}_p)\hat{\mathbf{e}}_i\hat{\mathbf{e}}_q$$

and $$(\mathsf{D}_c\cdot\mathsf{D})_c = D_{ji}D_{pq}(\hat{\mathbf{e}}_j\cdot\hat{\mathbf{e}}_p)\hat{\mathbf{e}}_q\hat{\mathbf{e}}_i = D_{pq}\hat{\mathbf{e}}_q(\hat{\mathbf{e}}_p\cdot\hat{\mathbf{e}}_j)\hat{\mathbf{e}}_i D_{ji} = D_{pq}\hat{\mathbf{e}}_q\hat{\mathbf{e}}_p\cdot D_{ji}\hat{\mathbf{e}}_j\hat{\mathbf{e}}_i = \mathsf{D}_c\cdot\mathsf{D}$$

1.9. Show that $(\mathsf{D}\times\mathbf{v})_c = -\mathbf{v}\times\mathsf{D}_c$.

$$\mathsf{D}\times\mathbf{v} = \mathbf{a}_1(\mathbf{b}_1\times\mathbf{v}) + \mathbf{a}_2(\mathbf{b}_2\times\mathbf{v}) + \cdots + \mathbf{a}_N(\mathbf{b}_N\times\mathbf{v})$$

$$(\mathsf{D}\times\mathbf{v})_c = (\mathbf{b}_1\times\mathbf{v})\mathbf{a}_1 + (\mathbf{b}_2\times\mathbf{v})\mathbf{a}_2 + \cdots + (\mathbf{b}_N\times\mathbf{v})\mathbf{a}_N$$
$$= -(\mathbf{v}\times\mathbf{b}_1)\mathbf{a}_1 - (\mathbf{v}\times\mathbf{b}_2)\mathbf{a}_2 - \cdots - (\mathbf{v}\times\mathbf{b}_N)\mathbf{a}_N = -\mathbf{v}\times\mathsf{D}_c$$

1.10. If $\mathsf{D} = a\hat{\mathbf{i}}\hat{\mathbf{i}} + b\hat{\mathbf{j}}\hat{\mathbf{j}} + c\hat{\mathbf{k}}\hat{\mathbf{k}}$ and $\mathbf{r}$ is the position vector $\mathbf{r} = x\hat{\mathbf{i}} + y\hat{\mathbf{j}} + z\hat{\mathbf{k}}$, show that $\mathbf{r}\cdot\mathsf{D}\cdot\mathbf{r} = 1$ represents the ellipsoid $ax^2 + by^2 + cz^2 = 1$.

$$\mathbf{r}\cdot\mathsf{D}\cdot\mathbf{r} = (x\hat{\mathbf{i}} + y\hat{\mathbf{j}} + z\hat{\mathbf{k}})\cdot(a\hat{\mathbf{i}}\hat{\mathbf{i}} + b\hat{\mathbf{j}}\hat{\mathbf{j}} + c\hat{\mathbf{k}}\hat{\mathbf{k}})\cdot(x\hat{\mathbf{i}} + y\hat{\mathbf{j}} + z\hat{\mathbf{k}})$$
$$= (x\hat{\mathbf{i}} + y\hat{\mathbf{j}} + z\hat{\mathbf{k}})\cdot(ax\hat{\mathbf{i}} + by\hat{\mathbf{j}} + cz\hat{\mathbf{k}}) = ax^2 + by^2 + cz^2 = 1$$

1.11. For the dyadics $\mathsf{D} = 3\hat{\mathbf{i}}\hat{\mathbf{i}} + 2\hat{\mathbf{j}}\hat{\mathbf{j}} - \hat{\mathbf{j}}\hat{\mathbf{k}} + 5\hat{\mathbf{k}}\hat{\mathbf{k}}$ and $\mathsf{F} = 4\hat{\mathbf{i}}\hat{\mathbf{k}} + 6\hat{\mathbf{j}}\hat{\mathbf{j}} - 3\hat{\mathbf{k}}\hat{\mathbf{j}} + \hat{\mathbf{k}}\hat{\mathbf{k}}$, compute and compare the double dot products $\mathsf{D}:\mathsf{F}$ and $\mathsf{D}\cdot\cdot\mathsf{F}$.

From the definition $\mathbf{ab}:\mathbf{cd} = (\mathbf{a}\cdot\mathbf{c})(\mathbf{b}\cdot\mathbf{d})$ it is seen that $\mathsf{D}:\mathsf{F} = 12 + 5 = 17$. Also, from $\mathbf{ab}\cdot\cdot\mathbf{cd} = (\mathbf{b}\cdot\mathbf{c})(\mathbf{a}\cdot\mathbf{d})$ it follows that $\mathsf{D}\cdot\cdot\mathsf{F} = 12 + 3 + 5 = 20$.

1.12. Determine the dyadics $\mathsf{G} = \mathsf{D}\cdot\mathsf{F}$ and $\mathsf{H} = \mathsf{F}\cdot\mathsf{D}$ if D and F are the dyadics given in Problem 1.11.

From the definition $\mathbf{ab}\cdot\mathbf{cd} = (\mathbf{b}\cdot\mathbf{c})\mathbf{ad}$,

$$\mathsf{G} = (3\hat{\mathbf{i}}\hat{\mathbf{i}} + 2\hat{\mathbf{j}}\hat{\mathbf{j}} - \hat{\mathbf{j}}\hat{\mathbf{k}} + 5\hat{\mathbf{k}}\hat{\mathbf{k}})\cdot(4\hat{\mathbf{i}}\hat{\mathbf{k}} + 6\hat{\mathbf{j}}\hat{\mathbf{j}} - 3\hat{\mathbf{k}}\hat{\mathbf{j}} + \hat{\mathbf{k}}\hat{\mathbf{k}})$$
$$= 12\hat{\mathbf{i}}\hat{\mathbf{k}} + 12\hat{\mathbf{j}}\hat{\mathbf{j}} + 3\hat{\mathbf{j}}\hat{\mathbf{j}} - \hat{\mathbf{j}}\hat{\mathbf{k}} - 15\hat{\mathbf{k}}\hat{\mathbf{j}} + 5\hat{\mathbf{k}}\hat{\mathbf{k}}$$

Similarly,
$$\mathsf{H} = (4\hat{\mathbf{i}}\hat{\mathbf{k}} + 6\hat{\mathbf{j}}\hat{\mathbf{j}} - 3\hat{\mathbf{k}}\hat{\mathbf{j}} + \hat{\mathbf{k}}\hat{\mathbf{k}})\cdot(3\hat{\mathbf{i}}\hat{\mathbf{i}} + 2\hat{\mathbf{j}}\hat{\mathbf{j}} - \hat{\mathbf{j}}\hat{\mathbf{k}} + 5\hat{\mathbf{k}}\hat{\mathbf{k}})$$
$$= 20\hat{\mathbf{i}}\hat{\mathbf{k}} + 12\hat{\mathbf{j}}\hat{\mathbf{j}} - 6\hat{\mathbf{j}}\hat{\mathbf{k}} - 6\hat{\mathbf{k}}\hat{\mathbf{j}} + 8\hat{\mathbf{k}}\hat{\mathbf{k}}$$

1.13. Show directly from the nonion form of the dyadic D that $\mathsf{D} = (\mathsf{D}\cdot\hat{\mathbf{i}})\hat{\mathbf{i}} + (\mathsf{D}\cdot\hat{\mathbf{j}})\hat{\mathbf{j}} + (\mathsf{D}\cdot\hat{\mathbf{k}})\hat{\mathbf{k}}$ and also $\hat{\mathbf{i}}\cdot\mathsf{D}\cdot\hat{\mathbf{i}} = D_{xx}$, $\hat{\mathbf{i}}\cdot\mathsf{D}\cdot\hat{\mathbf{j}} = D_{xy}$, etc.

Writing D in nonion form and regrouping terms,

$$\mathsf{D} = (D_{xx}\hat{\mathbf{i}} + D_{yx}\hat{\mathbf{j}} + D_{zx}\hat{\mathbf{k}})\hat{\mathbf{i}} + (D_{xy}\hat{\mathbf{i}} + D_{yy}\hat{\mathbf{j}} + D_{zy}\hat{\mathbf{k}})\hat{\mathbf{j}} + (D_{xz}\hat{\mathbf{i}} + D_{yz}\hat{\mathbf{j}} + D_{zz}\hat{\mathbf{k}})\hat{\mathbf{k}}$$
$$= \mathbf{d}_1\hat{\mathbf{i}} + \mathbf{d}_2\hat{\mathbf{j}} + \mathbf{d}_3\hat{\mathbf{k}} = (\mathsf{D}\cdot\hat{\mathbf{i}})\hat{\mathbf{i}} + (\mathsf{D}\cdot\hat{\mathbf{j}})\hat{\mathbf{j}} + (\mathsf{D}\cdot\hat{\mathbf{k}})\hat{\mathbf{k}}$$

Also now
$$\hat{\mathbf{i}}\cdot\mathbf{d}_1 = \hat{\mathbf{i}}\cdot(\mathsf{D}\cdot\hat{\mathbf{i}}) = \hat{\mathbf{i}}\cdot(D_{xx}\hat{\mathbf{i}} + D_{yx}\hat{\mathbf{j}} + D_{zx}\hat{\mathbf{k}}) = D_{xx}$$
$$\hat{\mathbf{j}}\cdot\mathbf{d}_1 = \hat{\mathbf{j}}\cdot\mathsf{D}\cdot\hat{\mathbf{i}} = D_{yx}, \qquad \hat{\mathbf{j}}\cdot\mathbf{d}_2 = \hat{\mathbf{j}}\cdot\mathsf{D}\cdot\hat{\mathbf{j}} = D_{yy}, \text{ etc.}$$

1.14. For an antisymmetric dyadic A and the arbitrary vector $\mathbf{b}$, show that $2\mathbf{b}\cdot\mathsf{A} = \mathsf{A}_v\times\mathbf{b}$.

From Problem 1.6(*a*), $\mathsf{A} = \hat{\mathbf{e}}_1\mathbf{c}_1 + \hat{\mathbf{e}}_2\mathbf{c}_2 + \hat{\mathbf{e}}_3\mathbf{c}_3$; and because it is antisymmetric, $2\mathsf{A} = (\mathsf{A} - \mathsf{A}_c)$ or

$$2\mathsf{A} = (\hat{\mathbf{e}}_1\mathbf{c}_1 + \hat{\mathbf{e}}_2\mathbf{c}_2 + \hat{\mathbf{e}}_3\mathbf{c}_3 - \mathbf{c}_1\hat{\mathbf{e}}_1 - \mathbf{c}_2\hat{\mathbf{e}}_2 - \mathbf{c}_3\hat{\mathbf{e}}_3)$$
$$= (\hat{\mathbf{e}}_1\mathbf{c}_1 - \mathbf{c}_1\hat{\mathbf{e}}_1 + \hat{\mathbf{e}}_2\mathbf{c}_2 - \mathbf{c}_2\hat{\mathbf{e}}_2 + \hat{\mathbf{e}}_3\mathbf{c}_3 - \mathbf{c}_3\hat{\mathbf{e}}_3)$$

and so
$$2\mathbf{b}\cdot\mathsf{A} = [(\mathbf{b}\cdot\hat{\mathbf{e}}_1)\mathbf{c}_1 - (\mathbf{b}\cdot\mathbf{c}_1)\hat{\mathbf{e}}_1] + [(\mathbf{b}\cdot\hat{\mathbf{e}}_2)\mathbf{c}_2 - (\mathbf{b}\cdot\mathbf{c}_2)\hat{\mathbf{e}}_2] + [(\mathbf{b}\cdot\hat{\mathbf{e}}_3)\mathbf{c}_3 - (\mathbf{b}\cdot\mathbf{c}_3)\hat{\mathbf{e}}_3]$$
$$= [(\hat{\mathbf{e}}_1\times\mathbf{c}_1)\times\mathbf{b} + (\hat{\mathbf{e}}_2\times\mathbf{c}_2)\times\mathbf{b} + (\hat{\mathbf{e}}_3\times\mathbf{c}_3)\times\mathbf{b}] = (\mathsf{A}_v\times\mathbf{b})$$

1.15. If $\mathsf{D} = 6\hat{\mathbf{i}}\hat{\mathbf{i}} + 3\hat{\mathbf{i}}\hat{\mathbf{j}} + 4\hat{\mathbf{k}}\hat{\mathbf{k}}$ and $\mathbf{u} = 2\hat{\mathbf{i}} + \hat{\mathbf{k}}$, $\mathbf{v} = 5\hat{\mathbf{j}}$, show by direct calculation that $\mathsf{D}\cdot(\mathbf{u}\times\mathbf{v}) = (\mathsf{D}\times\mathbf{u})\cdot\mathbf{v}$.

Since $\mathbf{u}\times\mathbf{v} = (2\hat{\mathbf{i}} + \hat{\mathbf{k}})\times 5\hat{\mathbf{j}} = 10\hat{\mathbf{k}} - 5\hat{\mathbf{i}}$,

$$\mathsf{D}\cdot(\mathbf{u}\times\mathbf{v}) = (6\hat{\mathbf{i}}\hat{\mathbf{i}} + 3\hat{\mathbf{i}}\hat{\mathbf{j}} + 4\hat{\mathbf{k}}\hat{\mathbf{k}})\cdot(-5\hat{\mathbf{i}} + 10\hat{\mathbf{k}}) = -30\hat{\mathbf{i}} + 40\hat{\mathbf{k}}$$

Next,
$$\mathsf{D}\times\mathbf{u} = (6\hat{\mathbf{i}}\hat{\mathbf{i}} + 3\hat{\mathbf{i}}\hat{\mathbf{j}} + 4\hat{\mathbf{k}}\hat{\mathbf{k}})\times(2\hat{\mathbf{i}} + \hat{\mathbf{k}}) = -6\hat{\mathbf{i}}\hat{\mathbf{k}} + 8\hat{\mathbf{k}}\hat{\mathbf{j}} - 6\hat{\mathbf{i}}\hat{\mathbf{j}} + 3\hat{\mathbf{i}}\hat{\mathbf{i}}$$

and
$$(\mathsf{D}\times\mathbf{u})\cdot\mathbf{v} = (3\hat{\mathbf{i}}\hat{\mathbf{i}} - 6\hat{\mathbf{i}}\hat{\mathbf{j}} - 6\hat{\mathbf{i}}\hat{\mathbf{k}} + 8\hat{\mathbf{k}}\hat{\mathbf{j}})\cdot 5\hat{\mathbf{j}} = -30\hat{\mathbf{i}} + 40\hat{\mathbf{k}}$$

1.16. Considering the dyadic

$$\mathbf{D} = 3\hat{\mathbf{i}}\hat{\mathbf{i}} - 4\hat{\mathbf{i}}\hat{\mathbf{j}} + 2\hat{\mathbf{j}}\hat{\mathbf{i}} + \hat{\mathbf{j}}\hat{\mathbf{j}} + \hat{\mathbf{k}}\hat{\mathbf{k}}$$

as a linear vector operator, determine the vector $\mathbf{r}'$ produced when $\mathbf{D}$ operates on $\mathbf{r} = 4\hat{\mathbf{i}} + 2\hat{\mathbf{j}} + 5\hat{\mathbf{k}}$.

$$\mathbf{r}' = \mathbf{D}\cdot\mathbf{r} = 12\hat{\mathbf{i}} + 8\hat{\mathbf{j}} - 8\hat{\mathbf{i}} + 2\hat{\mathbf{j}} + 5\hat{\mathbf{k}} = 4\hat{\mathbf{i}} + 10\hat{\mathbf{j}} + 5\hat{\mathbf{k}}$$

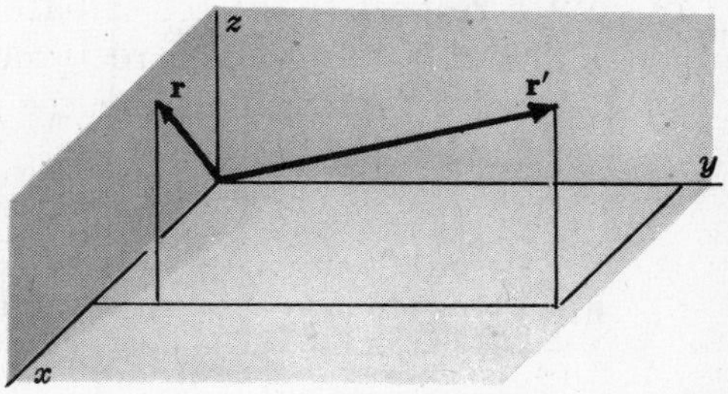

Fig. 1-14

1.17. Determine the dyadic $\mathbf{D}$ which serves as a linear vector operator for the vector function $\mathbf{a} = \mathbf{f}(\mathbf{b}) = \mathbf{b} + \mathbf{b}\times\mathbf{r}$ where $\mathbf{r} = x\hat{\mathbf{i}} + y\hat{\mathbf{j}} + z\hat{\mathbf{k}}$ and $\mathbf{b}$ is a constant vector.

In accordance with (*1.59*) and (*1.60*), construct the vectors

$$\mathbf{u} = \mathbf{f}(\hat{\mathbf{i}}) = \hat{\mathbf{i}} + \hat{\mathbf{i}}\times\mathbf{r} = \hat{\mathbf{i}} - z\hat{\mathbf{j}} + y\hat{\mathbf{k}}$$

$$\mathbf{v} = \mathbf{f}(\hat{\mathbf{j}}) = \hat{\mathbf{j}} + \hat{\mathbf{j}}\times\mathbf{r} = z\hat{\mathbf{i}} + \hat{\mathbf{j}} - x\hat{\mathbf{k}}$$

$$\mathbf{w} = \mathbf{f}(\hat{\mathbf{k}}) = \hat{\mathbf{k}} + \hat{\mathbf{k}}\times\mathbf{r} = -y\hat{\mathbf{i}} + x\hat{\mathbf{j}} + \hat{\mathbf{k}}$$

Then $\mathbf{D} = \mathbf{u}\hat{\mathbf{i}} + \mathbf{v}\hat{\mathbf{j}} + \mathbf{w}\hat{\mathbf{k}} = (\hat{\mathbf{i}} - z\hat{\mathbf{j}} + y\hat{\mathbf{k}})\hat{\mathbf{i}} + (z\hat{\mathbf{i}} + \hat{\mathbf{j}} - x\hat{\mathbf{k}})\hat{\mathbf{j}} + (-y\hat{\mathbf{i}} + x\hat{\mathbf{j}} + \hat{\mathbf{k}})\hat{\mathbf{k}}$

and $\mathbf{a} = \mathbf{D}\cdot\mathbf{b} = (b_x + b_y z - b_z y)\hat{\mathbf{i}} + (-b_x z + b_y + b_z x)\hat{\mathbf{j}} + (b_x y - b_y x + b_z)\hat{\mathbf{k}}$

As a check the same result may be obtained by direct expansion of the vector function,

$$\mathbf{a} = \mathbf{b} + \mathbf{b}\times\mathbf{r} = b_x\hat{\mathbf{i}} + b_y\hat{\mathbf{j}} + b_z\hat{\mathbf{k}} + (b_y z - b_z y)\hat{\mathbf{i}} + (b_z x - b_x z)\hat{\mathbf{j}} + (b_x y - b_y x)\hat{\mathbf{k}}$$

1.18. Express the unit triad $\hat{\mathbf{e}}_\phi, \hat{\mathbf{e}}_\theta, \hat{\mathbf{e}}_r$ in terms of $\hat{\mathbf{i}}, \hat{\mathbf{j}}, \hat{\mathbf{k}}$ and confirm that the curvilinear triad is right-handed by showing that $\hat{\mathbf{e}}_\phi \times \hat{\mathbf{e}}_\theta = \hat{\mathbf{e}}_r$.

By direct projection from Fig. 1-15,

$$\hat{\mathbf{e}}_\phi = (\cos\phi\cos\theta)\hat{\mathbf{i}} + (\cos\phi\sin\theta)\hat{\mathbf{j}} - (\sin\phi)\hat{\mathbf{k}}$$

$$\hat{\mathbf{e}}_\theta = (-\sin\theta)\hat{\mathbf{i}} + (\cos\theta)\hat{\mathbf{j}}$$

$$\hat{\mathbf{e}}_r = (\sin\phi\cos\theta)\hat{\mathbf{i}} + (\sin\phi\sin\theta)\hat{\mathbf{j}} + (\cos\phi)\hat{\mathbf{k}}$$

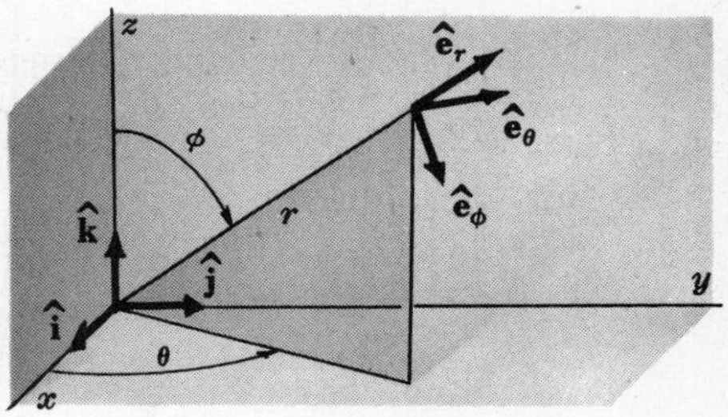

Fig. 1-15

and so

$$\hat{\mathbf{e}}_\phi \times \hat{\mathbf{e}}_\theta = \begin{vmatrix} \hat{\mathbf{i}} & \hat{\mathbf{j}} & \hat{\mathbf{k}} \\ \cos\phi\cos\theta & \cos\phi\sin\theta & -\sin\phi \\ -\sin\theta & \cos\theta & 0 \end{vmatrix}$$

$$= (\sin\phi\cos\theta)\hat{\mathbf{i}} + (\sin\phi\sin\theta)\hat{\mathbf{j}} + [(\cos^2\theta + \sin^2\theta)\cos\phi]\hat{\mathbf{k}} = \hat{\mathbf{e}}_r$$

1.19. Resolve the dyadic $\mathbf{D} = 3\hat{\mathbf{i}}\hat{\mathbf{i}} + 4\hat{\mathbf{i}}\hat{\mathbf{k}} + 6\hat{\mathbf{j}}\hat{\mathbf{i}} + 7\hat{\mathbf{j}}\hat{\mathbf{j}} + 10\hat{\mathbf{k}}\hat{\mathbf{i}} + 2\hat{\mathbf{k}}\hat{\mathbf{j}}$ into its symmetric and antisymmetric parts.

Let $\mathbf{D} = \mathbf{E} + \mathbf{F}$ where $\mathbf{E} = \mathbf{E}_c$ and $\mathbf{F} = -\mathbf{F}_c$. Then

$$\mathbf{E} = (1/2)(\mathbf{D} + \mathbf{D}_c) = (1/2)(6\hat{\mathbf{i}}\hat{\mathbf{i}} + 4\hat{\mathbf{i}}\hat{\mathbf{k}} + 4\hat{\mathbf{k}}\hat{\mathbf{i}} + 6\hat{\mathbf{j}}\hat{\mathbf{i}} + 6\hat{\mathbf{i}}\hat{\mathbf{j}} + 14\hat{\mathbf{j}}\hat{\mathbf{j}} + 10\hat{\mathbf{k}}\hat{\mathbf{i}} + 10\hat{\mathbf{i}}\hat{\mathbf{k}} + 2\hat{\mathbf{k}}\hat{\mathbf{j}} + 2\hat{\mathbf{j}}\hat{\mathbf{k}})$$

$$= 3\hat{\mathbf{i}}\hat{\mathbf{i}} + 3\hat{\mathbf{i}}\hat{\mathbf{j}} + 7\hat{\mathbf{i}}\hat{\mathbf{k}} + 3\hat{\mathbf{j}}\hat{\mathbf{i}} + 7\hat{\mathbf{j}}\hat{\mathbf{j}} + \hat{\mathbf{j}}\hat{\mathbf{k}} + 7\hat{\mathbf{k}}\hat{\mathbf{i}} + \hat{\mathbf{k}}\hat{\mathbf{j}} = \mathbf{E}_c$$

$$\mathbf{F} = (1/2)(\mathbf{D} - \mathbf{D}_c) = (1/2)(4\hat{\mathbf{i}}\hat{\mathbf{k}} - 4\hat{\mathbf{k}}\hat{\mathbf{i}} + 6\hat{\mathbf{j}}\hat{\mathbf{i}} - 6\hat{\mathbf{i}}\hat{\mathbf{j}} + 10\hat{\mathbf{k}}\hat{\mathbf{i}} - 10\hat{\mathbf{i}}\hat{\mathbf{k}} + 2\hat{\mathbf{k}}\hat{\mathbf{j}} - 2\hat{\mathbf{j}}\hat{\mathbf{k}})$$

$$= -3\hat{\mathbf{i}}\hat{\mathbf{j}} - 3\hat{\mathbf{i}}\hat{\mathbf{k}} + 3\hat{\mathbf{j}}\hat{\mathbf{i}} - \hat{\mathbf{j}}\hat{\mathbf{k}} + 3\hat{\mathbf{k}}\hat{\mathbf{i}} + \hat{\mathbf{k}}\hat{\mathbf{j}} = -\mathbf{F}_c$$

1.20. With respect to the set of base vectors $\mathbf{a}_1, \mathbf{a}_2, \mathbf{a}_3$ (not necessarily unit vectors), the set $\mathbf{a}^1, \mathbf{a}^2, \mathbf{a}^3$ is said to be a reciprocal basis if $\mathbf{a}_i \cdot \mathbf{a}^j = \delta_{ij}$. Determine the necessary relationships for constructing the reciprocal base vectors and carry out the calculations for the basis

$$\mathbf{b}_1 = 3\hat{\mathbf{i}} + 4\hat{\mathbf{j}}, \quad \mathbf{b}_2 = -\hat{\mathbf{i}} + 2\hat{\mathbf{j}} + 2\hat{\mathbf{k}}, \quad \mathbf{b}_3 = \hat{\mathbf{i}} + \hat{\mathbf{j}} + \hat{\mathbf{k}}$$

By definition, $\mathbf{a}_1 \cdot \mathbf{a}^1 = 1$, $\mathbf{a}_2 \cdot \mathbf{a}^1 = 0$, $\mathbf{a}_3 \cdot \mathbf{a}^1 = 0$. Hence $\mathbf{a}^1$ is perpendicular to both $\mathbf{a}_2$ and $\mathbf{a}_3$. Therefore it is parallel to $\mathbf{a}_2 \times \mathbf{a}_3$, i.e. $\mathbf{a}^1 = \lambda(\mathbf{a}_2 \times \mathbf{a}_3)$. Since $\mathbf{a}_1 \cdot \mathbf{a}^1 = 1$, $\mathbf{a}_1 \cdot \lambda \mathbf{a}_2 \times \mathbf{a}_3 = 1$ and $\lambda = 1/(\mathbf{a}_1 \cdot \mathbf{a}_2 \times \mathbf{a}_3) = 1/[\mathbf{a}_1\mathbf{a}_2\mathbf{a}_3]$. Thus, in general,

$$\mathbf{a}^1 = \frac{\mathbf{a}_2 \times \mathbf{a}_3}{[\mathbf{a}_1\mathbf{a}_2\mathbf{a}_3]}, \quad \mathbf{a}^2 = \frac{\mathbf{a}_3 \times \mathbf{a}_1}{[\mathbf{a}_1\mathbf{a}_2\mathbf{a}_3]}, \quad \mathbf{a}^3 = \frac{\mathbf{a}_1 \times \mathbf{a}_2}{[\mathbf{a}_1\mathbf{a}_2\mathbf{a}_3]}$$

For the basis $\mathbf{b}_1, \mathbf{b}_2, \mathbf{b}_3$, $1/\lambda = \mathbf{b}_1 \cdot \mathbf{b}_2 \times \mathbf{b}_3 = 12$ and so

$$\mathbf{b}^1 = (\mathbf{b}_2 \times \mathbf{b}_3)/12 = (\hat{\mathbf{j}} - \hat{\mathbf{k}})/4$$
$$\mathbf{b}^2 = (\mathbf{b}_3 \times \mathbf{b}_1)/12 = -\hat{\mathbf{i}}/3 + \hat{\mathbf{j}}/4 + \hat{\mathbf{k}}/12$$
$$\mathbf{b}^3 = (\mathbf{b}_1 \times \mathbf{b}_2)/12 = 2\hat{\mathbf{i}}/3 - \hat{\mathbf{j}}/2 + 5\hat{\mathbf{k}}/6$$

INDICIAL NOTATION — CARTESIAN TENSORS (Sec. 1.9-1.16)

1.21. For a range of three on the indices, give the meaning of the following Cartesian tensor symbols: $A_{ii}, B_{ijj}, R_{ij}, a_iT_{ij}, a_ib_jS_{ij}$.

A_{ii} represents the single sum $A_{ii} = A_{11} + A_{22} + A_{33}$.

B_{ijj} represents three sums: (1) For $i = 1$, $B_{111} + B_{122} + B_{133}$.
(2) For $i = 2$, $B_{211} + B_{222} + B_{233}$.
(3) For $i = 3$, $B_{311} + B_{322} + B_{333}$.

R_{ij} represents the nine components $R_{11}, R_{12}, R_{13}, R_{21}, R_{22}, R_{23}, R_{31}, R_{32}, R_{33}$.

a_iT_{ij} represents three sums: (1) For $j = 1$, $a_1T_{11} + a_2T_{21} + a_3T_{31}$.
(2) For $j = 2$, $a_1T_{12} + a_2T_{22} + a_3T_{32}$.
(3) For $j = 3$, $a_1T_{13} + a_2T_{23} + a_3T_{33}$.

$a_ib_jS_{ij}$ represents a single sum of nine terms. Summing first on i, $a_ib_jS_{ij} = a_1b_jS_{1j} + a_2b_jS_{2j} + a_3b_jS_{3j}$. Now summing each of these three terms on j,

$$a_ib_jS_{ij} = a_1b_1S_{11} + a_1b_2S_{12} + a_1b_3S_{13} + a_2b_1S_{21} + a_2b_2S_{22} + a_2b_3S_{23} + a_3b_1S_{31} + a_3b_2S_{32} + a_3b_3S_{33}$$

1.22. Evaluate the following expressions involving the Kronecker delta δ_{ij} for a range of three on the indices.

(a) $\delta_{ii} = \delta_{11} + \delta_{22} + \delta_{33} = 3$

(b) $\delta_{ij}\delta_{ij} = \delta_{1j}\delta_{1j} + \delta_{2j}\delta_{2j} + \delta_{3j}\delta_{3j} = 3$

(c) $\delta_{ij}\delta_{ik}\delta_{jk} = \delta_{1j}\delta_{1k}\delta_{jk} + \delta_{2j}\delta_{2k}\delta_{jk} + \delta_{3j}\delta_{3k}\delta_{jk} = 3$

(d) $\delta_{ij}\delta_{jk} = \delta_{i1}\delta_{1k} + \delta_{i2}\delta_{2k} + \delta_{i3}\delta_{3k} = \delta_{ik}$

(e) $\delta_{ij}A_{ik} = \delta_{1j}A_{1k} + \delta_{2j}A_{2k} + \delta_{3j}A_{3k} = A_{jk}$

1.23. For the permutation symbol ϵ_{ijk} show by direct expansion that (a) $\epsilon_{ijk}\epsilon_{kij} = 6$, (b) $\epsilon_{ijk}a_ja_k = 0$.

(a) First sum on i, $\epsilon_{ijk}\epsilon_{kij} = \epsilon_{1jk}\epsilon_{k1j} + \epsilon_{2jk}\epsilon_{k2j} + \epsilon_{3jk}\epsilon_{k3j}$

Next sum on j. The nonzero terms are

$$\epsilon_{ijk}\epsilon_{kij} = \epsilon_{12k}\epsilon_{k12} + \epsilon_{13k}\epsilon_{k13} + \epsilon_{21k}\epsilon_{k21} + \epsilon_{23k}\epsilon_{k23} + \epsilon_{31k}\epsilon_{k31} + \epsilon_{32k}\epsilon_{k32}$$

Finally summing on k, the nonzero terms are

$$\begin{aligned}\epsilon_{ijk}\epsilon_{kij} &= \epsilon_{123}\epsilon_{312} + \epsilon_{132}\epsilon_{213} + \epsilon_{213}\epsilon_{321} + \epsilon_{231}\epsilon_{123} + \epsilon_{312}\epsilon_{231} + \epsilon_{321}\epsilon_{132} \\ &= (1)(1) + (-1)(-1) + (-1)(-1) + (1)(1) + (1)(1) + (-1)(-1) = 6\end{aligned}$$

(*b*) Summing on j and k in turn,

$$\begin{aligned}\epsilon_{ijk}a_j a_k &= \epsilon_{i1k}a_1 a_k + \epsilon_{i2k}a_2 a_k + \epsilon_{i3k}a_3 a_k \\ &= \epsilon_{i12}a_1 a_2 + \epsilon_{i13}a_1 a_3 + \epsilon_{i21}a_2 a_1 + \epsilon_{i23}a_2 a_3 + \epsilon_{i31}a_3 a_1 + \epsilon_{i32}a_3 a_2\end{aligned}$$

From this expression,

when $i = 1$, $\epsilon_{1jk}a_j a_k = a_2 a_3 - a_3 a_2 = 0$

when $i = 2$, $\epsilon_{2jk}a_j a_k = a_1 a_3 - a_3 a_1 = 0$

when $i = 3$, $\epsilon_{3jk}a_j a_k = a_1 a_2 - a_2 a_1 = 0$

Note that $\epsilon_{ijk}a_j a_k$ is the indicial form of the vector a crossed into itself, and so $\mathbf{a} \times \mathbf{a} = 0$.

1.24. Determine the component f_2 for the vector expressions given below.

(*a*) $f_i = \epsilon_{ijk}T_{jk}$

$f_2 = \epsilon_{2jk}T_{jk} = \epsilon_{213}T_{13} + \epsilon_{231}T_{31} = -T_{13} + T_{31}$

(*b*) $f_i = c_{i,j}b_j - c_{j,i}b_j$

$$\begin{aligned}f_2 &= c_{2,1}b_1 + c_{2,2}b_2 + c_{2,3}b_3 - c_{1,2}b_1 - c_{2,2}b_2 - c_{3,2}b_3 \\ &= (c_{2,1} - c_{1,2})b_1 + (c_{2,3} - c_{3,2})b_3\end{aligned}$$

(*c*) $f_i = B_{ij}f_j^*$

$f_2 = B_{21}f_1^* + B_{22}f_2^* + B_{23}f_3^*$

1.25. Expand and simplify where possible the expression $D_{ij}x_i x_j$ for (*a*) $D_{ij} = D_{ji}$, (*b*) $D_{ij} = -D_{ji}$.

Expanding,
$$\begin{aligned}D_{ij}x_i x_j &= D_{1j}x_1 x_j + D_{2j}x_2 x_j + D_{3j}x_3 x_j \\ &= D_{11}x_1 x_1 + D_{12}x_1 x_2 + D_{13}x_1 x_3 + D_{21}x_2 x_1 + D_{22}x_2 x_2 \\ &\quad + D_{23}x_2 x_3 + D_{31}x_3 x_1 + D_{32}x_3 x_2 + D_{33}x_3 x_3\end{aligned}$$

(*a*) $D_{ij}x_i x_j = D_{11}(x_1)^2 + D_{22}(x_2)^2 + D_{33}(x_3)^2 + 2D_{12}x_1 x_2 + 2D_{23}x_2 x_3 + 2D_{13}x_1 x_3$

(*b*) $D_{ij}x_i x_j = 0$ since $D_{11} = -D_{11}$, $D_{12} = -D_{21}$, etc.

1.26. Show that $\epsilon_{ijk}\epsilon_{kpq} = \delta_{ip}\delta_{jq} - \delta_{iq}\delta_{jp}$ for (*a*) $i = 1$, $j = q = 2$, $p = 3$ and for (*b*) $i = q = 1$, $j = p = 2$. (It is shown in Problem 1.59 that this identity holds for every choice of indices.)

(*a*) Introduce $i = 1$, $j = 2$, $p = 3$, $q = 2$ and note that since k is a summed index it takes on all values. Then

$$\epsilon_{ijk}\epsilon_{kpq} = \epsilon_{12k}\epsilon_{k32} = \epsilon_{121}\epsilon_{132} + \epsilon_{122}\epsilon_{232} + \epsilon_{123}\epsilon_{332} = 0$$

and

$$\delta_{ip}\delta_{jq} - \delta_{iq}\delta_{jp} = \delta_{13}\delta_{22} - \delta_{12}\delta_{23} = 0$$

(*b*) Introduce $i = 1$, $j = 2$, $p = 2$, $q = 1$. Then $\epsilon_{ijk}\epsilon_{kpq} = \epsilon_{123}\epsilon_{321} = -1$ and $\delta_{ip}\delta_{jq} - \delta_{iq}\delta_{jp} = \delta_{12}\delta_{21} - \delta_{11}\delta_{22} = -1$.

1.27. Show that the tensor $B_{ik} = \epsilon_{ijk}a_j$ is skew-symmetric.

Since by definition of ϵ_{ijk} an interchange of two indices causes a sign change,

$$B_{ik} = \epsilon_{ijk}a_j = -(\epsilon_{kji}a_j) = -(B_{ki}) = -B_{ki}$$

1.28. If B_{ij} is a skew-symmetric Cartesian tensor for which the vector $b_i = (\frac{1}{2})\epsilon_{ijk}B_{jk}$, show that $B_{pq} = \epsilon_{pqi}b_i$.

Multiply the given equation by ϵ_{pqi} and use the identity given in Problem 1.26.

$$\epsilon_{pqi}b_i = \tfrac{1}{2}\epsilon_{pqi}\epsilon_{ijk}B_{jk} = \tfrac{1}{2}(\delta_{pj}\delta_{qk} - \delta_{pk}\delta_{qj})B_{jk} = \tfrac{1}{2}(B_{pq} - B_{qp}) = \tfrac{1}{2}(B_{pq} + B_{pq}) = B_{pq}$$

1.29. Determine directly the components of the metric tensor for spherical polar coordinates as shown in Fig. 1-7(*b*).

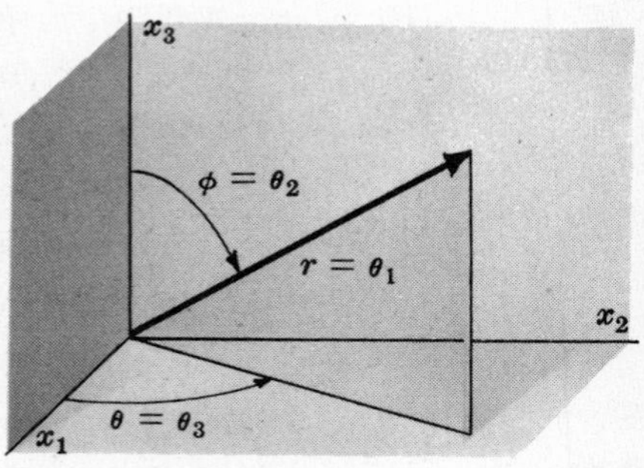

Fig. 1-16

Write (*1.87*) as $g_{pq} = \dfrac{\partial x_i}{\partial \theta_p}\dfrac{\partial x_i}{\partial \theta_q}$ and label the coordinates as shown in Fig. 1-16 ($r = \theta_1$, $\phi = \theta_2$, $\theta = \theta_3$). Then

$$x_1 = \theta_1 \sin\theta_2 \cos\theta_3$$
$$x_2 = \theta_1 \sin\theta_2 \sin\theta_3$$
$$x_3 = \theta_1 \cos\theta_2$$

Hence

$$\frac{\partial x_1}{\partial \theta_1} = \sin\theta_2\cos\theta_3 \qquad \frac{\partial x_1}{\partial \theta_2} = \theta_1\cos\theta_2\cos\theta_3 \qquad \frac{\partial x_1}{\partial \theta_3} = -\theta_1\sin\theta_2\sin\theta_3$$

$$\frac{\partial x_2}{\partial \theta_1} = \sin\theta_2\sin\theta_3 \qquad \frac{\partial x_2}{\partial \theta_2} = \theta_1\cos\theta_2\sin\theta_3 \qquad \frac{\partial x_2}{\partial \theta_3} = \theta_1\sin\theta_2\cos\theta_3$$

$$\frac{\partial x_3}{\partial \theta_1} = \cos\theta_2 \qquad \frac{\partial x_3}{\partial \theta_2} = -\theta_1\sin\theta_2 \qquad \frac{\partial x_3}{\partial \theta_3} = 0$$

from which $$g_{11} = \frac{\partial x_i}{\partial \theta_1}\frac{\partial x_i}{\partial \theta_1} = \sin^2\theta_2\cos^2\theta_3 + \sin^2\theta_2\sin^2\theta_3 + \cos^2\theta_2 = 1$$

$$g_{22} = \frac{\partial x_i}{\partial \theta_2}\frac{\partial x_i}{\partial \theta_2} = \theta_1^2\cos^2\theta_2\cos^2\theta_3 + \theta_1^2\cos^2\theta_2\sin^2\theta_3 + \theta_1^2\sin^2\theta_2 = \theta_1^2$$

$$g_{33} = \frac{\partial x_i}{\partial \theta_3}\frac{\partial x_i}{\partial \theta_3} = \theta_1^2\sin^2\theta_2\sin^2\theta_3 + \theta_1^2\sin^2\theta_2\cos^2\theta_3 = \theta_1^2\sin^2\theta_2$$

Also, $g_{pq} = 0$ for $p \neq q$. For example,

$$\begin{aligned} g_{12} = \frac{\partial x_i}{\partial \theta_1}\frac{\partial x_i}{\partial \theta_2} &= (\sin\theta_2\cos\theta_3)(\theta_1\cos\theta_2\cos\theta_3) \\ &\quad + (\sin\theta_2\sin\theta_3)(\theta_1\cos\theta_2\sin\theta_3) - (\cos\theta_2)(\theta_1\sin\theta_2) \\ &= 0 \end{aligned}$$

Thus for spherical coordinates, $(ds)^2 = (d\theta_1)^2 + (\theta_1)^2(d\theta_2)^2 + (\theta_1\sin\theta_2)^2(d\theta_3)^2$.

1.30. Show that the length of the line element ds resulting from the curvilinear coordinate increment $d\theta_i$ is given by $ds = \sqrt{g_{ii}}\,d\theta_i$ (no sum). Apply this result to the spherical coordinate system of Problem 1.29.

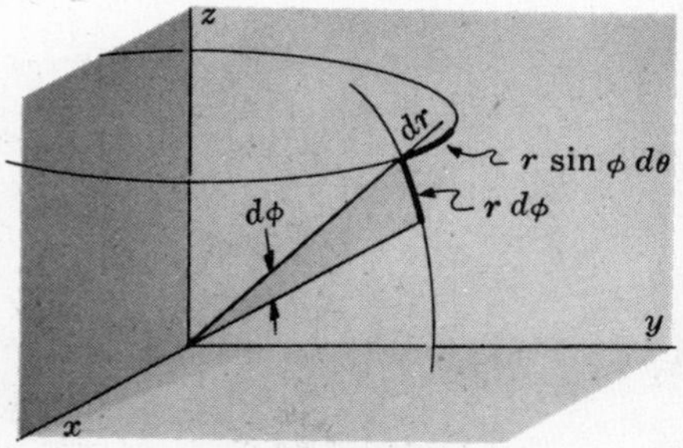

Fig. 1-17

Write (*1.86*) as $(ds)^2 = g_{pq}\,d\theta_p\,d\theta_q$. Thus for the line element $(d\theta_1, 0, 0)$, it follows that $(ds)^2 = g_{11}(d\theta_1)^2$ and $ds = \sqrt{g_{11}}\,d\theta_1$. Similarly for $(0, d\theta_2, 0)$, $ds = \sqrt{g_{22}}\,d\theta_2$; and for $(0, 0, d\theta_3)$, $ds = \sqrt{g_{33}}\,d\theta_3$. Therefore (Fig. 1-17),

(1) For $(d\theta_1, 0, 0)$, $ds = d\theta_1 = dr$

(2) For $(0, d\theta_2, 0)$, $ds = \theta_1\, d\theta_2 = r\, d\phi$

(3) For $(0, 0, d\theta_3)$, $ds = \theta_1 \sin\theta_2\, d\theta_3 = r \sin\phi\, d\theta$

1.31. If the angle between the line elements represented by $(d\theta_1, 0, 0)$ and $(0, d\theta_2, 0)$ is denoted by β_{12}, show that $\cos\beta_{12} = \dfrac{g_{12}}{\sqrt{g_{11}}\sqrt{g_{22}}}$.

Let $ds_1 = \sqrt{g_{11}}\, d\theta_1$ be the length of line element represented by $(d\theta_1, 0, 0)$ and $ds_2 = \sqrt{g_{22}}\, d\theta_2$ be that of $(0, d\theta_2, 0)$. Write *(1.85)* as $dx_i = \dfrac{\partial x_i}{\partial \theta_k} d\theta_k$, and since $(ds)^2 = \cos\beta_{12}\, ds_1\, ds_2$,

$$\begin{aligned}(ds)^2 &= dx_i\, dx_i = \delta_{ij}\, dx_i\, dx_j \\ &= \frac{\partial x_1}{\partial \theta_1}\frac{\partial x_1}{\partial \theta_2} d\theta_1\, d\theta_2 + \frac{\partial x_2}{\partial \theta_1}\frac{\partial x_2}{\partial \theta_2} d\theta_1\, d\theta_2 + \frac{\partial x_3}{\partial \theta_1}\frac{\partial x_3}{\partial \theta_2} d\theta_1\, d\theta_2 = g_{12}\, d\theta_1\, d\theta_2\end{aligned}$$

Hence using the result of Problem 1.30, $\cos\beta_{12} = g_{12}\dfrac{d\theta_1}{ds_1}\dfrac{d\theta_2}{ds_2} = \dfrac{g_{12}}{\sqrt{g_{11}}\sqrt{g_{22}}}$.

1.32. A primed set of Cartesian axes $Ox_1'x_2'x_3'$ is obtained by a rotation through an angle θ about the x_3 axis. Determine the transformation coefficients a_{ij} relating the axes, and give the primed components of the vector $\mathbf{v} = v_1\hat{\mathbf{e}}_1 + v_2\hat{\mathbf{e}}_2 + v_3\hat{\mathbf{e}}_3$.

From the definition (see Section 1.13) $a_{ij} = \cos(x_i', x_j)$ and Fig. 1-18, the table of direction cosines is

	x_1	x_2	x_3
x_1'	$\cos\theta$	$\sin\theta$	0
x_2'	$-\sin\theta$	$\cos\theta$	0
x_3'	0	0	1

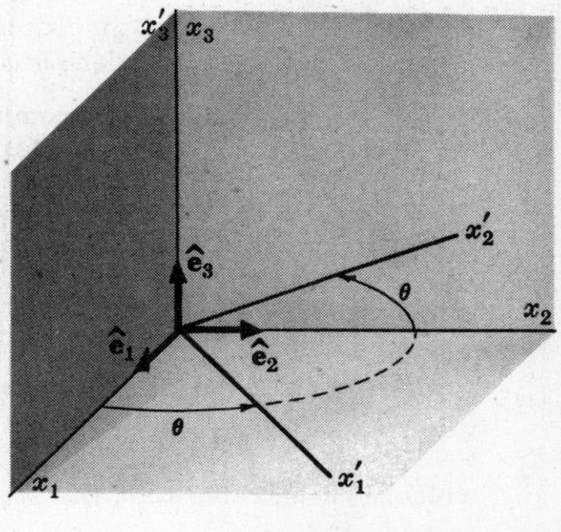

Fig. 1-18

Thus the transformation tensor is

$$\mathbf{A} = \begin{pmatrix} \cos\theta & \sin\theta & 0 \\ -\sin\theta & \cos\theta & 0 \\ 0 & 0 & 1 \end{pmatrix}$$

By the transformation law for vectors *(1.94)*,

$$\begin{aligned} v_1' &= a_{1j}v_j = v_1\cos\theta + v_2\sin\theta \\ v_2' &= a_{2j}v_j = -v_1\sin\theta + v_2\cos\theta \\ v_3' &= a_{3j}v_j = v_3 \end{aligned}$$

1.33. The table of direction cosines relating two sets of rectangular Cartesian axes is partially given as shown on the right. Determine the entries for the bottom row of the table so that $Ox_1'x_2'x_3'$ is a right-handed system.

	x_1	x_2	x_3
x_1'	3/5	−4/5	0
x_2'	0	0	1
x_3'			

The unit vector $\hat{\mathbf{e}}'_1$ along the x'_1 axis is given by the first row of the table as $\hat{\mathbf{e}}'_1 = (3/5)\hat{\mathbf{e}}_1 - (4/5)\hat{\mathbf{e}}_2$. Also from the table $\hat{\mathbf{e}}'_2 = \hat{\mathbf{e}}_3$. For a right-handed primed system $\hat{\mathbf{e}}'_3 = \hat{\mathbf{e}}'_1 \times \hat{\mathbf{e}}'_2$, or $\hat{\mathbf{e}}'_3 = [(3/5)\hat{\mathbf{e}}_1 - (4/5)\hat{\mathbf{e}}_2] \times \hat{\mathbf{e}}_3 = (-3/5)\hat{\mathbf{e}}_2 - (4/5)\hat{\mathbf{e}}_1$ and the third row is

x'_3	$-4/5$	$-3/5$	0

.

1.34. Let the *angles* between the primed and unprimed coordinate directions be given by the table shown on the right. Determine the transformation coefficients a_{ij} and show that the orthogonality conditions are satisfied.

	x_1	x_2	x_3
x'_1	135°	60°	120°
x'_2	90°	45°	45°
x'_3	45°	60°	120°

The coefficients a_{ij} are direction cosines and may be calculated directly from the table. Thus

$$a_{ij} = \begin{pmatrix} -1/\sqrt{2} & 1/2 & -1/2 \\ 0 & 1/\sqrt{2} & 1/\sqrt{2} \\ 1/\sqrt{2} & 1/2 & -1/2 \end{pmatrix}$$

The orthogonality conditions $a_{ij}a_{ik} = \delta_{jk}$ require:

1. For $j = k = 1$ that $a_{11}a_{11} + a_{21}a_{21} + a_{31}a_{31} = 1$ which is seen to be the sum of squares of the elements in the first column.
2. For $j = 2,\ k = 3$ that $a_{12}a_{13} + a_{22}a_{23} + a_{32}a_{33} = 0$ which is seen to be the sum of products of corresponding elements of the second and third columns.
3. Any two columns "multiplied together element by element and summed" to be zero. The sum of squares of elements of any column to be unity.

For orthogonality conditions in the form $a_{ji}a_{ki} = \delta_{jk}$, the rows are multiplied together instead of the columns. All of these conditions are satisfied by the above solution.

1.35. Show that the sum $\lambda A_{ij} + \mu B_{ij}$ represents the components of a second-order tensor if A_{ij} and B_{ij} are known second-order tensors.

By (*1.103*) and the statement of the problem, $A_{ij} = a_{pi}a_{qj}A'_{pq}$ and $B_{ij} = a_{pi}a_{qj}B'_{pq}$. Hence

$$\lambda A_{ij} + \mu B_{ij} = \lambda(a_{pi}a_{qj}A'_{pq}) + \mu(a_{pi}a_{qj}B'_{pq}) = a_{pi}a_{qj}(\lambda A'_{pq} + \mu B'_{pq})$$

which demonstrates that the sum transforms as a second-order Cartesian tensor.

1.36. Show that $(P_{ijk} + P_{jki} + P_{jik})\, x_i x_j x_k = 3P_{ijk}\, x_i x_j x_k$.

Since all indices are dummy indices and the order of the variables x_i is unimportant, each term of the sum is equivalent to the others. This may be readily shown by introducing new dummy variables. Thus replacing i, j, k in the second and third terms by p, q, r, the sum becomes

$$P_{ijk}x_ix_jx_k + P_{qrp}x_px_qx_r + P_{qpr}x_px_qx_r$$

Now change dummy indices in these same terms again so that the form is

$$P_{ijk}\,x_ix_jx_k + P_{ijk}\,x_kx_ix_j + P_{ijk}\,x_jx_ix_k = 3P_{ijk}\,x_ix_jx_k$$

1.37. If B_{ij} is skew-symmetric and A_{ij} is symmetric, show that $A_{ij}B_{ij} = 0$.

Since $A_{ij} = A_{ji}$ and $B_{ij} = -B_{ji}$, $A_{ij}B_{ij} = -A_{ji}B_{ji}$ or $A_{ij}B_{ij} + A_{ji}B_{ji} = A_{ij}B_{ij} + A_{pq}B_{pq} = 0$. Since all indices are dummy indices, $A_{pq}B_{pq} = A_{ij}B_{ij}$ and so $2A_{ij}B_{ij} = 0$, or $A_{ij}B_{ij} = 0$.

1.38. Show that the quadratic form $D_{ij}x_ix_j$ is unchanged if D_{ij} is replaced by its symmetric part $D_{(ij)}$.

Resolving D_{ij} into its symmetric and anti-symmetric parts,

$$D_{ij} = D_{(ij)} + D_{[ij]} = \tfrac{1}{2}(D_{ij} + D_{ji}) + \tfrac{1}{2}(D_{ij} - D_{ji})$$

Then $$D_{(ij)}x_ix_j = \tfrac{1}{2}(D_{ij} + D_{ji})x_ix_j = \tfrac{1}{2}(D_{ij}x_ix_j + D_{pq}x_qx_p) = D_{ij}x_ix_j$$

1.39. Use indicial notation to prove the vector identities

$$(1)\ \mathbf{a} \times (\mathbf{b} \times \mathbf{c}) = (\mathbf{a}\cdot\mathbf{c})\mathbf{b} - (\mathbf{a}\cdot\mathbf{b})\mathbf{c}, \qquad (2)\ \mathbf{a} \times \mathbf{b}\cdot\mathbf{a} = 0$$

(1) Let $\mathbf{v} = \mathbf{b} \times \mathbf{c}$. Then $v_i = \epsilon_{ijk}b_jc_k$; and if $\mathbf{a} \times \mathbf{v} = \mathbf{w}$, then

$$\begin{aligned} w_p &= \epsilon_{pqi}a_q\epsilon_{ijk}b_jc_k \\ &= (\delta_{pj}\delta_{qk} - \delta_{pk}\delta_{qj})a_qb_jc_k \quad \text{(see Problem 1.26)} \\ &= a_qb_pc_q - a_qb_qc_p \\ &= (a_qc_q)b_p - (a_qb_q)c_p \end{aligned}$$

Transcribing this expression into symbolic notation,

$$\mathbf{w} = \mathbf{a} \times (\mathbf{b} \times \mathbf{c}) = (\mathbf{a}\cdot\mathbf{c})\mathbf{b} - (\mathbf{a}\cdot\mathbf{b})\mathbf{c}$$

(2) Let $\mathbf{a} \times \mathbf{b} = \mathbf{v}$. Thus $v_i = \epsilon_{ijk}a_jb_k$; and if $\lambda = \mathbf{v}\cdot\mathbf{a}$, then $\lambda = \epsilon_{ijk}(a_ia_jb_k)$. But ϵ_{ijk} is skew-symmetric in i and j, while $(a_ia_jb_k)$ is symmetric in i and j. Hence the product $\epsilon_{ijk}a_ia_jb_k$ vanishes as may also be shown by direct expansion.

$$\begin{aligned} \lambda &= \epsilon_{ij1}a_ia_jb_1 + \epsilon_{ij2}a_ia_jb_2 + \epsilon_{ij3}a_ia_jb_3 \\ &= (\epsilon_{321}a_3a_2 + \epsilon_{231}a_2a_3)b_1 + \cdots \\ &= (-a_2a_3 + a_2a_3)b_1 + (0)b_2 + (0)b_3 = 0 \end{aligned}$$

1.40. Show that the determinant

$$\det A_{ij} = \begin{vmatrix} A_{11} & A_{12} & A_{13} \\ A_{21} & A_{22} & A_{23} \\ A_{31} & A_{32} & A_{33} \end{vmatrix}$$

may be expressed in the form $\epsilon_{ijk}A_{1i}A_{2j}A_{3k}$.

From (*1.52*) and (*1.109*) the box product [**abc**] may be written

$$\lambda = \mathbf{a}\cdot\mathbf{b}\times\mathbf{c} = [\mathbf{abc}] = \epsilon_{ijk}a_ib_jc_k = \begin{vmatrix} a_1 & a_2 & a_3 \\ b_1 & b_2 & b_3 \\ c_1 & c_2 & c_3 \end{vmatrix}$$

If now the substitutions $a_i = A_{1i}$, $b_i = A_{2i}$ and $c_i = A_{3i}$ are introduced,

$$\lambda = \epsilon_{ijk}a_ib_jc_k = \epsilon_{ijk}A_{1i}A_{2j}A_{3k}$$

This result may also be obtained by direct expansion of the determinant. An equivalent expression for the determinant is $\epsilon_{ijk}A_{i1}A_{j2}A_{k3}$.

1.41. If the vector v_i is given in terms of base vectors **a, b, c** by $v_i = \alpha a_i + \beta b_i + \gamma c_i$, show that $\alpha = \dfrac{\epsilon_{ijk}v_ib_jc_k}{\epsilon_{pqr}a_pb_qc_r}$.

$$\begin{aligned} v_1 &= \alpha a_1 + \beta b_1 + \gamma c_1 \\ v_2 &= \alpha a_2 + \beta b_2 + \gamma c_2 \\ v_3 &= \alpha a_3 + \beta b_3 + \gamma c_3 \end{aligned}$$

By Cramer's rule, $\alpha = \dfrac{\begin{vmatrix} v_1 & b_1 & c_1 \\ v_2 & b_2 & c_2 \\ v_3 & b_3 & c_3 \end{vmatrix}}{\begin{vmatrix} a_1 & b_1 & c_1 \\ a_2 & b_2 & c_2 \\ a_3 & b_3 & c_3 \end{vmatrix}}$ and by (*1.52*) and (*1.109*), $\alpha = \dfrac{\epsilon_{ijk}v_i b_j c_k}{\epsilon_{pqr}a_p b_q c_r}$.

Likewise $\beta = \dfrac{\epsilon_{ijk}a_i v_j c_k}{\epsilon_{pqr}a_p b_q c_r}$, $\gamma = \dfrac{\epsilon_{ijk}a_i b_j v_k}{\epsilon_{pqr}a_p b_q c_r}$.

MATRICES AND MATRIX METHODS (Sec. 1.17-1.20)

1.42. For the vectors $\mathbf{a} = 3\hat{\mathbf{i}} + 4\hat{\mathbf{k}}$, $\mathbf{b} = 2\hat{\mathbf{j}} - 6\hat{\mathbf{k}}$ and the dyadic $\mathbf{D} = 3\hat{\mathbf{i}}\hat{\mathbf{i}} + 2\hat{\mathbf{i}}\hat{\mathbf{k}} - 4\hat{\mathbf{j}}\hat{\mathbf{j}} - 5\hat{\mathbf{k}}\hat{\mathbf{j}}$, compute by matrix multiplication the products $\mathbf{a}\cdot\mathbf{D}$, $\mathbf{D}\cdot\mathbf{b}$ and $\mathbf{a}\cdot\mathbf{D}\cdot\mathbf{b}$.

Let $\mathbf{a}\cdot\mathbf{D} = \mathbf{v}$; then $[v_1, v_2, v_3] = [3, 0, 4]\begin{bmatrix} 3 & 0 & 2 \\ 0 & -4 & 0 \\ 0 & -5 & 0 \end{bmatrix} = [9, -20, 6]$.

Let $\mathbf{D}\cdot\mathbf{b} = \mathbf{w}$; then $\begin{bmatrix} w_1 \\ w_2 \\ w_3 \end{bmatrix} = \begin{bmatrix} 3 & 0 & 2 \\ 0 & -4 & 0 \\ 0 & -5 & 0 \end{bmatrix}\begin{bmatrix} 0 \\ 2 \\ -6 \end{bmatrix} = \begin{bmatrix} -12 \\ -8 \\ -10 \end{bmatrix}$.

Let $\mathbf{a}\cdot\mathbf{D}\cdot\mathbf{b} = \mathbf{v}\cdot\mathbf{b} = \lambda$; then $[\lambda] = [9, -20, 6]\begin{bmatrix} 0 \\ 2 \\ -6 \end{bmatrix} = [-76]$.

1.43. Determine the principal directions and principal values of the second-order Cartesian tensor $\mathbf{T}$ whose matrix representation is

$$[T_{ij}] = \begin{vmatrix} 3 & -1 & 0 \\ -1 & 3 & 0 \\ 0 & 0 & 1 \end{vmatrix}$$

From (*1.132*), for principal values λ,

$$\begin{vmatrix} 3-\lambda & -1 & 0 \\ -1 & 3-\lambda & 0 \\ 0 & 0 & 1-\lambda \end{vmatrix} = (1-\lambda)[(3-\lambda)^2 - 1] = 0$$

which results in the cubic equation $\lambda^3 - 7\lambda^2 + 14\lambda - 8 = (\lambda-1)(\lambda-2)(\lambda-4) = 0$ whose principal values are $\lambda_{(1)} = 1$, $\lambda_{(2)} = 2$, $\lambda_{(3)} = 4$.

Next let $n_i^{(1)}$ be the components of the unit normal in the principal direction associated with $\lambda_{(1)} = 1$. Then the first two equations of (*1.131*) give $2n_1^{(1)} - n_2^{(1)} = 0$ and $-n_1^{(1)} + 2n_2^{(1)} = 0$, from which $n_1^{(1)} = n_2^{(1)} = 0$; and from $n_i n_i = 1$, $n_3^{(1)} = \pm 1$.

For $\lambda_{(2)} = 2$, (*1.131*) yields $n_1^{(2)} - n_2^{(2)} = 0$, $-n_1^{(2)} + n_2^{(2)} = 0$, and $-n_3^{(2)} = 0$. Thus $n_1^{(2)} = n_2^{(2)} = \pm 1/\sqrt{2}$, since $n_i n_i = 1$ and $n_3^{(2)} = 0$.

For $\lambda_{(3)} = 4$, (*1.131*) yields $-n_1^{(3)} - n_2^{(3)} = 0$, $-n_1^{(3)} - n_2^{(3)} = 0$, and $3n_3^{(3)} = 0$. Thus $n_3^{(3)} = 0$, $n_1^{(3)} = -n_2^{(3)} = \mp 1/\sqrt{2}$.

The principal axes x_i^* may be referred to the original axes x_i through the table of direction cosines

	x_1	x_2	x_3
x_1^*	0	0	± 1
x_2^*	$\pm 1/\sqrt{2}$	$\pm 1/\sqrt{2}$	0
x_3^*	$\mp 1/\sqrt{2}$	$\pm 1/\sqrt{2}$	0

from which the transformation matrix (tensor) may be written:

$$\mathcal{A} = \begin{bmatrix} 0 & 0 & \pm 1 \\ \pm 1/\sqrt{2} & \pm 1/\sqrt{2} & 0 \\ \mp 1/\sqrt{2} & \pm 1/\sqrt{2} & 0 \end{bmatrix} \quad \text{or} \quad a_{ij} = \begin{pmatrix} 0 & 0 & \pm 1 \\ \pm 1/\sqrt{2} & \pm 1/\sqrt{2} & 0 \\ \mp 1/\sqrt{2} & \pm 1/\sqrt{2} & 0 \end{pmatrix}$$

1.44. Show that the principal axes determined in Problem 1.43 form a right-handed set of orthogonal axes.

Orthogonality requires that the conditions $a_{ij}a_{ik} = \delta_{jk}$ be satisfied. Since the condition $n_i n_i = 1$ was used in determining the a_{ij}, orthogonality is automatically satisfied for $j = k$. Multiplying the corresponding elements of any row (column) by those of any other row (column) and adding the products demonstrates that the conditions for $j \neq k$ are satisfied by the solution in Problem 1.43.

Finally for the system to be right-handed, $\hat{\mathbf{n}}^{(2)} \times \hat{\mathbf{n}}^{(3)} = \hat{\mathbf{n}}^{(1)}$. Thus

$$\begin{vmatrix} \hat{\mathbf{e}}_1 & \hat{\mathbf{e}}_2 & \hat{\mathbf{e}}_3 \\ 1/\sqrt{2} & 1/\sqrt{2} & 0 \\ -1/\sqrt{2} & 1/\sqrt{2} & 0 \end{vmatrix} = (\tfrac{1}{2} + \tfrac{1}{2})\,\hat{\mathbf{e}}_3 = \hat{\mathbf{e}}_3$$

As indicated by the plus and minus values of a_{ij} in Problem 1.43, there are two sets of principal axes, x_i^* and x_i^{**}. As shown by the sketch both sets are along the principal directions with x_i^* being a right-handed system, x_i^{**} a left-handed system.

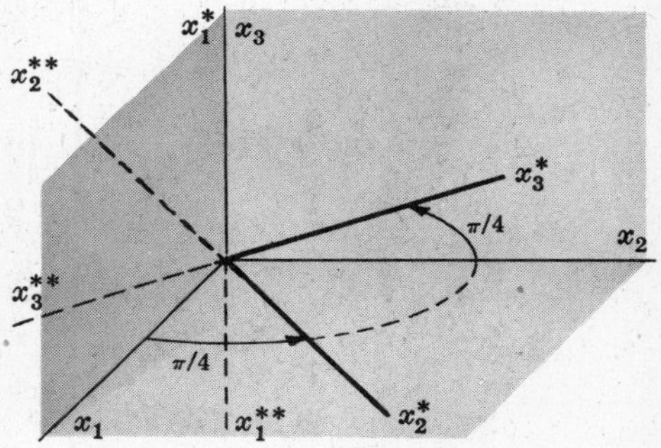

Fig. 1-19

1.45. Show that the matrix of the tensor T_{ij} of Problem 1.43 may be put into diagonal (principal) form by the transformation law $T^*_{ij} = a_{ip}a_{jq}T_{pq}$, (or in matrix symbols $\mathcal{T}^* = \mathcal{A}\mathcal{T}\mathcal{A}^T$).

$$[T^*_{ij}] = \begin{bmatrix} 0 & 0 & 1 \\ 1/\sqrt{2} & 1/\sqrt{2} & 0 \\ -1/\sqrt{2} & 1/\sqrt{2} & 0 \end{bmatrix} \begin{bmatrix} 3 & -1 & 0 \\ -1 & 3 & 0 \\ 0 & 0 & 1 \end{bmatrix} \begin{bmatrix} 0 & 1/\sqrt{2} & -1/\sqrt{2} \\ 0 & 1/\sqrt{2} & 1/\sqrt{2} \\ 1 & 0 & 0 \end{bmatrix}$$

$$= \begin{bmatrix} 0 & 0 & 1 \\ \sqrt{2} & \sqrt{2} & 0 \\ -2\sqrt{2} & 2\sqrt{2} & 0 \end{bmatrix} \begin{bmatrix} 0 & 1/\sqrt{2} & -1/\sqrt{2} \\ 0 & 1/\sqrt{2} & 1/\sqrt{2} \\ 1 & 0 & 0 \end{bmatrix} = \begin{bmatrix} 1 & 0 & 0 \\ 0 & 2 & 0 \\ 0 & 0 & 4 \end{bmatrix}$$

1.46. Prove that if the principal values $\lambda_{(1)}, \lambda_{(2)}, \lambda_{(3)}$ of a symmetric second-order tensor are all distinct, the principal directions are mutually orthogonal.

The proof is made for $\lambda_{(2)}$ and $\lambda_{(3)}$. For each of these (*1.129*) is satisfied, so that $T_{ij}n_j^{(2)} = \lambda_{(2)}n_i^{(2)}$ and $T_{ij}n_j^{(3)} = \lambda_{(3)}n_i^{(3)}$. Multiplying the first of these equations by $n_i^{(3)}$ and the second by $n_i^{(2)}$,

$$T_{ij}n_j^{(2)}n_i^{(3)} = \lambda_{(2)}n_i^{(2)}n_i^{(3)}$$

$$T_{ij}n_j^{(3)}n_i^{(2)} = \lambda_{(3)}n_i^{(3)}n_i^{(2)}$$

Since T_{ij} is symmetric, the dummy indices i and j may be interchanged on the left-hand side of the second of these equations and that equation subtracted from the first to yield

$$(\lambda_{(2)} - \lambda_{(3)})n_i^{(2)}n_i^{(3)} = 0$$

Since $\lambda_{(2)} \neq \lambda_{(3)}$, their difference is not zero. Hence $n_i^{(2)}n_i^{(3)} = 0$, the condition for the two directions to be perpendicular.

1.47. Compute the principal values of $(\mathbf{T})^2$ of Problem 1.43 and verify that its principal axes coincide with those of $\mathbf{T}$.

$$[T_{ij}]^2 = \begin{bmatrix} 3 & -1 & 0 \\ -1 & 3 & 0 \\ 0 & 0 & 1 \end{bmatrix}\begin{bmatrix} 3 & -1 & 0 \\ -1 & 3 & 0 \\ 0 & 0 & 1 \end{bmatrix} = \begin{bmatrix} 10 & -6 & 0 \\ -6 & 10 & 0 \\ 0 & 0 & 1 \end{bmatrix}$$

The characteristic equation for this matrix is

$$\begin{vmatrix} 10-\lambda & -6 & 0 \\ -6 & 10-\lambda & 0 \\ 0 & 0 & 1-\lambda \end{vmatrix} = (1-\lambda)[(10-\lambda)^2 - 36] = (1-\lambda)(\lambda-4)(\lambda-16) = 0$$

from which $\lambda_{(1)} = 1$, $\lambda_{(2)} = 4$, $\lambda_{(3)} = 16$. Substituting these into (*1.131*) and using the condition $n_in_i = 1$,

For $\lambda_{(1)} = 1$,
$$\left.\begin{aligned} 9n_1^{(1)} - 6n_2^{(1)} &= 0 \\ -6n_1^{(1)} + 9n_2^{(1)} &= 0 \end{aligned}\right\} \quad \text{or} \quad n_1^{(1)} = n_2^{(1)} = 0,\ n_3^{(1)} = \pm 1$$

For $\lambda_{(2)} = 4$,
$$\left.\begin{aligned} 6n_1^{(2)} - 6n_2^{(2)} &= 0 \\ -6n_1^{(2)} + 6n_2^{(2)} &= 0 \\ -3n_3^{(2)} &= 0 \end{aligned}\right\} \quad \text{or} \quad n_1^{(2)} = n_2^{(2)} = \pm 1\sqrt{2},\ n_3^{(2)} = 0$$

For $\lambda_{(3)} = 16$,
$$\left.\begin{aligned} -6n_1^{(3)} - 6n_2^{(3)} &= 0 \\ -6n_1^{(3)} - 6n_2^{(3)} &= 0 \\ -15n_3^{(3)} &= 0 \end{aligned}\right\} \quad \text{or} \quad n_1^{(3)} = -n_2^{(3)} = \mp 1/\sqrt{2},\ n_3^{(3)} = 0$$

which are the same as the principal directions of $\mathbf{T}$.

1.48. Use the fact that $(\mathbf{T})^2$ has the same principal directions as the symmetrical tensor $\mathbf{T}$ to obtain $\sqrt{\mathbf{T}}$ when

$$\mathbf{T} = \begin{pmatrix} 5 & -1 & -1 \\ -1 & 4 & 0 \\ -1 & 0 & 4 \end{pmatrix}$$

First, the principal values and principal directions of $\mathbf{T}$ are determined. Following the procedure of Problem 1.43, the diagonal form of $\mathbf{T}$ is given by

$$\mathbf{T}^* = \begin{pmatrix} 3 & 0 & 0 \\ 0 & 4 & 0 \\ 0 & 0 & 6 \end{pmatrix}$$

with the transformation matrix being

$$[a_{ij}] = \begin{bmatrix} 1/\sqrt{3} & 1/\sqrt{3} & 1/\sqrt{3} \\ 0 & 1/\sqrt{2} & -1/\sqrt{2} \\ -2/\sqrt{6} & 1/\sqrt{6} & 1/\sqrt{6} \end{bmatrix}$$

Therefore $\sqrt{\mathbf{T}^*} = \begin{pmatrix} \sqrt{3} & 0 & 0 \\ 0 & 2 & 0 \\ 0 & 0 & \sqrt{6} \end{pmatrix}$ and using $[a_{ij}]$ to relate this to the original axes by the transformation $\sqrt{\mathbf{T}} = \mathbf{A}_c\sqrt{\mathbf{T}^*}\,\mathbf{A}$, the matrix equation is

$$[\sqrt{T_{ij}}] = \begin{bmatrix} 1/\sqrt{3} & 0 & -2/\sqrt{6} \\ 1/\sqrt{3} & 1/\sqrt{2} & 1/\sqrt{6} \\ 1/\sqrt{3} & -1/\sqrt{2} & 1/\sqrt{6} \end{bmatrix} \begin{bmatrix} \sqrt{3} & 0 & 0 \\ 0 & 2 & 0 \\ 0 & 0 & \sqrt{6} \end{bmatrix} \begin{bmatrix} 1/\sqrt{3} & 1/\sqrt{3} & 1/\sqrt{3} \\ 0 & 1/\sqrt{2} & -1/\sqrt{2} \\ -2/\sqrt{6} & 1/\sqrt{6} & 1/\sqrt{6} \end{bmatrix}$$

$$= \frac{1}{\sqrt{6}} \begin{bmatrix} \sqrt{2}+4 & \sqrt{2}-2 & \sqrt{2}-2 \\ \sqrt{2}-2 & \sqrt{2}+\sqrt{6}+1 & \sqrt{2}-\sqrt{6}+1 \\ \sqrt{2}-2 & \sqrt{2}-\sqrt{6}+1 & \sqrt{2}+\sqrt{6}+1 \end{bmatrix} = .402 \begin{bmatrix} 5.414 & -0.586 & -0.586 \\ -0.586 & 4.863 & -0.035 \\ -0.586 & -0.035 & 4.863 \end{bmatrix}$$

CARTESIAN TENSOR CALCULUS (Sec. 1.21-1.23)

1.49. For the function $\lambda = A_{ij}x_ix_j$ where A_{ij} is constant, show that $\partial\lambda/\partial x_i = (A_{ij}+A_{ji})x_j$ and $\partial^2\lambda/\partial x_i\partial x_j = A_{ij} + A_{ji}$. Simplify these derivatives for the case $A_{ij} = A_{ji}$.

Consider $\dfrac{\partial\lambda}{\partial x_k} = A_{ij}\dfrac{\partial x_i}{\partial x_k}x_j + A_{ij}x_i\dfrac{\partial x_j}{\partial x_k}$. Since $\dfrac{\partial x_i}{\partial x_k} \equiv \delta_{ik}$, it is seen that $\dfrac{\partial\lambda}{\partial x_k} = A_{kj}x_j + A_{ik}x_i = (A_{kj}+A_{jk})x_j$. Continuing the differentiation, $\dfrac{\partial^2\lambda}{\partial x_p\,\partial x_k} = (A_{kj}+A_{jk})\dfrac{\partial x_j}{\partial x_p} = A_{kp} + A_{pk}$. If $A_{ij} = A_{ji}$, $\dfrac{\partial\lambda}{\partial x_k} = 2A_{kj}x_j$ and $\dfrac{\partial^2\lambda}{\partial x_p\,\partial x_k} = 2A_{pk}$.

1.50. Use indicial notation to prove the vector identities (*a*) $\nabla\times\nabla\phi = 0$, (*b*) $\nabla\cdot\nabla\times\mathbf{a} = 0$.

(*a*) By (*1.147*), $\nabla\phi$ is written $\phi_{,i}$ and so $\mathbf{v} = \nabla\times\nabla\phi$ has components $v_i = \epsilon_{ijk}\partial_j\phi_{,k} = \epsilon_{ijk}\phi_{,kj}$. But ϵ_{ijk} is anti-symmetric in j and k, whereas $\phi_{,kj}$ is symmetric in j and k; hence the product $\epsilon_{ijk}\phi_{,kj}$ vanishes. The same result may be found by computing individually the components of $\mathbf{v}$. For example, by expansion $v_1 = \epsilon_{123}\phi_{,23} + \epsilon_{132}\phi_{,32} = (\phi_{,23} - \phi_{,32}) = 0$.

(*b*) $\nabla\cdot\nabla\times\mathbf{a} = \lambda = (\epsilon_{ijk}a_{k,j})_{,i} = \epsilon_{ijk}a_{k,ji} = 0$ since $a_{k,ij} = a_{k,ji}$ and $\epsilon_{ijk} = -\epsilon_{jik}$.

1.51. Determine the derivative of the function $\lambda = (x_1)^2 + 2x_1x_2 - (x_3)^2$ in the direction of the unit normal $\hat{\mathbf{n}} = (2/7)\hat{\mathbf{e}}_1 - (3/7)\hat{\mathbf{e}}_2 - (6/7)\hat{\mathbf{e}}_3$ or $\hat{\mathbf{n}} = (2\hat{\mathbf{e}}_1 - 3\hat{\mathbf{e}}_2 - 6\hat{\mathbf{e}}_3)/7$.

The required derivative is $\dfrac{\partial\lambda}{\partial n} = \nabla\lambda\cdot\hat{\mathbf{n}} = \lambda_{,i}n_i$. Thus

$$\frac{\partial\lambda}{\partial n} = (2x_1+2x_2)\tfrac{2}{7} - (2x_1)\tfrac{3}{7} + (2x_3)\tfrac{6}{7} = \tfrac{2}{7}(-x_1+2x_2+6x_3)$$

1.52. If A_{ij} is a second-order Cartesian tensor, show that its derivative with respect to x_k, namely $A_{ij,k}$, is a third-order Cartesian tensor.

For the Cartesian coordinate systems x_i and x'_i, $x_i = a_{ji}x'_j$ and $\partial x_i/\partial x'_j = a_{ji}$. Hence

$$A'_{ij,k} \;=\; \frac{\partial(A'_{ij})}{\partial x'_k} \;=\; \frac{\partial}{\partial x'_k}(a_{ip}a_{jq}A_{pq}) \;=\; a_{ip}a_{jq}\frac{\partial A_{pq}}{\partial x_m}\frac{\partial x_m}{\partial x'_k} \;=\; a_{ip}a_{jq}a_{km}A_{pq,m}$$

which is the transformation law for a third-order Cartesian tensor.

1.53. If $r^2 = x_ix_i$ and $f(r)$ is an arbitrary function of r, show that (*a*) $\nabla(f(r)) = f'(r)\mathbf{x}/r$, and (*b*) $\nabla^2(f(r)) = f''(r) + 2f'/r$, where primes denote derivatives with respect to r.

(*a*) The components of ∇f are simply $f_{,i}$. Thus $f_{,i} = \dfrac{\partial f}{\partial r}\dfrac{\partial r}{\partial x_i}$; and since $\dfrac{\partial(r^2)}{\partial x_j} = 2r\dfrac{\partial r}{\partial x_j} = 2\delta_{ij}x_i$ it follows that $\dfrac{\partial r}{\partial x_j} = \dfrac{x_j}{r}$. Thus $f_{,i} = \dfrac{\partial f}{\partial r}\dfrac{\partial r}{\partial x_i} = f'x_i/r$.

(*b*) $\nabla^2 f = f_{,ii} = (f'x_i/r)_{,i} = f''\dfrac{x_ix_i}{r^2} + f'\left(\dfrac{3}{r} - \dfrac{x_ix_i}{r^3}\right) = f'' + \dfrac{2f'}{r}$.

1.54. Use the divergence theorem of Gauss to show that $\int_S x_in_j\,dS = V\delta_{ij}$ where $n_j\,dS$ represents the surface element of S, the bounding surface of the volume V shown in Fig. 1-20. x_i is the position vector of $n_j\,dS$, and n_i its outward normal.

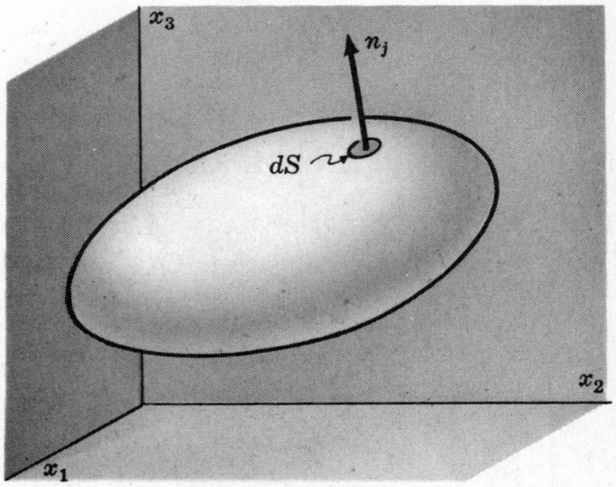

Fig. 1-20

By (*1.157*),

$$\int_S x_in_j\,dS \;=\; \int_V x_{i,j}\,dV \;=\; \int_V \delta_{ij}\,dV \;=\; \delta_{ij}V$$

1.55. If the vector $\mathbf{b} = \nabla\times\mathbf{v}$, show that $\int_S \lambda b_in_i\,dS = \int_V \lambda_{,i}b_i\,dV$ where $\lambda = \lambda(x_i)$ is a scalar function of the coordinates.

Since $\mathbf{b} = \nabla\times\mathbf{v}$, $b_i = \epsilon_{ijk}v_{k,j}$ and so

$$\begin{aligned}\int_S \lambda b_in_i\,dS &= \int_S \epsilon_{ijk}\lambda v_{k,j}n_i\,dS \\ &= \int_V \epsilon_{ijk}(\lambda v_{k,j})_{,i}\,dV \qquad \text{by (1.157)} \\ &= \int_V (\epsilon_{ijk}\lambda_{,i}v_{k,j} + \epsilon_{ijk}\lambda v_{k,ji})\,dV \\ &= \int_V \lambda_{,i}b_i\,dV \qquad \text{since } \lambda\epsilon_{ijk}v_{k,ji} = 0\end{aligned}$$

MISCELLANEOUS PROBLEMS

1.56. For the arbitrary vectors **a** and **b**, show that

$$\lambda = (\mathbf{a}\times\mathbf{b})\cdot(\mathbf{a}\times\mathbf{b}) + (\mathbf{a}\cdot\mathbf{b})^2 = (ab)^2$$

Interchange the dot and cross in the first term. Then

$$\begin{aligned}\lambda &= \mathbf{a}\cdot\mathbf{b}\times(\mathbf{a}\times\mathbf{b}) + (\mathbf{a}\cdot\mathbf{b})(\mathbf{a}\cdot\mathbf{b})\\ &= \mathbf{a}\cdot[(\mathbf{b}\cdot\mathbf{b})\mathbf{a} - (\mathbf{b}\cdot\mathbf{a})\mathbf{b}] + (\mathbf{a}\cdot\mathbf{b})(\mathbf{a}\cdot\mathbf{b})\\ &= (\mathbf{a}\cdot\mathbf{a})(\mathbf{b}\cdot\mathbf{b}) - (\mathbf{b}\cdot\mathbf{a})(\mathbf{a}\cdot\mathbf{b}) + (\mathbf{a}\cdot\mathbf{b})(\mathbf{a}\cdot\mathbf{b})\\ &= (ab)^2\end{aligned}$$

since the second and third terms cancel.

1.57. If $\dot{\mathbf{u}} = \boldsymbol{\omega}\times\mathbf{u}$ and $\dot{\mathbf{v}} = \boldsymbol{\omega}\times\mathbf{v}$, show that $\dfrac{d}{dt}(\mathbf{u}\times\mathbf{v}) = \boldsymbol{\omega}\times(\mathbf{u}\times\mathbf{v})$.

(*a*) In symbolic notation,

$$\begin{aligned}\frac{d}{dt}(\mathbf{u}\times\mathbf{v}) &= \dot{\mathbf{u}}\times\mathbf{v} + \mathbf{u}\times\dot{\mathbf{v}} = (\boldsymbol{\omega}\times\mathbf{u})\times\mathbf{v} + \mathbf{u}\times(\boldsymbol{\omega}\times\mathbf{v})\\ &= (\mathbf{v}\cdot\boldsymbol{\omega})\mathbf{u} - (\mathbf{v}\cdot\mathbf{u})\boldsymbol{\omega} + (\mathbf{u}\cdot\mathbf{v})\boldsymbol{\omega} - (\mathbf{u}\cdot\boldsymbol{\omega})\mathbf{v}\\ &= (\mathbf{v}\cdot\boldsymbol{\omega})\mathbf{u} - (\mathbf{u}\cdot\boldsymbol{\omega})\mathbf{v} = \boldsymbol{\omega}\times(\mathbf{u}\times\mathbf{v})\end{aligned}$$

(*b*) In indicial notation, let $\dfrac{d}{dt}(\mathbf{u}\times\mathbf{v}) = \mathbf{w}$. Then

$$w_i = \frac{d}{dt}(\epsilon_{ijk}u_jv_k) = \epsilon_{ijk}\dot{u}_jv_k + \epsilon_{ijk}u_j\dot{v}_k$$

and since $\dot{u}_j = \epsilon_{jpq}\omega_pu_q$ and $\dot{v}_k = \epsilon_{kmn}\omega_mv_n$,

$$w_i = \epsilon_{ijk}\epsilon_{jpq}\omega_pu_qv_k + \epsilon_{ijk}\epsilon_{kmn}u_j\omega_mv_n = (\epsilon_{ijk}\epsilon_{kmn} - \epsilon_{ink}\epsilon_{kmj})u_j\omega_mv_n$$

and using the result of Problem *1.59(a)* below,

$$\begin{aligned}w_i &= (\delta_{im}\delta_{jn} - \delta_{in}\delta_{jm} - \delta_{im}\delta_{jn} + \delta_{ij}\delta_{mn})u_j\omega_mv_n\\ &= (\delta_{ij}\delta_{mn} - \delta_{in}\delta_{jm})u_j\omega_mv_n = \epsilon_{imk}\epsilon_{kjn}u_j\omega_mv_n\end{aligned}$$

which is the indicial form of $\boldsymbol{\omega}\times(\mathbf{u}\times\mathbf{v})$.

1.58. Establish the identity $\epsilon_{pqs}\epsilon_{mnr} = \begin{vmatrix}\delta_{mp} & \delta_{mq} & \delta_{ms}\\ \delta_{np} & \delta_{nq} & \delta_{ns}\\ \delta_{rp} & \delta_{rq} & \delta_{rs}\end{vmatrix}$.

Let the determinant of A_{ij} be given by $\det A = \begin{vmatrix}A_{11} & A_{12} & A_{13}\\ A_{21} & A_{22} & A_{23}\\ A_{31} & A_{32} & A_{33}\end{vmatrix}$. An interchange of rows or columns causes a sign change. Thus

$$\begin{vmatrix}A_{21} & A_{22} & A_{23}\\ A_{11} & A_{12} & A_{13}\\ A_{31} & A_{32} & A_{33}\end{vmatrix} = \begin{vmatrix}A_{12} & A_{11} & A_{13}\\ A_{22} & A_{21} & A_{23}\\ A_{32} & A_{31} & A_{33}\end{vmatrix} = -\det A$$

and for an arbitrary number of row changes,

$$\begin{vmatrix}A_{m1} & A_{m2} & A_{m3}\\ A_{n1} & A_{n2} & A_{n3}\\ A_{r1} & A_{r2} & A_{r3}\end{vmatrix} = \epsilon_{mnr}\det A$$

or column changes,

$$\begin{vmatrix} A_{1p} & A_{1q} & A_{1s} \\ A_{2p} & A_{2q} & A_{2s} \\ A_{3p} & A_{3q} & A_{3s} \end{vmatrix} = \epsilon_{pqs} \det A$$

Hence for an arbitrary row and column interchange sequence,

$$\begin{vmatrix} A_{mp} & A_{mq} & A_{ms} \\ A_{np} & A_{nq} & A_{ns} \\ A_{rp} & A_{rq} & A_{rs} \end{vmatrix} = \epsilon_{mnr}\epsilon_{pqs} \det A$$

When $A_{ij} \equiv \delta_{ij}$, $\det A = 1$ and the identity is established.

1.59. Use the results of Problem 1.58 to prove (*a*) $\epsilon_{pqs}\epsilon_{snr} = \delta_{pn}\delta_{qr} - \delta_{pr}\delta_{qn}$, (*b*) $\epsilon_{pqs}\epsilon_{sqr} = -2\delta_{pr}$.

Expanding the determinant of the identity in Problem 1.58,

$$\epsilon_{pqs}\epsilon_{mnr} = \delta_{mp}(\delta_{nq}\delta_{rs} - \delta_{ns}\delta_{rq}) + \delta_{mq}(\delta_{ns}\delta_{rp} - \delta_{np}\delta_{rs}) + \delta_{ms}(\delta_{np}\delta_{rq} - \delta_{nq}\delta_{rp})$$

(*a*) Identifying s with m yields,

$$\begin{aligned} \epsilon_{pqs}\epsilon_{snr} &= \delta_{sp}(\delta_{nq}\delta_{rs} - \delta_{ns}\delta_{rq}) + \delta_{sq}(\delta_{ns}\delta_{rp} - \delta_{np}\delta_{rs}) + \delta_{ss}(\delta_{np}\delta_{rq} - \delta_{nq}\delta_{rp}) \\ &= \delta_{rp}\delta_{nq} - \delta_{pn}\delta_{rq} + \delta_{qn}\delta_{rp} - \delta_{np}\delta_{qr} + 3\delta_{np}\delta_{rq} - 3\delta_{nq}\delta_{rp} \\ &= \delta_{np}\delta_{rq} - \delta_{nq}\delta_{rp} \end{aligned}$$

(*b*) Identifying q with n in (*a*),

$$\epsilon_{pqs}\epsilon_{sqr} = \delta_{qp}\delta_{rq} - \delta_{qq}\delta_{rp} = \delta_{pr} - 3\delta_{pr} = -2\delta_{pr}$$

1.60. If the dyadic $\mathbf{B}$ is skew-symmetric $\mathbf{B} = -\mathbf{B}_c$, show that $\mathbf{B}_v \times \mathbf{a} = 2\mathbf{a}\cdot\mathbf{B}$.

Writing $\mathbf{B} = \mathbf{b}_1\hat{\mathbf{e}}_1 + \mathbf{b}_2\hat{\mathbf{e}}_2 + \mathbf{b}_3\hat{\mathbf{e}}_3$ (see Problem 1.6),

$$\mathbf{B}_v = \mathbf{b}_1 \times \hat{\mathbf{e}}_1 + \mathbf{b}_2 \times \hat{\mathbf{e}}_2 + \mathbf{b}_3 \times \hat{\mathbf{e}}_3$$

and

$$\begin{aligned} \mathbf{B}_v \times \mathbf{a} &= (\mathbf{b}_1 \times \hat{\mathbf{e}}_1) \times \mathbf{a} + (\mathbf{b}_2 \times \hat{\mathbf{e}}_2) \times \mathbf{a} + (\mathbf{b}_3 \times \hat{\mathbf{e}}_3) \times \mathbf{a} \\ &= (\mathbf{a}\cdot\mathbf{b}_1)\hat{\mathbf{e}}_1 - (\mathbf{a}\cdot\hat{\mathbf{e}}_1)\mathbf{b}_1 + (\mathbf{a}\cdot\mathbf{b}_2)\hat{\mathbf{e}}_2 - (\mathbf{a}\cdot\hat{\mathbf{e}}_2)\mathbf{b}_2 + (\mathbf{a}\cdot\mathbf{b}_3)\hat{\mathbf{e}}_3 - (\mathbf{a}\cdot\hat{\mathbf{e}}_3)\mathbf{b}_3 \\ &= \mathbf{a}\cdot(\mathbf{b}_1\hat{\mathbf{e}}_1 + \mathbf{b}_2\hat{\mathbf{e}}_2 + \mathbf{b}_3\hat{\mathbf{e}}_3) - \mathbf{a}\cdot(\hat{\mathbf{e}}_1\mathbf{b}_1 + \hat{\mathbf{e}}_2\mathbf{b}_2 + \hat{\mathbf{e}}_3\mathbf{b}_3) \\ &= \mathbf{a}\cdot\mathbf{B} - \mathbf{a}\cdot\mathbf{B}_c = 2\mathbf{a}\cdot\mathbf{B} \end{aligned}$$

1.61. Use the Hamilton-Cayley equation to obtain $(\mathbf{B})^4$ for the tensor $\mathbf{B} = \begin{pmatrix} 1 & 0 & -1 \\ 0 & 3 & 0 \\ -1 & 0 & -2 \end{pmatrix}$. Check the result directly by squaring $(\mathbf{B})^2$.

The characteristic equation for $\mathbf{B}$ is given by

$$\begin{vmatrix} 1-\lambda & 0 & -1 \\ 0 & 3-\lambda & 0 \\ -1 & 0 & -2-\lambda \end{vmatrix} = -(\lambda^3 - 2\lambda^2 - 6\lambda + 9) = 0$$

By the Hamilton-Cayley theorem the tensor satisfies its own characteristic equation. Hence $(\mathbf{B})^3 - 2(\mathbf{B})^2 - 6\mathbf{B} + 9\mathbf{I} = 0$, and multiplying this equation by $\mathbf{B}$ yields $(\mathbf{B})^4 = 2(\mathbf{B})^3 + 6(\mathbf{B})^2 - 9\mathbf{B}$ or $(\mathbf{B})^4 = 10(\mathbf{B})^2 + 3\mathbf{B} - 18\mathbf{I}$. Hence

$$(\mathbf{B})^4 = 10\begin{pmatrix}2 & 0 & 1\\ 0 & 9 & 0\\ 1 & 0 & 5\end{pmatrix} + \begin{pmatrix}3 & 0 & -3\\ 0 & 9 & 0\\ -3 & 0 & -6\end{pmatrix} - \begin{pmatrix}18 & 0 & 0\\ 0 & 18 & 0\\ 0 & 0 & 18\end{pmatrix} = \begin{pmatrix}5 & 0 & 7\\ 0 & 81 & 0\\ 7 & 0 & 26\end{pmatrix}$$

Checking by direct matrix multiplication of $(\mathbf{B})^2$,

$$(\mathcal{B})^4 = \begin{bmatrix}2 & 0 & 1\\ 0 & 9 & 0\\ 1 & 0 & 5\end{bmatrix}\begin{bmatrix}2 & 0 & 1\\ 0 & 9 & 0\\ 1 & 0 & 5\end{bmatrix} = \begin{bmatrix}5 & 0 & 7\\ 0 & 81 & 0\\ 7 & 0 & 26\end{bmatrix}$$

1.62. Prove that (*a*) A_{ii}, (*b*) $A_{ij}A_{ij}$, (*c*) $\epsilon_{ijk}\epsilon_{kjp}A_{ip}$ are invariant under the coordinate transformation represented by (*1.103*), i.e. show that $A_{ii} = A'_{ii}$, etc.

(*a*) By (*1.103*), $A_{ij} = a_{pi}a_{qj}A'_{pq}$. Hence $A_{ii} = a_{pi}a_{qi}A'_{pq} = \delta_{pq}A'_{pq} = A'_{pp} = A'_{ii}$.

(*b*) $A_{ij}A_{ij} = a_{pi}a_{qj}A'_{pq}a_{mi}a_{nj}A'_{mn} = \delta_{pm}\delta_{qn}A'_{pq}A'_{mn} = A'_{pq}A'_{pq} = A'_{ij}A'_{ij}$

(*c*) $$\begin{aligned}\epsilon_{ijk}\epsilon_{kjp}A_{ip} &= \epsilon_{ijk}\epsilon_{kjp}a_{mi}a_{np}A'_{mn} = (\delta_{ij}\delta_{jp} - \delta_{ip}\delta_{jj})a_{mi}a_{np}A'_{mn}\\ &= (\delta_{mn} - \delta_{mn}\delta_{jj})A'_{mn} = (\delta_{mj}\delta_{nj} - \delta_{mn}\delta_{jj})A'_{mn} = \epsilon_{mjk}\epsilon_{kjn}A'_{mn}\end{aligned}$$

1.63. Show that the dual vector of the arbitrary tensor T_{ij} depends only upon $T_{[ij]}$ but that the product $T_{ij}S_{ij}$ of T_{ij} with the symmetric tensor S_{ij} is independent of $T_{[ij]}$.

By (*1.110*) the dual vector of T_{ij} is $v_i = \epsilon_{ijk}T_{jk}$, or $v_i = \epsilon_{ijk}(T_{(jk)} + T_{[jk]}) = \epsilon_{ijk}T_{[jk]}$ since $\epsilon_{ijk}T_{(jk)} = 0$ (ϵ_{ijk} is anti-symmetric in j and k, $T_{(jk)}$ symmetric in j and k).

For the product $T_{ij}S_{ij} = T_{(ij)}S_{ij} + T_{[ij]}S_{ij}$. Here $T_{[ij]}S_{ij} = 0$ and $T_{ij}S_{ij} = T_{(ij)}S_{ij}$.

1.64. Show that $\mathbf{D} : \mathbf{E}$ is equal to $\mathbf{D} \cdot\cdot \mathbf{E}$ if $\mathbf{E}$ is a symmetric dyadic.

Write $\mathbf{D} = D_{ij}\hat{\mathbf{e}}_i\hat{\mathbf{e}}_j$ and $\mathbf{E} = E_{pq}\hat{\mathbf{e}}_p\hat{\mathbf{e}}_q$. By (*1.31*), $\mathbf{D}:\mathbf{E} = D_{ij}E_{pq}(\hat{\mathbf{e}}_i \cdot \hat{\mathbf{e}}_p)(\hat{\mathbf{e}}_j \cdot \hat{\mathbf{e}}_q)$. By (*1.35*), $\mathbf{D}\cdot\cdot\mathbf{E} = D_{ij}E_{pq}(\hat{\mathbf{e}}_j \cdot \hat{\mathbf{e}}_p)(\hat{\mathbf{e}}_i \cdot \hat{\mathbf{e}}_q) = D_{ij}E_{qp}(\hat{\mathbf{e}}_j \cdot \hat{\mathbf{e}}_p)(\hat{\mathbf{e}}_i \cdot \hat{\mathbf{e}}_q)$ since $E_{pq} = E_{qp}$. Now interchanging the role of dummy indices p and q in this last expression, $\mathbf{D}\cdot\cdot\mathbf{E} = D_{ij}E_{pq}(\hat{\mathbf{e}}_j \cdot \hat{\mathbf{e}}_q)(\hat{\mathbf{e}}_i \cdot \hat{\mathbf{e}}_p)$.

1.65. Use the indicial notation to prove the vector identity $\nabla \times (\mathbf{a} \times \mathbf{b}) = \mathbf{b} \cdot \nabla\mathbf{a} - \mathbf{b}(\nabla \cdot \mathbf{a}) + \mathbf{a}(\nabla \cdot \mathbf{b}) - \mathbf{a} \cdot \nabla\mathbf{b}$.

Let $\nabla \times (\mathbf{a} \times \mathbf{b}) = \mathbf{v}$; then $v_p = \epsilon_{pqi}\epsilon_{ijk}\partial_q a_j b_k$ or

$$\begin{aligned}v_p &= \epsilon_{pqi}\epsilon_{ijk}(a_j b_k)_{,q} = \epsilon_{pqi}\epsilon_{ijk}(a_{j,q}b_k + a_j b_{k,q})\\ &= (\delta_{pj}\delta_{qk} - \delta_{pk}\delta_{qj})(a_{j,q}b_k + a_j b_{k,q}) = a_{p,q}b_q - a_{q,q}b_p + a_p b_{q,q} - a_q b_{p,q}\end{aligned}$$

Hence $\mathbf{v} = \mathbf{b} \cdot \nabla\mathbf{a} - \mathbf{b}(\nabla \cdot \mathbf{a}) + \mathbf{a}(\nabla \cdot \mathbf{b}) - \mathbf{a} \cdot \nabla\mathbf{b}$.

1.66. By means of the divergence theorem of Gauss show that $\int_S \hat{\mathbf{n}} \times (\mathbf{a} \times \mathbf{x})\, dS = 2\mathbf{a}V$ where V is the volume enclosed by the surface S having the outward normal $\mathbf{n}$. The position vector to any point in V is $\mathbf{x}$, and $\mathbf{a}$ is an arbitrary constant vector.

In the indicial notation the surface integral is $\int_S \epsilon_{qpi}n_p\epsilon_{ijk}a_j x_k\, dS$. By (*1.157*) this becomes the volume integral $\int_V (\epsilon_{qpi}\epsilon_{ijk}a_j x_k)_{,p}\, dV$ and since $\mathbf{a}$ is constant, the last expression becomes

$$\begin{aligned}\int_V \epsilon_{qpi}\epsilon_{ijk}a_j x_{k,p}\, dV &= \int_V (\delta_{qj}\delta_{pk} - \delta_{qk}\delta_{pj})a_j x_{k,p}\, dV = \int_V (a_q x_{p,p} - a_p x_{q,p})\, dV\\ &= \int_V (a_q\delta_{pp} - a_p\delta_{qp})\, dV = \int_V (3a_q - a_q)\, dV = 2a_q V\end{aligned}$$

1.67. For the reflection of axes shown in Fig. 1-21 show that the transformation is orthogonal.

From the figure the transformation matrix is

$$[a_{ij}] = \begin{bmatrix} 1 & 0 & 0 \\ 0 & -1 & 0 \\ 0 & 0 & 1 \end{bmatrix}$$

The orthogonality conditions $a_{ij}a_{ik} = \delta_{jk}$ or $a_{ji}a_{ki} = \delta_{jk}$ are clearly satisfied. In matrix form, by (*1.117*),

$$\begin{bmatrix} 1 & 0 & 0 \\ 0 & -1 & 0 \\ 0 & 0 & 1 \end{bmatrix}\begin{bmatrix} 1 & 0 & 0 \\ 0 & -1 & 0 \\ 0 & 0 & 1 \end{bmatrix} = \begin{bmatrix} 1 & 0 & 0 \\ 0 & 1 & 0 \\ 0 & 0 & 1 \end{bmatrix}$$

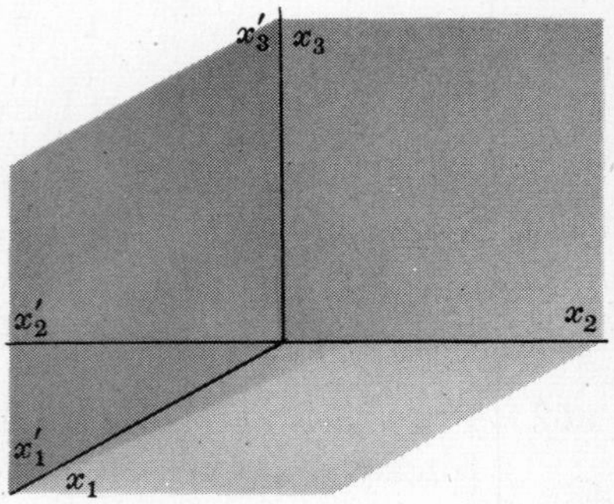

Fig. 1-21

1.68. Show that $(\mathbf{I} \times \mathbf{v}) \cdot \mathbf{D} = \mathbf{v} \times \mathbf{D}$.

$$\begin{aligned} \mathbf{I} \times \mathbf{v} &= (\hat{\mathbf{i}}\hat{\mathbf{i}} + \hat{\mathbf{j}}\hat{\mathbf{j}} + \hat{\mathbf{k}}\hat{\mathbf{k}}) \times (v_x\hat{\mathbf{i}} + v_y\hat{\mathbf{j}} + v_z\hat{\mathbf{k}}) \\ &= \hat{\mathbf{i}}(v_y\hat{\mathbf{k}} - v_z\hat{\mathbf{j}}) + \hat{\mathbf{j}}(-v_x\hat{\mathbf{k}} + v_z\hat{\mathbf{i}}) + \hat{\mathbf{k}}(v_x\hat{\mathbf{j}} - v_y\hat{\mathbf{i}}) \\ &= (\mathbf{v} \times \hat{\mathbf{i}})\hat{\mathbf{i}} + (\mathbf{v} \times \hat{\mathbf{j}})\hat{\mathbf{j}} + (\mathbf{v} \times \hat{\mathbf{k}})\hat{\mathbf{k}} = \mathbf{v} \times \mathbf{I} \end{aligned}$$

Hence $(\mathbf{I} \times \mathbf{v}) \cdot \mathbf{D} = \mathbf{v} \times \mathbf{I} \cdot \mathbf{D} = \mathbf{v} \times \mathbf{D}$.

Supplementary Problems

1.69. Show that $\mathbf{u} = \hat{\mathbf{i}} + \hat{\mathbf{j}} - \hat{\mathbf{k}}$ and $\mathbf{v} = \hat{\mathbf{i}} - \hat{\mathbf{j}}$ are perpendicular to one another. Determine $\hat{\mathbf{w}}$ so that $\mathbf{u}, \mathbf{v}, \hat{\mathbf{w}}$ forms a right-handed triad. *Ans.* $\mathbf{w} = (-1/\sqrt{6})(\hat{\mathbf{i}} + \hat{\mathbf{j}} + 2\hat{\mathbf{k}})$

1.70. Determine the transformation matrix between the $\mathbf{u}, \mathbf{v}, \hat{\mathbf{w}}$ axes of Problem 1.69 and the coordinate directions.

Ans. $$[a_{ij}] = \begin{bmatrix} 1/\sqrt{3} & 1/\sqrt{3} & -1/\sqrt{3} \\ 1/\sqrt{2} & -1/\sqrt{2} & 0 \\ -1/\sqrt{6} & -1/\sqrt{6} & -2/\sqrt{6} \end{bmatrix}$$

1.71. Use indicial notation to prove (*a*) $\nabla \cdot \mathbf{x} = 3$, (*b*) $\nabla \times \mathbf{x} = 0$, (*c*) $\mathbf{a} \cdot \nabla\mathbf{x} = \mathbf{a}$ where $\mathbf{x}$ is the position vector and $\mathbf{a}$ is a constant vector.

1.72. Determine the principal values of the symmetric part of the tensor $T_{ij} = \begin{pmatrix} 5 & -1 & 3 \\ 1 & -6 & -6 \\ -3 & -18 & 1 \end{pmatrix}$.
Ans. $\lambda_{(1)} = -15$, $\lambda_{(2)} = 5$, $\lambda_{(3)} = 10$

1.73. For the symmetric tensor $T_{ij} = \begin{pmatrix} 7 & 3 & 0 \\ 3 & 7 & 4 \\ 0 & 4 & 7 \end{pmatrix}$ determine the principal values and the directions of the principal axes.

Ans. $\lambda_{(1)} = 2,\ \lambda_{(2)} = 7,\ \lambda_{(3)} = 12,$

	x_1	x_2	x_3
x_1^*	$-3/5\sqrt{2}$	$1/\sqrt{2}$	$-4/5\sqrt{2}$
x_2^*	$4/5$	0	$-3/5$
x_3^*	$3/5\sqrt{2}$	$1/\sqrt{2}$	$4/5\sqrt{2}$

1.74. Given the arbitrary vector $\mathbf{v}$ and any unit vector $\hat{\mathbf{e}}$, show that $\mathbf{v}$ may be resolved into a component parallel and a component perpendicular to $\hat{\mathbf{e}}$, i.e. $\mathbf{v} = (\mathbf{v}\cdot\hat{\mathbf{e}})\hat{\mathbf{e}} + \hat{\mathbf{e}}\times(\mathbf{v}\times\hat{\mathbf{e}})$.

1.75. If $\nabla\cdot\mathbf{v} = 0$, $\nabla\times\mathbf{v} = \dot{\mathbf{w}}$ and $\nabla\times\mathbf{w} = -\dot{\mathbf{v}}$, show that $\nabla^2\mathbf{v} = \ddot{\mathbf{v}}$.

1.76. Check the result of Problem 1.48 by direct multiplication to show that $\sqrt{\mathbf{T}}\sqrt{\mathbf{T}} = \mathbf{T}$.

1.77. Determine the square root of the tensor $\mathbf{B} = \begin{pmatrix} 3 & 2 & 0 \\ 2 & 3 & 0 \\ 0 & 0 & 9 \end{pmatrix}$.

Ans. $\sqrt{\mathbf{B}} = \begin{pmatrix} \frac{1}{2}(\sqrt{5}+1) & \frac{1}{2}(\sqrt{5}-1) & 0 \\ \frac{1}{2}(\sqrt{5}-1) & \frac{1}{2}(\sqrt{5}+1) & 0 \\ 0 & 0 & 3 \end{pmatrix}$

1.78. Using the result of Problem 1.40, $\det\mathbf{A} = \epsilon_{ijk}A_{1i}A_{2j}A_{3k}$, show that $\det(\mathbf{AB}) = \det\mathbf{A}\det\mathbf{B}$.

1.79. Verify that (*a*) $\delta_{3p}v_p = v_3$, (*b*) $\delta_{3i}A_{ji} = A_{j3}$, (*c*) $\delta_{ij}\epsilon_{ijk} = 0$, (*d*) $\delta_{i2}\delta_{j3}A_{ij} = A_{23}$.

1.80. Let the axes $Ox_1'x_2'x_3'$ be related to $Ox_1x_2x_3$ by the table

	x_1	x_2	x_3
x_1'	$3/5\sqrt{2}$	$1/\sqrt{2}$	$4/5\sqrt{2}$
x_2'	$4/5$	0	$-3/5$
x_3'	$-3/5\sqrt{2}$	$1/\sqrt{2}$	$-4/5\sqrt{2}$

(*a*) Show that the orthogonality conditions $a_{ij}a_{ik} = \delta_{jk}$ and $a_{pq}a_{sq} = \delta_{ps}$ are satisfied.

(*b*) What are the primed coordinates of the point having position vector $\mathbf{x} = 2\hat{\mathbf{e}}_1 - \hat{\mathbf{e}}_3$?

(*c*) What is the equation of the plane $x_1 - x_2 + 3x_3 = 1$ in the primed system?

Ans. (*b*) $(2/5\sqrt{2},\ 11/5,\ -2/5\sqrt{2})$ (*c*) $\sqrt{2}\,x_1' - x_2' - 2\sqrt{2}\,x_3' = 1$

1.81. Show that the volume V enclosed by the surface S may be given as $V = \frac{1}{6}\int_S \nabla(\mathbf{x}\cdot\mathbf{x})\cdot\hat{\mathbf{n}}\,dS$ where $\mathbf{x}$ is the position vector and $\mathbf{n}$ the unit normal to the surface. *Hint*: Write $V = (1/6)\int_S (x_ix_i)_{,j}n_j\,dS$ and use (*1.157*).

Chapter 2

Analysis of Stress

2.1 THE CONTINUUM CONCEPT

The molecular nature of the structure of matter is well established. In numerous investigations of material behavior, however, the individual molecule is of no concern and only the behavior of the material as a whole is deemed important. For these cases the observed macroscopic behavior is usually explained by disregarding molecular considerations and, instead, by assuming the material to be continuously distributed throughout its volume and to completely fill the space it occupies. This *continuum concept* of matter is the fundamental postulate of Continuum Mechanics. Within the limitations for which the continuum assumption is valid, this concept provides a framework for studying the behavior of solids, liquids and gases alike.

Adoption of the continuum viewpoint as the basis for the mathematical description of material behavior means that field quantities such as stress and displacement are expressed as piecewise continuous functions of the space coordinates and time.

2.2 HOMOGENEITY. ISOTROPY. MASS-DENSITY

A *homogeneous* material is one having identical properties at all points. With respect to some property, a material is *isotropic* if that property is the same in all directions at a point. A material is called *anisotropic* with respect to those properties which are directional at a point.

The concept of *density* is developed from the *mass-volume ratio* in the neighborhood of a point in the continuum. In Fig. 2-1 the mass in the small element of volume ΔV is denoted by ΔM. The *average density* of the material within ΔV is therefore

$$\rho_{(av)} = \frac{\Delta M}{\Delta V} \qquad (2.1)$$

The *density* at some interior point P of the volume element ΔV is given mathematically in accordance with the continuum concept by the limit,

$$\rho = \lim_{\Delta V \to 0} \frac{\Delta M}{\Delta V} = \frac{dM}{dV} \qquad (2.2)$$

Mass-density ρ is a scalar quantity.

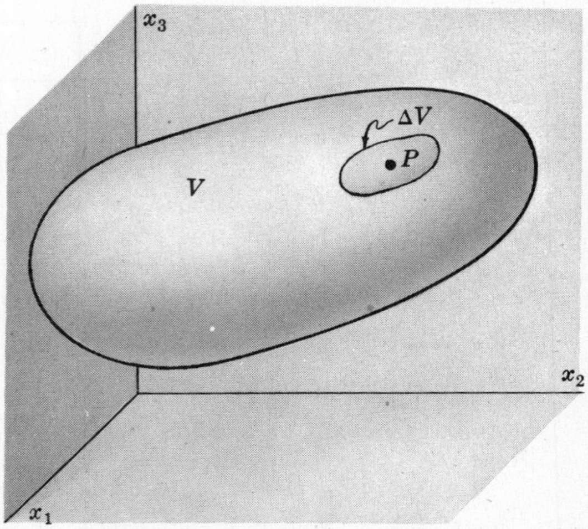

Fig. 2-1

2.3 BODY FORCES. SURFACE FORCES

Forces are vector quantities which are best described by intuitive concepts such as push or pull. Those forces which act on all elements of volume of a continuum are known as *body forces*. Examples are gravity and inertia forces. These forces are represented by the symbol b_i (force per unit mass), or as p_i (force per unit volume). They are related through the density by the equation

$$\rho b_i = p_i \quad \text{or} \quad \rho \mathbf{b} = \mathbf{p} \tag{2.3}$$

Those forces which act on a surface element, whether it is a portion of the bounding surface of the continuum or perhaps an arbitrary internal surface, are known as *surface forces*. These are designated by f_i (force per unit area). Contact forces between bodies are a type of surface forces.

2.4 CAUCHY'S STRESS PRINCIPLE. THE STRESS VECTOR

A material continuum occupying the region R of space, and subjected to surface forces f_i and body forces b_i, is shown in Fig. 2-2. As a result of forces being transmitted from one portion of the continuum to another, the material within an arbitrary volume V enclosed by the surface S interacts with the material outside of this volume. Taking n_i as the outward unit normal at point P of a small element of surface ΔS of S, let Δf_i be the resultant force exerted across ΔS upon the material within V by the material outside of V. Clearly the force element Δf_i will depend upon the choice of ΔS and upon n_i. It should also be noted that the distribution of force on ΔS is not necessarily uniform. Indeed the force distribution is, in general, equipollent to a force and a moment at P, as shown in Fig. 2-2 by the vectors Δf_i and ΔM_i.

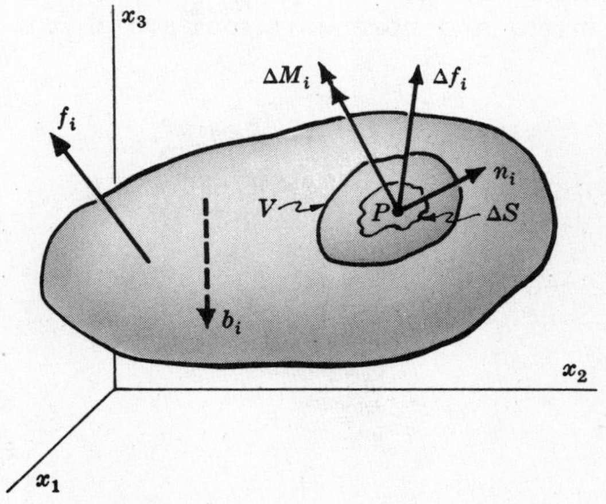

Fig. 2-2

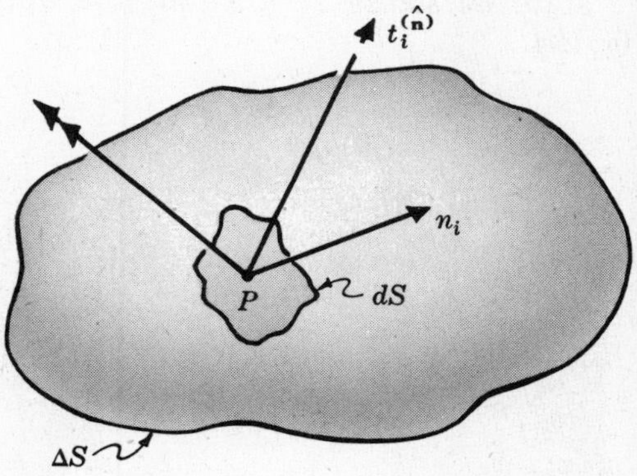

Fig. 2-3

The average force per unit area on ΔS is given by $\Delta f_i/\Delta S$. The *Cauchy stress principle* asserts that this ratio $\Delta f_i/\Delta s$ tends to a definite limit df_i/dS as ΔS approaches zero at the point P, while at the same time the moment of Δf_i about the point P vanishes in the limiting process. The resulting vector df_i/dS (force per unit area) is called the *stress vector* $t_i^{(\hat{n})}$ and is shown in Fig. 2-3. If the moment at P were not to vanish in the limiting process, a *couple-stress vector*, shown by the double-headed arrow in Fig. 2-3, would also be defined at the point. One branch of the theory of elasticity considers such couple stresses but they are not considered in this text.

Mathematically the stress vector is defined by

$$t_i^{(\hat{n})} = \lim_{\Delta S \to 0} \frac{\Delta f_i}{\Delta S} = \frac{df_i}{dS} \quad \text{or} \quad \mathbf{t}^{(\hat{n})} = \lim_{\Delta S \to 0} \frac{\Delta \mathbf{f}}{\Delta S} = \frac{d\mathbf{f}}{dS} \tag{2.4}$$

The notation $t_i^{(\hat{n})}$ (or $\mathbf{t}^{(\hat{n})}$) is used to emphasize the fact that the stress vector at a given point P in the continuum depends explicitly upon the particular surface element ΔS chosen there, as represented by the unit normal n_i (or $\hat{\mathbf{n}}$). For some differently oriented surface element, having a different unit normal, the associated stress vector at P will also be different. The stress vector arising from the action across ΔS at P of the material within V upon the material outside is the vector $-t_i^{(\hat{n})}$. Thus by Newton's law of action and reaction,

$$-t_i^{(\hat{n})} = t_i^{(-\hat{n})} \quad \text{or} \quad -\mathbf{t}^{(\hat{n})} = \mathbf{t}^{(-\hat{n})} \tag{2.5}$$

The *stress vector* is very often referred to as the *traction vector.*

2.5 STATE OF STRESS AT A POINT. STRESS TENSOR

At an arbitrary point P in a continuum, Cauchy's stress principle associates a stress vector $t_i^{(\hat{n})}$ with each unit normal vector n_i, representing the orientation of an infinitesimal surface element having P as an interior point. This is illustrated in Fig. 2-3. The totality of all possible pairs of such vectors $t_i^{(\hat{n})}$ and n_i at P defines the *state of stress* at that point. Fortunately it is not necessary to specify every pair of stress and normal vectors to completely describe the state of stress at a given point. This may be accomplished by giving the stress vector on each of three mutually perpendicular planes at P. Coordinate transformation equations then serve to relate the stress vector on any other plane at the point to the given three.

Adopting planes perpendicular to the coordinate axes for the purpose of specifying the state of stress at a point, the appropriate stress and normal vectors are shown in Fig. 2-4.

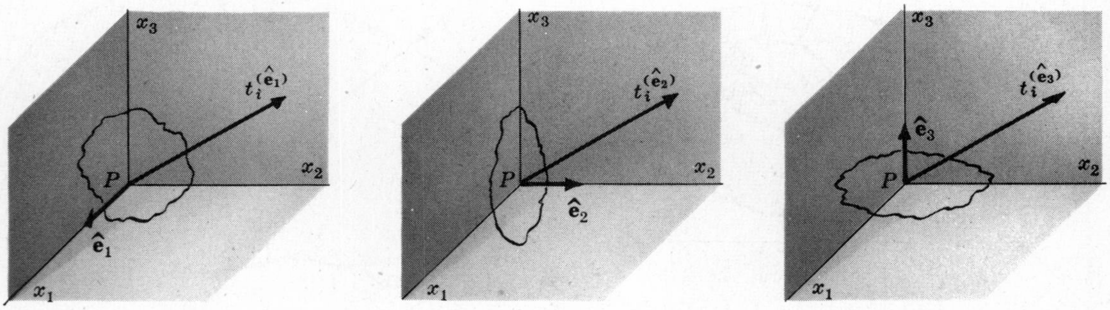

Fig. 2-4

For convenience, the three separate diagrams in Fig. 2-4 are often combined into a single schematic representation as shown in Fig. 2-5 below.

Each of the three coordinate-plane stress vectors may be written according to (*1.69*) in terms of its Cartesian components as

$$\begin{aligned} \mathbf{t}^{(\hat{e}_1)} &= t_1^{(\hat{e}_1)}\hat{\mathbf{e}}_1 + t_2^{(\hat{e}_1)}\hat{\mathbf{e}}_2 + t_3^{(\hat{e}_1)}\hat{\mathbf{e}}_3 = t_j^{(\hat{e}_1)}\hat{\mathbf{e}}_j \\ \mathbf{t}^{(\hat{e}_2)} &= t_1^{(\hat{e}_2)}\hat{\mathbf{e}}_1 + t_2^{(\hat{e}_2)}\hat{\mathbf{e}}_2 + t_3^{(\hat{e}_2)}\hat{\mathbf{e}}_3 = t_j^{(\hat{e}_2)}\hat{\mathbf{e}}_j \\ \mathbf{t}^{(\hat{e}_3)} &= t_1^{(\hat{e}_3)}\hat{\mathbf{e}}_1 + t_2^{(\hat{e}_3)}\hat{\mathbf{e}}_2 + t_3^{(\hat{e}_3)}\hat{\mathbf{e}}_3 = t_j^{(\hat{e}_3)}\hat{\mathbf{e}}_j \end{aligned} \tag{2.6}$$

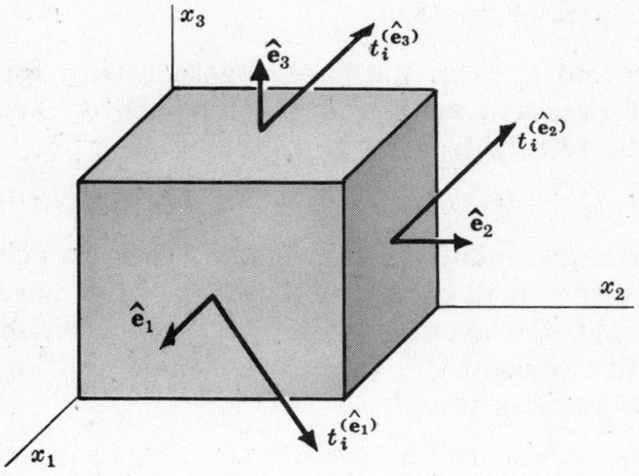

Fig. 2-5

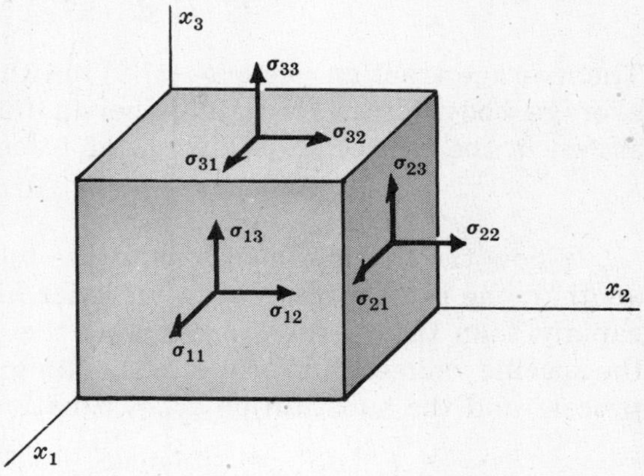

Fig. 2-6

The nine stress vector components,

$$t_j^{(\hat{\mathbf{e}}_i)} \equiv \sigma_{ij} \tag{2.7}$$

are the components of a second-order Cartesian tensor known as the *stress tensor*. The equivalent stress dyadic is designated by $\mathbf{\Sigma}$, so that explicit component and matrix representations of the stress tensor, respectively, take the forms

$$\mathbf{\Sigma} = \begin{pmatrix} \sigma_{11} & \sigma_{12} & \sigma_{13} \\ \sigma_{21} & \sigma_{22} & \sigma_{23} \\ \sigma_{31} & \sigma_{32} & \sigma_{33} \end{pmatrix} \quad \text{or} \quad [\sigma_{ij}] = \begin{bmatrix} \sigma_{11} & \sigma_{12} & \sigma_{13} \\ \sigma_{21} & \sigma_{22} & \sigma_{23} \\ \sigma_{31} & \sigma_{32} & \sigma_{33} \end{bmatrix} \tag{2.8}$$

Pictorially, the stress tensor components may be displayed with reference to the coordinate planes as shown in Fig. 2-6. The components perpendicular to the planes ($\sigma_{11}, \sigma_{22}, \sigma_{33}$) are called *normal stresses*. Those acting in (tangent to) the planes ($\sigma_{12}, \sigma_{13}, \sigma_{21}, \sigma_{23}, \sigma_{31}, \sigma_{32}$) are called *shear stresses*. A stress component is *positive* when it acts in the positive direction of the coordinate axes, and on a plane whose outer normal points in one of the positive coordinate directions. The component σ_{ij} acts in the direction of the jth coordinate axis and on the plane whose outward normal is parallel to the ith coordinate axis. The stress components shown in Fig. 2-6 are all positive.

2.6 THE STRESS TENSOR — STRESS VECTOR RELATIONSHIP

The relationship between the stress tensor σ_{ij} at a point P and the stress vector $t_i^{(\hat{\mathbf{n}})}$ on a plane of arbitrary orientation at that point may be established through the force equilibrium or momentum balance of a small tetrahedron of the continuum, having its vertex at P. The base of the tetrahedron is taken perpendicular to n_i, and the three faces are taken perpendicular to the coordinate planes as shown by Fig. 2-7. Designating the area of the base ABC as dS, the areas of the faces are the projected areas, $dS_1 = dS\, n_1$ for face CPB, $dS_2 = dS\, n_2$ for face APC, $dS_3 = dS\, n_3$ for face BPA or

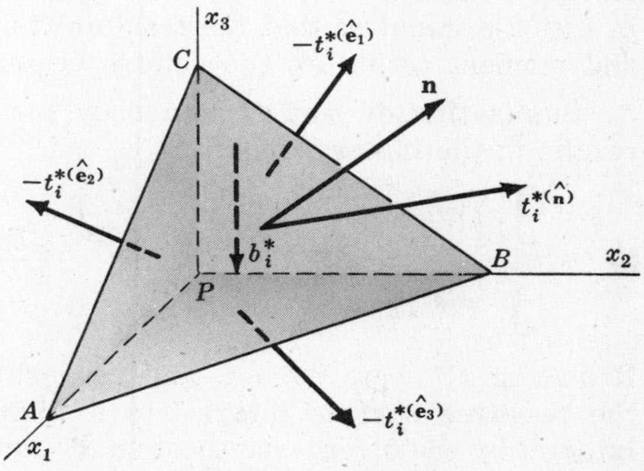

Fig. 2-7

$$dS_i = dS\,(\hat{\mathbf{n}}\cdot\hat{\mathbf{e}}_i) = dS\cos(\hat{\mathbf{n}},\hat{\mathbf{e}}_i) = dS\,n_i \tag{2.9}$$

The average traction vectors $-t_i^{*(\hat{\mathbf{e}}_j)}$ on the faces and $t_i^{*(\hat{\mathbf{n}})}$ on the base, together with the average body forces (including inertia forces, if present), acting on the tetrahedron are shown in the figure. Equilibrium of forces on the tetrahedron requires that

$$t_i^{*(\hat{\mathbf{n}})}\,dS - t_i^{*(\hat{\mathbf{e}}_1)}\,dS_1 - t_i^{*(\hat{\mathbf{e}}_2)}\,dS_2 - t_i^{*(\hat{\mathbf{e}}_3)}\,dS_3 + \rho b_i^*\,dV = 0 \tag{2.10}$$

If now the linear dimensions of the tetrahedron are reduced in a constant ratio to one another, the body forces, being an order higher in the small dimensions, tend to zero more rapidly than the surface forces. At the same time, the average stress vectors approach the specific values appropriate to the designated directions at P. Therefore by this limiting process and the substitution (*2.9*), equation (*2.10*) reduces to

$$t_i^{(\hat{\mathbf{n}})}\,dS = t_i^{(\hat{\mathbf{e}}_1)}\,n_1\,dS + t_i^{(\hat{\mathbf{e}}_2)}\,n_2\,dS + t_i^{(\hat{\mathbf{e}}_3)}\,n_3\,dS = t_i^{(\hat{\mathbf{e}}_j)}\,n_j\,dS \tag{2.11}$$

Cancelling the common factor dS and using the identity $t_i^{(\hat{\mathbf{e}}_j)} \equiv \sigma_{ji}$, (*2.11*) becomes

$$t_i^{(\hat{\mathbf{n}})} = \sigma_{ji}n_j \quad \text{or} \quad \mathbf{t}^{(\hat{\mathbf{n}})} = \hat{\mathbf{n}}\cdot\boldsymbol{\Sigma} \tag{2.12}$$

Equation (*2.12*) is also often expressed in the matrix form

$$[t_{1j}^{(\hat{\mathbf{n}})}] = [n_{1k}]\,[\sigma_{kj}] \tag{2.13}$$

which is written explicitly

$$[t_1^{(\hat{\mathbf{n}})}, t_2^{(\hat{\mathbf{n}})}, t_3^{(\hat{\mathbf{n}})}] = [n_1, n_2, n_3]\begin{bmatrix}\sigma_{11} & \sigma_{12} & \sigma_{13}\\ \sigma_{21} & \sigma_{22} & \sigma_{23}\\ \sigma_{31} & \sigma_{32} & \sigma_{33}\end{bmatrix} \tag{2.14}$$

The matrix form (*2.14*) is equivalent to the component equations

$$\begin{aligned} t_1^{(\hat{\mathbf{n}})} &= n_1\sigma_{11} + n_2\sigma_{21} + n_3\sigma_{31}\\ t_2^{(\hat{\mathbf{n}})} &= n_1\sigma_{12} + n_2\sigma_{22} + n_3\sigma_{32}\\ t_3^{(\hat{\mathbf{n}})} &= n_1\sigma_{13} + n_2\sigma_{23} + n_3\sigma_{33} \end{aligned} \tag{2.15}$$

2.7 FORCE AND MOMENT EQUILIBRIUM. STRESS TENSOR SYMMETRY

Equilibrium of an arbitrary volume V of a continuum, subjected to a system of surface forces $t_i^{(\hat{\mathbf{n}})}$ and body forces b_i (including inertia forces, if present) as shown in Fig. 2-8, requires that the resultant force and moment acting on the volume be zero.

Summation of surface and body forces results in the integral relation,

$$\int_S t_i^{(\hat{\mathbf{n}})}\,dS + \int_V \rho b_i\,dV = 0$$

or

$$\int_S \mathbf{t}^{(\hat{\mathbf{n}})}\,dS + \int_V \rho\mathbf{b}\,dV = 0 \tag{2.16}$$

Replacing $t_i^{(\hat{\mathbf{n}})}$ here by $\sigma_{ji}n_j$ and converting the resulting surface integral to a volume integral by the divergence theorem of Gauss (*1.157*), equation (*2.16*) becomes

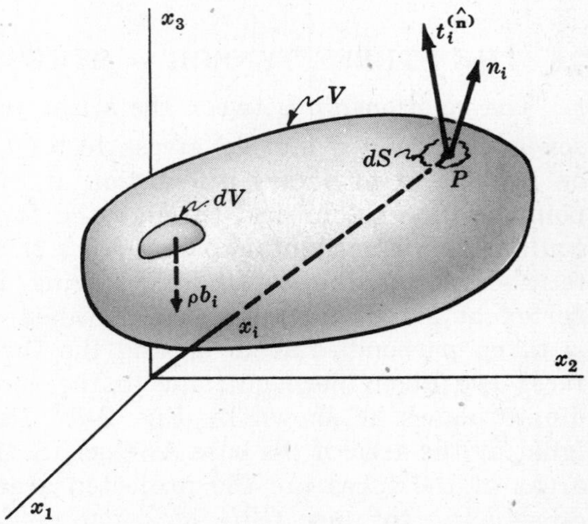

Fig. 2-8

$$\int_V (\sigma_{ji,j} + \rho b_i)\, dV = 0 \quad \text{or} \quad \int_V (\nabla \cdot \mathbf{\Sigma} + \rho \mathbf{b})\, dV = 0 \tag{2.17}$$

Since the volume V is arbitrary, the integrand in (*2.17*) must vanish, so that

$$\sigma_{ji,j} + \rho b_i = 0 \quad \text{or} \quad \nabla \cdot \mathbf{\Sigma} + \rho \mathbf{b} = 0 \tag{2.18}$$

which are called the *equilibrium equations.*

In the absence of distributed moments or couple-stresses, the equilibrium of moments about the origin requires that

$$\int_S \epsilon_{ijk} x_j t_k^{(\hat{\mathbf{n}})}\, dS + \int_V \epsilon_{ijk} x_j \rho b_k\, dV = 0$$

or

$$\int_S \mathbf{x} \times \mathbf{t}^{(\hat{\mathbf{n}})}\, dS + \int_V \mathbf{x} \times \rho \mathbf{b}\, dV = 0 \tag{2.19}$$

in which x_i is the position vector of the elements of surface and volume. Again, making the substitution $t_i^{(\hat{\mathbf{n}})} = \sigma_{ji} n_j$, applying the theorem of Gauss and using the result expressed in (*2.18*), the integrals of (*2.19*) are combined and reduced to

$$\int_V \epsilon_{ijk} \sigma_{jk}\, dV = 0 \quad \text{or} \quad \int_V \mathbf{\Sigma}_v\, dV = 0 \tag{2.20}$$

For the arbitrary volume V, (*2.20*) requires

$$\epsilon_{ijk} \sigma_{jk} = 0 \quad \text{or} \quad \mathbf{\Sigma}_v = 0 \tag{2.21}$$

Equation (*2.21*) represents the equations $\sigma_{12} = \sigma_{21}$, $\sigma_{23} = \sigma_{32}$, $\sigma_{13} = \sigma_{31}$, or in all

$$\sigma_{ij} = \sigma_{ji} \tag{2.22}$$

which shows that the *stress tensor is symmetric.* In view of (*2.22*), the equilibrium equations (*2.18*) are often written

$$\sigma_{ij,j} + \rho b_i = 0 \tag{2.23}$$

which appear in expanded form as

$$\begin{aligned}
\frac{\partial \sigma_{11}}{\partial x_1} + \frac{\partial \sigma_{12}}{\partial x_2} + \frac{\partial \sigma_{13}}{\partial x_3} + \rho b_1 &= 0 \\
\frac{\partial \sigma_{21}}{\partial x_1} + \frac{\partial \sigma_{22}}{\partial x_2} + \frac{\partial \sigma_{23}}{\partial x_3} + \rho b_2 &= 0 \\
\frac{\partial \sigma_{31}}{\partial x_1} + \frac{\partial \sigma_{32}}{\partial x_2} + \frac{\partial \sigma_{33}}{\partial x_3} + \rho b_3 &= 0
\end{aligned} \tag{2.24}$$

2.8 STRESS TRANSFORMATION LAWS

At the point P let the rectangular Cartesian coordinate systems $Px_1x_2x_3$ and $Px_1'x_2'x_3'$ of Fig. 2-9 be related to one another by the table of direction cosines

	x_1	x_2	x_3
x_1'	a_{11}	a_{12}	a_{13}
x_2'	a_{21}	a_{22}	a_{23}
x_3'	a_{31}	a_{32}	a_{33}

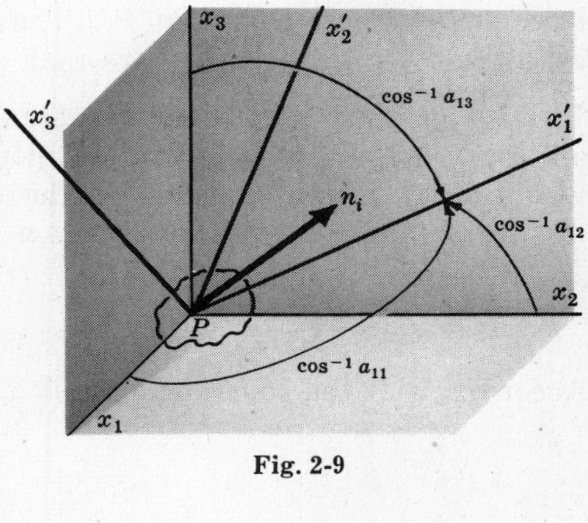

Fig. 2-9

or by the equivalent alternatives, the transformation matrix $[a_{ij}]$, or the transformation dyadic

$$\mathbf{A} = a_{ij}\hat{\mathbf{e}}_i\hat{\mathbf{e}}_j \tag{2.25}$$

According to the transformation law for Cartesian tensors of order one (*1.93*), the components of the stress vector $t_i^{(\hat{n})}$ referred to the unprimed axes are related to the primed axes components $t_i'^{(\hat{n})}$ by the equation

$$t_i'^{(\hat{n})} = a_{ij}t_j^{(\hat{n})} \quad \text{or} \quad \mathbf{t}'^{(\hat{n})} = \mathbf{A}\cdot\mathbf{t}^{(\hat{n})} \tag{2.26}$$

Likewise, by the transformation law (*1.102*) for second-order Cartesian tensors, the stress tensor components in the two systems are related by

$$\sigma'_{ij} = a_{ip}a_{jq}\sigma_{pq} \quad \text{or} \quad \mathbf{\Sigma}' = \mathbf{A}\cdot\mathbf{\Sigma}\cdot\mathbf{A}_c \tag{2.27}$$

In matrix form, the stress vector transformation is written

$$[t_{i1}'^{(\hat{n})}] = [a_{ij}][t_{j1}^{(\hat{n})}] \tag{2.28}$$

and the stress tensor transformation as

$$[\sigma_{ij}] = [a_{ip}][\sigma_{pq}][a_{qj}] \tag{2.29}$$

Explicitly, the matrix multiplications in (*2.28*) and (*2.29*) are given respectively by

$$\begin{bmatrix} t_1'^{(\hat{n})} \\ t_2'^{(\hat{n})} \\ t_3'^{(\hat{n})} \end{bmatrix} = \begin{bmatrix} a_{11} & a_{12} & a_{13} \\ a_{21} & a_{22} & a_{23} \\ a_{31} & a_{32} & a_{33} \end{bmatrix} \begin{bmatrix} t_1^{(\hat{n})} \\ t_2^{(\hat{n})} \\ t_3^{(\hat{n})} \end{bmatrix} \tag{2.30}$$

and

$$\begin{bmatrix} \sigma'_{11} & \sigma'_{12} & \sigma'_{13} \\ \sigma'_{21} & \sigma'_{22} & \sigma'_{23} \\ \sigma'_{31} & \sigma'_{32} & \sigma'_{33} \end{bmatrix} = \begin{bmatrix} a_{11} & a_{12} & a_{13} \\ a_{21} & a_{22} & a_{23} \\ a_{31} & a_{32} & a_{33} \end{bmatrix} \begin{bmatrix} \sigma_{11} & \sigma_{12} & \sigma_{13} \\ \sigma_{21} & \sigma_{22} & \sigma_{23} \\ \sigma_{31} & \sigma_{32} & \sigma_{33} \end{bmatrix} \begin{bmatrix} a_{11} & a_{21} & a_{31} \\ a_{12} & a_{22} & a_{32} \\ a_{13} & a_{23} & a_{33} \end{bmatrix} \tag{2.31}$$

2.9 STRESS QUADRIC OF CAUCHY

At the point P in a continuum, let the stress tensor have the values σ_{ij} when referred to directions parallel to the local Cartesian axes $P\zeta_1\zeta_2\zeta_3$ shown in Fig. 2-10. The equation

$$\sigma_{ij}\zeta_i\zeta_j = \pm k^2 \quad \text{(a constant)} \tag{2.32}$$

represents geometrically similar quadric surfaces having a common center at P. The plus or minus choice assures the surfaces are real.

The position vector $\mathbf{r}$ of an arbitrary point lying on the quadric surface has components $\zeta_i = rn_i$, where n_i is the unit normal in the direction of $\mathbf{r}$. At the point P the normal component $\sigma_N n_i$ of the stress vector $t_i^{(\hat{n})}$ has a magnitude

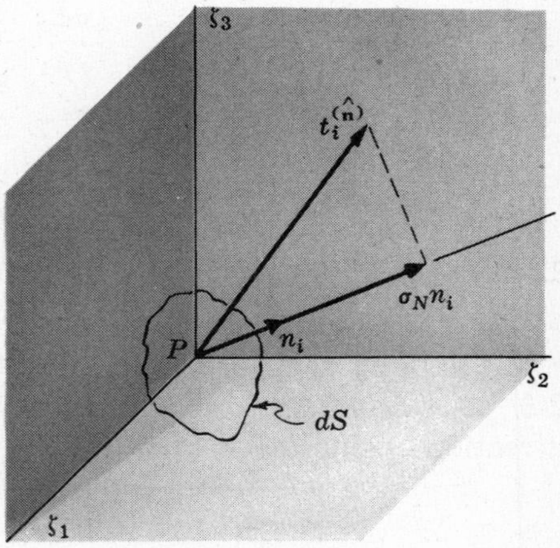

Fig. 2-10

$$\sigma_N = t_i^{(\hat{n})}n_i = \mathbf{t}^{(\hat{n})}\cdot\mathbf{n} = \sigma_{ij}n_in_j \tag{2.33}$$

Accordingly if the constant k^2 of (*2.32*) is set equal to $\sigma_N r^2$, the resulting quadric

$$\sigma_{ij}\zeta_i\zeta_j = \pm\sigma_N r^2 \tag{2.34}$$

is called the *stress quadric of Cauchy*. From this definition it follows that the magnitude σ_N of the normal stress component on the surface element dS perpendicular to the position vector $\mathbf{r}$ of a point on Cauchy's stress quadric, is inversely proportional to r^2, i.e. $\sigma_N = \pm k^2/r^2$. Furthermore it may be shown that the stress vector $t_i^{(\hat{n})}$ acting on dS at P is parallel to the normal of the tangent plane of the Cauchy quadric at the point identified by $\mathbf{r}$.

2.10 PRINCIPAL STRESSES. STRESS INVARIANTS. STRESS ELLIPSOID

At the point P for which the stress tensor components are σ_{ij}, the equation (*2.12*), $t_i^{(\hat{n})} = \sigma_{ji}n_j$, associates with each direction n_i a stress vector $t_i^{(\hat{n})}$. Those directions for which $t_i^{(\hat{n})}$ and n_i are collinear as shown in Fig. 2-11 are called *principal stress directions*. For a principal stress direction,

$$t_i^{(\hat{n})} = \sigma n_i \quad \text{or} \quad \mathbf{t}^{(\hat{n})} = \sigma \hat{\mathbf{n}} \tag{2.35}$$

in which σ, the magnitude of the stress vector, is called a *principal stress value*. Substituting (*2.35*) into (*2.12*) and making use of the identities $n_i = \delta_{ij}n_j$ and $\sigma_{ij} = \sigma_{ji}$, results in the equations

$$(\sigma_{ij} - \delta_{ij}\sigma)n_j = 0 \quad \text{or} \quad (\mathbf{\Sigma} - \mathbf{I}\sigma)\cdot\hat{\mathbf{n}} = 0 \tag{2.36}$$

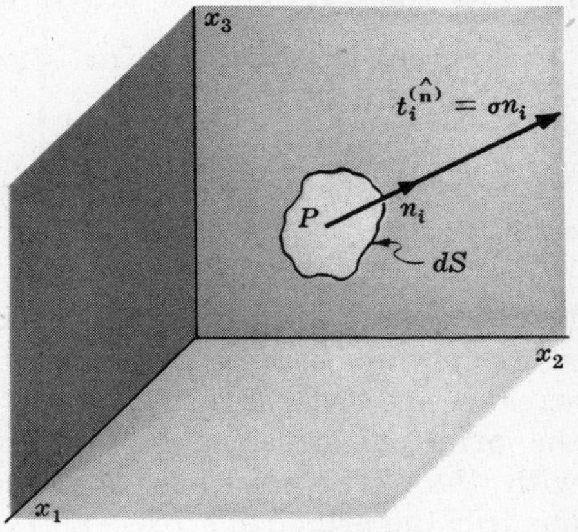

Fig. 2-11

In the three equations (*2.36*), there are four unknowns, namely, the three direction cosines n_i and the principal stress value σ.

For solutions of (*2.36*) other than the trivial one $n_j = 0$, the determinant of coefficients, $|\sigma_{ij} - \delta_{ij}\sigma|$, must vanish. Explicitly,

$$|\sigma_{ij} - \delta_{ij}\sigma| = 0 \quad \text{or} \quad \begin{vmatrix} \sigma_{11}-\sigma & \sigma_{12} & \sigma_{13} \\ \sigma_{21} & \sigma_{22}-\sigma & \sigma_{23} \\ \sigma_{31} & \sigma_{32} & \sigma_{33}-\sigma \end{vmatrix} = 0 \tag{2.37}$$

which upon expansion yields the cubic polynomial in σ,

$$\sigma^3 - \mathrm{I}_\Sigma\sigma^2 + \mathrm{II}_\Sigma\sigma - \mathrm{III}_\Sigma = 0 \tag{2.38}$$

where

$$\mathrm{I}_\Sigma = \sigma_{ii} = \operatorname{tr}\mathbf{\Sigma} \tag{2.39}$$

$$\mathrm{II}_\Sigma = \tfrac{1}{2}(\sigma_{ii}\sigma_{jj} - \sigma_{ij}\sigma_{ij}) \tag{2.40}$$

$$\mathrm{III}_\Sigma = |\sigma_{ij}| = \det\mathbf{\Sigma} \tag{2.41}$$

are known respectively as the *first, second* and *third stress invariants*.

The three roots of (*2.38*), $\sigma_{(1)}, \sigma_{(2)}, \sigma_{(3)}$ are the three principal stress values. Associated with each principal stress $\sigma_{(k)}$, there is a principal stress direction for which the direction cosines $n_i^{(k)}$ are solutions of the equations

$$(\sigma_{ij} - \sigma_{(k)}\delta_{ij})n_j^{(k)} = 0 \quad \text{or} \quad (\mathbf{\Sigma} - \sigma_{(k)}\mathbf{I})\cdot\hat{\mathbf{n}}^{(k)} = 0 \quad (k = 1,2,3) \tag{2.42}$$

In (*2.42*) letter subscripts or superscripts enclosed by parentheses are merely labels and as such do not participate in any summation process. The expanded form of (*2.42*) for the *second* principal direction, for example, is therefore

$$\begin{aligned}(\sigma_{11}-\sigma_{(2)})n_1^{(2)}+\sigma_{12}n_2^{(2)}+\sigma_{13}n_3^{(2)} &= 0\\ \sigma_{21}n_1^{(2)}+(\sigma_{22}-\sigma_{(2)})n_2^{(2)}+\sigma_{23}n_3^{(2)} &= 0\\ \sigma_{31}n_1^{(2)}+\sigma_{32}n_2^{(2)}+(\sigma_{33}-\sigma_{(2)})n_3^{(2)} &= 0\end{aligned} \tag{2.43}$$

Because the stress tensor is real and symmetric, the principal stress values are also real.

When referred to principal stress directions, the stress matrix $[\sigma_{ij}]$ is diagonal,

$$[\sigma_{ij}] \equiv \begin{bmatrix}\sigma_{(1)} & 0 & 0\\ 0 & \sigma_{(2)} & 0\\ 0 & 0 & \sigma_{(3)}\end{bmatrix} \quad \text{or} \quad [\sigma_{ij}] = \begin{bmatrix}\sigma_{\mathrm{I}} & 0 & 0\\ 0 & \sigma_{\mathrm{II}} & 0\\ 0 & 0 & \sigma_{\mathrm{III}}\end{bmatrix} \tag{2.44}$$

in the second form of which Roman numeral subscripts are used to show that the principal stresses are ordered, i.e. $\sigma_{\mathrm{I}} > \sigma_{\mathrm{II}} > \sigma_{\mathrm{III}}$. Since the principal stress directions are coincident with the principal axes of Cauchy's stress quadric, the principal stress values include both the maximum and minimum normal stress components at a point.

In a *principal stress space,* i.e. a space whose axes are in the principal stress directions and whose coordinate unit of measure is stress $(t_1^{(\hat{n})}, t_2^{(\hat{n})}, t_3^{(\hat{n})})$ as shown in Fig. 2-12, the arbitrary stress vector $t_i^{(\hat{n})}$ has components

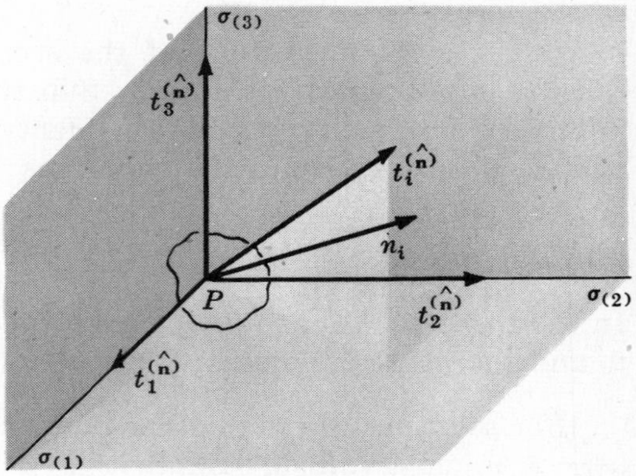

Fig. 2-12

$$t_1^{(\hat{n})} = \sigma_{(1)}n_1, \qquad t_2^{(\hat{n})} = \sigma_{(2)}n_2, \qquad t_3^{(\hat{n})} = \sigma_{(3)}n_3 \tag{2.45}$$

according to (*2.12*). But inasmuch as $(n_1)^2 + (n_2)^2 + (n_3)^2 = 1$ for the unit vector n_i, (*2.45*) requires the stress vector $t_i^{(\hat{n})}$ to satisfy the equation

$$\frac{(t_1^{(\hat{n})})^2}{(\sigma_{(1)})^2} + \frac{(t_2^{(\hat{n})})^2}{(\sigma_{(2)})^2} + \frac{(t_3^{(\hat{n})})^2}{(\sigma_{(3)})^2} = 1 \tag{2.46}$$

in stress space. This equation is an ellipsoid known as the *Lamé stress ellipsoid.*

2.11 MAXIMUM AND MINIMUM SHEAR STRESS VALUES

If the stress vector $t_i^{(\hat{n})}$ is resolved into orthogonal components normal and tangential to the surface element dS upon which it acts, the magnitude of the normal component may be determined from (*2.33*) and the magnitude of the tangential or *shearing component* is given by

$$\sigma_S^2 = t_i^{(\hat{n})}t_i^{(\hat{n})} - \sigma_N^2 \tag{2.47}$$

This resolution is shown in Fig. 2-13 where the axes are chosen in the principal stress directions and it is assumed the principal stresses are ordered according to $\sigma_I > \sigma_{II} > \sigma_{III}$. Hence from (2.12), the components of $t_i^{(\hat{n})}$ are

$$t_1^{(\hat{n})} = \sigma_I n_1$$
$$t_2^{(\hat{n})} = \sigma_{II} n_2 \qquad (2.48)$$
$$t_3^{(\hat{n})} = \sigma_{III} n_3$$

and from (2.33), the normal component magnitude is

$$\sigma_N = \sigma_I n_1^2 + \sigma_{II} n_2^2 + \sigma_{III} n_3^2 \qquad (2.49)$$

Substituting (2.48) and (2.49) into (2.47), the squared magnitude of the shear stress as a function of the direction cosines n_i is given by

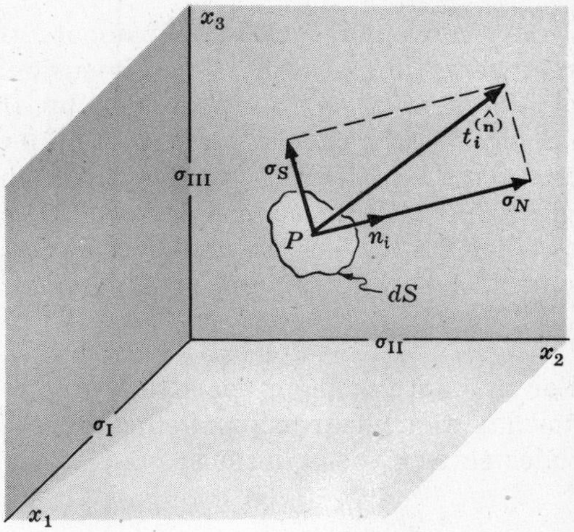

Fig. 2-13

$$\sigma_S^2 = \sigma_I^2 n_1^2 + \sigma_{II}^2 n_2^2 + \sigma_{III}^2 n_3^2 - (\sigma_I n_1^2 + \sigma_{II} n_2^2 + \sigma_{III} n_3^2)^2 \qquad (2.50)$$

The maximum and minimum values of σ_S may be obtained from (2.50) by the method of *Lagrangian multipliers*. The procedure is to construct the function

$$F = \sigma_S^2 - \lambda n_i n_i \qquad (2.51)$$

in which the scalar λ is called a Lagrangian multiplier. Equation (2.51) is clearly a function of the direction cosines n_i, so that the conditions for stationary (maximum or minimum) values of F are given by $\partial F/\partial n_i = 0$. Setting these partials equal to zero yields the equations

$$n_1\{\sigma_I^2 - 2\sigma_I(\sigma_I n_1^2 + \sigma_{II} n_2^2 + \sigma_{III} n_3^2) + \lambda\} = 0 \qquad (2.52a)$$
$$n_2\{\sigma_{II}^2 - 2\sigma_{II}(\sigma_I n_1^2 + \sigma_{II} n_2^2 + \sigma_{III} n_3^2) + \lambda\} = 0 \qquad (2.52b)$$
$$n_3\{\sigma_{III}^2 - 2\sigma_{III}(\sigma_I n_1^2 + \sigma_{II} n_2^2 + \sigma_{III} n_3^2) + \lambda\} = 0 \qquad (2.52c)$$

which, together with the condition $n_i n_i = 1$, may be solved for λ and the direction cosines n_1, n_2, n_3, conjugate to the extremum values of shear stress.

One set of solutions to (2.52), and the associated shear stresses from (2.50), are

$$n_1 = \pm 1, \quad n_2 = 0, \quad n_3 = 0; \quad \text{for which } \sigma_S = 0 \qquad (2.53a)$$
$$n_1 = 0, \quad n_2 = \pm 1, \quad n_3 = 0; \quad \text{for which } \sigma_S = 0 \qquad (2.53b)$$
$$n_1 = 0, \quad n_2 = 0, \quad n_3 = \pm 1; \quad \text{for which } \sigma_S = 0 \qquad (2.53c)$$

The shear stress values in (2.53) are obviously minimum values. Furthermore, since (2.35) indicates that shear components vanish on principal planes, the directions given by (2.53) are recognized as principal stress directions.

A second set of solutions to (2.52) may be verified to be given by

$$n_1 = 0, \quad n_2 = \pm 1/\sqrt{2}, \quad n_3 = \pm 1/\sqrt{2}; \quad \text{for which } \sigma_S = (\sigma_{II} - \sigma_{III})/2 \qquad (2.54a)$$
$$n_1 = \pm 1/\sqrt{2}, \quad n_2 = 0, \quad n_3 = \pm 1/\sqrt{2}; \quad \text{for which } \sigma_S = (\sigma_{III} - \sigma_I)/2 \qquad (2.54b)$$
$$n_1 = \pm 1/\sqrt{2}, \quad n_2 = \pm 1/\sqrt{2}, \quad n_3 = 0; \quad \text{for which } \sigma_S = (\sigma_I - \sigma_{II})/2 \qquad (2.54c)$$

Equation (2.54b) gives the maximum shear stress value, which is equal to half the difference of the largest and smallest principal stresses. Also from (2.54b), the maximum shear stress component acts in the plane which bisects the right angle between the directions of the maximum and minimum principal stresses.

2.12 MOHR'S CIRCLES FOR STRESS

A convenient two-dimensional graphical representation of the three-dimensional state of stress at a point is provided by the well-known *Mohr's stress circles*. In developing these, the coordinate axes are again chosen in the principal stress directions at P as shown by Fig. 2-14. The principal stresses are assumed to be distinct and ordered according to

$$\sigma_{\text{I}} > \sigma_{\text{II}} > \sigma_{\text{III}} \tag{2.55}$$

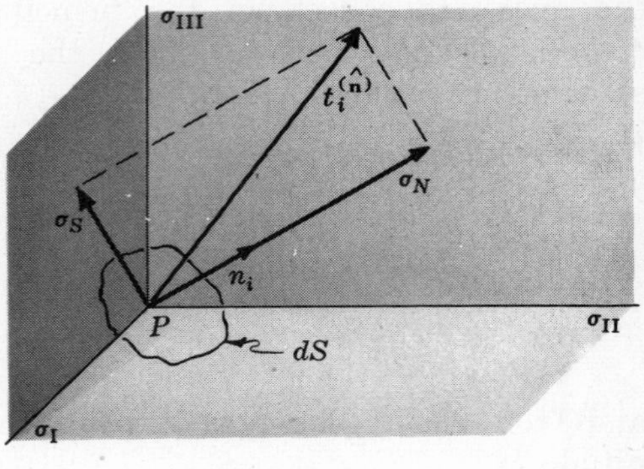

Fig. 2-14

For this arrangement the stress vector $t_i^{(\hat{n})}$ has normal and shear components whose magnitudes satisfy the equations

$$\sigma_N = \sigma_{\text{I}} n_1^2 + \sigma_{\text{II}} n_2^2 + \sigma_{\text{III}} n_3^2 \tag{2.56}$$

$$\sigma_N^2 + \sigma_S^2 = \sigma_{\text{I}}^2 n_1^2 + \sigma_{\text{II}}^2 n_2^2 + \sigma_{\text{III}}^2 n_3^2 \tag{2.57}$$

Combining these two expressions with the identity $n_i n_i = 1$ and solving for the direction cosines n_i, results in the equations

$$(n_1)^2 = \frac{(\sigma_N - \sigma_{\text{II}})(\sigma_N - \sigma_{\text{III}}) + (\sigma_S)^2}{(\sigma_{\text{I}} - \sigma_{\text{II}})(\sigma_{\text{I}} - \sigma_{\text{III}})} \tag{2.58a}$$

$$(n_2)^2 = \frac{(\sigma_N - \sigma_{\text{III}})(\sigma_N - \sigma_{\text{I}}) + (\sigma_S)^2}{(\sigma_{\text{II}} - \sigma_{\text{III}})(\sigma_{\text{II}} - \sigma_{\text{I}})} \tag{2.58b}$$

$$(n_3)^2 = \frac{(\sigma_N - \sigma_{\text{I}})(\sigma_N - \sigma_{\text{II}}) + (\sigma_S)^2}{(\sigma_{\text{III}} - \sigma_{\text{I}})(\sigma_{\text{III}} - \sigma_{\text{II}})} \tag{2.58c}$$

These equations serve as the basis for Mohr's stress circles, shown in the "stress plane" of Fig. 2-15, for which the σ_N axis is the abscissa, and the σ_S axis is the ordinate.

In (*2.58a*), since $\sigma_{\text{I}} - \sigma_{\text{II}} > 0$ and $\sigma_{\text{I}} - \sigma_{\text{III}} > 0$ from (*2.55*), and since $(n_1)^2$ is non-negative, the numerator of the right-hand side satisfies the relationship

$$(\sigma_N - \sigma_{\text{II}})(\sigma_N - \sigma_{\text{III}}) + (\sigma_S)^2 \geqq 0 \tag{2.59}$$

which represents stress points in the (σ_N, σ_S) plane that are *on* or *exterior* to the circle

$$[\sigma_N - (\sigma_{\text{II}} + \sigma_{\text{III}})/2]^2 + (\sigma_S)^2 = [(\sigma_{\text{II}} - \sigma_{\text{III}})/2]^2 \tag{2.60}$$

In Fig. 2-15, this circle is labeled C_1.

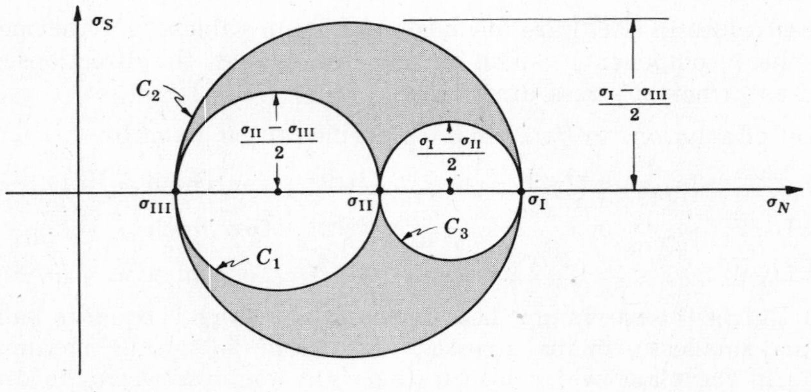

Fig. 2-15

Similarly, for (*2.58b*), since $\sigma_{II} - \sigma_{III} > 0$ and $\sigma_{II} - \sigma_I < 0$ from (*2.55*), and since $(n_2)^2$ is non-negative, the right hand numerator satisfies

$$(\sigma_N - \sigma_{III})(\sigma_N - \sigma_I) + (\sigma_S)^2 \leqq 0 \tag{2.61}$$

which represents points *on* or *interior* to the circle

$$[\sigma_N - (\sigma_{III} + \sigma_I)/2]^2 + (\sigma_S)^2 = [(\sigma_{III} - \sigma_I)/2]^2 \tag{2.62}$$

labeled C_2 in Fig. 2-15. Finally, for (*2.58c*), since $\sigma_{III} - \sigma_I < 0$ and $\sigma_{III} - \sigma_{II} < 0$ from (*2.55*), and since $(n_3)^2$ is non-negative,

$$(\sigma_N - \sigma_I)(\sigma_N - \sigma_{II}) + (\sigma_S)^2 \geqq 0 \tag{2.63}$$

which represents points *on* or *exterior* to the circle

$$[\sigma_N - (\sigma_I + \sigma_{II})/2]^2 + (\sigma_S)^2 = [(\sigma_I - \sigma_{II})/2]^2 \tag{2.64}$$

labeled C_3 in Fig. 2-15.

Since each "stress point" (pair of values of σ_N and σ_S) in the (σ_N, σ_S) plane represents a particular stress vector $t_i^{(\hat{n})}$, the state of stress at P expressed by (*2.58*) is represented in Fig. 2-15 as the shaded area bounded by the Mohr's stress circles. The diagram confirms a maximum shear stress of $(\sigma_I - \sigma_{III})/2$ as was determined analytically in Section 2.11. Frequently, because the sign of the shear stress is not of critical importance, only the top half of this symmetrical diagram is drawn.

The relationship between Mohr's stress diagram and the physical state of stress may be established through consideration of Fig. 2-16, which shows the first octant of a sphere of the continuum centered at point P. The normal n_i at the arbitrary point Q of the spherical surface ABC simulates the normal to the surface element dS at point P. Because of the symmetry properties of the stress tensor and the fact that principal stress axes are used in Fig. 2-16, the state of stress at P is completely represented through the totality of locations Q can occupy on the surface ABC. In the figure, circle arcs KD, GE and FH designate locations for Q along which one direction cosine of n_i has a constant value. Specifically,

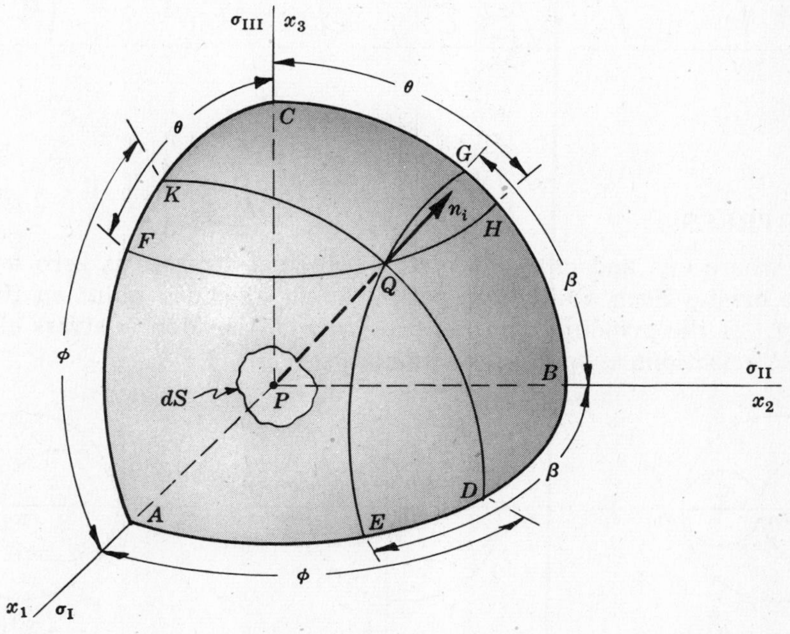

Fig. 2-16

$$n_1 = \cos\phi \text{ on } KD, \qquad n_2 = \cos\beta \text{ on } GE, \qquad n_3 = \cos\theta \text{ on } FH$$

and, on the bounding circle arcs BC, CA and AB,

$$n_1 = \cos\pi/2 = 0 \text{ on } BC, \qquad n_2 = \cos\pi/2 = 0 \text{ on } CA, \qquad n_3 = \cos\pi/2 = 0 \text{ on } AB$$

According to the first of these and the equation (*2.58a*), stress vectors for Q located on BC will have components given by stress points on the circle C_1 in Fig. 2-15. Likewise, CA in Fig. 2-16 corresponds to the circle C_2, and AB to the circle C_3 in Fig. 2-15.

The stress vector components σ_N and σ_S for an arbitrary location of Q may be determined by the construction shown in Fig. 2-17. Thus point e may be located on C_3 by drawing the radial line from the center of C_3 at the angle 2β. Note that angles in the physical space of Fig. 2-16 are doubled in the stress space of Fig. 2-17 (arc AB subtends 90° in Fig. 2-16 whereas the conjugate stress points σ_{I} and σ_{II} are 180° apart on C_3). In the same way, points g, h and f are located in Fig. 2-17 and the appropriate pairs joined by circle arcs having their centers on the σ_N axis. The intersection of circle arcs ge and hf represents the components σ_N and σ_S of the stress vector $t_i^{(\hat{n})}$ on the plane having the normal direction n_i at Q in Fig. 2-16.

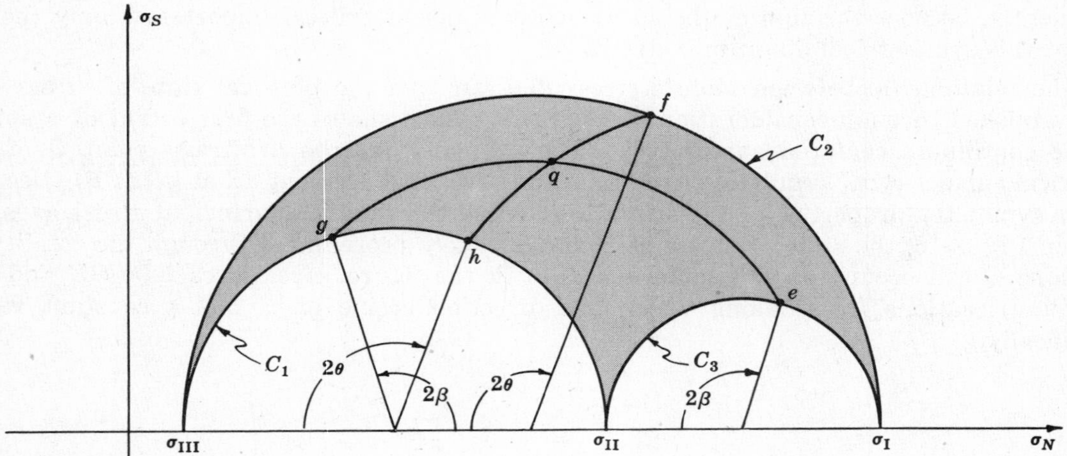

Fig. 2-17

2.13 PLANE STRESS

In the case where one and only one of the principal stresses is zero a state of *plane stress* is said to exist. Such a situation occurs at an unloaded point on the free surface bounding a body. If the principal stresses are ordered, the Mohr's stress circles will have one of the characterizations appearing in Fig. 2-18.

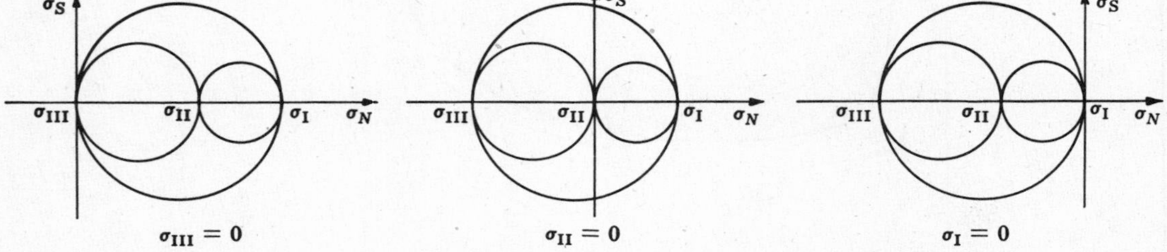

Fig. 2-18

If the principal stresses are not ordered and the direction of the zero principal stress is taken as the x_3 direction, the state of stress is termed plane stress parallel to the x_1x_2 plane. For arbitrary choice of orientation of the orthogonal axes x_1 and x_2 in this case, the stress matrix has the form

$$[\sigma_{ij}] = \begin{bmatrix} \sigma_{11} & \sigma_{12} & 0 \\ \sigma_{12} & \sigma_{22} & 0 \\ 0 & 0 & 0 \end{bmatrix} \tag{2.65}$$

The stress quadric for this plane stress is a cylinder with its base lying in the x_1x_2 plane and having the equation

$$\sigma_{11}x_1^2 + 2\sigma_{12}x_1x_2 + \sigma_{22}x_2^2 = \pm k^2 \tag{2.66}$$

Frequently in elementary books on Strength of Materials a state of plane stress is represented by a single Mohr's circle. As seen from Fig. 2-18 this representation is necessarily incomplete since all three circles are required to show the complete stress picture. In particular, the maximum shear stress value at a point will not be given if the single circle presented happens to be one of the inner circles of Fig. 2-18. A single circle Mohr's diagram is able, however, to display the stress points for all those planes at the point P which include the zero principal stress axis. For such planes, if the coordinate axes are chosen in accordance with the stress representation given in (*2.65*), the single plane stress Mohr's circle has the equation

$$[\sigma_N - (\sigma_{11} + \sigma_{22})/2]^2 + (\sigma_S)^2 = [(\sigma_{11} - \sigma_{22})/2]^2 + (\sigma_{12})^2 \tag{2.67}$$

The essential features in the construction of this circle are illustrated in Fig. 2-19. The circle is drawn by locating the center C at $\sigma_N = (\sigma_{11} + \sigma_{22})/2$ and using the radius $R = \sqrt{[(\sigma_{11} - \sigma_{22})/2]^2 + (\sigma_{12})^2}$ given in (*2.67*). Point A on the circle represents the stress state on the surface element whose normal is n_1 (the right-hand face of the rectangular parallelepiped shown in Fig. 2-19). Point B on the circle represents the stress state on the top surface of the parallelepiped with normal n_2. Principal stress points σ_I and σ_{II} are so labeled, and points E and D on the circle are points of maximum shear stress value.

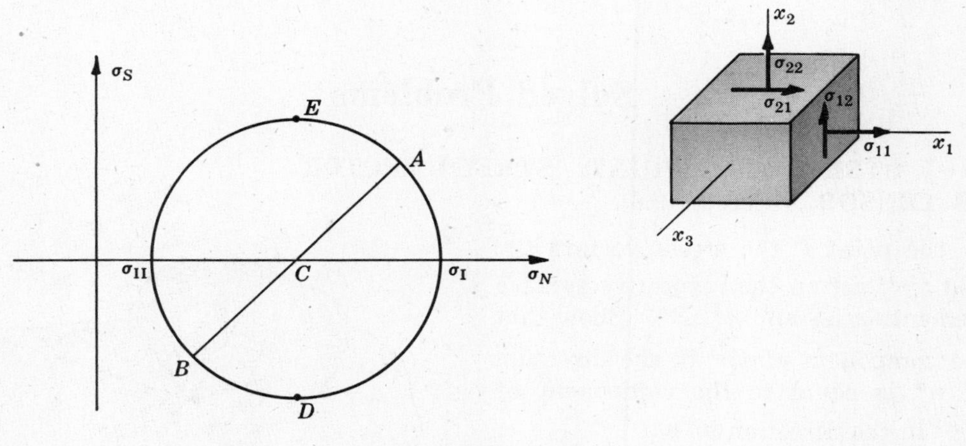

Fig. 2-19

2.14 DEVIATOR AND SPHERICAL STRESS TENSORS

It is very often useful to split the stress tensor σ_{ij} into two component tensors, one of which (the *spherical* or *hydrostatic stress tensor*) has the form

$$\Sigma_M = \sigma_M \mathbf{I} = \begin{pmatrix} \sigma_M & 0 & 0 \\ 0 & \sigma_M & 0 \\ 0 & 0 & \sigma_M \end{pmatrix} \quad (2.68)$$

where $\sigma_M = -p = \sigma_{kk}/3$ is the mean normal stress, and the second (the *deviator stress tensor*) has the form

$$\Sigma_D = \begin{pmatrix} \sigma_{11}-\sigma_M & \sigma_{12} & \sigma_{13} \\ \sigma_{21} & \sigma_{22}-\sigma_M & \sigma_{23} \\ \sigma_{31} & \sigma_{32} & \sigma_{33}-\sigma_M \end{pmatrix} \equiv \begin{pmatrix} s_{11} & s_{12} & s_{13} \\ s_{21} & s_{22} & s_{23} \\ s_{31} & s_{32} & s_{33} \end{pmatrix} \quad (2.69)$$

This decomposition is expressed by the equations

$$\sigma_{ij} = \delta_{ij}\sigma_{kk}/3 + s_{ij} \quad \text{or} \quad \Sigma = \sigma_M \mathbf{I} + \Sigma_D \quad (2.70)$$

The principal directions of the deviator stress tensor s_{ij} are the same as those of the stress tensor σ_{ij}. Thus *principal deviator stress* values are

$$s_{(k)} = \sigma_{(k)} - \sigma_M \quad (2.71)$$

The characteristic equation for the deviator stress tensor, comparable to (*2.38*) for the stress tensor, is the cubic

$$s^3 + II_{\Sigma_D} s - III_{\Sigma_D} = 0 \quad \text{or} \quad s^3 + (s_I s_{II} + s_{II} s_{III} + s_{III} s_I)s - s_I s_{II} s_{III} = 0 \quad (2.72)$$

It is easily shown that the first invariant of the deviator stress tensor I_{Σ_D} is identically zero, which accounts for its absence in (*2.72*).

Solved Problems

STATE OF STRESS AT A POINT. STRESS VECTOR. STRESS TENSOR (Sec. 2.1-2.6)

2.1. At the point P the stress vectors $t_i^{(\hat{n})}$ and $t_i^{(\hat{n}*)}$ act on the respective surface elements $n_i\,\Delta S$ and $n_i^*\,\Delta S^*$. Show that the component of $t_i^{(\hat{n})}$ in the direction of n_i^* is equal to the component of $t_i^{(\hat{n}*)}$ in the direction of n_i.

It is required to show that

$$t_i^{(\hat{n}*)} n_i = t_i^{(\hat{n})} n_i^*$$

From (*2.12*) $t_i^{(\hat{n}*)} n_i = \sigma_{ji} n_j^* n_i$, and by (*2.22*) $\sigma_{ji} = \sigma_{ij}$, so that

$$\sigma_{ji} n_j^* n_i = (\sigma_{ij} n_i) n_j^* = t_j^{(\hat{n})} n_j^*$$

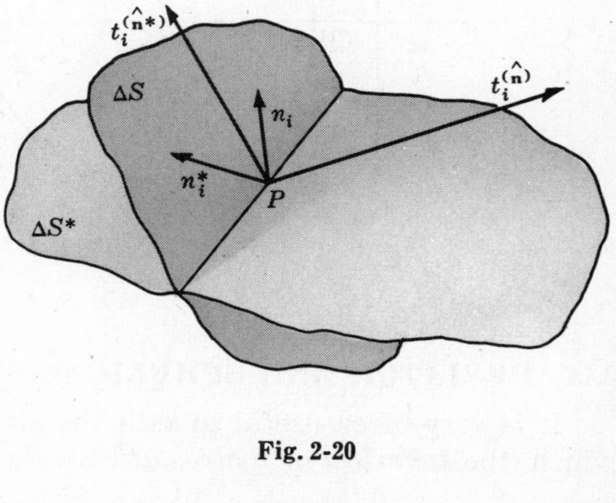

Fig. 2-20

2.2. The stress tensor values at a point P are given by the array

$$\mathbf{\Sigma} = \begin{pmatrix} 7 & 0 & -2 \\ 0 & 5 & 0 \\ -2 & 0 & 4 \end{pmatrix}$$

Determine the traction (stress) vector on the plane at P whose unit normal is $\hat{\mathbf{n}} = (2/3)\hat{\mathbf{e}}_1 - (2/3)\hat{\mathbf{e}}_2 + (1/3)\hat{\mathbf{e}}_3$.

From (*2.12*), $\mathbf{t}^{(\hat{\mathbf{n}})} = \hat{\mathbf{n}} \cdot \mathbf{\Sigma}$. The multiplication is best carried out in the matrix form of (*2.13*):

$$[t_1^{(\hat{\mathbf{n}})}, t_2^{(\hat{\mathbf{n}})}, t_3^{(\hat{\mathbf{n}})}] = [2/3, -2/3, 1/3]\begin{bmatrix} 7 & 0 & -2 \\ 0 & 5 & 0 \\ -2 & 0 & 4 \end{bmatrix} = \left[\frac{14}{3} - \frac{2}{3}, \frac{-10}{3}, \frac{-4}{3} + \frac{4}{3}\right]$$

Thus $\mathbf{t}^{(\hat{\mathbf{n}})} = 4\hat{\mathbf{e}}_1 - \frac{10}{3}\hat{\mathbf{e}}_2$.

2.3. For the traction vector of Problem 2.2, determine (*a*) the component perpendicular to the plane, (*b*) the magnitude of $t_i^{(\hat{\mathbf{n}})}$, (*c*) the angle between $t_i^{(\hat{\mathbf{n}})}$ and $\hat{\mathbf{n}}$.

(*a*) $t_i^{(\hat{\mathbf{n}})} \cdot \hat{\mathbf{n}} = (4\hat{\mathbf{e}}_1 - \frac{10}{3}\hat{\mathbf{e}}_2) \cdot (\frac{2}{3}\hat{\mathbf{e}}_1 - \frac{2}{3}\hat{\mathbf{e}}_2 + \frac{1}{3}\hat{\mathbf{e}}_3) = 44/9$

(*b*) $|t_i^{(\hat{\mathbf{n}})}| = \sqrt{16 + 100/9} = 5.2$

(*c*) Since $t_i^{(\hat{\mathbf{n}})} \cdot \hat{\mathbf{n}} = |t_i^{(\hat{\mathbf{n}})}| \cos\theta$, $\cos\theta = (44/9)/5.2 = 0.94$ and $\theta = 20°$.

2.4. The stress vectors acting on the three coordinate planes are given by $t_i^{(\hat{\mathbf{e}}_1)}$, $t_i^{(\hat{\mathbf{e}}_2)}$ and $t_i^{(\hat{\mathbf{e}}_3)}$. Show that the sum of the squares of the magnitudes of these vectors is independent of the orientation of the coordinate planes.

Let S be the sum in question. Then

$$S = t_i^{(\hat{\mathbf{e}}_1)} t_i^{(\hat{\mathbf{e}}_1)} + t_i^{(\hat{\mathbf{e}}_2)} t_i^{(\hat{\mathbf{e}}_2)} + t_i^{(\hat{\mathbf{e}}_3)} t_i^{(\hat{\mathbf{e}}_3)}$$

which from (*2.7*) becomes $S = \sigma_{1i}\sigma_{1i} + \sigma_{2i}\sigma_{2i} + \sigma_{3i}\sigma_{3i} = \sigma_{ji}\sigma_{ji}$, an invariant.

2.5. The state of stress at a point is given by the stress tensor

$$\sigma_{ij} = \begin{pmatrix} \sigma & a\sigma & b\sigma \\ a\sigma & \sigma & c\sigma \\ b\sigma & c\sigma & \sigma \end{pmatrix}$$

where a, b, c are constants and σ is some stress value. Determine the constants a, b and c so that the stress vector on the *octahedral* plane $(\hat{\mathbf{n}} = (1/\sqrt{3})\hat{\mathbf{e}}_1 + (1/\sqrt{3})\hat{\mathbf{e}}_2 + (1/\sqrt{3})\hat{\mathbf{e}}_3)$ vanishes.

In matrix form, $t_i^{(\hat{\mathbf{n}})} = \sigma_{ij}n_j$ must be zero for the given stress tensor and normal vector.

$$\begin{bmatrix} \sigma & a\sigma & b\sigma \\ a\sigma & \sigma & c\sigma \\ b\sigma & c\sigma & \sigma \end{bmatrix}\begin{bmatrix} 1/\sqrt{3} \\ 1/\sqrt{3} \\ 1/\sqrt{3} \end{bmatrix} = \begin{bmatrix} 0 \\ 0 \\ 0 \end{bmatrix} \qquad \text{hence} \qquad \begin{aligned} a + b &= -1 \\ a + c &= -1 \\ b + c &= -1 \end{aligned}$$

Solving these equations, $a = b = c = -1/2$. Therefore the solution tensor is

$$\sigma_{ij} = \begin{pmatrix} \sigma & -\sigma/2 & -\sigma/2 \\ -\sigma/2 & \sigma & -\sigma/2 \\ -\sigma/2 & -\sigma/2 & \sigma \end{pmatrix}$$

2.6. The stress tensor at point P is given by the array

$$\boldsymbol{\Sigma} = \begin{pmatrix} 7 & -5 & 0 \\ -5 & 3 & 1 \\ 0 & 1 & 2 \end{pmatrix}$$

Determine the stress vector on the plane passing through P and parallel to the plane ABC shown in Fig. 2-21.

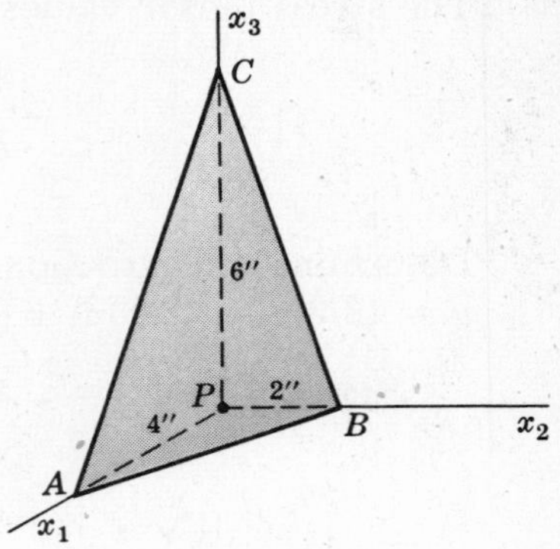

Fig. 2-21

The equation of the plane ABC is $3x_1+6x_2+2x_3=12$, and the unit normal to the plane is therefore (see Problem 1.2)

$$\hat{\mathbf{n}} = \tfrac{3}{7}\hat{\mathbf{e}}_1 + \tfrac{6}{7}\hat{\mathbf{e}}_2 + \tfrac{2}{7}\hat{\mathbf{e}}_3$$

From (*2.14*), the stress vector may be determined by matrix multiplication,

$$[3/7, 6/7, 2/7]\begin{bmatrix} 7 & -5 & 0 \\ -5 & 3 & 1 \\ 0 & 1 & 2 \end{bmatrix} = \frac{1}{7}[-9, 5, 10]$$

Thus $\mathbf{t}^{(\hat{\mathbf{n}})} = \dfrac{-9}{7}\hat{\mathbf{e}}_1 + \dfrac{5}{7}\hat{\mathbf{e}}_2 + \dfrac{10}{7}\hat{\mathbf{e}}_3.$

2.7. The state of stress throughout a continuum is given with respect to the Cartesian axes $Ox_1x_2x_3$ by the array

$$\boldsymbol{\Sigma} = \begin{pmatrix} 3x_1x_2 & 5x_2^2 & 0 \\ 5x_2^2 & 0 & 2x_3 \\ 0 & 2x_3 & 0 \end{pmatrix}$$

Determine the stress vector acting at the point $P(2,1,\sqrt{3})$ of the plane that is tangent to the cylindrical surface $x_2^2+x_3^2=4$ at P.

At P the stress components are given by

$$\boldsymbol{\Sigma} = \begin{pmatrix} 6 & 5 & 0 \\ 5 & 0 & 2\sqrt{3} \\ 0 & 2\sqrt{3} & 0 \end{pmatrix}$$

The unit normal to the surface at P is determined from $\text{grad}\,\phi = \nabla\phi = \nabla(x_2^2+x_3^2-4)$. Thus $\nabla\phi = 2x_2\hat{\mathbf{e}}_2 + 2x_3\hat{\mathbf{e}}_3$ and so

$$\nabla\phi = 2\hat{\mathbf{e}}_2 + 2\sqrt{3}\,\hat{\mathbf{e}}_3 \quad \text{at } P$$

Therefore the unit normal at P is $\hat{\mathbf{n}} = \dfrac{\hat{\mathbf{e}}_2}{2} + \dfrac{\sqrt{3}}{2}\hat{\mathbf{e}}_3$. This may also be seen in Fig. 2-22. Finally the stress vector at P on the plane $\perp$ to $\hat{\mathbf{n}}$ is given by

$$\begin{bmatrix} 6 & 5 & 0 \\ 5 & 0 & 2\sqrt{3} \\ 0 & 2\sqrt{3} & 0 \end{bmatrix}\begin{bmatrix} 0 \\ 1/2 \\ \sqrt{3}/2 \end{bmatrix} = \begin{bmatrix} 5/2 \\ 3 \\ \sqrt{3} \end{bmatrix}$$

or $\mathbf{t}^{(\hat{\mathbf{n}})} = 5\hat{\mathbf{e}}_1/2 + 3\hat{\mathbf{e}}_2 + \sqrt{3}\,\hat{\mathbf{e}}_3.$

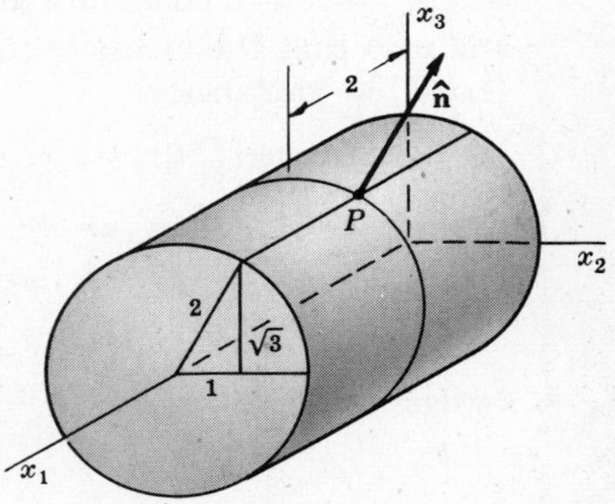

Fig. 2-22

EQUILIBRIUM EQUATIONS (Sec. 2.7)

2.8. For the distribution of the state of stress given in Problem 2.7, what form must the body force components have if the equilibrium equations (*2.24*) are to be satisfied everywhere.

Computing (*2.24*) directly from $\mathbf{\Sigma}$ given in Problem 2.7,

$$\begin{aligned} 3x_2 + 10x_2 + 0 + \rho b_1 &= 0 \\ 0 + 0 + 2 + \rho b_2 &= 0 \\ 0 + 0 + 0 + \rho b_3 &= 0 \end{aligned}$$

These equations are satisfied when $b_1 = -13x_2/\rho$, $b_2 = -2/\rho$, $b_3 = 0$.

2.9. Derive (*2.20*) from (*2.19*), page 49.

Starting with equation (*2.19*),

$$\int_S \epsilon_{ijk} x_j t_k^{(\hat{n})}\, dS + \int_V \epsilon_{ijk} x_j \rho b_k\, dV = 0$$

substitute $t_i^{(\hat{n})} = \sigma_{ji} n_j$ in the surface integral and convert the result to a volume integral by (*1.157*):

$$\int_S (\epsilon_{ijk} x_j \sigma_{pk}) n_p\, dS = \int_V (\epsilon_{ijk} x_j \sigma_{pk})_{,p}\, dV$$

Carrying out the indicated differentiation in this volume integral and combining with the first volume integral gives

$$\int_V \epsilon_{ijk}\{x_{j,p}\sigma_{pk} + x_j(\sigma_{pk,p} + \rho b_k)\}\, dV = 0$$

But from equilibrium equations, $\sigma_{pk,p} + \rho b_k = 0$; and since $x_{j,p} = \delta_{jp}$, this volume integral reduces to (*2.20*), $\int_V \epsilon_{ijk}\sigma_{jk}\, dV = 0$.

STRESS TRANSFORMATIONS (Sec. 2.8)

2.10. The state of stress at a point is given with respect to the Cartesian axes $Ox_1x_2x_3$ by the array

$$\mathbf{\Sigma} = \begin{pmatrix} 2 & -2 & 0 \\ -2 & \sqrt{2} & 0 \\ 0 & 0 & -\sqrt{2} \end{pmatrix}$$

Determine the stress tensor $\mathbf{\Sigma}'$ for the rotated axes $Ox_1'x_2'x_3'$ related to the unprimed axes by the transformation tensor

$$\mathbf{A} = \begin{pmatrix} 0 & 1/\sqrt{2} & 1/\sqrt{2} \\ 1/\sqrt{2} & 1/2 & -1/2 \\ -1/\sqrt{2} & 1/2 & -1/2 \end{pmatrix}$$

The stress transformation law is given by (*2.27*) as $\sigma'_{ij} = a_{ip}a_{jq}\sigma_{pq}$ or $\mathbf{\Sigma}' = \mathbf{A}\cdot\mathbf{\Sigma}\cdot\mathbf{A}_c$. The detailed calculation is best carried out by the matrix multiplication $[\sigma'_{ij}] = [a_{ip}][\sigma_{pq}][a_{qj}]$ given in (*2.29*). Thus

$$[\sigma'_{ij}] = \begin{bmatrix} 0 & 1/\sqrt{2} & 1/\sqrt{2} \\ 1/\sqrt{2} & 1/2 & -1/2 \\ -1/\sqrt{2} & 1/2 & -1/2 \end{bmatrix} \begin{bmatrix} 2 & -2 & 0 \\ -2 & \sqrt{2} & 0 \\ 0 & 0 & -\sqrt{2} \end{bmatrix} \begin{bmatrix} 0 & 1/\sqrt{2} & -1/\sqrt{2} \\ 1/\sqrt{2} & 1/2 & 1/2 \\ 1/\sqrt{2} & -1/2 & -1/2 \end{bmatrix}$$

$$= \begin{bmatrix} 0 & 0 & 2 \\ 0 & 1-\sqrt{2} & -1 \\ 2 & -1 & 1+\sqrt{2} \end{bmatrix}$$

2.11. Show that the stress transformation law may be derived from (*2.33*), the equation $\sigma_N = \sigma_{ij} n_i n_j$ expressing the normal stress value on an arbitrary plane having the unit normal vector n_i.

Since σ_N is a zero-order tensor, it is given with respect to an arbitrary set of primed or unprimed axes in the same form as

$$\sigma_N = \sigma'_{ij} n'_i n'_j = \sigma_{ij} n_i n_j$$

and since by (*1.94*) $n'_i = a_{ij} n_j$,

$$\sigma'_{ij} n'_i n'_j = \sigma'_{ij} a_{ip} n_p a_{jq} n_q = \sigma_N = \sigma_{pq} n_p n_q$$

where new dummy indices have been introduced in the last term. Therefore

$$(\sigma'_{ij} a_{ip} a_{jq} - \sigma_{pq}) n_p n_q = 0$$

and since the directions of the unprimed axes are arbitrary,

$$\sigma'_{ij} a_{ip} a_{jq} = \sigma_{pq}$$

2.12. For the unprimed axes in Fig. 2-23, the stress tensor is given by

$$\sigma_{ij} = \begin{pmatrix} \tau & 0 & 0 \\ 0 & \tau & 0 \\ 0 & 0 & \tau \end{pmatrix}$$

Determine the stress tensor for the primed axes specified as shown in the figure.

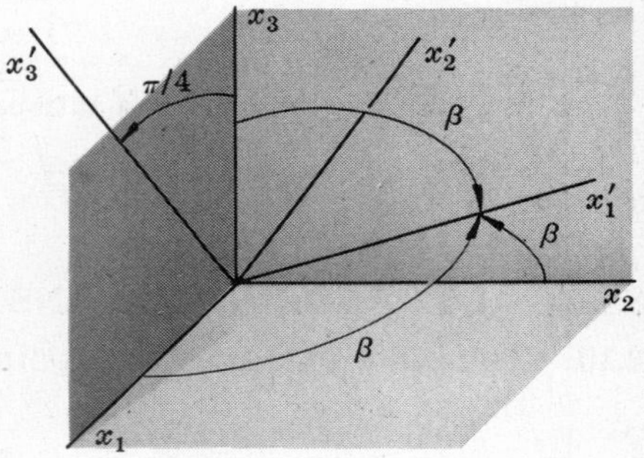

Fig. 2-23

It is first necessary to determine completely the transformation matrix **A**. Since x'_1 makes equal angles with the x_i axes, the first row of the transformation table together with a_{33} is known. Thus

	x_1	x_2	x_3
x'_1	$1/\sqrt{3}$	$1/\sqrt{3}$	$1/\sqrt{3}$
x'_2			
x'_3			$1/\sqrt{2}$

Using the orthogonality equations $a_{ij} a_{ik} = \delta_{jk}$, the transformation matrix is determined by computing the missing entries in the table. It is left as an exercise for the student to show that

$$[a_{ij}] = \begin{bmatrix} 1/\sqrt{3} & 1/\sqrt{3} & 1/\sqrt{3} \\ -2/\sqrt{6} & 1/\sqrt{6} & 1/\sqrt{6} \\ 0 & -1/\sqrt{2} & 1/\sqrt{2} \end{bmatrix}$$

Therefore
$$[\sigma'_{ij}] = \begin{bmatrix} 1/\sqrt{3} & 1/\sqrt{3} & 1/\sqrt{3} \\ -2/\sqrt{6} & 1/\sqrt{6} & 1/\sqrt{6} \\ 0 & -1/\sqrt{2} & 1/\sqrt{2} \end{bmatrix} \begin{bmatrix} \tau & 0 & 0 \\ 0 & \tau & 0 \\ 0 & 0 & \tau \end{bmatrix} \begin{bmatrix} 1/\sqrt{3} & -2/\sqrt{6} & 0 \\ 1/\sqrt{3} & 1/\sqrt{6} & -1/\sqrt{2} \\ 1/\sqrt{3} & 1/\sqrt{6} & 1/\sqrt{2} \end{bmatrix}$$

$$= \begin{bmatrix} \tau/\sqrt{3} & \tau/\sqrt{3} & \tau/\sqrt{3} \\ -2\tau/\sqrt{6} & \tau/\sqrt{6} & \tau/\sqrt{6} \\ 0 & -\tau/\sqrt{2} & \tau/\sqrt{2} \end{bmatrix} \begin{bmatrix} 1/\sqrt{3} & -2/\sqrt{6} & 0 \\ 1/\sqrt{3} & 1/\sqrt{6} & -1/\sqrt{2} \\ 1/\sqrt{3} & 1/\sqrt{6} & 1/\sqrt{2} \end{bmatrix} = \begin{bmatrix} \tau & 0 & 0 \\ 0 & \tau & 0 \\ 0 & 0 & \tau \end{bmatrix}$$

The result obtained here is not surprising when one considers the Mohr's circles for the state of stress having three equal principal stress values.

CAUCHY'S STRESS QUADRIC (Sec. 2.9)

2.13. Determine the Cauchy stress quadric at P for the following states of stress:

(a) uniform tension $\sigma_{11} = \sigma_{22} = \sigma_{33} = \sigma;\ \sigma_{12} = \sigma_{13} = \sigma_{23} = 0$

(b) uniaxial tension $\sigma_{11} = \sigma;\ \sigma_{22} = \sigma_{33} = \sigma_{12} = \sigma_{13} = \sigma_{23} = 0$

(c) simple shear $\sigma_{12} = \sigma_{21} = \tau;\ \sigma_{11} = \sigma_{22} = \sigma_{33} = \sigma_{13} = \sigma_{23} = 0$

(d) plane stress with $\sigma_{11} = \sigma = \sigma_{22};\ \sigma_{12} = \sigma_{21} = \tau;\ \sigma_{33} = \sigma_{31} = \sigma_{23} = 0.$

From (*2.32*), the quadric surface is given in symbolic notation by the equation $\boldsymbol{\zeta} \cdot \boldsymbol{\Sigma} \cdot \boldsymbol{\zeta} = \pm k^2$. Thus using the matrix form,

(a) $$[\zeta_1, \zeta_2, \zeta_3] \begin{bmatrix} \sigma & 0 & 0 \\ 0 & \sigma & 0 \\ 0 & 0 & \sigma \end{bmatrix} \begin{bmatrix} \zeta_1 \\ \zeta_2 \\ \zeta_3 \end{bmatrix} = \sigma\zeta_1^2 + \sigma\zeta_2^2 + \sigma\zeta_3^2 = \pm k^2$$

Hence the quadric surface for uniform tension is the sphere $\zeta_1^2 + \zeta_2^2 + \zeta_3^2 = \pm k^2/\sigma$.

(b) $$[\zeta_1, \zeta_2, \zeta_3] \begin{bmatrix} \sigma & 0 & 0 \\ 0 & 0 & 0 \\ 0 & 0 & 0 \end{bmatrix} \begin{bmatrix} \zeta_1 \\ \zeta_2 \\ \zeta_3 \end{bmatrix} = \sigma\zeta_1^2 = \pm k^2$$

Hence the quadric surface for uniaxial tension is a circular cylinder along the tension axis.

(c) $$[\zeta_1, \zeta_2, \zeta_3] \begin{bmatrix} 0 & \tau & 0 \\ \tau & 0 & 0 \\ 0 & 0 & 0 \end{bmatrix} \begin{bmatrix} \zeta_1 \\ \zeta_2 \\ \zeta_3 \end{bmatrix} = 2\tau\zeta_1\zeta_2 = \pm k^2$$

and so the quadric surface for simple shear is a hyperbolic cylinder parallel to the ζ_3 axis.

(d) $$[\zeta_1, \zeta_2, \zeta_3] \begin{bmatrix} \sigma & \tau & 0 \\ \tau & \sigma & 0 \\ 0 & 0 & 0 \end{bmatrix} \begin{bmatrix} \zeta_1 \\ \zeta_2 \\ \zeta_3 \end{bmatrix} = \sigma\zeta_1^2 + 2\tau\zeta_1\zeta_2 + \sigma\zeta_2^2 = \pm k^2$$

and so the quadric surface for plane stress is a general conic cylinder parallel to the zero principal axis.

2.14. Show that the Cauchy stress quadric for a state of stress represented by

$$\boldsymbol{\Sigma} = \begin{pmatrix} a & 0 & 0 \\ 0 & b & 0 \\ 0 & 0 & c \end{pmatrix}$$

is an ellipsoid (the stress ellipsoid) when a, b and c are all of the same sign.

The equation of the quadric is given by

$$[\zeta_1, \zeta_2, \zeta_3] \begin{bmatrix} a & 0 & 0 \\ 0 & b & 0 \\ 0 & 0 & c \end{bmatrix} \begin{bmatrix} \zeta_1 \\ \zeta_2 \\ \zeta_3 \end{bmatrix} = a\zeta_1^2 + b\zeta_2^2 + c\zeta_3^2 = \pm k^2$$

Therefore the quadric surface is the ellipsoid $\dfrac{\zeta_1^2}{bc} + \dfrac{\zeta_2^2}{ac} + \dfrac{\zeta_3^2}{ab} = \dfrac{\pm k^2}{abc}$.

PRINCIPAL STRESSES (Sec. 2.10-2.11)

2.15. The stress tensor at a point P is given with respect to the axes $Ox_1x_2x_3$ by the values

$$\sigma_{ij} = \begin{pmatrix} 3 & 1 & 1 \\ 1 & 0 & 2 \\ 1 & 2 & 0 \end{pmatrix}$$

Determine the principal stress values and the principal stress directions represented by the axes $Ox_1^*x_2^*x_3^*$.

From (*2.37*) the principal stress values σ are given by

$$\begin{vmatrix} 3-\sigma & 1 & 1 \\ 1 & -\sigma & 2 \\ 1 & 2 & -\sigma \end{vmatrix} = 0 \quad \text{or, upon expansion,} \quad (\sigma+2)(\sigma-4)(\sigma-1) = 0$$

The roots are the principal stress values $\sigma_{(1)} = -2$, $\sigma_{(2)} = 1$, $\sigma_{(3)} = 4$. Let the x_1^* axis be the direction of $\sigma_{(1)}$, and let $n_i^{(1)}$ be the direction cosines of this axis. Then from (*2.42*),

$$\begin{aligned} (3+2)n_1^{(1)} + n_2^{(1)} + n_3^{(1)} &= 0 \\ n_1^{(1)} + 2n_2^{(1)} + 2n_3^{(1)} &= 0 \\ n_1^{(1)} + 2n_2^{(1)} + 2n_3^{(1)} &= 0 \end{aligned}$$

Hence $n_1^{(1)} = 0$; $n_2^{(1)} = -n_3^{(1)}$ and since $n_in_i = 1$, $(n_2^{(1)})^2 = 1/2$. Therefore $n_1^{(1)} = 0$, $n_2^{(1)} = 1/\sqrt{2}$, $n_3^{(1)} = -1/\sqrt{2}$.

Likewise, let x_2^* be associated with $\sigma_{(2)}$. Then from (*2.42*),

$$\begin{aligned} 2n_1^{(2)} + n_2^{(2)} + n_3^{(2)} &= 0 \\ n_1^{(2)} - n_2^{(2)} + 2n_3^{(2)} &= 0 \\ n_1^{(2)} + 2n_2^{(2)} - n_3^{(2)} &= 0 \end{aligned}$$

so that $n_1^{(2)} = 1/\sqrt{3}$, $n_2^{(2)} = -1/\sqrt{3}$, $n_3^{(2)} = -1/\sqrt{3}$.

Finally, let x_3^* be associated with $\sigma_{(3)}$. Then from (*2.42*),

$$\begin{aligned} -n_1^{(3)} + n_2^{(3)} + n_3^{(3)} &= 0 \\ n_1^{(3)} - 4n_2^{(3)} + 2n_3^{(3)} &= 0 \\ n_1^{(3)} + 2n_2^{(3)} - 4n_3^{(3)} &= 0 \end{aligned}$$

so that $n_1^{(3)} = -2/\sqrt{6}$, $n_2^{(3)} = -1/\sqrt{6}$, $n_3^{(3)} = -1/\sqrt{6}$.

2.16. Show that the transformation tensor of direction cosines determined in Problem 2.15 transforms the original stress tensor into the diagonal principal axes stress tensor.

According to (*2.29*), $[\sigma_{ij}^*] = [a_{ip}][\sigma_{pq}][a_{qj}]$, which for the problem at hand becomes

$$[\sigma_{ij}^*] = \begin{bmatrix} 0 & 1/\sqrt{2} & -1/\sqrt{2} \\ 1/\sqrt{3} & -1/\sqrt{3} & -1/\sqrt{3} \\ -2/\sqrt{6} & -1/\sqrt{6} & -1/\sqrt{6} \end{bmatrix} \begin{bmatrix} 3 & 1 & 1 \\ 1 & 0 & 2 \\ 1 & 2 & 0 \end{bmatrix} \begin{bmatrix} 0 & 1/\sqrt{3} & -2/\sqrt{6} \\ 1/\sqrt{2} & -1/\sqrt{3} & -1/\sqrt{6} \\ -1/\sqrt{2} & -1/\sqrt{3} & -1/\sqrt{6} \end{bmatrix}$$

$$= \begin{bmatrix} 0 & -\sqrt{2} & \sqrt{2} \\ 1/\sqrt{3} & -1/\sqrt{3} & -1/\sqrt{3} \\ -8/\sqrt{6} & -4/\sqrt{6} & -4/\sqrt{6} \end{bmatrix} \begin{bmatrix} 0 & 1/\sqrt{3} & -2/\sqrt{6} \\ 1/\sqrt{2} & -1/\sqrt{3} & -1/\sqrt{6} \\ -1/\sqrt{2} & -1/\sqrt{3} & -1/\sqrt{6} \end{bmatrix} = \begin{bmatrix} -2 & 0 & 0 \\ 0 & 1 & 0 \\ 0 & 0 & 4 \end{bmatrix}$$

2.17. Determine the principal stress values and principal directions for the stress tensor

$$\sigma_{ij} = \begin{pmatrix} \tau & \tau & \tau \\ \tau & \tau & \tau \\ \tau & \tau & \tau \end{pmatrix}$$

From (*2.37*), $\begin{vmatrix} \tau-\sigma & \tau & \tau \\ \tau & \tau-\sigma & \tau \\ \tau & \tau & \tau-\sigma \end{vmatrix} = 0$ or $(\tau-\sigma)[-2\tau\sigma+\sigma^2]+2\tau^2\sigma = [3\tau-\sigma]\sigma^2 = 0.$

Hence $\sigma_{(1)}=0$, $\sigma_{(2)}=0$, $\sigma_{(3)}=3\tau$. For $\sigma_{(3)}=3\tau$, (*2.42*) yield

$$-2n_1^{(3)}+n_2^{(3)}+n_3^{(3)} = 0, \quad n_1^{(3)}-2n_2^{(3)}+n_3^{(3)} = 0, \quad n_1^{(3)}+n_2^{(3)}-2n_3^{(3)} = 0$$

and therefore $n_1^{(3)}=n_2^{(3)}=n_3^{(3)}=1/\sqrt{3}$. For $\sigma_{(1)}=\sigma_{(2)}=0$, (*2.42*) yield

$$n_1+n_2+n_3 = 0, \quad n_1+n_2+n_3 = 0, \quad n_1+n_2+n_3 = 0$$

which together with $n_in_i=1$ are insufficient to determine uniquely the first and second principal directions. Thus any pair of axes perpendicular to the $n_i^{(3)}$ direction and perpendicular to each other may serve as principal axes. For example, consider the axes determined in Problem 2.12, for which the transformation matrix is

$$[a_{ij}] = \begin{bmatrix} 1/\sqrt{3} & 1/\sqrt{3} & 1/\sqrt{3} \\ -2/\sqrt{6} & 1/\sqrt{6} & 1/\sqrt{6} \\ 0 & -1/\sqrt{2} & 1/\sqrt{2} \end{bmatrix}$$

According to the transformation law (*2.29*), the principal stress matrix $[\sigma_{ij}^*]$ is given by

$$[\sigma_{ij}^*] = \begin{bmatrix} 1/\sqrt{3} & 1/\sqrt{3} & 1/\sqrt{3} \\ -2/\sqrt{6} & 1/\sqrt{6} & 1/\sqrt{6} \\ 0 & -1/\sqrt{2} & 1/\sqrt{2} \end{bmatrix} \begin{bmatrix} \tau & \tau & \tau \\ \tau & \tau & \tau \\ \tau & \tau & \tau \end{bmatrix} \begin{bmatrix} 1/\sqrt{3} & -2/\sqrt{6} & 0 \\ 1/\sqrt{3} & 1/\sqrt{6} & -1/\sqrt{2} \\ 1/\sqrt{3} & 1/\sqrt{6} & 1/\sqrt{2} \end{bmatrix}$$

$$= \begin{bmatrix} \sqrt{3}\,\tau & \sqrt{3}\,\tau & \sqrt{3}\,\tau \\ 0 & 0 & 0 \\ 0 & 0 & 0 \end{bmatrix} \begin{bmatrix} 1/\sqrt{3} & -2/\sqrt{6} & 0 \\ 1/\sqrt{3} & 1/\sqrt{6} & -1/\sqrt{2} \\ 1/\sqrt{3} & 1/\sqrt{6} & 1/\sqrt{2} \end{bmatrix} = \begin{bmatrix} 3\tau & 0 & 0 \\ 0 & 0 & 0 \\ 0 & 0 & 0 \end{bmatrix}$$

2.18. Show that the axes $Ox_1^*x_2^*x_3^*$, (where x_2^*, x_3 and x_3^* are in the same vertical plane, and x_1^*, x_1 and x_2 are in the same horizontal plane) are also principal axes for the stress tensor of Problem 2.17.

The transformation matrix $[a_{ij}]$ relating the two sets of axes clearly has the known elements

$$[a_{ij}] = \begin{bmatrix} - & - & 0 \\ - & - & \sqrt{2}/\sqrt{3} \\ 1/\sqrt{3} & 1/\sqrt{3} & 1/\sqrt{3} \end{bmatrix}$$

as is evident from Fig. 2-24. From the orthogonality conditions $a_{ij}a_{ik}=\delta_{jk}$, the remaining four elements are determined so that

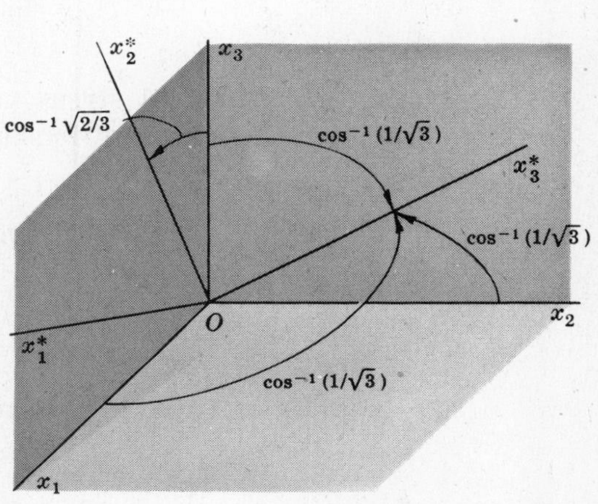

Fig. 2-24

$$[a_{ij}] = \begin{bmatrix} -1/\sqrt{2} & 1/\sqrt{2} & 0 \\ -1/\sqrt{6} & -1/\sqrt{6} & \sqrt{2}/\sqrt{3} \\ 1/\sqrt{3} & 1/\sqrt{3} & 1/\sqrt{3} \end{bmatrix}$$

As before,

$$[\sigma^*_{ij}] = \begin{bmatrix} -1/\sqrt{2} & 1/\sqrt{2} & 0 \\ -1/\sqrt{6} & -1/\sqrt{6} & \sqrt{2}/\sqrt{3} \\ 1/\sqrt{3} & 1/\sqrt{3} & 1/\sqrt{3} \end{bmatrix} \begin{bmatrix} \tau & \tau & \tau \\ \tau & \tau & \tau \\ \tau & \tau & \tau \end{bmatrix} \begin{bmatrix} -1/\sqrt{2} & -1/\sqrt{6} & 1/\sqrt{3} \\ 1/\sqrt{2} & -1/\sqrt{6} & 1/\sqrt{3} \\ 0 & \sqrt{2}/\sqrt{3} & 1/\sqrt{3} \end{bmatrix}$$

$$= \begin{bmatrix} 0 & 0 & 0 \\ 0 & 0 & 0 \\ \sqrt{3}\,\tau & \sqrt{3}\,\tau & \sqrt{3}\,\tau \end{bmatrix} \begin{bmatrix} -1/\sqrt{2} & -1/\sqrt{6} & 1/\sqrt{3} \\ 1/\sqrt{2} & -1/\sqrt{6} & 1/\sqrt{3} \\ 0 & \sqrt{2}/\sqrt{3} & 1/\sqrt{3} \end{bmatrix} = \begin{bmatrix} 0 & 0 & 0 \\ 0 & 0 & 0 \\ 0 & 0 & 3\tau \end{bmatrix}$$

2.19. Show that the principal stresses $\sigma^*_{(k)}$ and the stress components σ_{ij} for an arbitrary set of axes referred to the principal directions through the transformation coefficients a_{ij} are related by $\sigma_{ij} = \sum_{p=1}^{3} a_{pi} a_{pj} \sigma_p$.

From the transformation law for stress (*2.27*), $\sigma_{ij} = a_{pi}a_{qj}\sigma^*_{pq}$; but since σ^*_{pq} are principal stresses, there are only three terms on the right side of this equation, and in each $p = q$. Therefore the right hand side may be written in form $\sigma_{ij} = \sum_{p=1}^{3} a_{pi}a_{pj}\sigma^*_p$.

2.20. Prove that $\sigma_{ij}\sigma_{ik}\sigma_{kj}$ is an invariant of the stress tensor.

By the transformation law (*2.27*),

$$\begin{aligned} \sigma'_{ij}\sigma'_{ik}\sigma'_{kj} &= a_{ip}a_{jq}\sigma_{pq}a_{ir}a_{ks}\sigma_{rs}a_{km}a_{jn}\sigma_{mn} \\ &= (a_{ip}a_{ir})(a_{jq}a_{jn})(a_{ks}a_{km})\sigma_{pq}\sigma_{rs}\sigma_{mn} \\ &= \delta_{pr}\delta_{qn}\delta_{sm}\sigma_{pq}\sigma_{rs}\sigma_{mn} \\ &= (\delta_{pr}\sigma_{pq})(\delta_{qn}\sigma_{mn})(\delta_{sm}\sigma_{rs}) \\ &= \sigma_{rq}\sigma_{qm}\sigma_{rm} = \sigma_{ij}\sigma_{ik}\sigma_{kj} \end{aligned}$$

2.21. Evaluate directly the invariants $I_\Sigma, II_\Sigma, III_\Sigma$ for the stress tensor

$$\sigma_{ij} = \begin{pmatrix} 6 & -3 & 0 \\ -3 & 6 & 0 \\ 0 & 0 & 8 \end{pmatrix}$$

Determine the principal stress values for this state of stress and show that the diagonal form of the stress tensor yields the same values for the stress invariants.

From (*2.39*), $I_\Sigma = \sigma_{ii} = 6 + 6 + 8 = 20$.

From (*2.40*), $II_\Sigma = (1/2)(\sigma_{ii}\sigma_{jj} - \sigma_{ij}\sigma_{ij})$

$$= \sigma_{11}\sigma_{22} + \sigma_{22}\sigma_{33} + \sigma_{33}\sigma_{11} - \sigma_{12}\sigma_{12} - \sigma_{23}\sigma_{23} - \sigma_{31}\sigma_{31}$$

$$= 36 + 48 + 48 - 9 = 123.$$

From (*2.41*), $III_\Sigma = |\sigma_{ij}| = 6(48) + 3(-24) = 216$.

The principal stress values of σ_{ij} are $\sigma_{I} = 3$, $\sigma_{II} = 8$, $\sigma_{III} = 9$. In terms of principal values,

$$I_\Sigma = \sigma_I + \sigma_{II} + \sigma_{III} = 3 + 8 + 9 = 20$$

$$II_\Sigma = \sigma_I\sigma_{II} + \sigma_{II}\sigma_{III} + \sigma_{III}\sigma_I = 24 + 72 + 27 = 123$$

$$III_\Sigma = \sigma_I\sigma_{II}\sigma_{III} = (24)9 = 216$$

2.22. The *octahedral plane* is the plane which makes equal angles with the principal stress directions. Show that the shear stress on this plane, the so-called *octahedral shear stress,* is given by

$$\sigma_{OCT} = \tfrac{1}{3}\sqrt{(\sigma_I - \sigma_{II})^2 + (\sigma_{II} - \sigma_{III})^2 + (\sigma_{III} - \sigma_I)^2}$$

With respect to the principal axes, the normal to the octahedral plane is given by

$$\hat{\mathbf{n}} = \frac{1}{\sqrt{3}}(\hat{\mathbf{e}}_1 + \hat{\mathbf{e}}_2 + \hat{\mathbf{e}}_3)$$

Hence from (*2.12*) the stress vector on the octahedral plane is

$$\begin{aligned} \mathbf{t}^{(\hat{\mathbf{n}})} &= \frac{1}{\sqrt{3}}(\hat{\mathbf{e}}_1 + \hat{\mathbf{e}}_2 + \hat{\mathbf{e}}_3) \cdot (\sigma_I\hat{\mathbf{e}}_1\hat{\mathbf{e}}_1 + \sigma_{II}\hat{\mathbf{e}}_2\hat{\mathbf{e}}_2 + \sigma_{III}\hat{\mathbf{e}}_3\hat{\mathbf{e}}_3) \\ &= \frac{1}{\sqrt{3}}(\sigma_I\hat{\mathbf{e}}_1 + \sigma_{II}\hat{\mathbf{e}}_2 + \sigma_{III}\hat{\mathbf{e}}_3) \end{aligned}$$

and its normal component is

$$\sigma_N = \hat{\mathbf{n}} \cdot \mathbf{t}^{(\hat{\mathbf{n}})} = \tfrac{1}{3}(\sigma_I + \sigma_{II} + \sigma_{III})$$

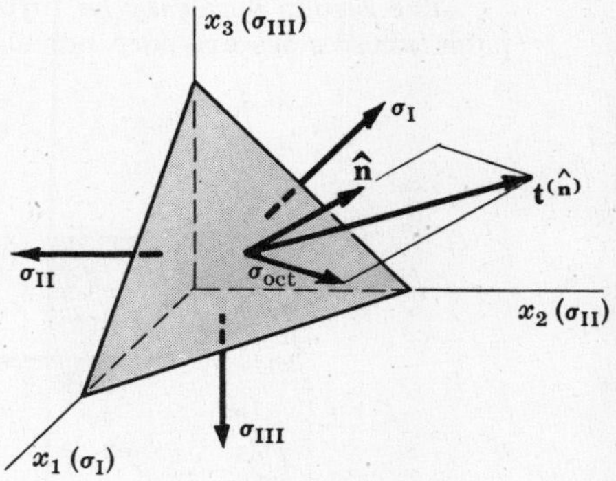

Fig. 2-25

Therefore the shear component is

$$\begin{aligned} \sigma_{OCT} &= \sqrt{\mathbf{t}^{(\hat{\mathbf{n}})} \cdot \mathbf{t}^{(\hat{\mathbf{n}})} - \sigma_N^2} = \{\tfrac{1}{3}(\sigma_I^2 + \sigma_{II}^2 + \sigma_{III}^2) - \tfrac{1}{9}(\sigma_I + \sigma_{II} + \sigma_{III})^2\}^{1/2} \\ &= \tfrac{1}{3}\{3(\sigma_I^2 + \sigma_{II}^2 + \sigma_{III}^2) - (\sigma_I^2 + \sigma_{II}^2 + \sigma_{III}^2 + 2\sigma_I\sigma_{II} + 2\sigma_{II}\sigma_{III} + 2\sigma_{III}\sigma_I)\}^{1/2} \\ &= \tfrac{1}{3}\{(\sigma_I^2 - 2\sigma_I\sigma_{II} + \sigma_{II}^2) + (\sigma_{II}^2 - 2\sigma_{II}\sigma_{III} + \sigma_{III}^2) + (\sigma_{III}^2 - 2\sigma_{III}\sigma_I + \sigma_I^2)\}^{1/2} \\ &= \tfrac{1}{3}\sqrt{(\sigma_I - \sigma_{II})^2 + (\sigma_{II} - \sigma_{III})^2 + (\sigma_{III} - \sigma_I)^2} \end{aligned}$$

2.23. The stress tensor at a point is given by $\sigma_{ij} = \begin{pmatrix} 5 & 0 & 0 \\ 0 & -6 & -12 \\ 0 & -12 & 1 \end{pmatrix}$. Determine the maximum shear stress at the point and show that it acts in the plane which bisects the maximum and minimum stress planes.

From (*2.38*) the reader should verify that the principal stresses are $\sigma_I = 10$, $\sigma_{II} = 5$, $\sigma_{III} = -15$. From (*2.54b*) the maximum shear value is $\sigma_S = (\sigma_{III} - \sigma_I)/2 = -12.5$. The principal axes $Ox_1^*x_2^*x_3^*$ are related to the axes of maximum shear $Ox_1'x_2'x_3'$ by the transformation table below and are situated as shown in Fig. 2-26.

	x_1^*	x_2^*	x_3^*
x_1'	$1/\sqrt{2}$	0	$1/\sqrt{2}$
x_2'	0	1	0
x_3'	$-1/\sqrt{2}$	0	$1/\sqrt{2}$

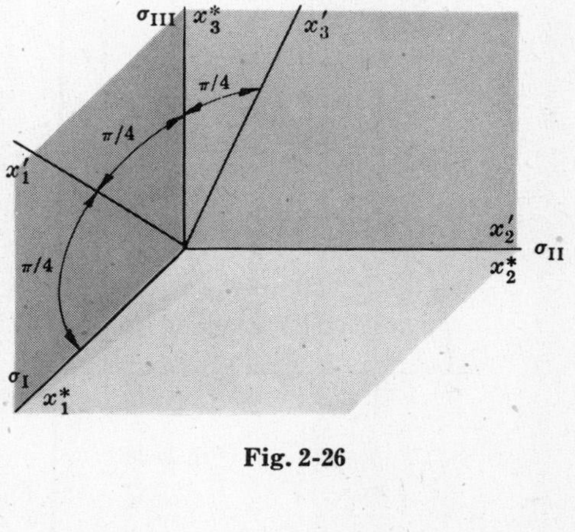

Fig. 2-26

The stress tensor referred to the primed axes is thus given by

$$[\sigma'_{ij}] = \begin{bmatrix} 1/\sqrt{2} & 0 & 1/\sqrt{2} \\ 0 & 1 & 0 \\ -1/\sqrt{2} & 0 & 1/\sqrt{2} \end{bmatrix} \begin{bmatrix} 10 & 0 & 0 \\ 0 & 5 & 0 \\ 0 & 0 & -15 \end{bmatrix} \begin{bmatrix} 1/\sqrt{2} & 0 & -1/\sqrt{2} \\ 0 & 1 & 0 \\ 1/\sqrt{2} & 0 & 1/\sqrt{2} \end{bmatrix} = \begin{bmatrix} -2.5 & 0 & -12.5 \\ 0 & 5 & 0 \\ -12.5 & 0 & -2.5 \end{bmatrix}$$

The results here may be further clarified by showing the stresses on infinitesimal cubes at the point whose sides are perpendicular to the coordinate axes (see Fig. 2-27).

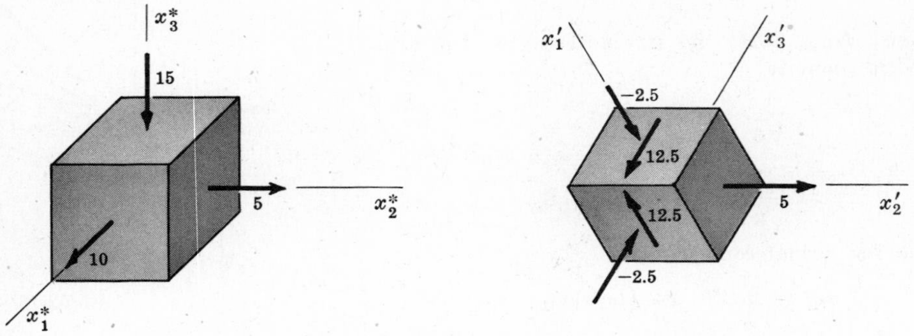

Fig. 2-27

MOHR'S CIRCLES (Sec. 2.12-2.13)

2.24. Draw the Mohr's circles for the state of stress discussed in Problem 2.23. Label important points. Relate the axes $Ox_1x_2x_3$ (conjugate to σ_{ij}) to the principal axes $Ox_1^*x_2^*x_3^*$ and locate on the Mohr's diagram the points giving the stress states on the coordinate planes of $Ox_1x_2x_3$.

The upper half of the symmetric Mohr's circles diagram is shown in Fig. 2-28 with the maximum shear point P and principal stresses labeled. The transformation table of direction cosines is

	x_1^*	x_2^*	x_3^*
x_1	0	1	0
x_2	−3/5	0	4/5
x_3	4/5	0	3/5

from which a diagram of the relative orientation of the axes is made as shown in Fig. 2-29. The x_1 and x_2^* axes are coincident. x_2 and x_3 are in the plane of $x_1^*x_3^*$ as shown. From the angles $\alpha = 36.8°$ and $\beta = 53.2°$ shown, the points $A(-6, 12)$ on the plane $\perp$ to x_2 and $B(1, 12)$ on the plane $\perp$ to x_3 are located. Point $C(5, 0)$ represents the stress state on the plane $\perp$ to x_1.

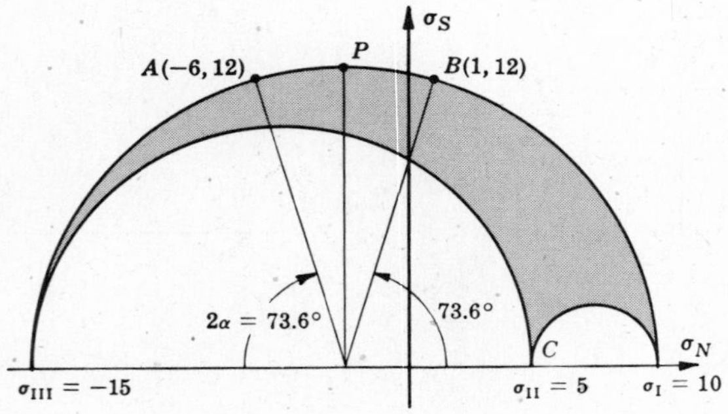

Fig. 2-28

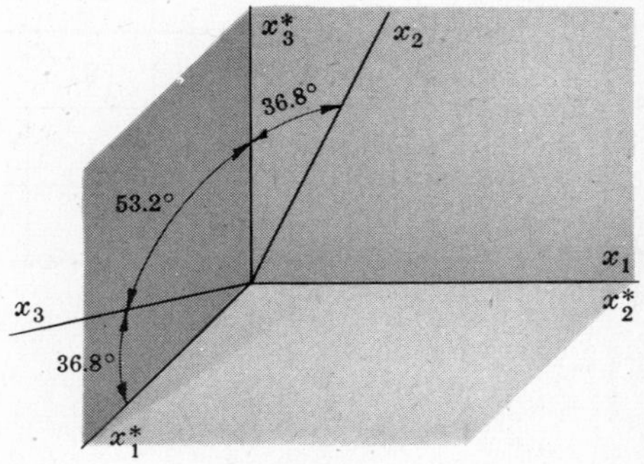

Fig. 2-29

2.25. The state of stress at a point referred to axes $Ox_1x_2x_3$ is given by

$$\sigma_{ij} = \begin{pmatrix} -5 & 0 & 0 \\ 0 & -6 & -12 \\ 0 & -12 & 1 \end{pmatrix}$$

Determine analytically the stress vector components on the plane whose unit normal is $\hat{\mathbf{n}} = (2/3)\hat{\mathbf{e}}_1 + (1/3)\hat{\mathbf{e}}_2 + (2/3)\hat{\mathbf{e}}_3$. Check the results by the Mohr's diagram for this problem.

From (*2.13*) and the symmetry property of the stress tensor, the stress vector on the plane of $\hat{\mathbf{n}}$ is given by the matrix product

$$\begin{bmatrix} -5 & 0 & 0 \\ 0 & -6 & -12 \\ 0 & -12 & 1 \end{bmatrix} \begin{bmatrix} 2/3 \\ 1/3 \\ 2/3 \end{bmatrix} = \begin{bmatrix} -10/3 \\ -10 \\ -10/3 \end{bmatrix}$$

Thus $\mathbf{t}^{(\hat{\mathbf{n}})} = -10\,\hat{\mathbf{e}}_1/3 - 10\,\hat{\mathbf{e}}_2 - 10\,\hat{\mathbf{e}}_3/3$; and from (*2.33*), $\sigma_N = \mathbf{t}^{(\hat{\mathbf{n}})} \cdot \hat{\mathbf{n}} = -70/9$. From (*2.47*), $\sigma_S = 70.7/9$.

For σ_{ij} the principal stress values are $\sigma_{\mathrm{I}} = 10$, $\sigma_{\mathrm{II}} = -5$, $\sigma_{\mathrm{III}} = -15$; and the principal axes are related to $Ox_1x_2x_3$ by the table

	x_1	x_2	x_3
x_1^*	0	−3/5	4/5
x_2^*	1	0	0
x_3^*	0	4/5	3/5

Thus in the principal axes frame, $n_i^* = a_{ij}n_j$ or $\begin{bmatrix} 0 & -3/5 & 4/5 \\ 1 & 0 & 0 \\ 0 & 4/5 & 3/5 \end{bmatrix} \begin{bmatrix} 2/3 \\ 1/3 \\ 2/3 \end{bmatrix} = \begin{bmatrix} 1/3 \\ 2/3 \\ 2/3 \end{bmatrix}$. Accordingly the angles of Fig. 2-16 are given by $\theta = \beta = \cos^{-1} 2/3 = 48.2°$ and $\phi = \cos^{-1} 1/3 = 70.5°$, and the Mohr's diagram comparable to Fig. 2-17 is as shown in Fig. 2-30.

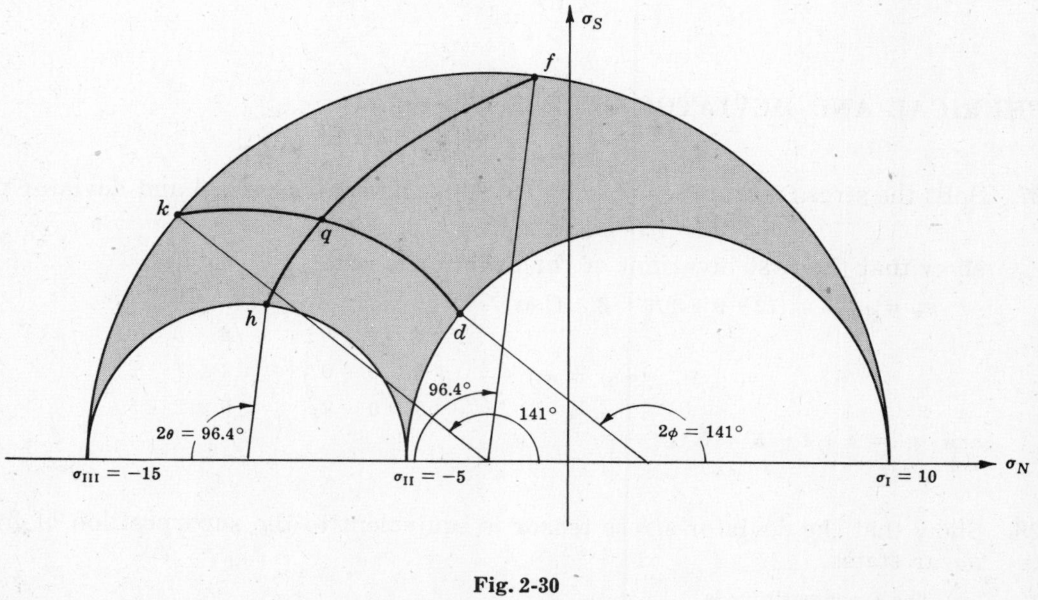

Fig. 2-30

2.26. Sketch the Mohr's circles for the three cases of plane stress depicted by the stresses on the small cube oriented along the coordinate axes shown in Fig. 2-31. Determine the maximum shear stress in each case.

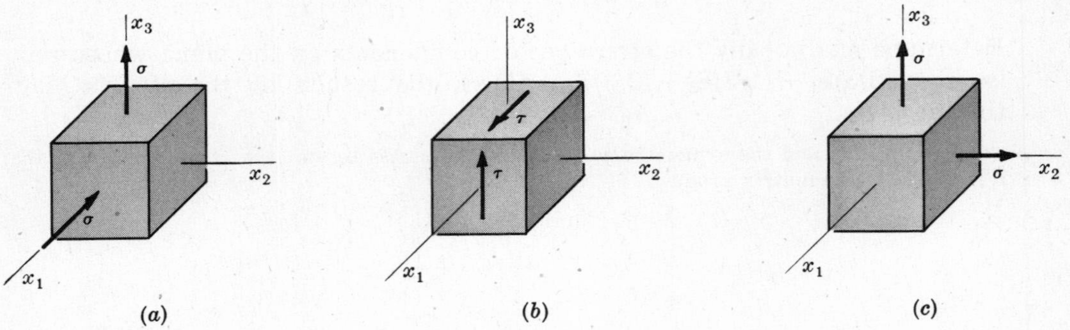

Fig. 2-31

The Mohr's circles are shown in Fig. 2-32.

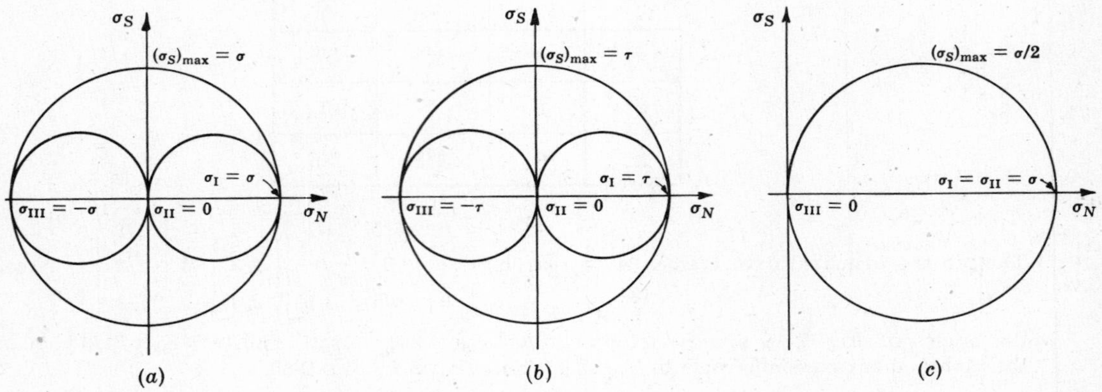

Fig. 2-32

SPHERICAL AND DEVIATOR STRESS (Sec. 2.14)

2.27. Split the stress tensor $\sigma_{ij} = \begin{pmatrix} 12 & 4 & 0 \\ 4 & 9 & -2 \\ 0 & -2 & 3 \end{pmatrix}$ into its spherical and deviator parts and show that the first invariant of the deviator is zero.

$\sigma_M = \sigma_{kk}/3 = (12+9+3)/3 = 8$. Thus

$$\sigma_{ij} \quad = \quad \sigma_M \delta_{ij} + s_{ij} \quad = \quad \begin{pmatrix} 8 & 0 & 0 \\ 0 & 8 & 0 \\ 0 & 0 & 8 \end{pmatrix} + \begin{pmatrix} 4 & 4 & 0 \\ 4 & 1 & -2 \\ 0 & -2 & -5 \end{pmatrix}$$

and $s_{ii} = 4 + 1 - 5 = 0$.

2.28. Show that the deviator stress tensor is equivalent to the superposition of five simple shear states.

The decomposition is

$$\begin{pmatrix} s_{11} & s_{12} & s_{13} \\ s_{21} & s_{22} & s_{23} \\ s_{31} & s_{32} & s_{33} \end{pmatrix} = \begin{pmatrix} 0 & s_{12} & 0 \\ s_{21} & 0 & 0 \\ 0 & 0 & 0 \end{pmatrix} + \begin{pmatrix} 0 & 0 & s_{13} \\ 0 & 0 & 0 \\ s_{31} & 0 & 0 \end{pmatrix} + \begin{pmatrix} 0 & 0 & 0 \\ 0 & 0 & s_{23} \\ 0 & s_{32} & 0 \end{pmatrix}$$

$$+ \begin{pmatrix} s_{11} & 0 & 0 \\ 0 & -s_{11} & 0 \\ 0 & 0 & 0 \end{pmatrix} + \begin{pmatrix} 0 & 0 & 0 \\ 0 & -s_{33} & 0 \\ 0 & 0 & s_{33} \end{pmatrix}$$

where the last two tensors are seen to be equivalent to simple shear states by comparison of cases (*a*) and (*b*) in Problem 2.26. Also note that since $s_{ii} = 0$, $-s_{11} - s_{33} = s_{22}$.

2.29. Determine the principal deviator stress values for the stress tensor

$$\sigma_{ij} = \begin{pmatrix} 10 & -6 & 0 \\ -6 & 10 & 0 \\ 0 & 0 & 1 \end{pmatrix}$$

The deviator of σ_{ij} is $s_{ij} = \begin{pmatrix} 3 & -6 & 0 \\ -6 & 3 & 0 \\ 0 & 0 & -6 \end{pmatrix}$ and its principal values may be determined from the determinant

$$\begin{vmatrix} 3-s & -6 & 0 \\ -6 & 3-s & 0 \\ 0 & 0 & -6-s \end{vmatrix} = (-6-s)(s+3)(s-9) = 0$$

Thus $s_{\rm I} = 9$, $s_{\rm II} = -3$, $s_{\rm III} = -6$. The same result is obtained by first calculating the principal stress values of σ_{ij} and then using (*2.71*). For σ_{ij}, as the reader should show, $\sigma_{\rm I} = 16$, $\sigma_{\rm II} = 4$, $\sigma_{\rm III} = 1$ and hence $s_{\rm I} = 16 - 7 = 9$, $s_{\rm II} = 4 - 7 = -3$, $s_{\rm III} = 1 - 7 = -6$.

2.30. Show that the second invariant of the stress deviator is given in terms of its principal stress values by $\text{II}_{\Sigma_D} = (s_{\rm I}s_{\rm II} + s_{\rm II}s_{\rm III} + s_{\rm III}s_{\rm I})$, or by the alternative form $\text{II}_{\Sigma_D} = -\frac{1}{2}(s_{\rm I}^2 + s_{\rm II}^2 + s_{\rm III}^2)$.

In terms of the principal deviator stresses the characteristic equation of the deviator stress tensor is given by the determinant

$$\begin{vmatrix} s_{\rm I}-s & 0 & 0 \\ 0 & s_{\rm II}-s & 0 \\ 0 & 0 & s_{\rm III}-s \end{vmatrix} = (s_{\rm I}-s)(s_{\rm II}-s)(s_{\rm III}-s) = 0$$

$$= s^3 + (s_{\rm I}s_{\rm II} + s_{\rm II}s_{\rm III} + s_{\rm III}s_{\rm I})s - s_{\rm I}s_{\rm II}s_{\rm III}$$

Hence from (2.72), $\text{II}_{\Sigma_D} = (s_{\rm I}s_{\rm II} + s_{\rm II}s_{\rm III} + s_{\rm III}s_{\rm I})$. Since $s_{\rm I} + s_{\rm II} + s_{\rm III} = 0$,

$$\text{II}_{\Sigma_D} = \tfrac{1}{2}(2s_{\rm I}s_{\rm II} + 2s_{\rm II}s_{\rm III} + 2s_{\rm III}s_{\rm I} - (s_{\rm I} + s_{\rm II} + s_{\rm III})^2) = -\tfrac{1}{2}(s_{\rm I}^2 + s_{\rm II}^2 + s_{\rm III}^2)$$

MISCELLANEOUS PROBLEMS

2.31. Prove that for any symmetric tensor such as the stress tensor σ_{ij}, the transformed tensor σ'_{ij} in any other coordinate system is also symmetric.

From (*2.27*), $\sigma'_{ij} = a_{ip}a_{jq}\sigma_{pq} = a_{jq}a_{ip}\sigma_{qp} = \sigma'_{ji}$.

2.32. At the point P the principal stresses are such that $2\sigma_{II} = \sigma_I + \sigma_{III}$. Determine the unit normal n_i for the plane upon which $\sigma_N = \sigma_{II}$ and $\sigma_S = (\sigma_I - \sigma_{III})/4$.

From (*2.33*), $\sigma_N = n_1^2\sigma_I + n_2^2(\sigma_I + \sigma_{III})/2 + n_3^2\sigma_{III} = (\sigma_I + \sigma_{III})/2$; and since $n_1^2 + n_2^2 + n_3^2 = 1$, these equations may be combined to yield $n_1 = n_3$. Next from (*2.47*),

$$\sigma_S^2 = n_1^2\sigma_I^2 + n_2^2(\sigma_I + \sigma_{III})^2/4 + n_3^2\sigma_{III}^2 - (\sigma_I + \sigma_{III})^2/4 = (\sigma_I - \sigma_{III})^2/16$$

Substituting $n_1 = n_3$ and $n_2^2 - 1 = -n_1^2 - n_3^2 = -2n_1^2$ into this equation and solving for n_1, the direction cosines are found to be $n_1 = 1/2\sqrt{2}$, $n_2 = \sqrt{3}/2$, $n_3 = 1/2\sqrt{2}$. The reader should apply these results to the stress tensor $\sigma_{ij} = \begin{pmatrix} 4 & 0 & 0 \\ 0 & 5 & 0 \\ 0 & 0 & 6 \end{pmatrix}$.

2.33. Show that the stress tensor σ_{ij} may be decomposed into a spherical and deviatoric part in one and only one way.

Assume two decompositions, $\sigma_{ij} = \delta_{ij}\lambda + s_{ij} = \delta_{ij}\lambda^* + s_{ij}^*$ with $s_{ii} = 0$ and $s_{ii}^* = 0$. Then $\sigma_{ii} = 3\lambda = 3\lambda^*$, so $\lambda = \lambda^*$; and from $\lambda\delta_{ij} + s_{ij} = \lambda\delta_{ij} + s_{ij}^*$ it follows that $s_{ij} = s_{ij}^*$.

2.34. Prove that the principal stress values are real if $\mathbf{\Sigma}$ is real and symmetric.

For real values of the stress components the stress invariants are real and hence the coefficients in (*2.38*) are all real. Thus by the theory of equations one root (principal stress) is real. Let $\sigma_{(3)}$ be this root and consider a set of primed axes x_i' of which x_3' is in the direction of $\sigma_{(3)}$. With respect to such axes the characteristic equation is given by the determinant $\begin{vmatrix} \sigma_{11}' - \sigma & \sigma_{12}' & 0 \\ \sigma_{21}' & \sigma_{22}' - \sigma & 0 \\ 0 & 0 & \sigma_{(3)} - \sigma \end{vmatrix} = 0$ or $(\sigma_{(3)} - \sigma)[(\sigma_{11}' - \sigma)(\sigma_{22}' - \sigma) - \sigma_{12}'^2] = 0$. Since the discriminant of the quadratic in square brackets $D = (\sigma_{11}' + \sigma_{22}')^2 - 4[\sigma_{11}'\sigma_{22}' - (\sigma_{12}')^2] = (\sigma_{11}' - \sigma_{22}')^2 + 4\sigma_{12}'^2 > 0$, the remaining roots must be real.

2.35. Use the method of Lagrangian multipliers to show that the extremal values (maximum and minimum) of the normal stress σ_N correspond to principal values.

From (*2.33*), $\sigma_N = \sigma_{ij}n_in_j$ with $n_in_i = 1$. Thus in analogy with (*2.51*) construct the function $H = \sigma_N - \lambda n_in_i$ for which $\partial H/\partial n_i = 0$. Then

$$\begin{aligned} \frac{\partial H}{\partial n_p} &= \sigma_{ij}n_{i,p}n_j + \sigma_{ij}n_in_{j,p} - 2\lambda n_{i,p}n_i \\ &= \sigma_{ij}\delta_{ip}n_j + \sigma_{ij}n_i\delta_{jp} - 2\lambda\delta_{ip}n_i \\ &= \sigma_{pj}n_j + \sigma_{ip}n_i - 2\lambda\delta_{ip}n_i = 2(\sigma_{pi} - \lambda\delta_{ip})n_i = 0 \end{aligned}$$

which is equivalent to equation (*2.36*) defining principal stress directions.

2.36. Assume that the stress components σ_{ij} are derivable from a symmetric tensor field ϕ_{ij} by the relationship $\sigma_{ij} = \epsilon_{ipq}\epsilon_{jmn}\phi_{qn,pm}$. Show that in the absence of body forces the equilibrium equations (*2.23*) are satisfied.

Using the results of Problem 1.58, the stress components are given by

$$\sigma_{ij} = \delta_{ij}(\phi_{qq,pp} - \phi_{qp,qp}) + \phi_{pi,pj} + \phi_{jp,pi} - \phi_{pp,ji} - \phi_{ji,pp}$$

or explicitly

$$\begin{aligned} \sigma_{11} &= \phi_{33,22} + \phi_{22,33} & \sigma_{12} &= \sigma_{21} = -\phi_{33,21} \\ \sigma_{22} &= \phi_{11,33} + \phi_{33,11} & \sigma_{23} &= \sigma_{32} = -\phi_{11,23} \\ \sigma_{33} &= \phi_{22,11} + \phi_{11,22} & \sigma_{31} &= \sigma_{13} = -\phi_{22,13} \end{aligned}$$

Substituting these values into $\sigma_{ij,j} = 0$,

$$\sigma_{11,1} + \sigma_{12,2} + \sigma_{13,3} = \phi_{33,221} + \phi_{22,331} - \phi_{33,212} - \phi_{22,133} = 0$$

$$\sigma_{21,1} + \sigma_{22,2} + \sigma_{23,3} = -\phi_{33,211} + \phi_{11,332} + \phi_{33,112} - \phi_{11,233} = 0$$

$$\sigma_{31,1} + \sigma_{32,2} + \sigma_{33,3} = -\phi_{22,131} - \phi_{11,232} + \phi_{22,113} + \phi_{11,223} = 0$$

2.37. Show that, as is asserted in Section 2.9, the normal to the Cauchy stress quadric at the point whose position vector is **r** is parallel to the stress vector $t_i^{(\hat{n})}$.

Let the quadric surface be given in the form $\phi = \sigma_{ij}\zeta_i\zeta_j \pm k^2 = 0$. The normal at any point is then $\nabla\phi$ or $\partial\phi/\partial\zeta_i = \phi_{,i}$. Hence $\phi_{,p} = \sigma_{ij}\delta_{ip}\zeta_j + \sigma_{ij}\zeta_i\delta_{jp} = 2\sigma_{pi}\zeta_i$. Now since $\zeta_i = rn_i$, this becomes $2\sigma_{pi}rn_i$ or $2r(\sigma_{pi}n_i) = 2t_p^{(\hat{n})}$.

2.38. At a point P the stress tensor referred to axes $Ox_1x_2x_3$ is given by

$$\sigma_{ij} = \begin{pmatrix} 15 & -10 & 0 \\ -10 & 5 & 0 \\ 0 & 0 & 20 \end{pmatrix}$$

If new axes $Ox_1'x_2'x_3'$ are chosen by a rotation about the origin for which the transformation matrix is $[a_{ij}] = \begin{bmatrix} 3/5 & 0 & -4/5 \\ 0 & 1 & 0 \\ 4/5 & 0 & 3/5 \end{bmatrix}$, determine the traction vectors on each of the primed coordinate planes by projecting the traction vectors of the original axes onto the primed directions. In this way determine σ_{ij}'. Check the result by the transformation formula (*2.27*).

From (*2.6*) and the identity $t_j^{(\hat{e}_i)} = \sigma_{ij}$ (*2.7*), the traction vectors on the unprimed coordinate axes are $\mathbf{t}^{(\hat{e}_1)} = 15\hat{\mathbf{e}}_1 - 10\hat{\mathbf{e}}_2$, $\mathbf{t}^{(\hat{e}_2)} = -10\hat{\mathbf{e}}_1 + 5\hat{\mathbf{e}}_2$, $\mathbf{t}^{(\hat{e}_3)} = 20\hat{\mathbf{e}}_3$ which correspond to the rows of the stress tensor. Projecting these vectors onto the primed axes by the vector form of (*2.12*), $\mathbf{t}^{(\hat{n})} = n_1\mathbf{t}^{(\hat{e}_1)} + n_2\mathbf{t}^{(\hat{e}_2)} + n_3\mathbf{t}^{(\hat{e}_3)}$, gives

$$\mathbf{t}^{(\hat{e}_1')} = \tfrac{3}{5}(15\hat{\mathbf{e}}_1 - 10\hat{\mathbf{e}}_2) - \tfrac{4}{5}(20\hat{\mathbf{e}}_3) = 9\hat{\mathbf{e}}_1 - 6\hat{\mathbf{e}}_2 - 16\hat{\mathbf{e}}_3$$

which by the transformation of the unit vectors becomes

$$\mathbf{t}^{(\hat{e}_1')} = 9(\tfrac{3}{5}\hat{\mathbf{e}}_1' + \tfrac{4}{5}\hat{\mathbf{e}}_3') - 6\hat{\mathbf{e}}_2' - 16(-\tfrac{4}{5}\hat{\mathbf{e}}_1' + \tfrac{3}{5}\hat{\mathbf{e}}_3') = 91\hat{\mathbf{e}}_1'/5 - 6\hat{\mathbf{e}}_2' - 12\hat{\mathbf{e}}_3'/5$$

Likewise
$$\mathbf{t}^{(\hat{e}_2')} = -6\hat{\mathbf{e}}_1' + 5\hat{\mathbf{e}}_2' - 8\hat{\mathbf{e}}_3'$$

and
$$\mathbf{t}^{(\hat{e}_3')} = -12\hat{\mathbf{e}}_1'/5 - 8\hat{\mathbf{e}}_2' + 84\hat{\mathbf{e}}_3'/5$$

so that
$$\sigma_{ij}' = \begin{pmatrix} 91/5 & -6 & -12/5 \\ -6 & 5 & -8 \\ -12/5 & -8 & 84/5 \end{pmatrix}$$

By (*2.27*),
$$\sigma_{ij}' = \begin{bmatrix} 3/5 & 0 & -4/5 \\ 0 & 1 & 0 \\ 4/5 & 0 & 3/5 \end{bmatrix}\begin{bmatrix} 15 & -10 & 0 \\ -10 & 5 & 0 \\ 0 & 0 & 20 \end{bmatrix}\begin{bmatrix} 3/5 & 0 & 4/5 \\ 0 & 1 & 0 \\ -4/5 & 0 & 3/5 \end{bmatrix}$$

$$= \begin{bmatrix} 9 & -6 & -16 \\ -10 & 5 & 0 \\ 12 & -8 & 12 \end{bmatrix}\begin{bmatrix} 3/5 & 0 & 4/5 \\ 0 & 1 & 0 \\ -4/5 & 0 & 3/5 \end{bmatrix} = \begin{bmatrix} 91/5 & -6 & -12/5 \\ -6 & 5 & -8 \\ -12/5 & -8 & 84/5 \end{bmatrix}$$

2.39. Show that the second invariant of the deviator stress tensor II_{Σ_D} is related to the octahedral shear stress by the equation $\sigma_{OCT} = \sqrt{-\frac{2}{3}\text{II}_{\Sigma_D}}$.

From Problem 2.22, $\sigma_{OCT} = \frac{1}{3}\sqrt{(\sigma_{\text{I}} - \sigma_{\text{II}})^2 + (\sigma_{\text{II}} - \sigma_{\text{III}})^2 + (\sigma_{\text{III}} - \sigma_{\text{I}})^2}$, and because $\sigma_{\text{I}} = \sigma_M + s_{\text{I}}$, $\sigma_{\text{II}} = \sigma_M + s_{\text{II}}$, etc.,

$$\begin{aligned}\sigma_{OCT} &= \tfrac{1}{3}\sqrt{(s_{\text{I}} - s_{\text{II}})^2 + (s_{\text{II}} - s_{\text{III}})^2 + (s_{\text{III}} - s_{\text{I}})^2} \\ &= \tfrac{1}{3}\sqrt{2(s_{\text{I}}^2 + s_{\text{II}}^2 + s_{\text{III}}^2) - 2(s_{\text{I}}s_{\text{II}} + s_{\text{II}}s_{\text{III}} + s_{\text{III}}s_{\text{I}})}\end{aligned}$$

Also $s_{\text{I}} + s_{\text{II}} + s_{\text{III}} = 0$ and so $(s_{\text{I}} + s_{\text{II}} + s_{\text{III}})^2 = 0$ or

$$s_{\text{I}}^2 + s_{\text{II}}^2 + s_{\text{III}}^2 = -2(s_{\text{I}}s_{\text{II}} + s_{\text{II}}s_{\text{III}} + s_{\text{III}}s_{\text{I}})$$

Hence

$$\sigma_{OCT} = \tfrac{1}{3}\sqrt{-6(s_{\text{I}}s_{\text{II}} + s_{\text{II}}s_{\text{III}} + s_{\text{III}}s_{\text{I}})} = \sqrt{-\tfrac{2}{3}\text{II}_{\Sigma_D}}$$

2.40. The state of stress throughout a body is given by the stress tensor

$$\sigma_{ij} = \begin{pmatrix} 0 & Cx_3 & 0 \\ Cx_3 & 0 & -Cx_1 \\ 0 & -Cx_1 & 0 \end{pmatrix}$$

where C is an arbitrary constant. (*a*) Show that the equilibrium equations are satisfied if body forces are zero. (*b*) At the point $P(4, -4, 7)$ calculate the stress vector on the plane $2x_1 + 2x_2 - x_3 = -7$, and on the sphere $(x_1)^2 + (x_2)^2 + (x_3)^2 = (9)^2$. (*c*) Determine the principal stresses, maximum shear stresses and principal deviator stresses at P. (*d*) Sketch the Mohr's circles for the state of stress at P.

(*a*) Substituting directly into (*2.24*) from σ_{ij}, the equilibrium equations are satisfied identically.

(*b*) From Problem 1.2, the unit normal to the plane $2x_1 + 2x_2 - x_3 = -7$ is $\hat{\mathbf{n}} = \frac{2}{3}\hat{\mathbf{e}}_1 + \frac{2}{3}\hat{\mathbf{e}}_2 - \frac{1}{3}\hat{\mathbf{e}}_3$. Thus from (*2.12*) the stress vector on the plane at P is

$$\begin{aligned}\mathbf{t}^{(\hat{\mathbf{n}})} &= (\tfrac{2}{3}\hat{\mathbf{e}}_1 + \tfrac{2}{3}\hat{\mathbf{e}}_2 - \tfrac{1}{3}\hat{\mathbf{e}}_3) \cdot (7C\,\hat{\mathbf{e}}_1\hat{\mathbf{e}}_2 + 7C\,\hat{\mathbf{e}}_2\hat{\mathbf{e}}_1 - 4C\,\hat{\mathbf{e}}_2\hat{\mathbf{e}}_3 - 4C\,\hat{\mathbf{e}}_3\hat{\mathbf{e}}_2) \\ &= C(\tfrac{14}{3}\hat{\mathbf{e}}_2 + \tfrac{14}{3}\hat{\mathbf{e}}_1 - \tfrac{8}{3}\hat{\mathbf{e}}_3 + \tfrac{4}{3}\hat{\mathbf{e}}_2) = \tfrac{1}{3}C(14\,\hat{\mathbf{e}}_1 + 18\,\hat{\mathbf{e}}_2 - 8\,\hat{\mathbf{e}}_3)\end{aligned}$$

The normal to the sphere $x_ix_i = (9)^2$ at P is $n_i = \phi_{,i}$ with $\phi = x_ix_i - 81$, or $\hat{\mathbf{n}} = \frac{4}{9}\hat{\mathbf{e}}_1 - \frac{4}{9}\hat{\mathbf{e}}_2 + \frac{7}{9}\hat{\mathbf{e}}_3$. In the matrix form of (*2.14*) the stress vector at P is

$$[4/9, -4/9, 7/9]\begin{bmatrix} 0 & 7C & 0 \\ 7C & 0 & -4C \\ 0 & -4C & 0 \end{bmatrix} = [-28C/9, 0, 16C/9]$$

(*c*) From (*2.37*), for principal stresses σ,

$$\begin{vmatrix} -\sigma & 7 & 0 \\ 7 & -\sigma & -4 \\ 0 & -4 & -\sigma \end{vmatrix} = \sigma(\sigma^2 - 65) = 0; \text{ hence}$$

$\sigma_{\text{I}} = \sqrt{65}$, $\sigma_{\text{II}} = 0$, $\sigma_{\text{III}} = -\sqrt{65}$. The maximum shear stress value is given by (*2.54b*) as $\sigma_S = (\sigma_{\text{III}} - \sigma_{\text{I}})/2 = \pm\sqrt{65}$. Since the mean normal stress at P is $\sigma_M = (\sigma_{\text{I}} + \sigma_{\text{II}} + \sigma_{\text{III}})/3 = 0$, the principal deviator stresses are the same as the principal stresses.

(*d*) The Mohr's circles are shown in Fig. 2-33.

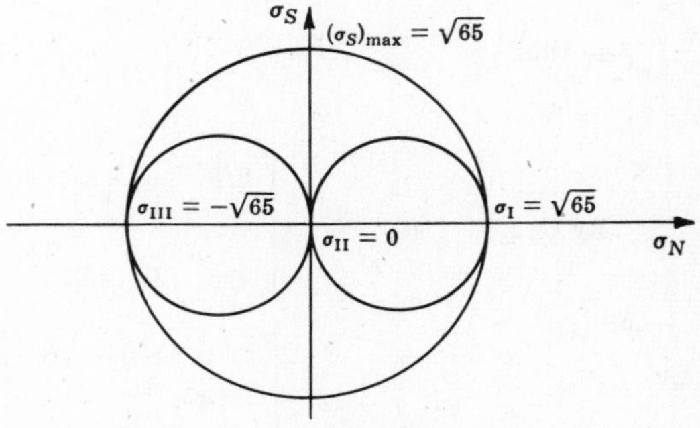

Fig. 2-33

Supplementary Problems

2.41. At point P the stress tensor is $\sigma_{ij} = \begin{pmatrix} 14 & 7 & -7 \\ 7 & 21 & 0 \\ -7 & 0 & 35 \end{pmatrix}$. Determine the stress vector on the plane at P parallel to plane (*a*) BGE, (*b*) $BGFC$ of the small parallelepiped shown in Fig. 2-34.

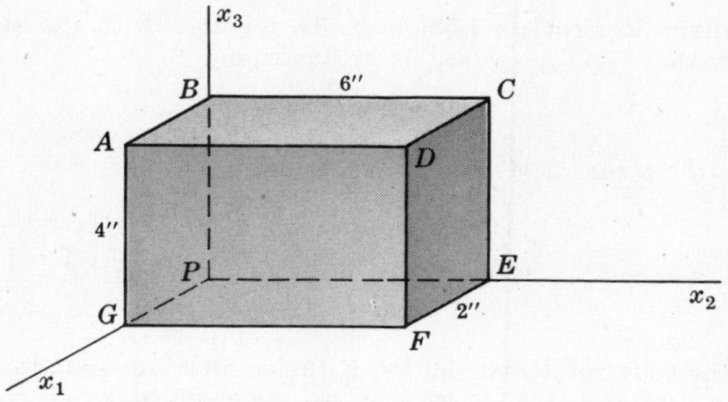

Fig. 2-34

Ans. (*a*) $\mathbf{t}^{(\hat{\mathbf{n}})} = 11\hat{\mathbf{e}}_1 + 12\hat{\mathbf{e}}_2 + 9\hat{\mathbf{e}}_3$, (*b*) $\mathbf{t}^{(\hat{\mathbf{n}})} = (21\hat{\mathbf{e}}_1 + 14\hat{\mathbf{e}}_2 + 21\hat{\mathbf{e}}_3)/\sqrt{5}$

2.42. Determine the normal and shear stress components on the plane $BGFC$ of Problem 2.41.
Ans. $\sigma_N = 63/5$, $\sigma_S = 37.7/5$

2.43. The principal stresses at point P are $\sigma_I = 12$, $\sigma_{II} = 3$, $\sigma_{III} = -6$. Determine the stress vector and its normal component on the octahedral plane at P.
Ans. $\mathbf{t}^{(\hat{\mathbf{n}})} = (12\hat{\mathbf{e}}_1 + 3\hat{\mathbf{e}}_2 - 6\hat{\mathbf{e}}_3)/\sqrt{3}$, $\sigma_N = 3$

2.44. Determine the principal stress values for

$$(a) \quad \sigma_{ij} = \begin{pmatrix} 0 & 1 & 1 \\ 1 & 0 & 1 \\ 1 & 1 & 0 \end{pmatrix} \qquad \text{and} \qquad (b) \quad \sigma_{ij} = \begin{pmatrix} 2 & 1 & 1 \\ 1 & 2 & 1 \\ 1 & 1 & 2 \end{pmatrix}$$

and show that both have the same principal directions.
Ans. (*a*) $\sigma_I = 2$, $\sigma_{II} = \sigma_{III} = -1$, (*b*) $\sigma_I = 4$, $\sigma_{II} = \sigma_{III} = 1$

2.45. Decompose the stress tensor $\sigma_{ij} = \begin{pmatrix} 3 & -10 & 0 \\ -10 & 0 & 30 \\ 0 & 30 & -27 \end{pmatrix}$ into its spherical and deviator parts and determine the principal deviator stresses. *Ans.* $s_I = 31$, $s_{II} = 8$, $s_{III} = -39$

2.46. Show that the normal component of the stress vector on the octahedral plane is equal to one third the first invariant of the stress tensor.

2.47. The stress tensor at a point is given as $\sigma_{ij} = \begin{pmatrix} 0 & 1 & 2 \\ 1 & \sigma_{22} & 1 \\ 2 & 1 & 0 \end{pmatrix}$ with σ_{22} unspecified. Determine σ_{22} so that the stress vector on some plane at the point will be zero. Give the unit normal for this traction-free plane.
Ans. $\sigma_{22} = 1$, $\hat{\mathbf{n}} = (\hat{\mathbf{e}}_1 - 2\hat{\mathbf{e}}_2 + \hat{\mathbf{e}}_3)/\sqrt{6}$

2.48. Sketch the Mohr's circles and determine the maximum shear stress for each of the following stress states:

$$(a)\quad \sigma_{ij} = \begin{pmatrix} \tau & \tau & 0 \\ \tau & \tau & 0 \\ 0 & 0 & 0 \end{pmatrix} \qquad (b)\quad \sigma_{ij} = \begin{pmatrix} \tau & 0 & 0 \\ 0 & -\tau & 0 \\ 0 & 0 & -2\tau \end{pmatrix}$$

Ans. (a) $\sigma_S = \tau$, (b) $\sigma_S = 3\tau/2$

2.49. Use the result given in Problem 1.58, page 39, together with the stress transformation law (2.27), page 50, to show that $\epsilon_{ijk}\epsilon_{pqm}\sigma_{ip}\sigma_{jq}\sigma_{km}$ is an invariant.

2.50. In a continuum, the stress field is given by the tensor

$$\sigma_{ij} = \begin{pmatrix} x_1^2 x_2 & (1-x_2^2)x_1 & 0 \\ (1-x_2^2)x_1 & (x_2^3 - 3x_2)/3 & 0 \\ 0 & 0 & 2x_3^2 \end{pmatrix}$$

Determine (a) the body force distribution if the equilibrium equations are to be satisfied throughout the field, (b) the principal stress values at the point $P(a, 0, 2\sqrt{a})$, (c) the maximum shear stress at P, (d) the principal deviator stresses at P.

Ans. (a) $b_3 = -4x_3$, (b) $a, -a, 8a$, (c) $\pm 4.5a$, (d) $-11a/3, -5a/3, 16a/3$

Chapter 3

Deformation and Strain

3.1 PARTICLES AND POINTS

In the kinematics of continua, the meaning of the word "point" must be clearly understood since it may be construed to refer either to a "point" in space, or to a "point" of a continuum. To avoid misunderstanding, the term "point" will be used exclusively to designate a location in fixed space. The word "particle" will denote a small volumetric element, or "material point", of a continuum. In brief, a *point* is a place in space, a *particle* is a small part of a material continuum.

3.2 CONTINUUM CONFIGURATION. DEFORMATION AND FLOW CONCEPTS

At any instant of time t, a continuum having a volume V and bounding surface S will occupy a certain region R of physical space. The identification of the particles of the continuum with the points of the space it occupies at time t by reference to a suitable set of coordinate axes is said to specify the *configuration* of the continuum at that instant.

The term *deformation* refers to a change in the shape of the continuum between some initial (undeformed) configuration and a subsequent (deformed) configuration. The emphasis in deformation studies is on the initial and final configurations. No attention is given to intermediate configurations or to the particular sequence of configurations by which the deformation occurs. By contrast, the word *flow* is used to designate the continuing state of motion of a continuum. Indeed, a configuration history is inherent in flow investigations for which the specification of a time-dependent velocity field is given.

3.3 POSITION VECTOR. DISPLACEMENT VECTOR

In Fig. 3-1 the undeformed configuration of a material continuum at time $t = 0$ is shown together with the deformed configuration of the same continuum at a later time $t = t$. For the present development it is useful to refer the initial and final configurations to separate coordinate axes as in the figure.

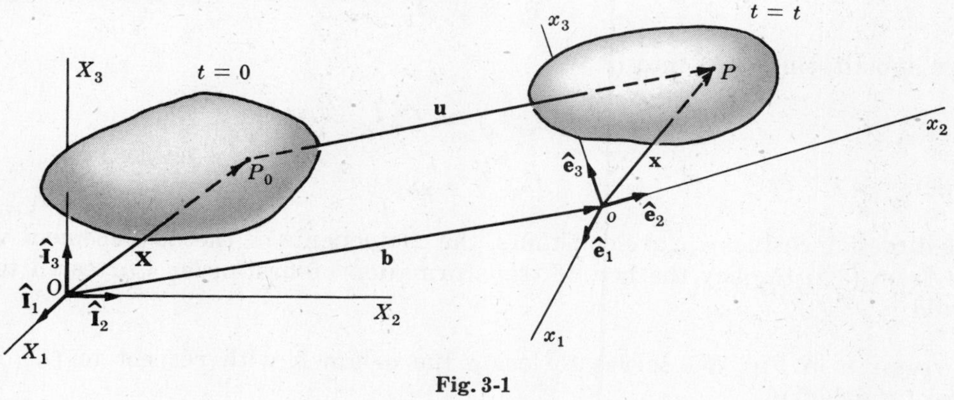

Fig. 3-1

Accordingly, in the initial configuration a representative particle of the continuum occupies a point P_0 in space and has the *position vector*

$$\mathbf{X} = X_1\hat{\mathbf{I}}_1 + X_2\hat{\mathbf{I}}_2 + X_3\hat{\mathbf{I}}_3 = X_K\hat{\mathbf{I}}_K \tag{3.1}$$

with respect to the rectangular Cartesian axes $OX_1X_2X_3$. Upper-case letters are used as indices in (*3.1*) and will appear as such in several equations that follow, but their use as summation indices is restricted to this section. In the remainder of the book upper-case subscripts or superscripts serve as labels only. Their use here is to emphasize the connection of certain expressions with the coordinates $(X_1X_2X_3)$, which are called the *material coordinates*. In the deformed configuration the particle originally at P_0 is located at the point P and has the position vector

$$\mathbf{x} = x_1\hat{\mathbf{e}}_1 + x_2\hat{\mathbf{e}}_2 + x_3\hat{\mathbf{e}}_3 = x_i\hat{\mathbf{e}}_i \tag{3.2}$$

when referred to the rectangular Cartesian axes $ox_1x_2x_3$. Here lower-case letters are used as subscripts to identify with the coordinates $(x_1x_2x_3)$ which give the current position of the particle and are frequently called the *spatial coordinates*.

The relative orientation of the material axes $OX_1X_2X_3$ and the spatial axes $ox_1x_2x_3$ is specified through direction cosines α_{kK} and α_{Kk}, which are defined by the dot products of unit vectors as

$$\hat{\mathbf{e}}_k \cdot \hat{\mathbf{I}}_K = \hat{\mathbf{I}}_K \cdot \hat{\mathbf{e}}_k = \alpha_{kK} = \alpha_{Kk} \tag{3.3}$$

No summation is implied by the indices in these expressions since k and K are distinct indices. Inasmuch as Kronecker deltas are designated by the equations $\hat{\mathbf{I}}_K \cdot \hat{\mathbf{I}}_P = \delta_{KP}$ and $\hat{\mathbf{e}}_k \cdot \hat{\mathbf{e}}_p = \delta_{kp}$, the *orthogonality conditions* between spatial and material axes take the form

$$\alpha_{Kk}\alpha_{Kp} = \alpha_{kK}\alpha_{pK} = \delta_{kp}; \quad \alpha_{Kp}\alpha_{Mp} = \alpha_{pK}\alpha_{pM} = \delta_{KM} \tag{3.4}$$

In Fig. 3-1 the vector $\mathbf{u}$ joining the points P_0 and P (the initial and final positions, respectively, of the particle), is known as the *displacement vector*. This vector may be expressed as

$$\mathbf{u} = u_k\hat{\mathbf{e}}_k \tag{3.5}$$

or alternatively as

$$\mathbf{U} = U_K\hat{\mathbf{I}}_K \tag{3.6}$$

in which the components U_K and u_k are interrelated through the direction cosines α_{kK}. From (*1.89*) the unit vector $\hat{\mathbf{e}}_k$ is expressed in terms of the material base vectors $\hat{\mathbf{I}}_K$ as

$$\hat{\mathbf{e}}_k = \alpha_{kK}\hat{\mathbf{I}}_K \tag{3.7}$$

Therefore substituting (*3.7*) into (*3.5*),

$$\mathbf{u} = u_k(\alpha_{kK}\hat{\mathbf{I}}_K) = U_K\hat{\mathbf{I}}_K = \mathbf{U} \tag{3.8}$$

from which

$$U_K = \alpha_{kK}u_k \tag{3.9}$$

Since the direction cosines α_{kK} are constants, the components of the displacement vector are observed from (*3.9*) to obey the law of transformation of first-order Cartesian tensors, as they should.

The vector $\mathbf{b}$ in Fig. 3-1 serves to locate the origin o with respect to O. From the geometry of the figure,

$$\mathbf{u} = \mathbf{b} + \mathbf{x} - \mathbf{X} \tag{3.10}$$

Very often in continuum mechanics it is possible to consider the coordinate systems $OX_1X_2X_3$ and $ox_1x_2x_3$ superimposed, with $\mathbf{b} \equiv \mathbf{0}$, so that (*3.10*) becomes

$$\mathbf{u} = \mathbf{x} - \mathbf{X} \qquad (3.11)$$

In Cartesian component form this equation is given by the general expression

$$u_k = x_k - \alpha_{kK} X_K \qquad (3.12)$$

However, for superimposed axes the unit triads of base vectors for the two systems are identical, which results in the direction cosine symbols α_{kK} becoming Kronecker deltas. Accordingly, (*3.12*) reduces to

$$u_k = x_k - X_k \qquad (3.13)$$

in which only lower-case subscripts appear. In the remainder of this book, unless specifically stated otherwise, the material and spatial axes are assumed *superimposed* and hence only lower-case indices will be used.

3.4 LAGRANGIAN AND EULERIAN DESCRIPTIONS

When a continuum undergoes deformation (or flow), the particles of the continuum move along various paths in space. This motion may be expressed by equations of the form

$$x_i = x_i(X_1, X_2, X_3, t) = x_i(\mathbf{X}, t) \quad \text{or} \quad \mathbf{x} = \mathbf{x}(\mathbf{X}, t) \qquad (3.14)$$

which give the present location x_i of the particle that occupied the point $(X_1X_2X_3)$ at time $t = 0$. Also, (*3.14*) may be interpreted as a mapping of the initial configuration into the current configuration. It is assumed that such a mapping is one-to-one and continuous, with continuous partial derivatives to whatever order is required. The description of motion or deformation expressed by (*3.14*) is known as the *Lagrangian* formulation.

If, on the other hand, the motion or deformation is given through equations of the form

$$X_i = X_i(x_1, x_2, x_3, t) = X_i(\mathbf{x}, t) \quad \text{or} \quad \mathbf{X} = \mathbf{X}(\mathbf{x}, t) \qquad (3.15)$$

in which the independent variables are the coordinates x_i and t, the description is known as the *Eulerian* formulation. This description may be viewed as one which provides a tracing to its original position of the particle that now occupies the location (x_1, x_2, x_3). If (*3.15*) is a continuous one-to-one mapping with continuous partial derivatives, as was also assumed for (*3.14*), the two mappings are the unique inverses of one another. A necessary and sufficient condition for the inverse functions to exist is that the Jacobian

$$J = \left| \frac{\partial x_i}{\partial X_j} \right| \qquad (3.16)$$

should not vanish.

As a simple example, the Lagrangian description given by the equations

$$\begin{aligned} x_1 &= X_1 + X_2(e^t - 1) \\ x_2 &= X_1(e^{-t} - 1) + X_2 \\ x_3 &= X_3 \end{aligned} \qquad (3.17)$$

has the inverse Eulerian formulation,

$$\begin{aligned} X_1 &= \frac{-x_1 + x_2(e^t - 1)}{1 - e^t - e^{-t}} \\ X_2 &= \frac{x_1(e^{-t} - 1) - x_2}{1 - e^t - e^{-t}} \\ X_3 &= x_3 \end{aligned} \qquad (3.18)$$

3.5 DEFORMATION GRADIENTS. DISPLACEMENT GRADIENTS

Partial differentiation of (*3.14*) with respect to X_j produces the tensor $\partial x_i/\partial X_j$ which is called the *material deformation gradient.* In symbolic notation, $\partial x_i/\partial X_j$ is represented by the dyadic

$$\mathbf{F} \;=\; \mathbf{x}\nabla_{\mathbf{X}} \;\equiv\; \frac{\partial \mathbf{x}}{\partial X_1}\hat{\mathbf{e}}_1 + \frac{\partial \mathbf{x}}{\partial X_2}\hat{\mathbf{e}}_2 + \frac{\partial \mathbf{x}}{\partial X_3}\hat{\mathbf{e}}_3 \tag{3.19}$$

in which the differential operator $\nabla_{\mathbf{X}} = \frac{\partial}{\partial X_i}\hat{\mathbf{e}}_i$ is applied from the right (as shown explicitly in the equation). The matrix form of $\mathbf{F}$ serves to further clarify this property of the operator $\nabla_{\mathbf{X}}$ when it appears as the consequent of a dyad. Thus

$$\mathcal{F} \;=\; \begin{bmatrix} x_1 \\ x_2 \\ x_3 \end{bmatrix}\left[\frac{\partial}{\partial X_1}, \frac{\partial}{\partial X_2}, \frac{\partial}{\partial X_3}\right] \;=\; \begin{bmatrix} \partial x_1/\partial X_1 & \partial x_1/\partial X_2 & \partial x_1/\partial X_3 \\ \partial x_2/\partial X_1 & \partial x_2/\partial X_2 & \partial x_2/\partial X_3 \\ \partial x_3/\partial X_1 & \partial x_3/\partial X_2 & \partial x_3/\partial X_3 \end{bmatrix} \;=\; [\partial x_i/\partial X_j] \tag{3.20}$$

Partial differentiation of (*3.15*) with respect to x_j produces the tensor $\partial X_i/\partial x_j$ which is called the *spatial deformation gradient.* This tensor is represented by the dyadic

$$\mathbf{H} \;=\; \mathbf{X}\nabla_{\mathbf{x}} \;\equiv\; \frac{\partial \mathbf{X}}{\partial x_1}\hat{\mathbf{e}}_1 + \frac{\partial \mathbf{X}}{\partial x_2}\hat{\mathbf{e}}_2 + \frac{\partial \mathbf{X}}{\partial x_3}\hat{\mathbf{e}}_3 \tag{3.21}$$

having a matrix form

$$\mathcal{H} \;=\; \begin{bmatrix} X_1 \\ X_2 \\ X_3 \end{bmatrix}\left[\frac{\partial}{\partial x_1}, \frac{\partial}{\partial x_2}, \frac{\partial}{\partial x_3}\right] \;=\; \begin{bmatrix} \partial X_1/\partial x_1 & \partial X_1/\partial x_2 & \partial X_1/\partial x_3 \\ \partial X_2/\partial x_1 & \partial X_2/\partial x_2 & \partial X_2/\partial x_3 \\ \partial X_3/\partial x_1 & \partial X_3/\partial x_2 & \partial X_3/\partial x_3 \end{bmatrix} \;=\; [\partial X_i/\partial x_j] \tag{3.22}$$

The material and spatial deformation tensors are interrelated through the well-known chain rule for partial differentiation,

$$\frac{\partial x_i}{\partial X_j}\frac{\partial X_j}{\partial x_k} \;=\; \frac{\partial X_i}{\partial x_j}\frac{\partial x_j}{\partial X_k} \;=\; \delta_{ik} \tag{3.23}$$

Partial differentiation of the displacement vector u_i with respect to the coordinates produces either the *material displacement gradient* $\partial u_i/\partial X_j$, or the *spatial displacement gradient* $\partial u_i/\partial x_j$. From (*3.13*), which expresses u_i as a difference of coordinates, these tensors are given in terms of the deformation gradients as the material gradient

$$\frac{\partial u_i}{\partial X_j} \;=\; \frac{\partial x_i}{\partial X_j} - \delta_{ij} \qquad \text{or} \qquad \mathbf{J} \equiv \mathbf{u}\nabla_{\mathbf{X}} \;=\; \mathbf{F} - \mathbf{I} \tag{3.24}$$

and the spatial gradient

$$\frac{\partial u_i}{\partial x_j} \;=\; \delta_{ij} - \frac{\partial X_i}{\partial x_j} \qquad \text{or} \qquad \mathbf{K} \equiv \mathbf{u}\nabla_{\mathbf{x}} \;=\; \mathbf{I} - \mathbf{H} \tag{3.25}$$

In the usual manner, the matrix forms of $\mathbf{J}$ and $\mathbf{K}$ are respectively

$$\mathcal{J} \;=\; \begin{bmatrix} u_1 \\ u_2 \\ u_3 \end{bmatrix}\left[\frac{\partial}{\partial X_1}, \frac{\partial}{\partial X_2}, \frac{\partial}{\partial X_3}\right] \;=\; \begin{bmatrix} \partial u_1/\partial X_1 & \partial u_1/\partial X_2 & \partial u_1/\partial X_3 \\ \partial u_2/\partial X_1 & \partial u_2/\partial X_2 & \partial u_2/\partial X_3 \\ \partial u_3/\partial X_1 & \partial u_3/\partial X_2 & \partial u_3/\partial X_3 \end{bmatrix} \;=\; [\partial u_i/\partial X_j] \tag{3.26}$$

and

$$\mathcal{K} \;=\; \begin{bmatrix} u_1 \\ u_2 \\ u_3 \end{bmatrix}\left[\frac{\partial}{\partial x_1}, \frac{\partial}{\partial x_2}, \frac{\partial}{\partial x_3}\right] \;=\; \begin{bmatrix} \partial u_1/\partial x_1 & \partial u_1/\partial x_2 & \partial u_1/\partial x_3 \\ \partial u_2/\partial x_1 & \partial u_2/\partial x_2 & \partial u_2/\partial x_3 \\ \partial u_3/\partial x_1 & \partial u_3/\partial x_2 & \partial u_3/\partial x_3 \end{bmatrix} \;=\; [\partial u_i/\partial x_j] \tag{3.27}$$

3.6 DEFORMATION TENSORS. FINITE STRAIN TENSORS

In Fig. 3-2 the initial (undeformed) and final (deformed) configurations of a continuum are referred to the superposed rectangular Cartesian coordinate axes $OX_1X_2X_3$ and $ox_1x_2x_3$. The neighboring particles which occupy points P_0 and Q_0 before deformation, move to points P and Q respectively in the deformed configuration.

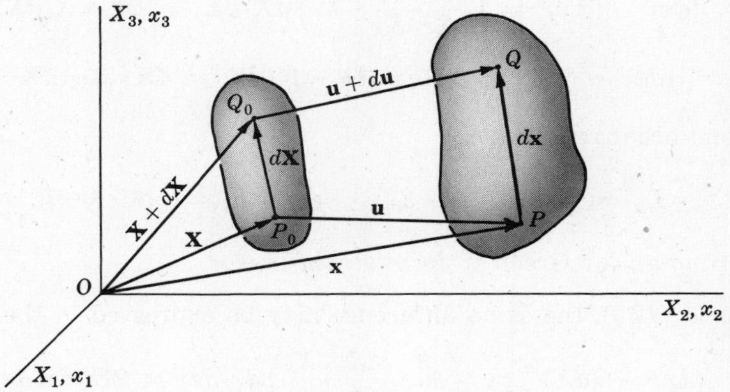

Fig. 3-2

The square of the differential element of length between P_0 and Q_0 is

$$(dX)^2 = d\mathbf{X} \cdot d\mathbf{X} = dX_i \, dX_i = \delta_{ij} \, dX_i \, dX_j \tag{3.28}$$

From (3.15), the distance differential dX_i is seen to be

$$dX_i = \frac{\partial X_i}{\partial x_j} dx_j \quad \text{or} \quad d\mathbf{X} = \mathbf{H} \cdot d\mathbf{x} \tag{3.29}$$

so that the squared length $(dX)^2$ in (3.28) may be written

$$(dX)^2 = \frac{\partial X_k}{\partial x_i} \frac{\partial X_k}{\partial x_j} dx_i \, dx_j = C_{ij} \, dx_i \, dx_j \quad \text{or} \quad (dX)^2 = d\mathbf{x} \cdot \mathbf{C} \cdot d\mathbf{x} \tag{3.30}$$

in which the second-order tensor

$$C_{ij} = \frac{\partial X_k}{\partial x_i} \frac{\partial X_k}{\partial x_j} \quad \text{or} \quad \mathbf{C} = \mathbf{H}_c \cdot \mathbf{H} \tag{3.31}$$

is known as *Cauchy's deformation tensor.*

In the deformed configuration, the square of the differential element of length between P and Q is

$$(dx)^2 = d\mathbf{x} \cdot d\mathbf{x} = dx_i \, dx_i = \delta_{ij} \, dx_i \, dx_j \tag{3.32}$$

From (3.14) the distance differential here is

$$dx_i = \frac{\partial x_i}{\partial X_j} dX_j \quad \text{or} \quad d\mathbf{x} = \mathbf{F} \cdot d\mathbf{X} \tag{3.33}$$

so that the squared length $(dx)^2$ in (3.32) may be written

$$(dx)^2 = \frac{\partial x_k}{\partial X_i} \frac{\partial x_k}{\partial X_j} dX_i \, dX_j = G_{ij} \, dX_i \, dX_j \quad \text{or} \quad (dx)^2 = d\mathbf{X} \cdot \mathbf{G} \cdot d\mathbf{X} \tag{3.34}$$

in which the second-order tensor

$$G_{ij} = \frac{\partial x_k}{\partial X_i} \frac{\partial x_k}{\partial X_j} \quad \text{or} \quad \mathbf{G} = \mathbf{F}_c \cdot \mathbf{F} \tag{3.35}$$

is known as *Green's deformation tensor.*

The difference $(dx)^2 - (dX)^2$ for two neighboring particles of a continuum is used as the *measure of deformation* that occurs in the neighborhood of the particles between the initial and final configurations. If this difference is identically zero for all neighboring particles of a continuum, a *rigid displacement* is said to occur. Using (*3.34*) and (*3.28*), this difference may be expressed in the form

$$(dx)^2 - (dX)^2 = \left(\frac{\partial x_k}{\partial X_i}\frac{\partial x_k}{\partial X_j} - \delta_{ij}\right) dX_i\,dX_j = 2L_{ij}\,dX_i\,dX_j$$

or

$$(dx)^2 - (dX)^2 = d\mathbf{X}\cdot(\mathbf{F}_c\cdot\mathbf{F} - \mathbf{I})\cdot d\mathbf{X} = d\mathbf{X}\cdot 2\mathbf{L}_G\cdot d\mathbf{X} \tag{3.36}$$

in which the second-order tensor

$$L_{ij} = \frac{1}{2}\left(\frac{\partial x_k}{\partial X_i}\frac{\partial x_k}{\partial X_j} - \delta_{ij}\right) \quad \text{or} \quad \mathbf{L}_G = \tfrac{1}{2}(\mathbf{F}_c\cdot\mathbf{F} - \mathbf{I}) \tag{3.37}$$

is called the *Lagrangian* (or Green's) *finite strain tensor.*

Using (*3.32*) and (*3.30*), the same difference may be expressed in the form

$$(dx)^2 - (dX)^2 = \left(\delta_{ij} - \frac{\partial X_k}{\partial x_i}\frac{\partial X_k}{\partial x_j}\right) dx_i\,dx_j = 2E_{ij}\,dx_i\,dx_j$$

or

$$(dx)^2 - (dX)^2 = d\mathbf{x}\cdot(\mathbf{I} - \mathbf{H}_c\cdot\mathbf{H})\cdot d\mathbf{x} = d\mathbf{x}\cdot 2\mathbf{E}_A\cdot d\mathbf{x} \tag{3.38}$$

in which the second-order tensor

$$E_{ij} = \frac{1}{2}\left(\delta_{ij} - \frac{\partial X_k}{\partial x_i}\frac{\partial X_k}{\partial x_j}\right) \quad \text{or} \quad \mathbf{E}_A = \tfrac{1}{2}(\mathbf{I} - \mathbf{H}_c\cdot\mathbf{H}) \tag{3.39}$$

is called the *Eulerian* (or Almansi's) *finite strain tensor.*

An especially useful form of the Lagrangian and Eulerian finite strain tensors is that in which these tensors appear as functions of the displacement gradients. Thus if $\partial x_i/\partial X_j$ from (*3.24*) is substituted into (*3.37*), the result after some simple algebraic manipulations is the Lagrangian finite strain tensor in the form

$$L_{ij} = \frac{1}{2}\left(\frac{\partial u_i}{\partial X_j} + \frac{\partial u_j}{\partial X_i} + \frac{\partial u_k}{\partial X_i}\frac{\partial u_k}{\partial X_j}\right) \quad \text{or} \quad \mathbf{L}_G = \tfrac{1}{2}(\mathbf{J} + \mathbf{J}_c + \mathbf{J}_c\cdot\mathbf{J}) \tag{3.40}$$

In the same manner, if $\partial X_i/\partial x_j$ from (*3.25*) is substituted into (*3.39*), the result is the Eulerian finite strain tensor in the form

$$E_{ij} = \frac{1}{2}\left(\frac{\partial u_i}{\partial x_j} + \frac{\partial u_j}{\partial x_i} - \frac{\partial u_k}{\partial x_i}\frac{\partial u_k}{\partial x_j}\right) \quad \text{or} \quad \mathbf{E}_A = \tfrac{1}{2}(\mathbf{K} + \mathbf{K}_c - \mathbf{K}_c\cdot\mathbf{K}) \tag{3.41}$$

The matrix representations of (*3.40*) and (*3.41*) may be written directly from (*3.26*) and (*3.27*) respectively.

3.7 SMALL DEFORMATION THEORY. INFINITESIMAL STRAIN TENSORS

The so-called *small deformation theory* of continuum mechanics has as its basic condition the requirement that the displacement gradients be small compared to unity. The fundamental measure of deformation is the difference $(dx)^2 - (dX)^2$, which may be expressed in terms of the displacement gradients by inserting (*3.40*) and (*3.41*) into (*3.36*) and (*3.38*) respectively. If the displacement gradients are small, the finite strain tensors in (*3.36*) and (*3.38*) reduce to infinitesimal strain tensors, and the resulting equations represent small deformations.

In (3.40), if the displacement gradient components $\partial u_i/\partial X_j$ are each small compared to unity, the product terms are negligible and may be dropped. The resulting tensor is the *Lagrangian infinitesimal strain tensor,* which is denoted by

$$l_{ij} = \frac{1}{2}\left(\frac{\partial u_i}{\partial X_j} + \frac{\partial u_j}{\partial X_i}\right) \quad \text{or} \quad \mathbf{L} = \tfrac{1}{2}(\mathbf{u}\nabla_{\mathbf{X}} + \nabla_{\mathbf{X}}\mathbf{u}) = \tfrac{1}{2}(\mathbf{J} + \mathbf{J}_c) \tag{3.42}$$

Likewise for $\partial u_i/\partial x_j \ll 1$ in (3.41), the product terms may be dropped to yield the *Eulerian infinitesimal strain tensor,* which is denoted by

$$\epsilon_{ij} = \frac{1}{2}\left(\frac{\partial u_i}{\partial x_j} + \frac{\partial u_j}{\partial x_i}\right) \quad \text{or} \quad \mathbf{E} = \tfrac{1}{2}(\mathbf{u}\nabla_{\mathbf{x}} + \nabla_{\mathbf{x}}\mathbf{u}) = \tfrac{1}{2}(\mathbf{K} + \mathbf{K}_c) \tag{3.43}$$

If both the displacement gradients and the displacements themselves are small, there is very little difference in the material and spatial coordinates of a continuum particle. Accordingly the material gradient components $\partial u_i/\partial X_j$ and spatial gradient components $\partial u_i/\partial x_j$ are very nearly equal, so that the Eulerian and Lagrangian infinitesimal strain tensors may be taken as equal. Thus

$$l_{ij} = \epsilon_{ij} \quad \text{or} \quad \mathbf{L} = \mathbf{E} \tag{3.44}$$

if both the displacements and displacement gradients are sufficiently small.

3.8 RELATIVE DISPLACEMENTS. LINEAR ROTATION TENSOR. ROTATION VECTOR

In Fig. 3-3 the displacements of two neighboring particles are represented by the vectors $u_i^{(P_0)}$ and $u_i^{(Q_0)}$ (see also Fig. 3-2). The vector

$$du_i = u_i^{(Q_0)} - u_i^{(P_0)} \quad \text{or} \quad d\mathbf{u} = \mathbf{u}^{(Q_0)} - \mathbf{u}^{(P_0)} \tag{3.45}$$

is called the *relative displacement vector* of the particle originally at Q_0 with respect to the particle originally at P_0. Assuming suitable continuity conditions on the displacement field, a Taylor series expansion for $u_i^{(P_0)}$ may be developed in the neighborhood of P_0. Neglecting higher-order terms in this expansion, the relative displacement vector can be written as

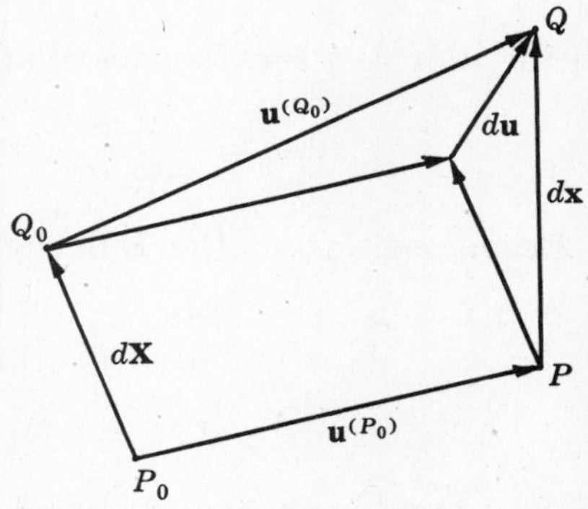

Fig. 3-3

$$du_i = (\partial u_i/\partial X_j)_{P_0}\, dX_j \quad \text{or} \quad d\mathbf{u} = (\mathbf{u}\nabla_{\mathbf{X}})_{P_0} \cdot d\mathbf{X} \tag{3.46}$$

Here the parentheses on the partial derivatives are to emphasize the requirement that the derivatives are to be evaluated at point P_0. These derivatives are actually the components of the material displacement gradient. Equation (3.46) is the Lagrangian form of the relative displacement vector.

It is also useful to define the *unit relative displacement vector* du_i/dX in which dX is the magnitude of the differential distance vector dX_i. Accordingly if ν_i is a unit vector in the direction of dX_i so that $dX_i = \nu_i\, dX$, then

$$\frac{du_i}{dX} = \frac{\partial u_i}{\partial X_j}\frac{dX_j}{dX} = \frac{\partial u_i}{\partial X_j}\nu_j \quad \text{or} \quad \frac{d\mathbf{u}}{dX} = \mathbf{u}\nabla_{\mathbf{X}} \cdot \hat{\boldsymbol{\nu}} = \mathbf{J} \cdot \hat{\boldsymbol{\nu}} \tag{3.47}$$

Since the material displacement gradient $\partial u_i/\partial X_j$ may be decomposed uniquely into a symmetric and an antisymmetric part, the relative displacement vector du_i may be written as

$$du_i = \left[\frac{1}{2}\left(\frac{\partial u_i}{\partial X_j} + \frac{\partial u_j}{\partial X_i}\right) + \frac{1}{2}\left(\frac{\partial u_i}{\partial X_j} - \frac{\partial u_j}{\partial X_i}\right)\right] dX_j$$

or

$$d\mathbf{u} = [\tfrac{1}{2}(\mathbf{u}\nabla_{\mathbf{X}} + \nabla_{\mathbf{X}}\mathbf{u}) + \tfrac{1}{2}(\mathbf{u}\nabla_{\mathbf{X}} - \nabla_{\mathbf{X}}\mathbf{u})] \cdot d\mathbf{X} \qquad (3.48)$$

The first term in the square brackets in (*3.48*) is recognized as the linear Lagrangian strain tensor l_{ij}. The second term is known as the *linear Lagrangian rotation tensor* and is denoted by

$$W_{ij} = \frac{1}{2}\left(\frac{\partial u_i}{\partial X_j} - \frac{\partial u_j}{\partial X_i}\right) \quad \text{or} \quad \mathbf{W} = \tfrac{1}{2}(\mathbf{u}\nabla_{\mathbf{X}} - \nabla_{\mathbf{X}}\mathbf{u}) \qquad (3.49)$$

In a displacement for which the strain tensor l_{ij} is identically zero in the vicinity of point P_0, the relative displacement at that point will be an infinitesimal rigid body rotation. This infinitesimal rotation may be represented by the *rotation vector*

$$w_i = \tfrac{1}{2}\epsilon_{ijk}W_{kj} \quad \text{or} \quad \mathbf{w} = \tfrac{1}{2}\nabla_{\mathbf{X}} \times \mathbf{u} \qquad (3.50)$$

in terms of which the relative displacement is given by the expression

$$du_i = \epsilon_{ijk}w_j\,dX_k \quad \text{or} \quad d\mathbf{u} = \mathbf{w} \times d\mathbf{X} \qquad (3.51)$$

The development of the Lagrangian description of the relative displacement vector, the linear rotation tensor and the linear rotation vector is paralleled completely by an analogous development for the Eulerian counterparts of these quantities. Accordingly the *Eulerian description* of the *relative displacement vector* is given by

$$du_i = \frac{\partial u_i}{\partial x_j}dx_j \quad \text{or} \quad d\mathbf{u} = \mathbf{K}\cdot d\mathbf{x} \qquad (3.52)$$

and the *unit relative displacement vector* by

$$du_i = \frac{\partial u_i}{\partial x_j}\frac{dx_j}{dx} = \frac{\partial u_i}{\partial x_j}\mu_j \quad \text{or} \quad \frac{d\mathbf{u}}{dx} = \mathbf{u}\nabla_{\mathbf{x}} \cdot \hat{\boldsymbol{\mu}} = \mathbf{K}\cdot\hat{\boldsymbol{\mu}} \qquad (3.53)$$

Decomposition of the Eulerian displacement gradient $\partial u_i/\partial x_j$ results in the expression

$$\frac{du_i}{dx} = \left[\frac{1}{2}\left(\frac{\partial u_i}{\partial x_j} + \frac{\partial u_j}{\partial x_i}\right) + \frac{1}{2}\left(\frac{\partial u_i}{\partial x_j} - \frac{\partial u_j}{\partial x_i}\right)\right] dx_j$$

or

$$d\mathbf{u} = [\tfrac{1}{2}(\mathbf{u}\nabla_{\mathbf{x}} + \nabla_{\mathbf{x}}\mathbf{u}) + \tfrac{1}{2}(\mathbf{u}\nabla_{\mathbf{x}} - \nabla_{\mathbf{x}}\mathbf{u})] \cdot d\mathbf{x} \qquad (3.54)$$

The first term in the square brackets of (*3.54*) is the Eulerian linear strain tensor ϵ_{ij}. The second term is the *linear Eulerian rotation tensor* and is denoted by

$$\omega_{ij} = \frac{1}{2}\left(\frac{\partial u_i}{\partial x_j} - \frac{\partial u_j}{\partial x_i}\right) \quad \text{or} \quad \mathbf{\Omega} = \tfrac{1}{2}(\mathbf{u}\nabla_{\mathbf{x}} - \nabla_{\mathbf{x}}\mathbf{u}) \qquad (3.55)$$

From (*3.55*), the *linear Eulerian rotation vector* is defined by

$$\omega_i = \tfrac{1}{2}\epsilon_{ijk}\omega_{kj} \quad \text{or} \quad \boldsymbol{\omega} = \tfrac{1}{2}\nabla_{\mathbf{x}} \times \mathbf{u} \qquad (3.56)$$

in terms of which the relative displacement is given by the expression

$$du_i = \epsilon_{ijk}\omega_j\,dx_k \quad \text{or} \quad d\mathbf{u} = \boldsymbol{\omega} \times d\mathbf{x} \qquad (3.57)$$

3.9 INTERPRETATION OF THE LINEAR STRAIN TENSORS

For small deformation theory, the finite Lagrangian strain tensor L_{ij} in (*3.36*) may be replaced by the linear Lagrangian strain tensor l_{ij}, and that expression may now be written

$$(dx)^2 - (dX)^2 = (dx - dX)(dx + dX) = 2l_{ij}\, dX_i\, dX_j$$

or

$$(dx)^2 - (dX)^2 = (dx - dX)(dx + dX) = d\mathbf{X} \cdot 2\mathbf{L} \cdot d\mathbf{X} \tag{3.58}$$

Since $dx \approx dX$ for small deformations, this equation may be put into the form

$$\frac{dx - dX}{dX} = l_{ij} \frac{dX_i}{dX} \frac{dX_j}{dX} = l_{ij}\nu_i\nu_j \quad \text{or} \quad \frac{dx - dX}{dX} = \hat{\boldsymbol{\nu}} \cdot \mathbf{L} \cdot \hat{\boldsymbol{\nu}} \tag{3.59}$$

The left-hand side of (*3.59*) is recognized as the change in length per unit original length of the differential element and is called the *normal strain* for the line element originally having direction cosines dX_i/dX.

When (*3.59*) is applied to the differential line element P_0Q_0, located with respect to the set of local axes at P_0 as shown in Fig. 3-4, the result will be the normal strain for that element. Because P_0Q_0 here lies along the X_2 axis,

$$dX_1/dX = dX_3/dX = 0, \quad dX_2/dX = 1$$

and therefore (*3.59*) becomes

$$\frac{dx - dX}{dX} = l_{22} = \frac{\partial u_2}{\partial X_2} \tag{3.60}$$

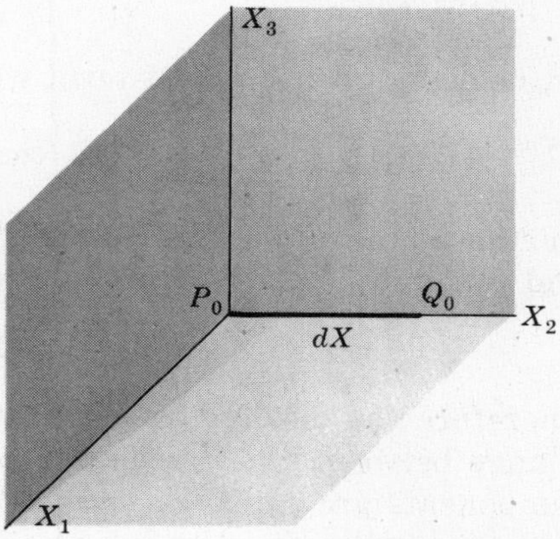

Fig. 3-4

Thus the normal strain for an element originally along the X_2 axis is seen to be the component l_{22}. Likewise for elements originally situated along the X_1 and X_3 axes, (*3.59*) yields normal strain values l_{11} and l_{33} respectively. In general, therefore, the diagonal terms of the linear strain tensor represent normal strains in the coordinate directions.

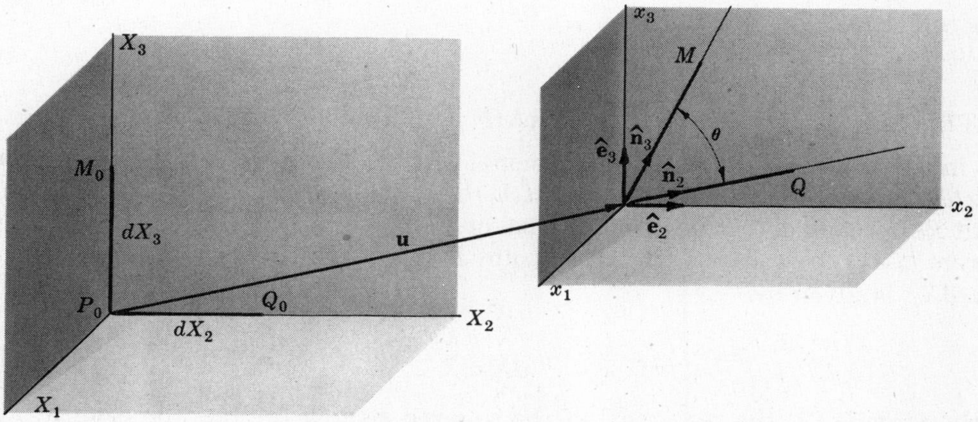

Fig. 3-5

The physical interpretation of the off-diagonal terms of l_{ij} may be obtained by a consideration of the line elements originally located along two of the coordinate axes. In Fig. 3-5 the line elements P_0Q_0 and P_0M_0 originally along the X_2 and X_3 axes, respectively, become after deformation the line elements PQ and PM with respect to the parallel set of local axes with origin at P. The original right angle between the line elements becomes the angle θ. From (*3.46*) and the assumption of small deformation theory, a first order approximation gives the unit vector at P in the direction of Q as

$$\hat{\mathbf{n}}_2 = \frac{\partial u_1}{\partial X_2}\hat{\mathbf{e}}_1 + \hat{\mathbf{e}}_2 + \frac{\partial u_3}{\partial X_2}\hat{\mathbf{e}}_3 \tag{3.61}$$

and, for the unit vector at P in the direction of M, as

$$\hat{\mathbf{n}}_3 = \frac{\partial u_1}{\partial X_3}\hat{\mathbf{e}}_1 + \frac{\partial u_2}{\partial X_3}\hat{\mathbf{e}}_2 + \hat{\mathbf{e}}_3 \tag{3.62}$$

Therefore

$$\cos\theta = \hat{\mathbf{n}}_2 \cdot \hat{\mathbf{n}}_3 = \frac{\partial u_1}{\partial X_3}\frac{\partial u_1}{\partial X_2} + \frac{\partial u_2}{\partial X_3} + \frac{\partial u_3}{\partial X_2} \tag{3.63}$$

or, neglecting the product term which is of higher order,

$$\cos\theta = \frac{\partial u_2}{\partial X_3} + \frac{\partial u_3}{\partial X_2} = 2l_{23} \tag{3.64}$$

Furthermore, taking the change in the right angle between the elements as $\gamma_{23} = \pi/2 - \theta$, and remembering that for the linear theory γ_{23} is very small, it follows that

$$\gamma_{23} \approx \sin\gamma_{23} = \sin(\pi/2 - \theta) = \cos\theta = 2l_{23} \tag{3.65}$$

Therefore the off-diagonal terms of the linear strain tensor represent one-half the angle change between two line elements originally at right angles to one another. These strain components are called *shearing strains,* and because of the factor 2 in (*3.65*) these tensor components are equal to one-half the familiar "engineering" shearing strains.

A development, essentially paralleling the one just presented for the interpretation of the components of l_{ij}, may also be made for the linear Eulerian strain tensor ϵ_{ij}. The essential difference in the derivations rests in the choice of line elements, which in the Eulerian description must be those that lie along the coordinate axes after deformation. The diagonal terms of ϵ_{ij} are the normal strains, and the off-diagonal terms the shearing strains. For those deformations in which the assumption $l_{ij} = \epsilon_{ij}$ is valid, no distinction is made between the Eulerian and Lagrangian interpretations.

3.10 STRETCH RATIO. FINITE STRAIN INTERPRETATION

An important measure of the extensional strain of a differential line element is the ratio dx/dX, known as the *stretch* or *stretch ratio.* This quantity may be defined at either the point P_0 in the undeformed configuration or at the point P in the deformed configuration. Thus from (*3.34*) the squared stretch at point P_0 for the line element along the unit vector $\hat{\mathbf{m}} = d\mathbf{X}/dX$, is given by

$$\left(\frac{dx}{dX}\right)^2_{P_0} = \Lambda^2_{(\hat{\mathbf{m}})} = G_{ij}\frac{dX_i}{dX}\frac{dX_j}{dX} \quad \text{or} \quad \Lambda^2_{(\hat{\mathbf{m}})} = \hat{\mathbf{m}} \cdot \mathbf{G} \cdot \hat{\mathbf{m}} \tag{3.66}$$

Similarly, from (*3.30*) the reciprocal of the squared stretch for the line element at P along the unit vector $\hat{\mathbf{n}} = d\mathbf{x}/dx$ is given by

$$\left(\frac{dX}{dx}\right)^2_{P} = \frac{1}{\lambda^2_{(\hat{\mathbf{n}})}} = C_{ij}\frac{dx_i}{dx}\frac{dx_j}{dx} \quad \text{or} \quad \frac{1}{\lambda^2_{(\hat{\mathbf{n}})}} = \hat{\mathbf{n}} \cdot \mathbf{C} \cdot \hat{\mathbf{n}} \tag{3.67}$$

For an element originally along the local X_2 axis shown in Fig. 3-4, $\hat{\mathbf{m}} \equiv \hat{\mathbf{e}}_2$ and therefore $dX_1/dX = dX_3/dX = 0$, $dX_2/dX = 1$ so that (*3.66*) yields for such an element

$$\Lambda^2_{(\hat{\mathbf{e}}_2)} = G_{22} = 1 + 2L_{22} \tag{3.68}$$

Similar results may be determined for $\Lambda^2_{(\hat{\mathbf{e}}_1)}$ and $\Lambda^2_{(\hat{\mathbf{e}}_3)}$.

For an element parallel to the x_2 axis after deformation, (*3.67*) yields the result

$$\frac{1}{\lambda^2_{(\hat{\mathbf{e}}_2)}} = 1 - 2E_{22} \tag{3.69}$$

with similar expressions for the quantities $1/\lambda^2_{(\hat{\mathbf{e}}_1)}$ and $1/\lambda^2_{(\hat{\mathbf{e}}_3)}$. In general, $\Lambda_{(\hat{\mathbf{e}}_2)}$ is not equal to $\lambda_{(\hat{\mathbf{e}}_2)}$ since the element originally along the X_2 axis will not likely lie along the x_2 axis after deformation.

The stretch ratio provides a basis for interpretation of the finite strain tensors. Thus the change of length per unit of original length is

$$\frac{dx - dX}{dX} = \frac{dx}{dX} - 1 = \Lambda_{(\hat{\mathbf{m}})} - 1 \tag{3.70}$$

and for the element P_0Q_0 along the X_2 axis (of Fig. 3-4), the *unit extension* is therefore

$$L_{(2)} = \Lambda_{(\hat{\mathbf{e}}_2)} - 1 = \sqrt{1 + 2L_{22}} - 1 \tag{3.71}$$

This result may also be derived directly from (*3.36*). For small deformation theory, (*3.71*) reduces to (*3.60*). Also, the unit extensions $L_{(1)}$ and $L_{(3)}$ are given by analogous equations in terms of L_{11} and L_{33} respectively.

For the two differential line elements shown in Fig. 3-5, the change in angle $\gamma_{23} = \pi/2 - \theta$ is given in terms of $\Lambda_{(\hat{\mathbf{e}}_2)}$ and $\Lambda_{(\hat{\mathbf{e}}_3)}$ by

$$\sin \gamma_{23} = \frac{2L_{23}}{\Lambda_{(\hat{\mathbf{e}}_2)}\Lambda_{(\hat{\mathbf{e}}_3)}} = \frac{2L_{23}}{\sqrt{1+2L_{22}}\sqrt{1+2L_{33}}} \tag{3.72}$$

When deformations are small, (*3.72*) reduces to (*3.65*).

3.11 STRETCH TENSORS. ROTATION TENSOR

The so-called *polar decomposition* of an arbitrary, nonsingular, second-order tensor is given by the product of a positive symmetric second-order tensor with an orthogonal second-order tensor. When such a multiplicative decomposition is applied to the deformation gradient **F**, the result may be written

$$F_{ij} \equiv \frac{\partial x_i}{\partial X_j} = R_{ik}S_{kj} = T_{ik}R_{kj} \quad \text{or} \quad \mathbf{F} = \mathbf{R}\cdot\mathbf{S} = \mathbf{T}\cdot\mathbf{R} \tag{3.73}$$

in which **R** is the *orthogonal rotation tensor,* and **S** and **T** are positive symmetric tensors known as the *right stretch tensor* and *left stretch tensor* respectively.

The interpretation of (*3.73*) is provided through the relationship $dx_i = (\partial x_i/\partial X_j)\,dX_j$ given by (*3.33*). Inserting the inner products of (*3.73*) into (*3.33*) results in the equations

$$dx_i = R_{ik}S_{kj}\,dX_j = T_{ik}R_{kj}\,dX_j \quad \text{or} \quad d\mathbf{x} = \mathbf{R}\cdot\mathbf{S}\cdot d\mathbf{X} = \mathbf{T}\cdot\mathbf{R}\cdot d\mathbf{X} \tag{3.74}$$

From these expressions the deformation of dX_i into dx_i as illustrated in Fig. 3-2 may be given either of two physical interpretations. In the first form of the right hand side of (*3.74*), the deformation consists of a sequential stretching (by **S**) and rotation to be followed by a rigid body displacement to the point P. In the second form, a rigid body translation to P is followed by a rotation and finally the stretching (by **T**). The translation, of course, does not alter the vector components relative to the axes X_i and x_i.

3.12 TRANSFORMATION PROPERTIES OF STRAIN TENSORS

The various strain tensors L_{ij}, E_{ij}, l_{ij} and ϵ_{ij} defined respectively by (*3.37*), (*3.39*), (*3.42*) and (*3.43*) are all second-order Cartesian tensors as indicated by the two free indices in each. Accordingly for a set of rotated axes X'_i having the transformation matrix $[b_{ij}]$ with respect to the set of local unprimed axes X_i at point P_0 as shown in Fig. 3-6(*a*), the components of L'_{ij} and l'_{ij} are given by

$$L'_{ij} = b_{ip}b_{jq}L_{pq} \quad \text{or} \quad \mathbf{L}'_G = \mathbf{B}\cdot\mathbf{L}_G\cdot\mathbf{B}_c \tag{3.75}$$

and

$$l'_{ij} = b_{ip}b_{jq}l_{pq} \quad \text{or} \quad \mathbf{L}' = \mathbf{B}\cdot\mathbf{L}\cdot\mathbf{B}_c \tag{3.76}$$

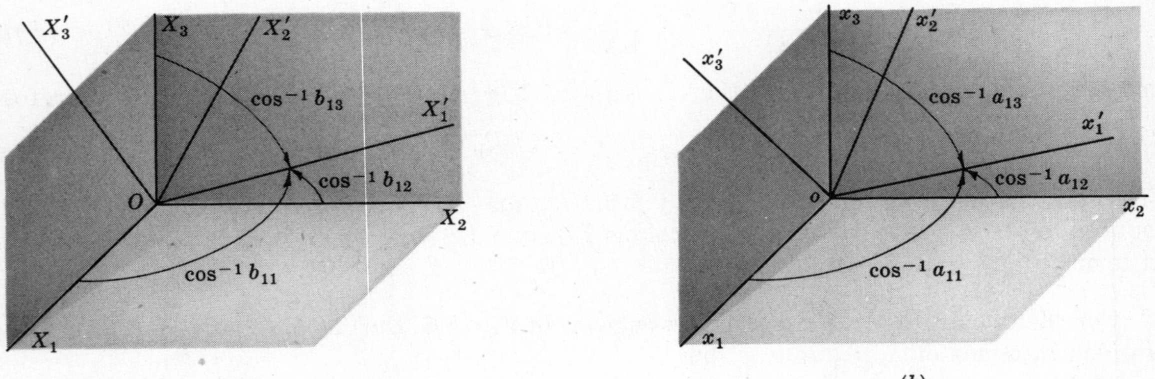

(*a*) (*b*)

Fig. 3-6

Likewise, for the rotated axes x'_i having the transformation matrix $[a_{ij}]$ in Fig. 3-6(*b*), the components of E'_{ij} and e'_{ij} are given by

$$E'_{ij} = a_{ip}a_{jq}E_{pq} \quad \text{or} \quad \mathbf{E}'_A = \mathbf{A}\cdot\mathbf{E}_A\cdot\mathbf{A}_c \tag{3.77}$$

and

$$\epsilon'_{ij} = a_{ip}a_{jq}\epsilon_{pq} \quad \text{or} \quad \mathbf{E}' = \mathbf{A}\cdot\mathbf{E}\cdot\mathbf{A}_c \tag{3.78}$$

By analogy with the stress quadric described in Section 2.9, page 50, the *Lagrangian* and *Eulerian linear strain quadrics* may be given with reference to local Cartesian coordinates η_i and ζ_i at the points P_0 and P respectively as shown in Fig. 3-7. Thus the equation of the *Lagrangian strain quadric* is given by

$$l_{ij}\eta_i\eta_j = \pm h^2 \quad \text{or} \quad \boldsymbol{\eta}\cdot\mathbf{L}\cdot\boldsymbol{\eta} = \pm h^2 \tag{3.79}$$

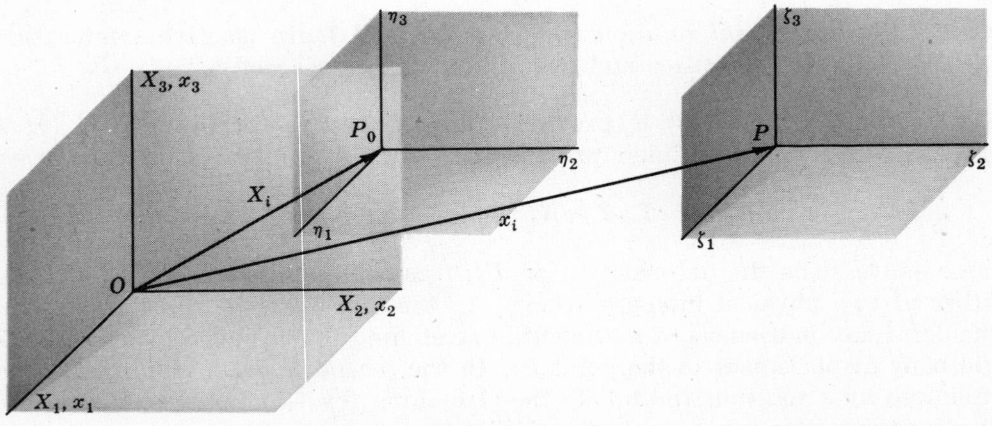

Fig. 3-7

and the equation of the *Eulerian strain quadric* is given by

$$\epsilon_{ij}\zeta_i\zeta_j = \pm g^2 \qquad \text{or} \qquad \boldsymbol{\zeta}\cdot\mathbf{E}\cdot\boldsymbol{\zeta} = \pm g^2 \tag{3.80}$$

Two important properties of the Lagrangian {Eulerian} linear strain quadric are:

1. The normal strain with respect to the original {final} length of a line element is inversely proportional to the distance squared from the origin of the quadric P_0 {P} to a point on its surface.
2. The relative displacement of the neighboring particle located at Q_0 {Q} per unit original {final} length is parallel to the normal of the quadric surface at the point of intersection with the line through P_0Q_0 {PQ}.

Additional insight into the nature of local deformations in the neighborhood of P_0 is provided by defining the *strain ellipsoid* at that point. Thus for the undeformed continuum, the equation of the bounding surface of an infinitesimal sphere of radius R is given in terms of local material coordinates by (*3.28*) as

$$(dX)^2 = \delta_{ij}\,dX_i\,dX_j = R^2 \qquad \text{or} \qquad (dX)^2 = d\mathbf{X}\cdot\mathbf{I}\cdot d\mathbf{X} = R^2 \tag{3.81}$$

After deformation, the equation of the surface of the same material particles is given by (*3.30*) as

$$(dX)^2 = C_{ij}\,dx_i\,dx_j = R^2 \qquad \text{or} \qquad (dX)^2 = d\mathbf{x}\cdot\mathbf{C}\cdot d\mathbf{x} = R^2 \tag{3.82}$$

which describes an ellipsoid, known as the *material strain ellipsoid*. Therefore a spherical volume of the continuum in the undeformed state is changed into an ellipsoid at P_0 by the deformation. By comparison, an infinitesimal spherical volume at P in the deformed continuum began as an ellipsoidal volume element in the undeformed state. For a sphere of radius r at P, the equations for these surfaces in terms of local coordinates are given by (*3.32*) for the sphere as

$$(dx)^2 = \delta_{ij}\,dx_i\,dx_j = r^2 \qquad \text{or} \qquad (dx)^2 = d\mathbf{x}\cdot\mathbf{I}\cdot d\mathbf{x} = r^2 \tag{3.83}$$

and by (*3.34*) for the ellipsoid as

$$(dx)^2 = G_{ij}\,dX_i\,dX_j = r^2 \qquad \text{or} \qquad (dx)^2 = d\mathbf{X}\cdot\mathbf{G}\cdot d\mathbf{X} = r^2 \tag{3.84}$$

The ellipsoid of (*3.84*) is called the *spatial strain ellipsoid*. Such strain ellipsoids as described here are frequently known as *Cauchy strain ellipsoids*.

3.13 PRINCIPAL STRAINS. STRAIN INVARIANTS. CUBICAL DILATATION

The Lagrangian and Eulerian linear strain tensors are symmetric second-order Cartesian tensors, and accordingly the determination of their principal directions and principal strain values follows the standard development presented in Section 1.19, page 20. Physically, a principal direction of the strain tensor is one for which the orientation of an element at a given point is not altered by a pure strain deformation. The principal strain value is simply the unit relative displacement (normal strain) that occurs in the principal direction.

For the Lagrangian strain tensor l_{ij}, the unit relative displacement vector is given by (*3.47*), which may be written

$$\frac{du_i}{dX} = (l_{ij} + W_{ij})\nu_j \qquad \text{or} \qquad \frac{d\mathbf{u}}{dX} = (\mathbf{L}+\mathbf{W})\cdot\hat{\boldsymbol{\nu}} \tag{3.85}$$

Calling $l_i^{(\hat{n})}$ the normal strain in the direction of the unit vector n_i, (*3.85*) yields for pure strain ($W_{ij} \equiv 0$) the relation

$$l_i^{(\hat{n})} = l_{ij}n_j \qquad \text{or} \qquad \mathbf{l}^{(\hat{n})} = \mathbf{L}\cdot\hat{\mathbf{n}} \tag{3.86}$$

If the direction n_i is a principal direction with a principal strain value l, then

$$l_i^{(\hat{n})} = l n_i = l \delta_{ij} n_j \quad \text{or} \quad \mathbf{l}^{(\hat{n})} = l\hat{\mathbf{n}} = l\mathbf{I}\cdot\hat{\mathbf{n}} \tag{3.87}$$

Equating the right-hand sides of (*3.86*) and (*3.87*) leads to the relationship

$$(l_{ij} - \delta_{ij} l) n_j = 0 \quad \text{or} \quad (\mathbf{L} - \mathbf{I}l)\cdot\hat{\mathbf{n}} = 0 \tag{3.88}$$

which together with the condition $n_i n_i = 1$ on the unit vectors n_i provide the necessary equations for determining the principal strain value l and its direction cosines n_i. Nontrivial solutions of (*3.88*) exist if and only if the determinant of coefficients vanishes. Therefore

$$|l_{ij} - \delta_{ij} l| = 0 \quad \text{or} \quad |\mathbf{L} - \mathbf{I}l| = 0 \tag{3.89}$$

which upon expansion yields the characteristic equation of l_{ij}, the cubic

$$l^3 - \mathrm{I}_L l^2 + \mathrm{II}_L l - \mathrm{III}_L = 0 \tag{3.90}$$

where $\quad \mathrm{I}_L = l_{ii} = \operatorname{tr}\mathbf{L}, \quad \mathrm{II}_L = \frac{1}{2}(l_{ii}l_{jj} - l_{ij}l_{ij}), \quad \mathrm{III}_L = |l_{ij}| = \det\mathbf{L} \qquad (3.91)$

are the first, second and third *Lagrangian strain invariants* respectively. The roots of (*3.90*) are the principal strain values denoted by $l_{(1)}$, $l_{(2)}$ and $l_{(3)}$.

The first invariant of the Lagrangian strain tensor may be expressed in terms of the principal strains as

$$\mathrm{I}_L = l_{ii} = l_{(1)} + l_{(2)} + l_{(3)} \tag{3.92}$$

and has an important physical interpretation. To see this, consider a differential rectangular parallelepiped whose edges are parallel to the principal strain directions as shown in Fig. 3-8. The change in volume per unit original volume of this element is called the *cubical dilatation* and is given by

$$D_0 = \frac{\Delta V_0}{V_0} = \frac{dX_1(1 + l_{(1)})\, dX_2(1 + l_{(2)})\, dX_3(1 + l_{(3)}) - dX_1\, dX_2\, dX_3}{dX_1\, dX_2\, dX_3} \tag{3.93}$$

For small strain theory, the first-order approximation of this ratio is the sum

$$D_0 = l_{(1)} + l_{(2)} + l_{(3)} = \mathrm{I}_L \tag{3.94}$$

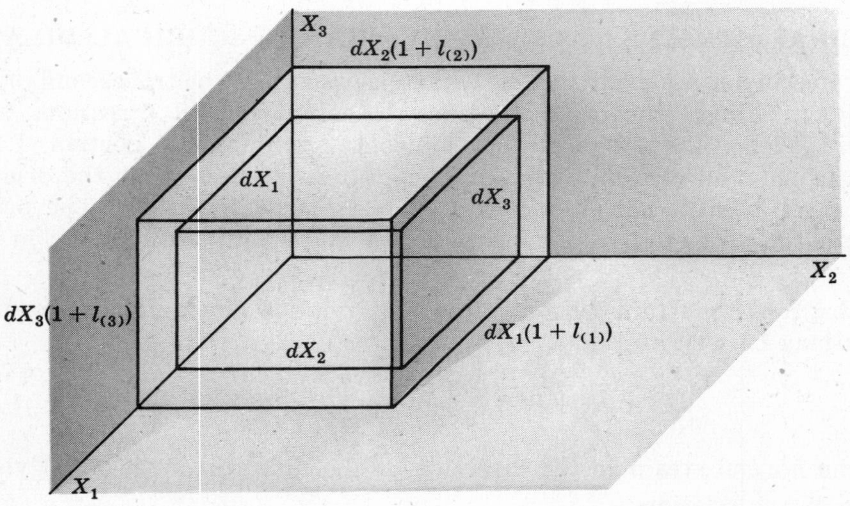

Fig. 3-8

With regard to the Eulerian strain tensor ϵ_{ij} and its associated unit relative displacement vector $\epsilon_i^{(\hat{n})}$, the principal directions and principal strain values $\epsilon_{(1)}, \epsilon_{(2)}, \epsilon_{(3)}$ are determined in exactly the same way as their Lagrangian counterparts. The Eulerian strain invariants may be expressed in terms of the principal strains as

$$\begin{aligned} \mathrm{I}_E &= \epsilon_{(1)} + \epsilon_{(2)} + \epsilon_{(3)} \\ \mathrm{II}_E &= \epsilon_{(1)}\epsilon_{(2)} + \epsilon_{(2)}\epsilon_{(3)} + \epsilon_{(3)}\epsilon_{(1)} \\ \mathrm{III}_E &= \epsilon_{(1)}\epsilon_{(2)}\epsilon_{(3)} \end{aligned} \qquad (3.95)$$

The cubical dilatation for the Eulerian description is given by

$$\Delta V/V = D = \epsilon_{(1)} + \epsilon_{(2)} + \epsilon_{(3)} \qquad (3.96)$$

3.14 SPHERICAL AND DEVIATOR STRAIN TENSORS

The Lagrangian and Eulerian *linear* strain tensors may each be split into a *spherical* and *deviator* tensor in the same manner in which the stress tensor decomposition was carried out in Chapter 2. As before, if Lagrangian and Eulerian deviator tensor components are denoted by d_{ij} and e_{ij} respectively, the resolution expressions are

$$l_{ij} = d_{ij} + \delta_{ij}\frac{l_{kk}}{3} \quad \text{or} \quad \mathbf{L} = \mathbf{L}_D + \frac{\mathbf{I}(\text{tr}\,\mathbf{L})}{3} \qquad (3.97)$$

and

$$\epsilon_{ij} = e_{ij} + \delta_{ij}\frac{\epsilon_{kk}}{3} \quad \text{or} \quad \mathbf{E} = \mathbf{E}_D + \frac{\mathbf{I}(\text{tr}\,\mathbf{E})}{3} \qquad (3.98)$$

The deviator tensors are associated with shear deformation for which the cubical dilatation vanishes. Therefore it is not surprising that the first invariants d_{ii} and e_{ii} of the deviator strain tensors are identically zero.

3.15 PLANE STRAIN. MOHR'S CIRCLES FOR STRAIN

When one and only one of the principal strains at a point in a continuum is zero, a state of *plane strain* is said to exist at that point. In the Eulerian description (the Lagrangian description follows exactly the same pattern), if x_3 is taken as the direction of the zero principal strain, a state of plane strain parallel to the x_1x_2 plane exists and the linear strain tensor is given by

$$\epsilon_{ij} = \begin{pmatrix} \epsilon_{11} & \epsilon_{12} & 0 \\ \epsilon_{12} & \epsilon_{22} & 0 \\ 0 & 0 & 0 \end{pmatrix} \quad \text{or} \quad [\epsilon_{ij}] = \begin{bmatrix} \epsilon_{11} & \epsilon_{12} & 0 \\ \epsilon_{12} & \epsilon_{22} & 0 \\ 0 & 0 & 0 \end{bmatrix} \qquad (3.99)$$

When x_1 and x_2 are also principal directions, the strain tensor has the form

$$\epsilon_{ij} = \begin{pmatrix} \epsilon_{(1)} & 0 & 0 \\ 0 & \epsilon_{(2)} & 0 \\ 0 & 0 & 0 \end{pmatrix} \quad \text{or} \quad [\epsilon_{ij}] = \begin{bmatrix} \epsilon_{(1)} & 0 & 0 \\ 0 & \epsilon_{(2)} & 0 \\ 0 & 0 & 0 \end{bmatrix} \qquad (3.100)$$

In many books on "Strength of Materials" and "Elasticity", plane strain is referred to as *plane deformation* since the deformation field is identical in all planes perpendicular to the direction of the zero principal strain. For plane strain perpendicular to the x_3 axis, the displacement vector may be taken as a function of x_1 and x_2 only. The appropriate displacement components for this case of plane strain are designated by

$$\begin{aligned} u_1 &= u_1(x_1, x_2) \\ u_2 &= u_2(x_1, x_2) \\ u_3 &= C \quad \text{(a constant, usually taken as zero)} \end{aligned} \tag{3.101}$$

Inserting these expressions into the definition of ϵ_{ij} given by (*3.43*) produces the plane strain tensor in the same form shown in (*3.99*).

A graphical description of the state of strain at a point is provided by the *Mohr's circles for strain* in a manner exactly like that presented in Chapter 2 for the Mohr's circles for stress. For this purpose the strain tensor is often displayed in the form

$$\epsilon_{ij} = \begin{pmatrix} \epsilon_{11} & \frac{1}{2}\gamma_{12} & \frac{1}{2}\gamma_{13} \\ \frac{1}{2}\gamma_{12} & \epsilon_{22} & \frac{1}{2}\gamma_{23} \\ \frac{1}{2}\gamma_{13} & \frac{1}{2}\gamma_{23} & \epsilon_{33} \end{pmatrix} \tag{3.102}$$

Here the γ_{ij} (with $i \neq j$) are the so-called "engineering" shear strain components, which are twice the tensorial shear strain components.

The state of strain at an unloaded point on the bounding surface of a continuum body is locally plane strain. Frequently in experimental studies involving strain measurements at such a surface point, Mohr's strain circles are useful for reporting the observed data. Usually three normal strains are measured at the given point by means of a strain rosette, and the Mohr's circles diagram constructed from these. Corresponding to the plane stress Mohr's circles, a typical case of plane strain diagram is shown in Fig. 3-9. The principal normal strains are labeled as such in the diagram, and the maximum shear strain values are represented by points D and E.

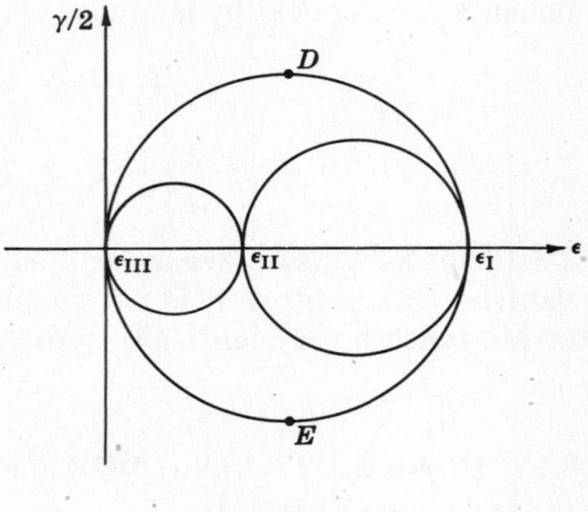

Fig. 3-9

3.16 COMPATIBILITY EQUATIONS FOR LINEAR STRAINS

If the strain components ϵ_{ij} are given explicitly as functions of the coordinates, the six independent equations (*3.43*)

$$\epsilon_{ij} = \frac{1}{2}\left(\frac{\partial u_i}{\partial x_j} + \frac{\partial u_j}{\partial x_i}\right)$$

may be viewed as a system of six partial differential equations for determining the three displacement components u_i. The system is over-determined and will not, in general, possess a solution for an arbitrary choice of the strain components ϵ_{ij}. Therefore if the displacement components u_i are to be single-valued and continuous, some conditions must be imposed upon the strain components. The necessary and sufficient conditions for such a displacement field are expressed by the equations

$$\frac{\partial^2 \epsilon_{ij}}{\partial x_k\, \partial x_m} + \frac{\partial^2 \epsilon_{km}}{\partial x_i\, \partial x_j} - \frac{\partial^2 \epsilon_{ik}}{\partial x_j\, \partial x_m} - \frac{\partial^2 \epsilon_{jm}}{\partial x_i\, \partial x_k} = 0 \tag{3.103}$$

There are eighty-one equations in all in (*3.103*) but only six are distinct. These six written in explicit and symbolic form appear as

$$\left.\begin{aligned}
&1. \quad \frac{\partial^2 \epsilon_{11}}{\partial x_2^2} + \frac{\partial^2 \epsilon_{22}}{\partial x_1^2} = 2\frac{\partial^2 \epsilon_{12}}{\partial x_1 \partial x_2}\\
&2. \quad \frac{\partial^2 \epsilon_{22}}{\partial x_3^2} + \frac{\partial^2 \epsilon_{33}}{\partial x_2^2} = 2\frac{\partial^2 \epsilon_{23}}{\partial x_2 \partial x_3}\\
&3. \quad \frac{\partial^2 \epsilon_{33}}{\partial x_1^2} + \frac{\partial^2 \epsilon_{11}}{\partial x_3^2} = 2\frac{\partial^2 \epsilon_{31}}{\partial x_3 \partial x_1}\\
&4. \quad \frac{\partial}{\partial x_1}\left(-\frac{\partial \epsilon_{23}}{\partial x_1} + \frac{\partial \epsilon_{31}}{\partial x_2} + \frac{\partial \epsilon_{12}}{\partial x_3}\right) = \frac{\partial^2 \epsilon_{11}}{\partial x_2 \partial x_3}\\
&5. \quad \frac{\partial}{\partial x_2}\left(\frac{\partial \epsilon_{23}}{\partial x_1} - \frac{\partial \epsilon_{31}}{\partial x_2} + \frac{\partial \epsilon_{12}}{\partial x_3}\right) = \frac{\partial^2 \epsilon_{22}}{\partial x_3 \partial x_1}\\
&6. \quad \frac{\partial}{\partial x_3}\left(\frac{\partial \epsilon_{23}}{\partial x_1} + \frac{\partial \epsilon_{31}}{\partial x_2} - \frac{\partial \epsilon_{12}}{\partial x_3}\right) = \frac{\partial^2 \epsilon_{33}}{\partial x_1 \partial x_2}
\end{aligned}\right\} \quad \text{or} \quad \nabla_{\mathbf{x}} \times \mathbf{E} \times \nabla_{\mathbf{x}} = 0 \qquad (3.104)$$

Compatibility equations in terms of the Lagrangian linear strain tensor l_{ij} may also be written down by an obvious correspondence to the Eulerian form given above. For plane strain parallel to the x_1x_2 plane, the six equations in (*3.104*) reduce to the single equation

$$\frac{\partial^2 \epsilon_{11}}{\partial x_2^2} + \frac{\partial^2 \epsilon_{22}}{\partial x_1^2} = 2\frac{\partial^2 \epsilon_{12}}{\partial x_1 \partial x_2} \quad \text{or} \quad \nabla_{\mathbf{x}} \times \mathbf{E} \times \nabla_{\mathbf{x}} = 0 \qquad (3.105)$$

where $\mathbf{E}$ is of the form given by (*3.99*).

Solved Problems

DISPLACEMENT AND DEFORMATION (Sec. 3.1-3.5)

3.1. With respect to superposed material axes X_i and spatial axes x_i, the displacement field of a continuum body is given by $x_1 = X_1$, $x_2 = X_2 + AX_3$, $x_3 = X_3 + AX_2$ where A is a constant. Determine the displacement vector components in both the material and spatial forms.

From (*3.13*) directly, the displacement components in material form are $u_1 = x_1 - X_1 = 0$, $u_2 = x_2 - X_2 = AX_3$, $u_3 = x_3 - X_3 = AX_2$. Inverting the given displacement relations to obtain $X_1 = x_1$, $X_2 = (x_2 - Ax_3)/(1 - A^2)$, $X_3 = (x_3 - Ax_2)/(1 - A^2)$, the spatial components of $\mathbf{u}$ are $u_1 = 0$, $u_2 = A(x_3 - Ax_2)/(1 - A^2)$, $u_3 = A(x_2 - Ax_3)/(1 - A^2)$.

From these results it is noted that the originally straight line of material particles expressed by $X_1 = 0$, $X_2 + X_3 = 1/(1 + A)$ occupies the location $x_1 = 0$, $x_2 + x_3 = 1$ after displacement. Likewise the particle line $X_1 = 0$, $X_2 = X_3$ becomes after displacement $x_1 = 0$, $x_2 = x_3$. (Interpret the physical meaning of this.)

3.2. For the displacement field of Problem 3.1 determine the displaced location of the material particles which originally comprise (*a*) the plane circular surface $X_1 = 0$, $X_2^2 + X_3^2 = 1/(1 - A^2)$, (*b*) the infinitesimal cube with edges along the coordinate axes of length $dX_i = dX$. Sketch the displaced configurations for (*a*) and (*b*) if $A = \frac{1}{2}$.

(*a*) By the direct substitutions $X_2 = (x_2 - Ax_3)/(1 - A^2)$ and $X_3 = (x_3 - Ax_2)/(1 - A^2)$, the circular surface becomes the elliptical surface $(1 + A^2)x_2^2 - 4Ax_2x_3 + (1 + A^2)x_3^2 = (1 - A^2)$. For $A = \frac{1}{2}$, this is bounded by the ellipse $5x_2^2 - 8x_2x_3 + 5x_3^2 = 3$ which when referred to its principal axes x_i^* (at 45° with x_i, $i = 2, 3$) has the equation $x_2^{*2} + 9x_3^{*2} = 3$. Fig. 3-10 below shows this displacement pattern.

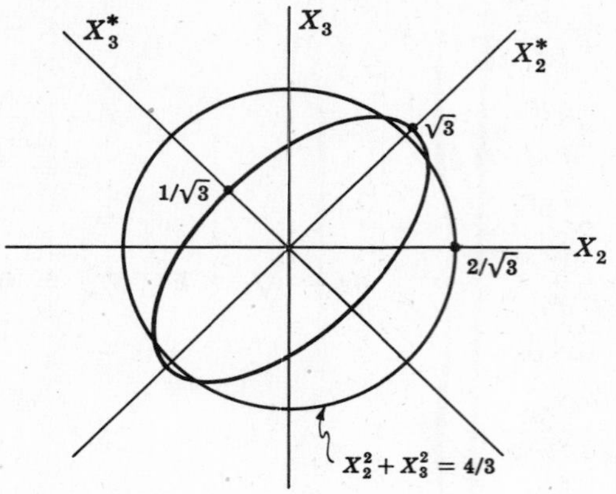

Fig. 3-10

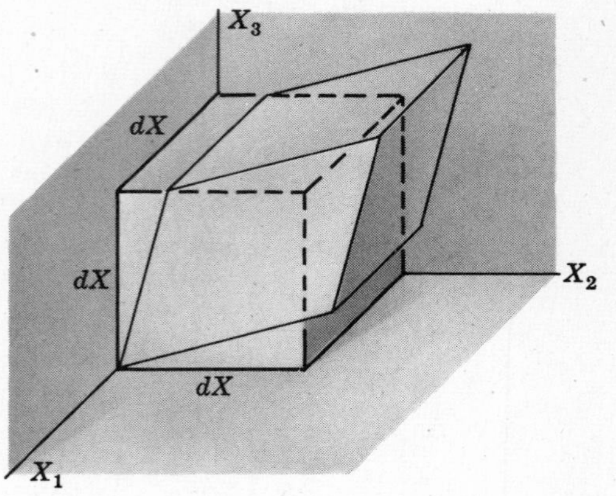

Fig. 3-11

(b) From Problem 3.1, the displacements of the edges of the cube are readily calculated. For the edge $X_1 = X_1$, $X_2 = X_3 = 0$, $u_1 = u_2 = u_3 = 0$. For the edge $X_1 = 0 = X_2$, $X_3 = X_3$, $u_1 = u_3 = 0$, $u_2 = AX_3$ and the particles on this edge are displaced in the X_2 direction proportionally to their distance from the origin. For the edge $X_1 = X_3 = 0$, $X_2 = X_2$, $u_1 = u_2 = 0$, $u_3 = AX_2$. The initial and displaced positions of the cube are shown in Fig. 3-11.

3.3. For superposed material and spatial axes, the displacement vector of a body is given by $\mathbf{u} = 4X_1^2\hat{\mathbf{e}}_1 + X_2X_3^2\hat{\mathbf{e}}_2 + X_1X_3^2\hat{\mathbf{e}}_3$. Determine the displaced location of the particle originally at $(1, 0, 2)$.

The original position vector of the particle is $\mathbf{X} = \hat{\mathbf{e}}_1 + 2\hat{\mathbf{e}}_3$. Its displacement is $\mathbf{u} = 4\hat{\mathbf{e}}_1 + 4\hat{\mathbf{e}}_3$ and since $\mathbf{x} = \mathbf{X} + \mathbf{u}$, its final position vector is $\mathbf{x} = 5\hat{\mathbf{e}}_1 + 6\hat{\mathbf{e}}_3$.

3.4. With respect to rectangular Cartesian material coordinates X_i, a displacement field is given by $U_1 = -AX_2X_3$, $U_2 = AX_1X_3$, $U_3 = 0$ where A is a constant. Determine the displacement components for cylindrical spatial coordinates x_i if the two systems have a common origin.

From the geometry of the axes (Fig. 3-12) the transformation tensor $\alpha_{pK} = \hat{\mathbf{e}}_p \cdot \hat{\mathbf{I}}_K$ is

$$\alpha_{pK} = \begin{pmatrix} \cos x_2 & \sin x_2 & 0 \\ -\sin x_2 & \cos x_2 & 0 \\ 0 & 0 & 1 \end{pmatrix}$$

and from the inverse form of (3.9) $u_p = \alpha_{pK}U_K$. Thus since Cartesian and cylindrical coordinates are related through the equations $X_1 = x_1 \cos x_2$, $X_2 = x_1 \sin x_2$, $X_3 = x_3$, equation (3.9) gives

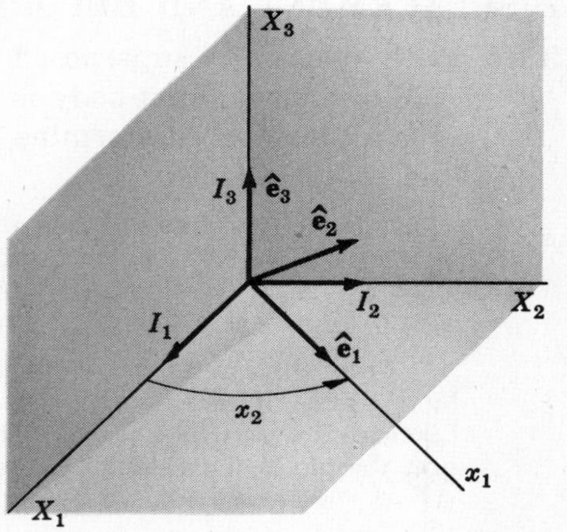

Fig. 3-12

$$u_1 = (-\cos x_2)AX_2X_3 + (\sin x_2)AX_1X_3$$
$$= (-\cos x_2)Ax_3x_1 \sin x_2 + (\sin x_2)Ax_3x_1 \cos x_2 = 0$$

$$u_2 = (\sin x_2)AX_2X_3 + (\cos x_2)AX_1X_3$$
$$= (\sin^2 x_2)Ax_1x_3 + (\cos^2 x_2)Ax_1x_3 = Ax_1x_3$$

$$u_3 = 0$$

This displacement is that of a circular shaft in torsion.

3.5. The Lagrangian description of a deformation is given by $x_1 = X_1 + X_3(e^2-1)$, $x_2 = X_2 + X_3(e^2 - e^{-2})$, $x_3 = e^2X_3$ where e is a constant. Show that the Jacobian J does not vanish and determine the Eulerian equations describing this motion.

From (*3.16*),
$$J = \begin{vmatrix} 1 & 0 & (e^2-1) \\ 0 & 1 & (e^2-e^{-2}) \\ 0 & 0 & (e^2) \end{vmatrix} = e^2 \neq 0.$$

Inverting the equations, $X_1 = x_1 + x_3(e^{-2}-1)$, $X_2 = x_2 + x_3(e^{-4}-1)$, $X_3 = e^{-2}x_3$.

3.6. A displacement field is given by $\mathbf{u} = X_1X_3^2\hat{\mathbf{e}}_1 + X_1^2X_2\hat{\mathbf{e}}_2 + X_2^2X_3\hat{\mathbf{e}}_3$. Determine independently the material deformation gradient $\mathbf{F}$ and the material displacement gradient $\mathbf{J}$ and verify (*3.24*), $\mathbf{J} = \mathbf{F} - \mathbf{I}$.

From the given displacement vector $\mathbf{u}$, $\mathbf{J}$ is found to be

$$\frac{\partial u_i}{\partial X_j} = \begin{pmatrix} X_3^2 & 0 & 2X_1X_3 \\ 2X_1X_2 & X_1^2 & 0 \\ 0 & 2X_2X_3 & X_2^2 \end{pmatrix}$$

Since $\mathbf{x} = \mathbf{u} + \mathbf{X}$, the displacement field may also be described by equations $x_1 = X_1(1+X_3^2)$, $x_2 = X_2(1+X_1^2)$, $x_3 = X_3(1+X_2^2)$ from which $\mathbf{F}$ is readily found to be

$$\partial x_i/\partial X_j = \begin{pmatrix} 1+X_3^2 & 0 & 2X_1X_3 \\ 2X_1X_2 & 1+X_1^2 & 0 \\ 0 & 2X_2X_3 & 1+X_2^2 \end{pmatrix}$$

Direct substitution of the calculated tensors $\mathbf{F}$ and $\mathbf{J}$ into (*3.24*) verifies that the equation is satisfied.

3.7. A continuum body undergoes the displacement $\mathbf{u} = (3X_2 - 4X_3)\hat{\mathbf{e}}_1 + (2X_1 - X_3)\hat{\mathbf{e}}_2 + (4X_2 - X_1)\hat{\mathbf{e}}_3$. Determine the displaced position of the vector joining particles $A(1,0,3)$ and $B(3,6,6)$, assuming superposed material and spatial axes.

From (*3.13*), the spatial coordinates for this displacement are $x_1 = X_1 + 3X_2 - 4X_3$, $x_2 = 2X_1 + X_2 - X_3$, $x_3 = -X_1 + 4X_2 + X_3$. Thus the displaced position of particle A is given by $x_1 = -11$, $x_2 = -1$, $x_3 = 2$; and of particle B, $x_1 = -3$, $x_2 = 6$, $x_3 = 27$. Therefore the displaced position of the vector joining A and B may be written $\mathbf{V} = 8\hat{\mathbf{e}}_1 + 7\hat{\mathbf{e}}_2 + 25\hat{\mathbf{e}}_3$.

3.8. For the displacement field of Problem 3.7 determine the displaced position of the position vector of particle $C(2,6,3)$ which is parallel to the vector joining particles A and B. Show that the two vectors remain parallel after deformation.

By the analysis of Problem 3.7 the position vector of C becomes $\mathbf{U} = 8\hat{\mathbf{e}}_1 + 7\hat{\mathbf{e}}_2 + 25\hat{\mathbf{e}}_3$ which is clearly parallel to $\mathbf{V}$. This is an example of so-called *homogeneous deformation.*

3.9. The general formulation of homogeneous deformation is given by the displacement field $u_i = A_{ij}X_j$ where the A_{ij} are constants or at most functions of time. Show that this deformation is such that (*a*) plane sections remain plane, (*b*) straight lines remain straight.

(*a*) From (*3.13*),
$$x_i = X_i + u_i = X_i + A_{ij}X_j = (\delta_{ij} + A_{ij})X_j$$

According to (3.16) the inverse equations $X_i = (\delta_{ij} + B_{ij})x_j$ exist provided the determinant $|\delta_{ij} + A_{ij}|$ does not vanish. Assuming this is the case, the material plane $\beta_i X_i + \alpha = 0$ becomes $\beta_i(\delta_{ij} + B_{ij})x_j + \alpha = 0$ which may be written in standard form as the plane $\lambda_j x_j + \alpha = 0$ where the coefficients $\lambda_j = \beta_i(\delta_{ij} + B_{ij})$.

(b) A straight line may be considered as the intersection of two planes. In the deformed geometry, planes remain plane as proven and hence the intersection of two planes remains a straight line.

3.10. An *infinitesimal* homogeneous deformation $u_i = A_{ij}X_j$ is one for which the coefficients A_{ij} are so small that their products may be neglected in comparison to the coefficients themselves. Show that the total deformation resulting from two successive infinitesimal homogeneous deformations may be considered as the sum of the two individual deformations, and that the order of applying the displacements does not alter the final configuration.

Let $x_i = (\delta_{ij} + A_{ij})X_j$ and $x_i' = (\delta_{ij} + B_{ij})x_j$ be successive infinitesimal homogeneous displacements. Then $x_i' = (\delta_{ij} + B_{ij})(\delta_{jk} + A_{jk})X_k = (\delta_{ik} + B_{ik} + A_{ik} + B_{ij}A_{jk})X_k$. Neglecting the higher order product terms $B_{ij}A_{jk}$, this becomes $x_i' = (\delta_{ik} + B_{ik} + A_{ik})X_k = (\delta_{ik} + C_{ik})X_k$ which represents the infinitesimal homogeneous deformation

$$u_i'' = x_i' - X_i = C_{ik}X_k = (B_{ik} + A_{ik})X_k = (A_{ik} + B_{ik})X_k = u_i + u_i'$$

DEFORMATION AND STRAIN TENSORS (Sec. 3.6-3.9)

3.11. A continuum body undergoes the deformation $x_1 = X_1$, $x_2 = X_2 + AX_3$, $x_3 = X_3 + AX_2$ where A is a constant. Compute the deformation tensor $\mathbf{G}$ and use this to determine the Lagrangian finite strain tensor $\mathbf{L}_G$.

From (3.35), $\mathbf{G} = \mathbf{F}_c \cdot \mathbf{F}$ and by (3.20) $\mathbf{F}$ is given in matrix form as

$$[\partial x_i/\partial X_j] = \begin{bmatrix} 1 & 0 & 0 \\ 0 & 1 & A \\ 0 & A & 1 \end{bmatrix} \quad \text{so that} \quad [G_{ij}] = \begin{bmatrix} 1 & 0 & 0 \\ 0 & 1+A^2 & 2A \\ 0 & 2A & 1+A^2 \end{bmatrix}$$

Therefore from (3.37),
$$\mathbf{L}_G = \tfrac{1}{2}(\mathbf{G} - \mathbf{I}) = \frac{1}{2}\begin{pmatrix} 0 & 0 & 0 \\ 0 & A^2 & 2A \\ 0 & 2A & A^2 \end{pmatrix}$$

3.12. For the displacement field of Problem 3.11 calculate the squared length $(dx)^2$ of the edges OA and OB, and the diagonal OC after deformation for the small rectangle shown in Fig. 3-13.

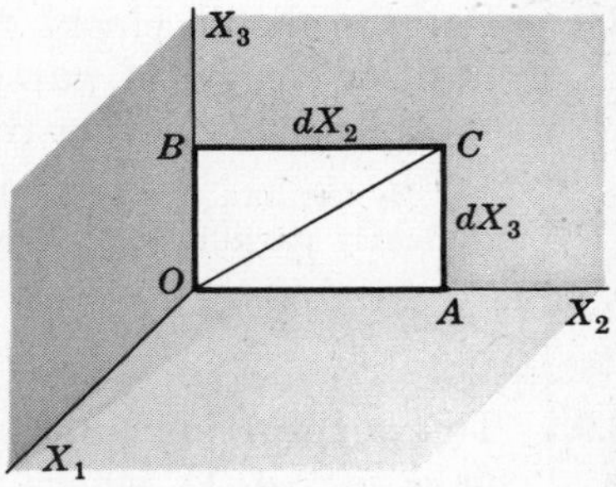

Fig. 3-13

Using $\mathbf{G}$ as determined in Problem 3.11 in (3.34), the squared length of the diagonal OC is given in matrix form by

$$(dx)^2 = [0, dX_2, dX_3]\begin{bmatrix} 1 & 0 & 0 \\ 0 & 1+A^2 & 2A \\ 0 & 2A & 1+A^2 \end{bmatrix}\begin{bmatrix} 0 \\ dX_2 \\ dX_3 \end{bmatrix}$$

$$= (1+A^2)(dX_2)^2 + 4A\,dX_2\,dX_3 + (1+A^2)(dX_3)^2$$

Similarly for OA, $(dx)^2 = (1+A^2)(dX_2)^2$; and for OB, $(dx)^2 = (1+A^2)(dX_3)^2$.

3.13. Calculate the *change* in squared length of the line elements of Problem 3.12 and check the result by use of (*3.36*) and the strain tensor $\mathbf{L}_G$ found in Problem 3.11.

Directly from the results of Problem 3.12, the changes are:

$$(a)\ \text{for } OC,\quad (dx)^2 - (dX)^2 = (1+A^2)(dX_2^2 + dX_3^2) + 4A\,dX_2\,dX_3 - (dX_2^2+dX_3^2) = A^2(dX_2^2+dX_3^2) + 4A\,dX_2\,dX_3$$

$$(b)\ \text{for } OB,\quad (dx)^2 - (dX)^2 = (1+A^2)\,dX_3^2 - dX_3^2 = A^2\,dX_3^2$$

$$(c)\ \text{for } OA,\quad (dx)^2 - (dX)^2 = (1+A^2)\,dX_2^2 - dX_2^2 = A^2\,dX_2^2.$$

By equation (*3.36*), for OC

$$(dx)^2 - (dX)^2 = [0, dX_2, dX_3]\begin{bmatrix} 0 & 0 & 0 \\ 0 & A^2 & 2A \\ 0 & 2A & A^2 \end{bmatrix}\begin{bmatrix} 0 \\ dX_2 \\ dX_3 \end{bmatrix} = A^2(dX_2^2 + dX_3^2) + 4A\,dX_2\,dX_3$$

The changes for OA and OB may also be confirmed in the same way.

3.14. For the displacement field of Problem 3.11 calculate the material displacement gradient $\mathbf{J}$ and use this tensor to determine the Lagrangian finite strain tensor $\mathbf{L}_G$. Compare with result of Problem 3.11.

From Problem 3.11 the displacement vector components are $u_1 = 0$, $u_2 = AX_3$, $u_3 = AX_2$ so that

$$\mathbf{J} = \begin{pmatrix} 0 & 0 & 0 \\ 0 & 0 & A \\ 0 & A & 0 \end{pmatrix} \quad \text{and} \quad \mathbf{J}_c \cdot \mathbf{J} = \begin{pmatrix} 0 & 0 & 0 \\ 0 & A^2 & 0 \\ 0 & 0 & A^2 \end{pmatrix}$$

Thus from (*3.40*)

$$2\mathbf{L}_G = \begin{pmatrix} 0 & 0 & 0 \\ 0 & 0 & A \\ 0 & A & 0 \end{pmatrix} + \begin{pmatrix} 0 & 0 & 0 \\ 0 & 0 & A \\ 0 & A & 0 \end{pmatrix} + \begin{pmatrix} 0 & 0 & 0 \\ 0 & A^2 & 0 \\ 0 & 0 & A^2 \end{pmatrix} = \begin{pmatrix} 0 & 0 & 0 \\ 0 & A^2 & 2A \\ 0 & 2A & A^2 \end{pmatrix}$$

the identical result obtained in Problem 3.11.

3.15. A displacement field is given by $x_1 = X_1 + AX_2$, $x_2 = X_2 + AX_3$, $x_3 = X_3 + AX_1$ where A is a constant. Calculate the Lagrangian linear strain tensor $\mathbf{L}$ and the Eulerian linear strain tensor $\mathbf{E}$. Compare $\mathbf{L}$ and $\mathbf{E}$ for the case when A is very small.

From (*3.42*),

$$2\mathbf{L} = (\mathbf{J} + \mathbf{J}_c) = \begin{pmatrix} 0 & A & 0 \\ 0 & 0 & A \\ A & 0 & 0 \end{pmatrix} + \begin{pmatrix} 0 & 0 & A \\ A & 0 & 0 \\ 0 & A & 0 \end{pmatrix} = \begin{pmatrix} 0 & A & A \\ A & 0 & A \\ A & A & 0 \end{pmatrix}$$

Inverting the displacement equations gives

$$u_1 = A(A^2x_1 + x_2 - Ax_3)/(1+A^3), \quad u_2 = A(-Ax_1 + A^2x_2 + x_3)/(1+A^3),$$
$$u_3 = A(x_1 - Ax_2 + A^2x_3)/(1+A^3)$$

from which by (*3.43*)

$$2\mathbf{E} = (\mathbf{K} + \mathbf{K}_c) = \frac{A}{1+A^3}\begin{pmatrix} A^2 & 1 & -A \\ -A & A^2 & 1 \\ 1 & -A & A^2 \end{pmatrix} + \frac{A}{1+A^3}\begin{pmatrix} A^2 & -A & 1 \\ 1 & A^2 & -A \\ -A & 1 & A^2 \end{pmatrix}$$

$$= \frac{A}{1+A^3}\begin{pmatrix} 2A^2 & 1-A & 1-A \\ 1-A & 2A^2 & 1-A \\ 1-A & 1-A & 2A^2 \end{pmatrix}$$

When A is very small, A^2 and higher powers may be neglected with the result that $\mathbf{E}$ reduces to $\mathbf{L}$.

3.16. A displacement field is specified by $\mathbf{u} = X_1^2 X_2 \hat{\mathbf{e}}_1 + (X_2 - X_3^2)\hat{\mathbf{e}}_2 + X_2^2 X_3 \hat{\mathbf{e}}_3$. Determine the relative displacement vector $d\mathbf{u}$ in the direction of the $-X_2$ axis at $P(1,2,-1)$. Determine the relative displacements $\mathbf{u}_{Q_i} - \mathbf{u}_P$ for $Q_1(1,1,-1)$, $Q_2(1,3/2,-1)$, $Q_3(1,7/4,-1)$ and $Q_4(1,15/8,-1)$ and compare their directions with the direction of $d\mathbf{u}$.

For the given $\mathbf{u}$, the displacement gradient J in matrix form is

$$[\partial u_i/\partial X_j] = \begin{bmatrix} 2X_1X_2 & X_1^2 & 0 \\ 0 & 1 & -2X_3 \\ 0 & 2X_2X_3 & X_2^2 \end{bmatrix}$$

so that from (*3.46*) at P in the $-X_2$ direction,

$$[du_i] = \begin{bmatrix} 4 & 1 & 0 \\ 0 & 1 & 2 \\ 0 & -4 & 4 \end{bmatrix}\begin{bmatrix} 0 \\ -1 \\ 0 \end{bmatrix} = \begin{bmatrix} -1 \\ -1 \\ 4 \end{bmatrix}$$

Next by direct calculation from $\mathbf{u}$, $\mathbf{u}_P = 2\hat{\mathbf{e}}_1 + \hat{\mathbf{e}}_2 - 4\hat{\mathbf{e}}_3$ and $\mathbf{u}_{Q_1} = \hat{\mathbf{e}}_1 - \hat{\mathbf{e}}_3$. Thus $\mathbf{u}_{Q_1} - \mathbf{u}_P = -\mathbf{e}_1 - \mathbf{e}_2 + 3\,\mathbf{e}_3$. Likewise, $\mathbf{u}_{Q_2} - \mathbf{u}_P = (-\hat{\mathbf{e}}_1 - \hat{\mathbf{e}}_2 + 3.5\,\hat{\mathbf{e}}_3)/2$, $\mathbf{u}_{Q_3} - \mathbf{u}_P = (-\hat{\mathbf{e}}_1 - \hat{\mathbf{e}}_2 + 3.75\,\hat{\mathbf{e}}_3)/4$, $\mathbf{u}_{Q_4} - \mathbf{u}_P = (-\hat{\mathbf{e}}_1 - \hat{\mathbf{e}}_2 + 3.875\,\hat{\mathbf{e}}_3)/8$. It is clear that as Q_i approaches P the direction of the relative displacement of the two particles approaches the limiting direction of $d\mathbf{u}$.

3.17. For the displacement field of Problem 3.16 determine the unit relative displacement vector at $P(1,2,-1)$ in the direction of $Q(4,2,3)$.

The unit vector at P in the direction of Q is $\hat{\boldsymbol{\nu}} = 3\hat{\mathbf{e}}_1/5 + 4\hat{\mathbf{e}}_3/5$, so that from (*3.47*) and the matrix of J as calculated in Problem 3.16,

$$[du_i/dX] = \begin{bmatrix} 4 & 1 & 0 \\ 0 & 1 & 2 \\ 0 & -4 & 4 \end{bmatrix}\begin{bmatrix} 3/5 \\ 0 \\ 4/5 \end{bmatrix} = \begin{bmatrix} 12/5 \\ 8/5 \\ 16/5 \end{bmatrix}$$

3.18. Under the restriction of small deformation theory, $\mathsf{L} = \mathsf{E}$. Accordingly for a displacement field given by $\mathbf{u} = (x_1 - x_3)^2\hat{\mathbf{e}}_1 + (x_2 + x_3)^2\hat{\mathbf{e}}_2 - x_1x_2\hat{\mathbf{e}}_3$, determine the linear strain tensor, the linear rotation tensor and the rotation vector at the point $P(0,2,-1)$.

Here the displacement gradient is given in matrix form by

$$[\partial u_i/\partial x_j] = \begin{bmatrix} 2(x_1 - x_3) & 0 & -2(x_1 - x_3) \\ 0 & 2(x_2 + x_3) & 2(x_2 + x_3) \\ -x_2 & -x_1 & 0 \end{bmatrix}$$

which at the point P becomes

$$[\partial u_i/\partial x_j]_P = \begin{bmatrix} 2 & 0 & -2 \\ 0 & 2 & 2 \\ -2 & 0 & 0 \end{bmatrix}$$

Decomposing this matrix into its symmetric and antisymmetric components gives

$$[\epsilon_{ij}] + [\omega_{ij}] = \begin{bmatrix} 2 & 0 & -2 \\ 0 & 2 & 1 \\ -2 & 1 & 0 \end{bmatrix} + \begin{bmatrix} 0 & 0 & 0 \\ 0 & 0 & 1 \\ 0 & -1 & 0 \end{bmatrix}$$

Therefore from (*3.56*) the rotation vector ω_i has components $\omega_1 = -1$, $\omega_2 = \omega_3 = 0$.

3.19. For the displacement field of Problem 3.18 determine the change in length per unit length (normal strain) in the direction of $\hat{\nu} = (8\hat{e}_1 - \hat{e}_2 + 4\hat{e}_3)/9$ at point $P(0, 2, -1)$.

From (*3.59*) and the strain tensor at P as computed in Problem 3.18, the normal strain at P in the direction of $\hat{\nu}$ is the matrix product

$$\epsilon_p^{(\hat{\nu})} = [8/9, -1/9, 4/9]\begin{bmatrix} 2 & 0 & -2 \\ 0 & 2 & 1 \\ -2 & 1 & 0 \end{bmatrix}\begin{bmatrix} 8/9 \\ -1/9 \\ 4/9 \end{bmatrix} = -6/81$$

3.20. Show that the change in the right angle between two orthogonal unit vectors $\hat{\nu}$ and $\hat{\mu}$ in the undeformed configuration is given by $\hat{\nu} \cdot 2\mathbf{L} \cdot \hat{\mu}$ for small deformation theory.

Assuming small displacement gradients, the unit vectors in the deformed directions of $\hat{\nu}$ and $\hat{\mu}$ are given by (*3.47*) as $(\hat{\nu} + \mathbf{J} \cdot \hat{\nu})$ and $(\hat{\mu} + \mathbf{J} \cdot \hat{\mu})$ respectively. (The student should check equations (*3.61*) and (*3.62*) by this method.) Writing $\mathbf{J} \cdot \hat{\nu}$ in the equivalent form $\hat{\nu} \cdot \mathbf{J}_c$ and dotting the two displaced unit vectors gives (as in (*3.63*)), $\cos\theta = \sin(\pi/2 - \theta) = \sin\gamma_{\nu\mu} = \gamma_{\nu\mu}$ or $\gamma_{\nu\mu} = [\hat{\nu} + \hat{\nu} \cdot \mathbf{J}_c] \cdot [\hat{\mu} + \mathbf{J} \cdot \hat{\mu}] = \hat{\nu} \cdot \hat{\mu} + \hat{\nu} \cdot (\mathbf{J} + \mathbf{J}_c) \cdot \hat{\mu} + \hat{\nu} \cdot \mathbf{J}_c \cdot \mathbf{J} \cdot \hat{\mu}$. Here $\mathbf{J}_c \cdot \mathbf{J}$ is of higher order for small displacement gradients and since $\hat{\nu} \perp \hat{\mu}$, $\hat{\nu} \cdot \hat{\mu} = 0$ so that finally by (*3.42*), $\gamma_{\nu\mu} = \hat{\nu} \cdot 2\mathbf{L} \cdot \hat{\mu}$.

3.21. Use the results of Problem 3.20 to compute the change in the right angle between $\hat{\nu} = (8\hat{e}_1 - \hat{e}_2 + 4\hat{e}_3)/9$ and $\hat{\mu} = (4\hat{e}_1 + 4\hat{e}_2 - 7\hat{e}_3)/9$ at the point $P(0, 2, -1)$ for the displacement field of Problem 3.18.

Since $\mathbf{L} = \mathbf{E}$ for small deformation theory, the strain tensor $\epsilon_{ij} = l_{ij}$ and so at P

$$\gamma_{\nu\mu} = [8/9, -1/9, 4/9]\begin{bmatrix} 4 & 0 & -4 \\ 0 & 4 & 2 \\ -4 & 2 & 0 \end{bmatrix}\begin{bmatrix} 4/9 \\ 4/9 \\ -7/9 \end{bmatrix} = 318/81$$

STRETCH AND ROTATION (Sec. 3.10-3.11)

3.22. For the shear deformation $x_1 = X_1$, $x_2 = X_2 + AX_3$, $x_3 = X_3 + AX_2$ of Problem 3.11 show that the stretch $\Lambda_{(\hat{m})}$ is unity (zero normal strain) for line elements parallel to the X_1 axis. For the diagonal directions OC and DB of the infinitesimal square $OBCD$ (Fig. 3-14), compute $\Lambda_{(\hat{m})}$ and check the results by direct calculation from the displacement field.

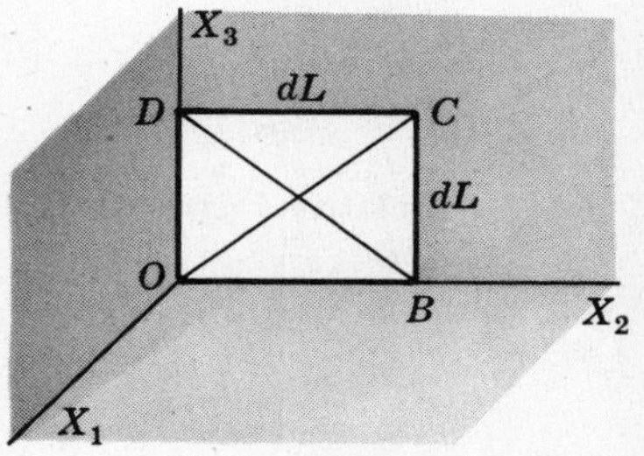

Fig. 3-14

From (*3.66*) and the matrix of $\mathbf{G}$ as determined in Problem 3.11, the squared stretch for $\hat{m} = \hat{e}_1$ is

$$\Lambda^2_{(\hat{e}_1)} = [1, 0, 0]\begin{bmatrix} 1 & 0 & 0 \\ 0 & 1 + A^2 & 2A \\ 0 & 2A & 1 + A^2 \end{bmatrix}\begin{bmatrix} 1 \\ 0 \\ 0 \end{bmatrix} = 1$$

Likewise for OC, $\hat{m} = (\hat{e}_2 + \hat{e}_3)/\sqrt{2}$ and so

$$\Lambda^2_{(\hat{m})} = [0, 1/\sqrt{2}, 1/\sqrt{2}]\begin{bmatrix} 1 & 0 & 0 \\ 0 & 1 + A^2 & 2A \\ 0 & 2A & 1 + A^2 \end{bmatrix}\begin{bmatrix} 0 \\ 1/\sqrt{2} \\ 1/\sqrt{2} \end{bmatrix} = (1 + A)^2$$

From the displacement equations the deformed location of C is $x_1 = 0$, $x_2 = dL + A\,dL$, $x_3 = dL + A\,dL$. Thus $(dx)^2 = 2(1+A)^2(dL)^2$ and since $dX = \sqrt{2}\,dL$, the squared stretch $(dx/dX)^2$ is $(1+A)^2$ as calculated from (3.66).

Similarly, for DB, $\hat{\mathbf{m}} = (-\hat{\mathbf{e}}_2 + \hat{\mathbf{e}}_3)/\sqrt{2}$ and so $\Lambda^2_{(\hat{m})} = (1-A)^2$.

3.23. **The stretch ratios $\Lambda_{(\hat{m})}$ and $\lambda_{(\hat{n})}$ are equal only if $\hat{\mathbf{n}}$ is the deformed direction of $\hat{\mathbf{m}}$. For the displacement field of Problem 3.22, calculate $\lambda_{(\hat{n})}$ for $\hat{\mathbf{n}} = (\hat{\mathbf{e}}_2 + \hat{\mathbf{e}}_3)/\sqrt{2}$ and show that it agrees with $\Lambda^2_{(\hat{m})}$ for the diagonal OC in Problem 3.22.**

Inverting the displacement equations of Problem 3.22 one obtains

$$X_1 = x_1, \quad X_2 = (x_2 - Ax_3)/(1-A^2), \quad X_3 = (x_3 - Ax_2)/(1-A^2)$$

from which the Cauchy deformation tensor $\mathbf{c}$ may be computed. Then using (3.67),

$$\frac{1}{\lambda^2_{(\hat{n})}} = [0, 1/\sqrt{2}, 1/\sqrt{2}]\begin{bmatrix} 1 & 0 & 0 \\ 0 & (1+A^2)/(1-A^2)^2 & -2A/(1-A^2)^2 \\ 0 & -2A/(1-A^2)^2 & (1+A^2)/(1-A^2)^2 \end{bmatrix}\begin{bmatrix} 0 \\ 1/\sqrt{2} \\ 1/\sqrt{2} \end{bmatrix} = (1-A)^2/(1-A^2)^2$$

Thus $\lambda^2_{(\hat{n})} = (1-A^2)^2/(1-A)^2 = (1+A)^2$ which is identical with $\Lambda^2_{(\hat{m})}$ calculated for OC. The diagonal element OC does not change direction under the given shear deformation.

3.24. **By a polar decomposition of the deformation gradient $\mathbf{F}$ for the shear deformation $x_1 = X_1$, $x_2 = X_2 + AX_3$, $x_3 = X_3 + AX_2$, determine the right stretch tensor $\mathbf{S}$ together with the rotation tensor $\mathbf{R}$. Show that the principal values of $\mathbf{S}$ are the stretch ratios of the diagonals OC and DB determined in Problem 3.22.**

In the polar decomposition of $\mathbf{F}$, the stretch tensor $\mathbf{S} = \sqrt{\mathbf{G}}$; and from (3.73), $\mathbf{R} = \mathbf{F}\mathbf{S}^{-1}$. By (3.35), $\mathbf{G} = \mathbf{F}_c \cdot \mathbf{F}$ or here $[G_{ij}] = \begin{bmatrix} 1 & 0 & 0 \\ 0 & 1+A^2 & 2A \\ 0 & 2A & 1+A^2 \end{bmatrix}$. The principal axes of $\mathbf{G}$ are given by a 45° rotation about X_1 with the tensor in principal form $[G^*_{ij}] = \begin{bmatrix} 1 & 0 & 0 \\ 0 & (1-A)^2 & 0 \\ 0 & 0 & (1+A)^2 \end{bmatrix}$.

$$\text{Therefore} \quad [S_{ij}] = [\sqrt{G^*_{ij}}] = \begin{bmatrix} 1 & 0 & 0 \\ 0 & (1-A) & 0 \\ 0 & 0 & (1+A) \end{bmatrix} = \begin{bmatrix} 1 & 0 & 0 \\ 0 & \Lambda_{(DB)} & 0 \\ 0 & 0 & \Lambda_{(OC)} \end{bmatrix}$$

Relative to the coordinate axes X_i, the decomposition is

$$[F_{ij}] = [R_{ik}][S_{kj}] = \begin{bmatrix} 1 & 0 & 0 \\ 0 & 1 & 0 \\ 0 & 0 & 1 \end{bmatrix}\begin{bmatrix} 1 & 0 & 0 \\ 0 & 1 & A \\ 0 & A & 1 \end{bmatrix}$$

In this example the deformation gradient $\mathbf{F}$ is its own stretch tensor $\mathbf{S}$ and $\mathbf{R} = \mathbf{I}$. This is the result of the coincidence of the principal axes of $\mathbf{L}_G$ and $\mathbf{E}_A$ for the given shear deformation.

3.25. **An infinitesimal rigid body rotation is given by $u_1 = -CX_2 + BX_3$, $u_2 = CX_1 - AX_3$, $u_3 = -BX_1 + AX_2$ where A, B, C are very small constants. Show that the stretch is zero ($\mathbf{S} = \mathbf{I}$) if terms involving squares and products of the constants are neglected.**

For this displacement,

$$[G_{ij}] = \begin{bmatrix} 1+C^2+B^2 & -AB & -AC \\ -AB & 1+A^2+C^2 & -BC \\ -AC & -BC & 1+A^2+B^2 \end{bmatrix}$$

Neglecting higher order terms, this becomes

$$[G_{ij}] = \begin{bmatrix} 1 & 0 & 0 \\ 0 & 1 & 0 \\ 0 & 0 & 1 \end{bmatrix} = [\sqrt{G_{ij}}\,] = [S_{ij}]$$

STRAIN TRANSFORMATIONS AND PRINCIPAL STRAINS (Sec. 3.12-3.14)

3.26. For the shear deformation $x_1 = X_1,\ x_2 = X_2 + \sqrt{2}\,X_3,\ x_3 = X_3 + \sqrt{2}\,X_2$ show that the principal directions of $\mathbf{L}_G$ and $\mathbf{E}_A$ coincide as was asserted in Problem 3.24.

From (*3.37*), $[L_{ij}] = \begin{bmatrix} 0 & 0 & 0 \\ 0 & 1 & \sqrt{2} \\ 0 & \sqrt{2} & 1 \end{bmatrix}$ which for principal axes given by the transformation

matrix $[a_{ij}] = \begin{bmatrix} 1 & 0 & 0 \\ 0 & 1/\sqrt{2} & 1/\sqrt{2} \\ 0 & -1/\sqrt{2} & 1/\sqrt{2} \end{bmatrix}$ becomes $[L^*_{ij}] = \begin{bmatrix} 0 & 0 & 0 \\ 0 & 1-\sqrt{2} & 0 \\ 0 & 0 & 1+\sqrt{2} \end{bmatrix}$.

Likewise from (*3.39*), $[E_{ij}] = \begin{bmatrix} 0 & 0 & 0 \\ 0 & -1 & \sqrt{2} \\ 0 & \sqrt{2} & -1 \end{bmatrix}$ which by the same transformation matrix $[a_{ij}]$

is converted into the principal-axes form $[E^*_{ij}] = \begin{bmatrix} 0 & 0 & 0 \\ 0 & -1-\sqrt{2} & 0 \\ 0 & 0 & -1+\sqrt{2} \end{bmatrix}$. The student should verify these calculations.

3.27. Using the definition (*3.37*), show that the Lagrangian finite strain tensor L_{ij} transforms as a second order Cartesian tensor under the coordinate transformations $x_i = b_{ji}x'_j$ and $X'_i = b_{ij}X_j$.

By (*3.37*), $L_{ij} = \frac{1}{2}\left(\frac{\partial x_k}{\partial X_i}\frac{\partial x_k}{\partial X_j} - \delta_{ij}\right)$ which by the stated transformation becomes

$$L_{ij} = \frac{1}{2}\left(\frac{\partial(b_{pk}x'_p)}{\partial X'_m}\frac{\partial X'_m}{\partial X_i}\frac{\partial(b_{qk}x'_q)}{\partial X'_n}\frac{\partial X'_n}{\partial X_j} - \frac{\partial x_i}{\partial x_j}\right)$$

$$= \frac{1}{2}\left(b_{mi}b_{nj}\delta_{pq}\frac{\partial x'_p}{\partial X'_m}\frac{\partial x'_q}{\partial X'_n} - \frac{\partial(b_{mi}x'_m)}{\partial x'_n}\frac{\partial x'_n}{\partial x_j}\right) \qquad \text{(since } b_{pk}b_{qk} = \delta_{pq}\text{)}$$

$$= b_{mi}b_{nj}\left[\frac{1}{2}\left(\frac{\partial x'_p}{\partial X'_m}\frac{\partial x'_p}{\partial X'_n} - \delta'_{mn}\right)\right] = b_{mi}b_{nj}L'_{mn}$$

3.28. A certain homogeneous deformation field results in the finite strain tensor $[L_{ij}] = \begin{bmatrix} 1 & 3 & -2 \\ 3 & 1 & -2 \\ -2 & -2 & 6 \end{bmatrix}$. Determine the principal strains and their directions for this deformation.

Being a symmetric second order Cartesian tensor, the principal strains are the roots of

$$\begin{vmatrix} 1-L & 3 & -2 \\ 3 & 1-L & -2 \\ -2 & -2 & 6-L \end{vmatrix} = L^3 - 8L^2 - 4L + 32 = 0$$

Thus $L_{(1)} = -2$, $L_{(2)} = 2$, $L_{(3)} = 8$. The transformation matrix for principal directions is

$$[a_{ij}] = \begin{bmatrix} 1/\sqrt{2} & -1/\sqrt{2} & 0 \\ 1/\sqrt{3} & 1/\sqrt{3} & 1/\sqrt{3} \\ -1/\sqrt{6} & -1/\sqrt{6} & 2/\sqrt{6} \end{bmatrix}$$

3.29. For the homogeneous deformation $x_1 = \sqrt{3}\,X_1$, $x_2 = 2X_2$, $x_3 = \sqrt{3}\,X_3 - X_2$ determine the material strain ellipsoid resulting from deformation of the spherical surface $X_1^2 + X_2^2 + X_3^2 = 1$. Show that this ellipsoid has the form $x_1^2/\Lambda_{(1)}^2 + x_2^2/\Lambda_{(2)}^2 + x_3^2/\Lambda_{(3)}^2 = 1$.

By (*3.82*), or alternatively by inverting the given displacement equations and substituting into $X_iX_i = 1$, the material strain ellipsoid is $x_1^2 + x_2^2 + x_3^2 + x_2x_3 = 3$. This equation is put into the principal-axes form $x_1^2/3 + x_2^2/6 + x_3^2/2 = 1$ by the transformation

$$[a_{ij}] = \begin{bmatrix} 1 & 0 & 0 \\ 0 & 1/\sqrt{2} & 1/\sqrt{2} \\ 0 & -1/\sqrt{2} & 1/\sqrt{2} \end{bmatrix}$$

From the deformation equations, the stretch tensor $\mathbf{S} = \sqrt{\mathbf{G}}$ is given (calculation is similar to that in Problem 3.24) as

$$[S_{ij}] = \begin{bmatrix} \sqrt{3} & 0 & 0 \\ 0 & \dfrac{3\sqrt{3}+1}{2\sqrt{2}} & \dfrac{\sqrt{3}-3}{2\sqrt{2}} \\ 0 & \dfrac{\sqrt{3}-3}{2\sqrt{2}} & \dfrac{+\sqrt{3}+3}{2\sqrt{2}} \end{bmatrix}$$

which by the transformation $[a_{ij}] = \begin{bmatrix} 1 & 0 & 0 \\ 0 & \sqrt{3}/2 & -1/2 \\ 0 & 1/2 & \sqrt{3}/2 \end{bmatrix}$ is put into the principal form

$$[S_{ij}^*] = \begin{bmatrix} \sqrt{3} & 0 & 0 \\ 0 & \sqrt{6} & 0 \\ 0 & 0 & \sqrt{2} \end{bmatrix}$$

with principal stretches $\Lambda_{(1)}^2 = 3$, $\Lambda_{(2)}^2 = 6$, $\Lambda_{(3)}^2 = 2$. Note also that the principal stretches may be calculated directly from (*3.66*) using $[a_{ij}]$ above.

3.30. For the deformation of Problem 3.29, determine the spatial strain ellipsoid and show that it is of the form $\Lambda_{(1)}^2 X_1^2 + \Lambda_{(2)}^2 X_2^2 + \Lambda_{(3)}^2 X_3^2 = 1$.

By (*3.84*) the sphere $x_ix_i = 1$ resulted from the ellipsoid $\mathbf{X}\cdot\mathbf{G}\cdot\mathbf{X} = 1$, or

$$[X_1, X_2, X_3]\begin{bmatrix} 3 & 0 & 0 \\ 0 & 5 & -\sqrt{3} \\ 0 & -\sqrt{3} & 3 \end{bmatrix}\begin{bmatrix} X_1 \\ X_2 \\ X_3 \end{bmatrix} = 3X_1^2 + 5X_2^2 + 3X_3^2 - 2\sqrt{3}\,X_2X_3 = 1$$

This ellipsoid is put into the principal-axes form $3X_1^2 + 6X_2^2 + 2X_3^2 = 1$ by the transformation

$$[a_{ij}] = \begin{bmatrix} 1 & 0 & 0 \\ 0 & \sqrt{3}/2 & 1/2 \\ 0 & -1/2 & \sqrt{3}/2 \end{bmatrix}$$

3.31. Verify by direct expansion that the second invariant II_L of the strain tensor may be expressed by

$$\text{II}_L = \begin{vmatrix} l_{11} & l_{12} \\ l_{21} & l_{22} \end{vmatrix} + \begin{vmatrix} l_{11} & l_{13} \\ l_{31} & l_{33} \end{vmatrix} + \begin{vmatrix} l_{22} & l_{23} \\ l_{32} & l_{33} \end{vmatrix}$$

Expansion of the given determinants results in $\text{II}_L = l_{11}l_{22} + l_{22}l_{33} + l_{33}l_{11} - (l_{12}^2 + l_{23}^2 + l_{31}^2)$. In comparison, direct expansion of the second equation of (*3.91*) yields

$$\begin{aligned} \text{II}_L &= \tfrac{1}{2}[(l_{11} + l_{22} + l_{33})l_{jj} - (l_{1j}l_{1j} + l_{2j}l_{2j} + l_{3j}l_{3j})] \\ &= \tfrac{1}{2}[(l_{11} + l_{22} + l_{33})(l_{11} + l_{22} + l_{33}) - (l_{11}l_{11} + l_{12}l_{12} + l_{13}l_{13} \\ &\qquad + l_{21}l_{21} + l_{22}l_{22} + l_{23}l_{23} + l_{31}l_{31} + l_{32}l_{32} + l_{33}l_{33})] \\ &= l_{11}l_{22} + l_{22}l_{33} + l_{33}l_{11} - (l_{12}^2 + l_{23}^2 + l_{31}^2) \end{aligned}$$

3.32. For the finite homogeneous deformation given by $u_i = A_{ij}X_j$ where A_{ij} are constants, determine an expression for the change of volume per unit original volume. If the A_{ij} are very small, show that the result reduces to the cubical dilatation.

Consider a rectangular element of volume having original dimensions dX_1, dX_2, dX_3 along the coordinate axes. For the given deformation, $x_i = (A_{ij} + \delta_{ij})X_j$. Thus by (*3.33*) the original volume dV_0 becomes a skewed parallelepiped having edge lengths $dx_i = (A_{i(n)} + \delta_{i(n)})\,dX_{(n)}$, $n = 1, 2, 3$. From (*1.109*) this deformed element has the volume $dV = \epsilon_{ijk}(A_{i1} + \delta_{i1})(A_{j2} + \delta_{j2})(A_{k3} + \delta_{k3})\,dX_1\,dX_2\,dX_3$. Then

$$\frac{dV}{dV_0} = \frac{dV_0 + \Delta V}{dV_0} = 1 + \frac{\Delta V}{dV_0} = \epsilon_{ijk}(A_{i1} + \delta_{i1})(A_{j2} + \delta_{j2})(A_{k3} + \delta_{k3})$$

If the A_{ij} are very small and their powers neglected,

$$\Delta V/dV_0 = \epsilon_{ijk}(A_{i1}\delta_{j2}\delta_{k3} + \delta_{i1}A_{j2}\delta_{k3} + \delta_{i1}\delta_{j2}A_{k3} + \delta_{i1}\delta_{j2}\delta_{k3}) - 1 = A_{11} + A_{22} + A_{33}$$

For linear theory the cubical dilatation $l_{ii} = \partial u_i/\partial X_i$, which for $u_i = A_{ij}X_j$ is $l_{ii} = A_{11} + A_{22} + A_{33}$.

3.33. A linear (small strain) deformation is specified by $u_1 = 4x_1 - x_2 + 3x_3$, $u_2 = x_1 + 7x_2$, $u_3 = -3x_1 + 4x_2 + 4x_3$. Determine the principal strains $\epsilon_{(n)}$ and the principal deviator strains $e_{(n)}$ for this deformation.

Since ϵ_{ij} is the symmetrical part of the displacement gradient $\partial u_i/\partial x_j$, it is given here by $\epsilon_{ij} = \begin{pmatrix} 4 & 0 & 0 \\ 0 & 7 & 2 \\ 0 & 2 & 4 \end{pmatrix}$ or in principal-axes form by $\epsilon_{ij}^* = \begin{pmatrix} 8 & 0 & 0 \\ 0 & 4 & 0 \\ 0 & 0 & 3 \end{pmatrix}$. Also, $\epsilon_{kk}/3 = 5$ and so the strain deviator is $e_{ij} = \begin{pmatrix} -1 & 0 & 0 \\ 0 & 2 & 2 \\ 0 & 2 & -1 \end{pmatrix}$ and its principal-axes form $e_{ij}^* = \begin{pmatrix} 3 & 0 & 0 \\ 0 & -1 & 0 \\ 0 & 0 & -2 \end{pmatrix}$. Note that $e_{(m)} = \epsilon_{(m)} - \epsilon_{kk}/3$.

PLANE STRAIN AND COMPATIBILITY (Sec. 3.15-3.16)

3.34. A 45° strain-rosette measures longitudinal strain along the axes shown in Fig. 3-15. At a point P, $\epsilon_{11} = 5 \times 10^{-4}$, $\epsilon'_{11} = 4 \times 10^{-4}$, $\epsilon_{22} = 7 \times 10^{-4}$ in/in. Determine the shear strain ϵ_{12} at the point.

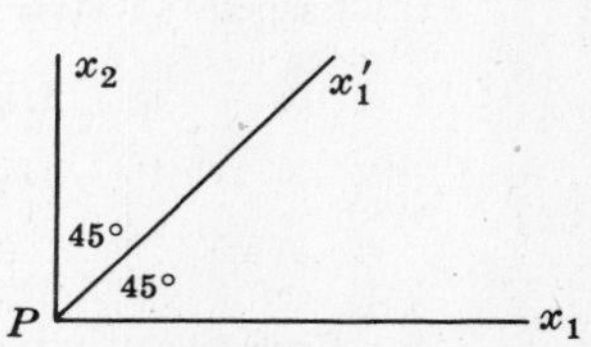

Fig. 3-15

By (*3.59*), with $\hat{\nu} = (\hat{\mathbf{e}}_1 + \hat{\mathbf{e}}_2)/\sqrt{2}$ as the unit vector in the x_1' direction,

$$[1/\sqrt{2}, 1/\sqrt{2}, 0]\begin{bmatrix} 5\times 10^{-4} & \epsilon_{12} & 0 \\ \epsilon_{12} & 7\times 10^{-4} & 0 \\ 0 & 0 & 0 \end{bmatrix}\begin{bmatrix} 1/\sqrt{2} \\ 1/\sqrt{2} \\ 0 \end{bmatrix} = 4\times 10^{-4}$$

Therefore $\dfrac{12\times 10^{-4} + 2\epsilon_{12}}{2} = 4\times 10^{-4}$ or $\epsilon_{12} = -2\times 10^{-4}$.

3.35. Construct the Mohr's circles for the case of plane strain

$$\epsilon_{ij} = \begin{pmatrix} 0 & 0 & 0 \\ 0 & 5 & \sqrt{3} \\ 0 & \sqrt{3} & 3 \end{pmatrix}$$

and determine the maximum shear strain. Verify the result analytically.

With the given state of strain referred to the x_i axes, the points $B(\epsilon_{22} = 5, \epsilon_{23} = \sqrt{3})$ and D are established as the diameter of the larger inner circle in Fig. 3-16. Since $\epsilon_{(1)} = 0$ is a principal value for plane strain, the other circles are drawn as shown.

A rotation of 30° about the x_1 axis (equivalent to 60° in the Mohr's diagram) results in the principal strain axes with the principal strain tensor ϵ^*_{ij} given by

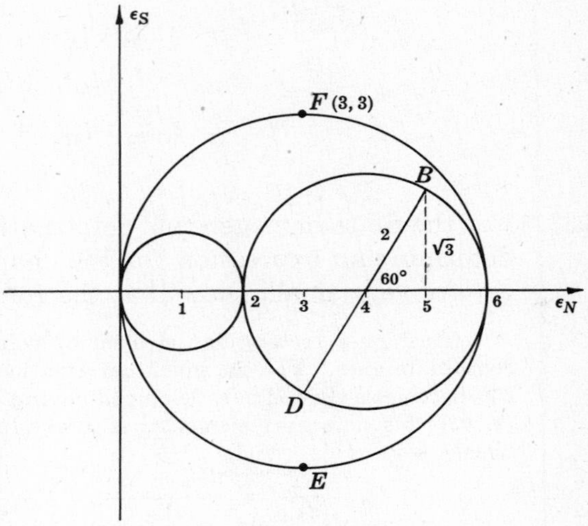

Fig. 3-16

$$\begin{bmatrix} 1 & 0 & 0 \\ 0 & \sqrt{3}/2 & 1/2 \\ 0 & -1/2 & \sqrt{3}/2 \end{bmatrix}\begin{bmatrix} 0 & 0 & 0 \\ 0 & 5 & \sqrt{3} \\ 0 & \sqrt{3} & 3 \end{bmatrix}\begin{bmatrix} 1 & 0 & 0 \\ 0 & \sqrt{3}/2 & -1/2 \\ 0 & 1/2 & \sqrt{3}/2 \end{bmatrix} = \begin{bmatrix} 0 & 0 & 0 \\ 0 & 6 & 0 \\ 0 & 0 & 2 \end{bmatrix}$$

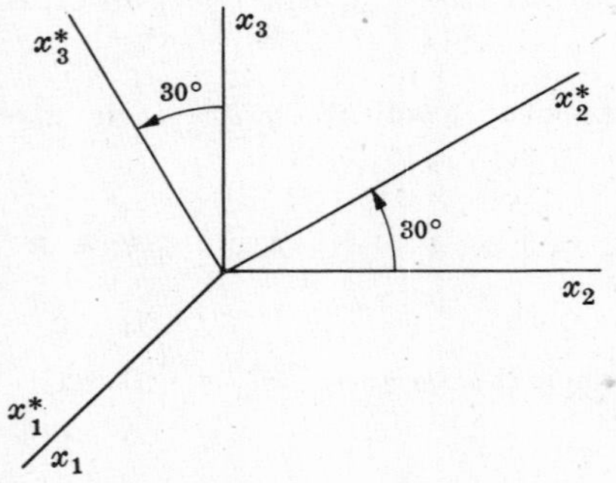

Fig. 3-17

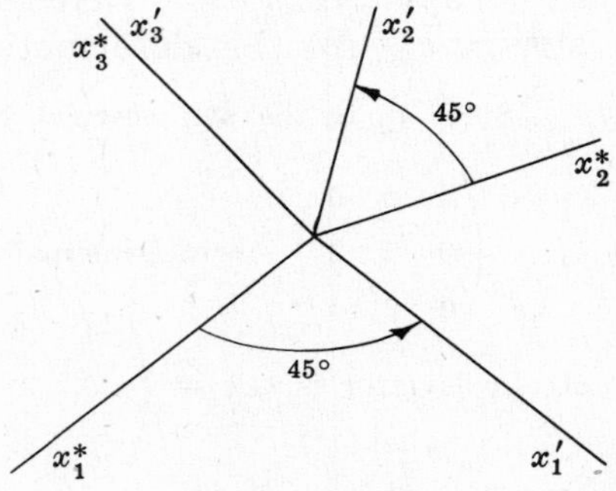

Fig. 3-18

Next a rotation of 45° about the x_3^* axis (90° in the Mohr diagram) results in the x_i' axes and the associated strain tensor ϵ'_{ij} given by

$$\begin{bmatrix} 1/\sqrt{2} & 1/\sqrt{2} & 0 \\ -1/\sqrt{2} & 1/\sqrt{2} & 0 \\ 0 & 0 & 1 \end{bmatrix}\begin{bmatrix} 0 & 0 & 0 \\ 0 & 6 & 0 \\ 0 & 0 & 2 \end{bmatrix}\begin{bmatrix} 1/\sqrt{2} & -1/\sqrt{2} & 0 \\ 1/\sqrt{2} & 1/\sqrt{2} & 0 \\ 0 & 0 & 1 \end{bmatrix} = \begin{bmatrix} 3 & 3 & 0 \\ 3 & 3 & 0 \\ 0 & 0 & 2 \end{bmatrix}$$

the first two rows of which represent the state of strain specified by point F in Fig. 3-16. Note that a rotation of $-45°$ about x_3^* would correspond to point E in Fig. 3-16.

3.36. The state of strain throughout a continuum is specified by

$$\epsilon_{ij} = \begin{pmatrix} x_1^2 & x_2^2 & x_1x_3 \\ x_2^2 & x_3 & x_3^2 \\ x_1x_3 & x_3^2 & 5 \end{pmatrix}$$

Are the compatibility equations for strain satisfied?

Substituting directly into (*3.104*), all equations are satisfied identically. The student should carry out the details.

MISCELLANEOUS PROBLEMS

3.37. Derive the indicial form of the Lagrangian finite strain tensor L_G of (*3.40*) from its definition (*3.37*).

From (*3.24*), $\partial x_i/\partial X_j = \delta_{ij} + \partial u_i/\partial X_j$. Thus (*3.37*) may be written

$$L_{ij} = \frac{1}{2}\left[\left(\delta_{ki} + \frac{\partial u_k}{\partial X_i}\right)\left(\delta_{kj} + \frac{\partial u_k}{\partial X_j}\right) - \delta_{ij}\right]$$

$$= \frac{1}{2}\left[\delta_{ki}\delta_{kj} + \delta_{ki}\frac{\partial u_k}{\partial X_j} + \delta_{kj}\frac{\partial u_k}{\partial X_i} + \frac{\partial u_k}{\partial X_i}\frac{\partial u_k}{\partial X_j} - \delta_{ij}\right]$$

$$= \frac{1}{2}\left[\frac{\partial u_i}{\partial X_j} + \frac{\partial u_j}{\partial X_i} + \frac{\partial u_k}{\partial X_i}\frac{\partial u_k}{\partial X_j}\right]$$

3.38. A displacement field is defined by $x_1 = X_1 - CX_2 + BX_3$, $x_2 = CX_1 + X_2 - AX_3$, $x_3 = -BX_1 + AX_2 + X_3$. Show that this displacement represents a rigid body rotation only if the constants A, B, C are very small. Determine the rotation vector $\mathbf{w}$ for the infinitesimal rigid body rotation.

For the given displacements, $\mathsf{F} = \begin{pmatrix} 1 & -C & B \\ C & 1 & -A \\ -B & A & 1 \end{pmatrix}$ and from (*3.37*),

$$\mathsf{L}_G = \frac{1}{2}\begin{pmatrix} B^2+C^2 & -AB & -AC \\ -AB & A^2+C^2 & -BC \\ -AC & -BC & A^2+B^2 \end{pmatrix}$$

If products of the constants are neglected, this strain tensor is zero and the displacement reduces to a rigid body rotation. From (*3.50*), the rotation vector is

$$\mathbf{w} = \frac{1}{2}\begin{vmatrix} \hat{\mathbf{e}}_1 & \hat{\mathbf{e}}_2 & \hat{\mathbf{e}}_3 \\ \partial/\partial X_1 & \partial/\partial X_2 & \partial/\partial X_3 \\ -CX_2 + BX_3 & CX_1 - AX_3 & -BX_1 + AX_2 \end{vmatrix} = A\,\hat{\mathbf{e}}_1 + B\,\hat{\mathbf{e}}_2 + C\,\hat{\mathbf{e}}_3$$

3.39. For the rigid body rotation represented by $u_1 = 0.02X_3$, $u_2 = -0.03X_3$, $u_3 = -0.02X_1 + 0.03X_2$, determine the relative displacement of $Q(3, 0.1, 4)$ with respect to $P(3, 0, 4)$.

From the displacement equations, $\mathbf{u}_Q = .08\,\hat{\mathbf{e}}_1 - .12\,\hat{\mathbf{e}}_2 - .057\,\hat{\mathbf{e}}_3$ and $\mathbf{u}_P = .08\,\hat{\mathbf{e}}_1 - .12\,\hat{\mathbf{e}}_2 - .06\,\hat{\mathbf{e}}_3$. Hence $d\mathbf{u} = \mathbf{U}_Q - \mathbf{U}_P = -.003\,\hat{\mathbf{e}}_3$. The same result is obtained by (*3.51*), with $\mathbf{w} = .03\,\hat{\mathbf{e}}_1 + .02\,\hat{\mathbf{e}}_2$:

$$d\mathbf{u} = \begin{vmatrix} \hat{\mathbf{e}}_1 & \hat{\mathbf{e}}_2 & \hat{\mathbf{e}}_3 \\ .03 & .02 & 0 \\ 0 & .1 & 0 \end{vmatrix} = -.003\,\hat{\mathbf{e}}_3$$

3.40. For a state of plane strain parallel to the x_2x_3 axes, determine expressions for the normal strain ϵ'_{22} and the shear strain ϵ'_{23} when the primed and unprimed axes are oriented as shown in Fig. 3-19.

By equation (*3.59*),

$$\epsilon'_{22} = [0, \cos\theta, \sin\theta]\begin{bmatrix} 0 & 0 & 0 \\ 0 & \epsilon_{22} & \epsilon_{23} \\ 0 & \epsilon_{23} & \epsilon_{33} \end{bmatrix}\begin{bmatrix} 0 \\ \cos\theta \\ \sin\theta \end{bmatrix}$$

$$= \epsilon_{22}\cos^2\theta + 2\epsilon_{23}\sin\theta\cos\theta + \epsilon_{33}\sin^2\theta$$

$$= \frac{\epsilon_{22}+\epsilon_{33}}{2} + \frac{\epsilon_{22}-\epsilon_{33}}{2}\cos 2\theta + \epsilon_{23}\sin 2\theta$$

Similarly from (*3.65*) and Problem 3.20,

$$\epsilon'_{23} = [0, \cos\theta, \sin\theta]\begin{bmatrix} 0 & 0 & 0 \\ 0 & \epsilon_{22} & \epsilon_{23} \\ 0 & \epsilon_{23} & \epsilon_{33} \end{bmatrix}\begin{bmatrix} 0 \\ -\sin\theta \\ \cos\theta \end{bmatrix}$$

$$= -\epsilon_{22}\sin c\cos\theta + \epsilon_{23}\cos^2\theta - \epsilon_{23}\sin^2\theta + \epsilon_{33}\sin\theta\cos\theta$$

$$= \epsilon_{23}\cos 2\theta - \frac{\epsilon_{22}-\epsilon_{33}}{2}\sin 2\theta$$

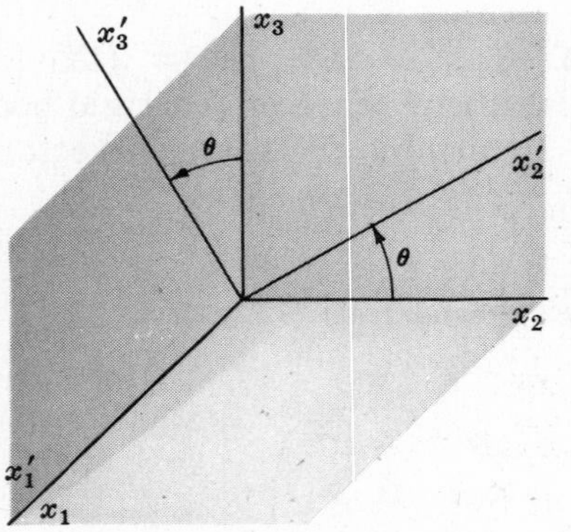

Fig. 3-19

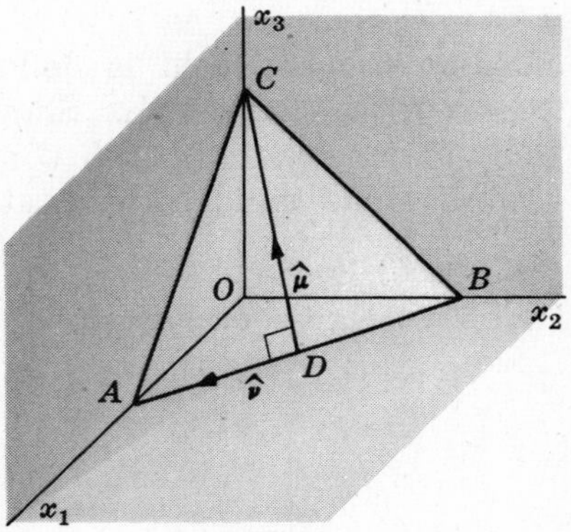

Fig. 3-20

3.41. For a homogeneous deformation the small strain tensor is given by

$$[\epsilon_{ij}] = \begin{bmatrix} 0.01 & -0.005 & 0 \\ -0.005 & 0.02 & 0.01 \\ 0 & 0.01 & -0.03 \end{bmatrix}$$

What is the change in the 90° angle *ADC* depicted by the small tetrahedron *OABC* in Fig. 3-20 if $OA = OB = OC$, and *D* is the midpoint of *AB*?

The unit vectors $\hat{\nu}$ and $\hat{\mu}$ at *D* are given by $\hat{\nu} = (\hat{\mathbf{e}}_1 - \hat{\mathbf{e}}_2)/\sqrt{2}$ and $\hat{\mu} = (2\hat{\mathbf{e}}_3 - \hat{\mathbf{e}}_2 - \hat{\mathbf{e}}_1)/\sqrt{6}$. From the result of Problem 3.20,

$$\gamma_{\mu\nu} = [1/\sqrt{2}, -1/\sqrt{2}, 0]\begin{bmatrix} .02 & -.01 & 0 \\ -.01 & .04 & .02 \\ 0 & .02 & -.06 \end{bmatrix}\begin{bmatrix} -1/\sqrt{6} \\ -1/\sqrt{6} \\ 2/\sqrt{6} \end{bmatrix} = -.01/\sqrt{3}$$

3.42. At a point the strain tensor is given by $\epsilon_{ij} = \begin{pmatrix} 5 & -1 & -1 \\ -1 & 4 & 0 \\ -1 & 0 & 4 \end{pmatrix}$ and in principal form by $\epsilon^*_{ij} = \begin{pmatrix} 6 & 0 & 0 \\ 0 & 4 & 0 \\ 0 & 0 & 3 \end{pmatrix}$. Calculate the strain invariants for each of these tensors and show their equivalence.

By (*3.95*) and Problem 3.31, $\text{I}_E = 5+4+4 = 13$, $\text{I}_{E*} = 6+4+3 = 13$. Likewise $\text{II}_E = 19+19+16 = 54$, $\text{II}_{E*} = 24+18+12 = 54$. Finally $\text{III}_E = 5(16)-4-4 = 72$, $\text{III}_{E*} = (6)(4)(3) = 72$. The student should check these calculations.

3.43. For the displacement field $x_1 = X_1 + AX_3$, $x_2 = X_2 - AX_3$, $x_3 = X_3 - AX_1 + AX_2$, determine the finite strain tensor $\mathbf{L}_G$. Show that if A is very small the displacement represents a rigid body rotation.

Since $u_1 = AX_3$, $u_2 = -AX_3$, $u_3 = -AX_1 + AX_2$, by (*3.40*),

$$2\mathbf{L}_G = \begin{pmatrix} 0 & 0 & A \\ 0 & 0 & -A \\ -A & A & 0 \end{pmatrix} + \begin{pmatrix} 0 & 0 & -A \\ 0 & 0 & A \\ A & -A & 0 \end{pmatrix} + \begin{pmatrix} A^2 & -A^2 & 0 \\ -A^2 & A^2 & 0 \\ 0 & 0 & 2A^2 \end{pmatrix} = \begin{pmatrix} A^2 & -A^2 & 0 \\ -A^2 & A^2 & 0 \\ 0 & 0 & 2A^2 \end{pmatrix}$$

If A is small so that A^2 may be neglected, $\mathbf{L}_G \equiv 0$; and by (*3.50*) the rotation vector $\mathbf{w} = A\,\hat{\mathbf{e}}_1 + A\,\hat{\mathbf{e}}_2$.

3.44. Show that the displacement field $u_1 = Ax_1 + 3x_2$, $u_2 = 3x_1 - Bx_2$, $u_3 = 5$ gives a state of plane strain and determine the relationship between A and B for which the deformation is *isochoric* (constant volume deformation).

From the displacement equations, by (*3.43*), $\epsilon_{ij} = \begin{bmatrix} A & 3 & 0 \\ 3 & -B & 0 \\ 0 & 0 & 0 \end{bmatrix}$ which is of the form of (*3.99*). From (*3.96*), the cubical dilatation is $D \equiv \epsilon_{ii} = A - B$, which is zero if $A = B$.

3.45. A so-called delta-rosette for measuring longitudinal surface strains has the shape of the equilateral triangle $\triangle$ and records normal strains $\epsilon_{11}, \epsilon'_{11}, \epsilon''_{11}$ in the directions shown in Fig. 3-21. If $\epsilon_{11} = a$, $\epsilon'_{11} = b$, $e''_{11} = c$, determine ϵ_{12} and ϵ_{22} at the point.

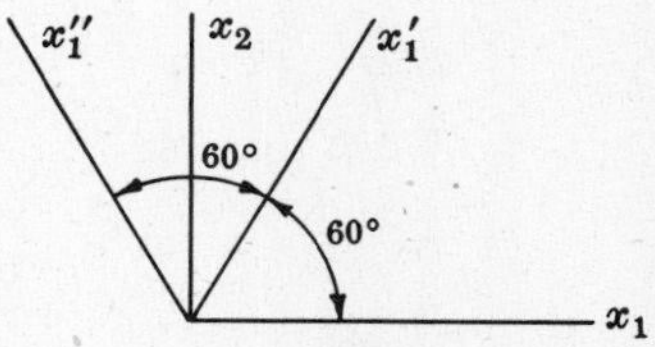

Fig. 3-21

By (*3.59*) with $\mathbf{L} = \mathbf{E}$, for the x_1' direction,

$$[1/2, \sqrt{3}/2, 0]\begin{bmatrix} a & \epsilon_{12} & 0 \\ \epsilon_{12} & \epsilon_{22} & 0 \\ 0 & 0 & 0 \end{bmatrix}\begin{bmatrix} 1/2 \\ \sqrt{3}/2 \\ 0 \end{bmatrix} = b \quad \text{or} \quad 2\sqrt{3}\,\epsilon_{12} + 3\epsilon_{22} = 4b - a$$

Also for the x_1'' direction

$$[-1/2, \sqrt{3}/2, 0]\begin{bmatrix} a & \epsilon_{12} & 0 \\ \epsilon_{12} & \epsilon_{22} & 0 \\ 0 & 0 & 0 \end{bmatrix}\begin{bmatrix} -1/2 \\ \sqrt{3}/2 \\ 0 \end{bmatrix} = c \quad \text{or} \quad -2\sqrt{3}\,\epsilon_{12} + 3\epsilon_{22} = 4c - a$$

Solving simultaneously for ϵ_{12} and ϵ_{22} yields $\epsilon_{12} = (b-c)/\sqrt{3}$ and $\epsilon_{22} = (-a+2b+2c)/3$.

3.46. Derive equation (3.72) expressing the change in angle between the coordinate directions X_2 and X_3 under a finite deformation. Show that (3.72) reduces to (3.65) when displacement gradients are small.

Let $\gamma_{23} = \pi/2 - \theta$ be the angle change as shown in Fig. 3-5. Then $\sin \gamma_{23} = \cos(\pi/2 - \theta) = \hat{\mathbf{n}}_2 \cdot \hat{\mathbf{n}}_3$, or by (3.33) and (3.34)

$$\sin \gamma_{23} = \frac{d\mathbf{x}_2}{|dx_2|} \cdot \frac{d\mathbf{x}_3}{|dx_2|} = \frac{d\mathbf{X}_2 \cdot \mathbf{F}_c \cdot \mathbf{F} \cdot d\mathbf{X}_3}{\sqrt{d\mathbf{X}_2 \cdot \mathbf{G} \cdot d\mathbf{X}_2}\sqrt{d\mathbf{X}_3 \cdot \mathbf{G} \cdot d\mathbf{X}_3}}$$

Now dividing the numerator and denominator of this equation by $|d\mathbf{X}_2|$ and $|d\mathbf{X}_3|$ and using (3.35) and (3.66) gives

$$\sin \gamma_{23} = \frac{\hat{\mathbf{e}}_2 \cdot \mathbf{G} \cdot \hat{\mathbf{e}}_3}{\sqrt{\hat{\mathbf{e}}_2 \cdot \mathbf{G} \cdot \hat{\mathbf{e}}_2}\sqrt{\hat{\mathbf{e}}_3 \cdot \mathbf{G} \cdot \hat{\mathbf{e}}_3}} = \frac{\hat{\mathbf{e}}_2 \cdot \mathbf{G} \cdot \hat{\mathbf{e}}_3}{\Lambda_{(\hat{\mathbf{e}}_2)}\Lambda_{(\hat{\mathbf{e}}_3)}}$$

Next from (3.37), $\hat{\mathbf{e}}_2 \cdot \mathbf{G} \cdot \hat{\mathbf{e}}_3 = \hat{\mathbf{e}}_2 \cdot (2\mathbf{L}_G + \mathbf{I}) \cdot \hat{\mathbf{e}}_3 = \hat{\mathbf{e}}_2 \cdot 2\mathbf{L}_G \cdot \hat{\mathbf{e}}_3 + \hat{\mathbf{e}}_2 \cdot \mathbf{I} \cdot \hat{\mathbf{e}}_3 = 2L_{23}$ since $\hat{\mathbf{e}}_2 \cdot \hat{\mathbf{e}}_3 = 0$. Also from (3.68), $\Lambda_{(\hat{\mathbf{e}}_2)} = \sqrt{1 + 2L_{22}}$, etc., and so

$$\sin \gamma_{23} = \frac{2L_{23}}{\sqrt{1+2L_{22}}\sqrt{1+2L_{33}}}$$

$$= \frac{\partial u_2/\partial X_3 + \partial u_3/\partial X_2 + (\partial u_k/\partial X_2)(\partial u_k/\partial X_3)}{\sqrt{1 + 2\partial u_2/\partial X_2 + (\partial u_k/\partial X_2)(\partial u_k/\partial X_2)}\sqrt{1 + 2\partial u_3/\partial X_3 + (\partial u_k/\partial X_3)(\partial u_k/\partial X_3)}}$$

If $\partial u_i/\partial X_j \ll 1$ this reduces to $\sin \gamma_{23} = \partial u_2/\partial X_3 + \partial u_3/\partial X_2 = 2l_{23}$.

3.47. For the simple shear displacement $x_1 = X_1$, $x_2 = X_2$, $x_3 = X_3 + 2X_2/\sqrt{3}$, determine the direction of the line element in the X_2X_3 plane for which the normal strain is zero.

Let $\hat{\mathbf{m}} = m_2\hat{\mathbf{e}}_2 + m_3\hat{\mathbf{e}}_3$ be the unit normal in the direction of zero strain. Then from (3.66), since $\Lambda^2_{(\hat{\mathbf{m}})} = 1$,

$$[0, m_2, m_3]\begin{bmatrix} 1 & 0 & 0 \\ 0 & 7/3 & 2/\sqrt{3} \\ 0 & 2/\sqrt{3} & 1 \end{bmatrix}\begin{bmatrix} 0 \\ m_2 \\ m_3 \end{bmatrix} = 1$$

or $7m_2^2 + 4\sqrt{3}\, m_2 m_3 + 3m_3^2 = 3$. Also $m_2^2 + m_3^2 = 1$, and solving simultaneously $m_2 = \pm\sqrt{3}/2$, $m_3 = \mp 1/2$, or $m_2 = 0$, $m_3 = \pm 1$. Thus there is zero strain along the X_3 axis and for the element at 60° to the X_3 axis.

The student should verify this result by using the relation $\hat{\mathbf{m}} \cdot 2\mathbf{L}_G \cdot \hat{\mathbf{m}} = 0$ derived from (3.36).

Supplementary Problems

3.48. For the shear displacement of Problem 3.47, determine the equation of the ellipse into which the circle $X_2^2 + X_3^2 = 1$ is deformed. *Ans.* $x_2^2 + 9x_3^2 = 3$

3.49. Determine the shear angle γ_{23} for the deformation of Problem 3.47 (Fig. 3-22). *Ans.* $\gamma_{23} = \sin^{-1} 2/\sqrt{7}$

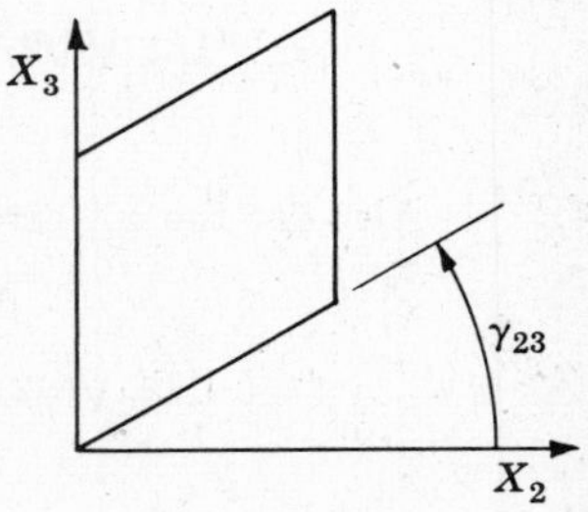

Fig. 3-22

3.50. Given the displacement field $x_1 = X_1 + 2X_3$, $x_2 = X_2 - 2X_3$, $x_3 = X_3 - 2X_1 + 2X_2$, determine the Lagrangian and Eulerian finite strain tensors $\mathbf{L}_G$ and $\mathbf{E}_A$.

Ans. $\mathbf{L}_G = \begin{pmatrix} 2 & -2 & 0 \\ -2 & 2 & 0 \\ 0 & 0 & 4 \end{pmatrix}$, $\mathbf{E}_A = \frac{1}{9}\begin{pmatrix} 2 & -2 & 0 \\ -2 & 2 & 0 \\ 0 & 0 & 4 \end{pmatrix}$

3.51. Determine the principal-axes form of the two tensors of Problem 3.50.

Ans. $\mathbf{L}_G^* = \begin{pmatrix} 4 & 0 & 0 \\ 0 & 0 & 0 \\ 0 & 0 & 4 \end{pmatrix}, \quad \mathbf{E}_A^* = \begin{pmatrix} 4/9 & 0 & 0 \\ 0 & 0 & 0 \\ 0 & 0 & 4/9 \end{pmatrix}$

3.52. For the displacement field of Problem 3.50 determine the deformation gradient $\mathbf{F}$, and by a polar decomposition of $\mathbf{F}$ find the rotation tensor $\mathbf{R}$ and the right stretch tensor $\mathbf{S}$.

Ans. $\mathbf{R} = \frac{1}{3}\begin{pmatrix} 2 & 1 & 2 \\ 1 & 2 & -2 \\ -2 & 2 & 1 \end{pmatrix}, \quad \mathbf{S} = \begin{pmatrix} 2 & -1 & 0 \\ -1 & 2 & 0 \\ 0 & 0 & 3 \end{pmatrix}, \quad \mathbf{F} = \begin{pmatrix} 1 & 0 & 2 \\ 0 & 1 & -2 \\ -2 & 2 & 1 \end{pmatrix}$

3.53. Show that the first invariant of $\mathbf{L}_G$ may be written in terms of the principal stretches as $I_{\mathbf{L}_G} = [(\Lambda^2_{(\hat{\mathbf{e}}_1)} - 1) + (\Lambda^2_{(\hat{\mathbf{e}}_2)} - 1) + (\Lambda^2_{(\hat{\mathbf{e}}_3)} - 1)]/2$. *Hint*: See equation (*3.68*).

3.54. The strain tensor at a point is given by $\epsilon_{ij} = \begin{pmatrix} 1 & -3 & \sqrt{2} \\ -3 & 1 & -\sqrt{2} \\ \sqrt{2} & -\sqrt{2} & 4 \end{pmatrix}$. Determine the normal strain in the direction of $\hat{\boldsymbol{\nu}} = \hat{\mathbf{e}}_1/2 - \hat{\mathbf{e}}_2/2 + \hat{\mathbf{e}}_3/\sqrt{2}$ and the shear strain between $\hat{\boldsymbol{\nu}}$ and $\hat{\boldsymbol{\mu}} = -\hat{\mathbf{e}}_1/2 + \hat{\mathbf{e}}_2/2 + \hat{\mathbf{e}}_3/\sqrt{2}$. *Ans.* $\epsilon_{(\hat{\boldsymbol{\nu}})} = 6, \ \gamma_{\nu\mu} = 0.$

3.55. Determine the principal-axes form of ϵ_{ij} given in Problem 3.54 and note that $\hat{\boldsymbol{\nu}}$ and $\hat{\boldsymbol{\mu}}$ of that problem are principal directions (hence $\gamma_{\nu\mu} = 0$).

Ans. $\epsilon^*_{ij} = \begin{pmatrix} 6 & 0 & 0 \\ 0 & 2 & 0 \\ 0 & 0 & -2 \end{pmatrix}$

3.56. Draw the Mohr's circle for the state of strain given in Problem 3.54 and determine the maximum shear strain value. Verify this result analytically. *Ans.* $\gamma_{max} = 4$

3.57. Using ϵ_{ij} of Problem 3.54 and ϵ^*_{ij} given in Problem 3.55, calculate the three strain invariants from each and compare the results. *Ans.* $I_E = 6, \ II_E = -4, \ III_E = -24.$

3.58. For ϵ_{ij} of Problem 3.54, determine the deviator tensor e_{ij} and calculate its principal values.

Ans. $e_{ij} = \begin{pmatrix} -1 & -3 & \sqrt{2} \\ -3 & -1 & -\sqrt{2} \\ \sqrt{2} & -\sqrt{2} & 2 \end{pmatrix}, \quad e_{ij} = \begin{pmatrix} 4 & 0 & 0 \\ 0 & 0 & 0 \\ 0 & 0 & -4 \end{pmatrix}$

3.59. A displacement field is given by $u_1 = 3x_1x_2^2, \ u_2 = 2x_3x_1, \ u_3 = x_3^2 - x_1x_2$. Determine the strain tensor ϵ_{ij} and check whether or not the compatibility conditions are satisfied.

Ans. $\epsilon_{ij} = \begin{bmatrix} 3x_2^2 & 3x_1x_2 + x_3 & -x_2/2 \\ 3x_1x_2 + x_3 & 0 & x_1/2 \\ -x_2/2 & x_1/2 & 2x_3 \end{bmatrix}$, Yes.

3.60. For a delta-strain-rosette the normal strains are found to be those shown in Fig. 3-23. Determine ϵ_{12} and ϵ_{22} at the location.
Ans. $\epsilon_{22} = 1 \times 10^{-4}, \ \epsilon_{12} = -0.2885 \times 10^{-4}$

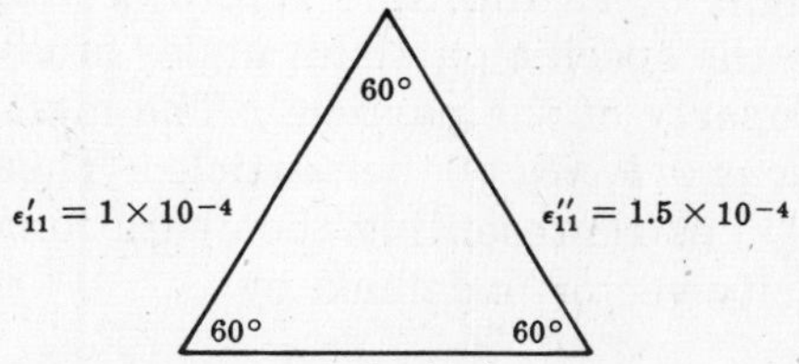

Fig. 3-23

3.61. For the displacement field $x_1 = X_1 + AX_3$, $x_2 = X_2$, $x_3 = X_3 - AX_1$, calculate the volume change and show that it is zero if A is a very small constant.

Chapter 4

Motion and Flow

4.1 MOTION. FLOW. MATERIAL DERIVATIVE

Motion and *flow* are terms used to describe the instantaneous or continuing change in configuration of a continuum. *Flow* sometimes carries the connotation of a motion leading to a permanent deformation as, for example, in plasticity studies. In fluid flow, however, the word denotes continuing motion. As indicated by (*3.14*) and (*3.15*), the motion of a continuum may be expressed either in terms of material coordinates (Lagrangian description) by

$$x_i = x_i(X_1, X_2, X_3, t) = x_i(\mathbf{X}, t) \quad \text{or} \quad \mathbf{x} = \mathbf{x}(\mathbf{X}, t) \tag{4.1}$$

or by the inverse of these equations in terms of the spatial coordinates (Eulerian description) as

$$X_i = X_i(x_1, x_2, x_3, t) = X_i(\mathbf{x}, t) \quad \text{or} \quad \mathbf{X} = \mathbf{X}(\mathbf{x}, t) \tag{4.2}$$

The necessary and sufficient condition for the inverse functions (*4.2*) to exist is that the Jacobian determinant

$$J = |\partial x_i / \partial X_j| \tag{4.3}$$

should not vanish. Physically, the Lagrangian description fixes attention on specific particles of the continuum, whereas the Eulerian description concerns itself with a particular region of the space occupied by the continuum.

Since (*4.1*) and (*4.2*) are the inverses of one another, any physical property of the continuum that is expressed with respect to a specific particle (Lagrangian, or material description) may also be expressed with respect to the particular location in space occupied by the particle (Eulerian, or spatial description). For example, if the material description of the density ρ is given by

$$\rho = \rho(X_i, t) \quad \text{or} \quad \rho = \rho(\mathbf{X}, t) \tag{4.4}$$

the spatial description is obtained by replacing $\mathbf{X}$ in this equation by the function given in (*4.2*). Thus the spatial description of the density is

$$\rho = \rho(X_i(\mathbf{x}, t), t) = \rho^*(x_i, t) \quad \text{or} \quad \rho = \rho(\mathbf{X}(\mathbf{x}, t), t) = \rho^*(\mathbf{x}, t) \tag{4.5}$$

where the symbol ρ^* is used here to emphasize that the functional form of the Eulerian description is not necessarily the same as the Lagrangian form.

The time rate of change of any property of a continuum with respect to specific particles of the moving continuum is called the *material derivative* of that property. The material derivative (also known as the *substantial*, or *comoving*, or *convective* derivative) may be thought of as the time rate of change that would be measured by an observer traveling with the specific particles under study. The instantaneous position x_i of a particle is itself a property of the particle. The material derivative of the particle's position is the *instantaneous velocity* of the particle. Therefore adopting the symbol d/dt or the superpositioned dot ($\cdot$) as representing the operation of material differentiation (some books use D/Dt), the velocity vector is defined by

$$v_i = dx_i/dt = \dot{x}_i \quad \text{or} \quad \mathbf{v} = d\mathbf{x}/dt = \dot{\mathbf{x}} \tag{4.6}$$

In general, if $P_{ij\ldots}$ is any scalar, vector or tensor property of a continuum that may be expressed as a point function of the coordinates, and if the Lagrangian description is given by

$$P_{ij\ldots} = P_{ij\ldots}(\mathbf{X}, t) \tag{4.7}$$

the material derivative of the property is expressed by

$$\frac{dP_{ij\ldots}}{dt} = \frac{\partial P_{ij\ldots}(\mathbf{X}, t)}{\partial t} \tag{4.8}$$

The right-hand side of (*4.8*) is sometimes written $\left[\frac{\partial P_{ij\ldots}(\mathbf{X}, t)}{\partial t}\right]_{\mathbf{X}}$ to emphasize that the **X** coordinates are held constant, i.e. the same particles are involved, in taking the derivative. When the property $P_{ij\ldots}$ is expressed by the spatial description in the form

$$P_{ij\ldots} = P_{ij\ldots}(\mathbf{x}, t) \tag{4.9}$$

the material derivative is given by

$$\frac{dP_{ij\ldots}(\mathbf{x}, t)}{dt} = \frac{\partial P_{ij\ldots}(\mathbf{x}, t)}{\partial t} + \frac{\partial P_{ij\ldots}(\mathbf{x}, t)}{\partial x_k}\frac{dx_k}{dt} \tag{4.10}$$

where the second term on the right arises because the specific particles are changing position in space. The first term on the right of (*4.10*) gives the rate of change at a particular location and is accordingly called the *local rate of change*. This term is sometimes written $\left[\frac{\partial P_{ij\ldots}(\mathbf{x}, t)}{\partial t}\right]_{\mathbf{x}}$ to emphasize that **x** is held constant in this differentiation. The second term on the right in (*4.10*) is called the *convective rate of change* since it expresses the contribution due to the motion of the particles in the variable field of the property.

From (*4.6*), the material derivative (*4.10*) may be written

$$\frac{dP_{ij\ldots}(\mathbf{x}, t)}{dt} = \frac{\partial P_{ij\ldots}(\mathbf{x}, t)}{\partial t} + v_k \frac{\partial P_{ij\ldots}(\mathbf{x}, t)}{\partial x_k} \tag{4.11}$$

which immediately suggests the introduction of the *material derivative operator*

$$\frac{d}{dt} = \frac{\partial}{\partial t} + v_k \frac{\partial}{\partial x_k} \quad \text{or} \quad \frac{d}{dt} = \frac{\partial}{\partial t} + \mathbf{v} \cdot \nabla_{\mathbf{x}} \tag{4.12}$$

which is used in taking the material derivatives of quantities expressed in spatial coordinates.

4.2 VELOCITY. ACCELERATION. INSTANTANEOUS VELOCITY FIELD

One definition of the velocity vector is given by (*4.6*) as $v_i = dx_i/dt$ (or $\mathbf{v} = d\mathbf{x}/dt$). An alternative definition of the same vector may be obtained from (*3.11*) which gives $x_i = u_i + X_i$ (or $\mathbf{x} = \mathbf{u} + \mathbf{X}$). Thus the velocity may be defined by

$$v_i \equiv \frac{dx_i}{dt} = \frac{d(u_i + X_i)}{dt} = \frac{du_i}{dt} \quad \text{or} \quad \mathbf{v} = \frac{d\mathbf{x}}{dt} = \frac{d(\mathbf{u} + \mathbf{X})}{dt} = \frac{d\mathbf{u}}{dt} \tag{4.13}$$

since **X** is independent of time. In (*4.13*), if the displacement is expressed in the Lagrangian form $u_i = u_i(\mathbf{X}, t)$, then

$$v_i \equiv \dot{u}_i = \frac{du_i(\mathbf{X}, t)}{dt} = \frac{\partial u_i(\mathbf{X}, t)}{\partial t} \quad \text{or} \quad \mathbf{v} = \dot{\mathbf{u}} = \frac{d\mathbf{u}(\mathbf{X}, t)}{dt} = \frac{\partial \mathbf{u}(\mathbf{X}, t)}{\partial t} \tag{4.14}$$

If, on the other hand, the displacement is in the Eulerian form $u_i = u_i(\mathbf{x}, t)$, then

$$v_i(\mathbf{x}, t) \equiv \dot{u}_i(\mathbf{x}, t) \equiv \frac{du_i(\mathbf{x}, t)}{dt} = \frac{\partial u_i(\mathbf{x}, t)}{\partial t} + v_k(\mathbf{x}, t)\frac{\partial u_i(\mathbf{x}, t)}{\partial x_k}$$

or

$$\mathbf{v}(\mathbf{x}, t) \equiv \dot{\mathbf{u}}(\mathbf{x}, t) \equiv \frac{d\mathbf{u}(\mathbf{x}, t)}{dt} = \frac{\partial \mathbf{u}(\mathbf{x}, t)}{\partial t} + \mathbf{v}(\mathbf{x}, t) \cdot \nabla_{\mathbf{x}}\, \mathbf{u}(\mathbf{x}, t) \tag{4.15}$$

In (*4.15*) the velocity is given implicitly since it appears as a factor of the second term on the right. The function

$$v_i = v_i(\mathbf{x}, t) \quad \text{or} \quad \mathbf{v} = \mathbf{v}(\mathbf{x}, t) \tag{4.16}$$

is said to specify the *instantaneous velocity field.*

The material derivative of the velocity is the *acceleration.* If the velocity is given in the Lagrangian form (*4.14*), then

$$a_i \equiv \dot{v}_i \equiv \frac{dv_i(\mathbf{X}, t)}{dt} = \frac{\partial v_i(\mathbf{X}, t)}{\partial t} \quad \text{or} \quad \mathbf{a} \equiv \dot{\mathbf{v}} \equiv \frac{d\mathbf{v}(\mathbf{X}, t)}{dt} = \frac{\partial \mathbf{v}(\mathbf{X}, t)}{\partial t} \tag{4.17}$$

If the velocity is given in the Eulerian form (*4.15*), then

$$a_i(x, t) \equiv \frac{dv_i(x, t)}{dt} = \frac{\partial v_i(\mathbf{x}, t)}{\partial t} + v_k(\mathbf{x}, t)\frac{\partial v_i(\mathbf{x}, t)}{\partial x_k}$$

or

$$\mathbf{a}(\mathbf{x}, t) \equiv \frac{d\mathbf{v}(\mathbf{x}, t)}{dt} = \frac{\partial \mathbf{v}(\mathbf{x}, t)}{\partial t} + \mathbf{v}(\mathbf{x}, t) \cdot \nabla_{\mathbf{x}}\, \mathbf{v}(\mathbf{x}, t) \tag{4.18}$$

4.3 PATH LINES. STREAM LINES. STEADY MOTION

A *path line* is the curve or path followed by a particle during motion or flow. A *stream line* is the curve whose tangent at any point is in the direction of the velocity at that point. The motion of a continuum is termed *steady motion* if the velocity field is independent of time so that $\partial v_i/\partial t = 0$. For steady motion, stream lines and path lines coincide.

4.4 RATE OF DEFORMATION. VORTICITY. NATURAL STRAIN INCREMENTS

The spatial gradient of the instantaneous velocity field defines the *velocity gradient tensor,* $\partial v_i/\partial x_j$ (or Y_{ij}). This tensor may be decomposed into its symmetric and skew-symmetric parts according to

$$Y_{ij} \equiv \frac{\partial v_i}{\partial x_j} = \frac{1}{2}\left(\frac{\partial v_i}{\partial x_j} + \frac{\partial v_j}{\partial x_i}\right) + \frac{1}{2}\left(\frac{\partial v_i}{\partial x_j} - \frac{\partial v_j}{\partial x_i}\right) = D_{ij} + V_{ij}$$

or

$$\mathbf{Y} = \tfrac{1}{2}(\mathbf{v}\nabla_{\mathbf{x}} + \nabla_{\mathbf{x}}\,\mathbf{v}) + \tfrac{1}{2}(\mathbf{v}\nabla_{\mathbf{x}} - \nabla_{\mathbf{x}}\,\mathbf{v}) = (\mathbf{D} + \mathbf{V}) \tag{4.19}$$

This decomposition is valid even if v_i and $\partial v_i/\partial x_j$ are finite quantities. The symmetric tensor

$$D_{ij} = D_{ji} = \frac{1}{2}\left(\frac{\partial v_i}{\partial x_j} + \frac{\partial v_j}{\partial x_i}\right) \quad \text{or} \quad \mathbf{D} = \tfrac{1}{2}(\mathbf{v}\nabla_{\mathbf{x}} + \nabla_{\mathbf{x}}\,\mathbf{v}) \tag{4.20}$$

is called the *rate of deformation tensor.* Many other names are used for this tensor; among them *rate of strain, stretching, strain rate* and *velocity strain* tensor. The skew-symmetric tensor

$$V_{ij} = -V_{ji} = \frac{1}{2}\left(\frac{\partial v_i}{\partial x_j} - \frac{\partial v_j}{\partial x_i}\right) \quad \text{or} \quad \mathbf{V} = \tfrac{1}{2}(\mathbf{v}\nabla_{\mathbf{x}} - \nabla_{\mathbf{x}}\,\mathbf{v}) \tag{4.21}$$

is called the *vorticity* or *spin* tensor.

The rate of deformation tensor is easily shown to be the material derivative of the Eulerian linear strain tensor. Thus if in the equation

$$\frac{d\epsilon_{ij}}{dt} = \frac{1}{2}\frac{d}{dt}\left(\frac{\partial u_i}{\partial x_j} + \frac{\partial u_j}{\partial x_i}\right) \quad \text{or} \quad \frac{d\mathbf{E}}{dt} = \frac{1}{2}\frac{d}{dt}(\mathbf{u}\nabla_{\mathbf{x}} + \nabla_{\mathbf{x}}\mathbf{u}) \tag{4.22}$$

the differentiations with respect to the coordinates and time are interchanged, i.e. if $\frac{d}{dt}\left(\frac{\partial u_i}{\partial x_j}\right)$ is replaced by $\frac{\partial}{\partial x_j}\left(\frac{du_i}{dt}\right)$, the equation takes the form

$$\frac{d\epsilon_{ij}}{dt} = \frac{1}{2}\left(\frac{\partial v_i}{\partial x_j} + \frac{\partial v_j}{\partial x_i}\right) = D_{ij} \quad \text{or} \quad \frac{d\mathbf{E}}{dt} = \tfrac{1}{2}(\mathbf{v}\nabla_{\mathbf{x}} + \nabla_{\mathbf{x}}\mathbf{v}) = \mathbf{D} \tag{4.23}$$

By the same procedure the vorticity tensor may be shown to be equal to the material derivative of the Eulerian linear rotation tensor. The result is expressed by the equation

$$\frac{d\omega_{ij}}{dt} = \frac{1}{2}\left(\frac{\partial v_i}{\partial x_j} - \frac{\partial v_j}{\partial x_i}\right) = V_{ij} \quad \text{or} \quad \frac{d\mathbf{\Omega}}{dt} = \tfrac{1}{2}(\mathbf{v}\nabla_{\mathbf{x}} - \nabla_{\mathbf{x}}\mathbf{v}) = \mathbf{V} \tag{4.24}$$

A rather interesting interpretation may be attached to (*4.23*) when that equation is rewritten in the form

$$d\epsilon_{ij} = D_{ij}\,dt \quad \text{or} \quad d\mathbf{E} = \mathbf{D}\,dt \tag{4.25}$$

The left hand side of (*4.25*) represents the components known as the *natural strain increments,* widely used in flow problems in the theory of plasticity (see Chapter 8).

4.5 PHYSICAL INTERPRETATION OF RATE OF DEFORMATION AND VORTICITY TENSORS

In Fig. 4-1 the velocities of the neighboring particles at points P and Q in a moving continuum are given by v_i and $v_i + dv_i$ respectively. The relative velocity of the particle at Q with respect to the one at P is therefore

$$dv_i = \frac{\partial v_i}{\partial x_j}dx_j \quad \text{or} \quad d\mathbf{v} = \mathbf{v}\nabla_{\mathbf{x}} \cdot d\mathbf{x} \tag{4.26}$$

in which the partial derivatives are to be evaluated at P. In terms of D_{ij} and V_{ij}, (*4.26*) becomes

$$dv_i = (D_{ij} + V_{ij})\,dx_j$$

or

$$d\mathbf{v} = (\mathbf{D} + \mathbf{V}) \cdot d\mathbf{x} \tag{4.27}$$

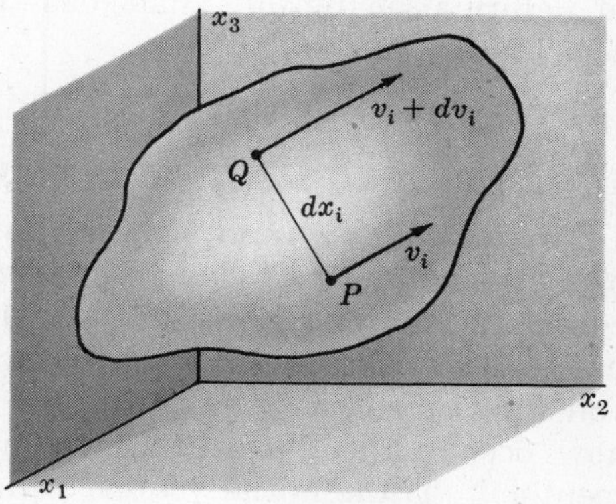

Fig. 4-1

If the rate of deformation tensor is identically zero ($D_{ij} \equiv 0$),

$$dv_i = V_{ij}\,dx_j \quad \text{or} \quad d\mathbf{v} = \mathbf{V} \cdot d\mathbf{x} \tag{4.28}$$

and the motion in the neighborhood of P is a *rigid body rotation.* For this reason a velocity field is said to be *irrotational* if the vorticity tensor vanishes everywhere within the field.

Associated with the vorticity tensor, the vector defined by

$$q_i = \epsilon_{ijk}v_{k,j} \quad \text{or} \quad \mathbf{q} = \nabla_{\mathbf{x}} \times \mathbf{v} \tag{4.29}$$

is known as the *vorticity vector.* The symbolic form of (*4.29*) shows that the vorticity vector is the curl of the velocity field. The vector defined as one-half the vorticity vector,

$$\Omega_i = \tfrac{1}{2}q_i = \tfrac{1}{2}\epsilon_{ijk}v_{k,j} \qquad \text{or} \qquad \mathbf{\Omega} = \tfrac{1}{2}\mathbf{q} = \tfrac{1}{2}\nabla_{\mathbf{x}} \times \mathbf{v} \tag{4.30}$$

is called the *rate of rotation* vector. For a rigid body rotation such as that described by (*4.28*), the relative velocity of a neighboring particle separated from P by dx_i is given as

$$dv_i = \epsilon_{ijk}\Omega_j\,dx_k \qquad \text{or} \qquad d\mathbf{v} = \mathbf{\Omega} \times d\mathbf{x} \tag{4.31}$$

The components of the rate of deformation tensor have the following physical interpretations. The diagonal elements of D_{ij} are known as the *stretching* or *rate of extension* components. Thus for pure deformation, from (*4.27*),

$$dv_i = D_{ij}\,dx_j \qquad \text{or} \qquad d\mathbf{v} = \mathbf{D} \cdot d\mathbf{x} \tag{4.32}$$

and, since the rate of change of length of the line element dx_i per unit instantaneous length is given by

$$d_i^{(\nu)} = \frac{dv_i}{dx} = D_{ij}\frac{dx_j}{dx} = D_{ij}\nu_j \qquad \text{or} \qquad \mathbf{d}^{(\hat{\nu})} = \mathbf{D} \cdot \hat{\boldsymbol{\nu}} \tag{4.33}$$

the *rate of deformation* in the direction of the unit vector ν_i is

$$d = d_i^{(\nu)}\nu_i = D_{ij}\nu_j\nu_i \qquad \text{or} \qquad d = \hat{\boldsymbol{\nu}} \cdot \mathbf{D} \cdot \hat{\boldsymbol{\nu}} \tag{4.34}$$

From (*4.34*), if ν_i is in the direction of a coordinate axis, say $\hat{\mathbf{e}}_2$,

$$d = d_{22} \qquad \text{or} \qquad d = \hat{\mathbf{e}}_2 \cdot \mathbf{D} \cdot \hat{\mathbf{e}}_2 = D_{22} \tag{4.35}$$

The off-diagonal elements of D_{ij} are *shear rates*, being a measure of the rate of change between directions at right angles (See Problem 4.18).

Since D_{ij} is a symmetric, second-order tensor, the concepts of *principal axes, principal values, invariants,* a *rate of deformation quadric,* and a *rate of deformation deviator tensor* may be associated with it. Also, *equations of compatibility* for the components of the rate of deformation tensor, analogous to those presented in Chapter 3 for the linear strain tensors may be developed.

4.6 MATERIAL DERIVATIVES OF VOLUME, AREA AND LINE ELEMENTS

In the motion from some initial configuration at time $t = 0$ to the present configuration at time t, the continuum particles which occupied the differential volume element dV_0 in the initial state now occupy the differential volume element dV. If the initial volume element is taken as the rectangular parallelepiped shown in Fig. 4-2, then by (*1.10*)

$$\begin{aligned} dV &= dX_1\hat{\mathbf{e}}_1 \times dX_2\hat{\mathbf{e}}_2 \cdot dX_3\hat{\mathbf{e}}_3 \\ &= dX_1\,dX_2\,dX_3 \end{aligned} \tag{4.36}$$

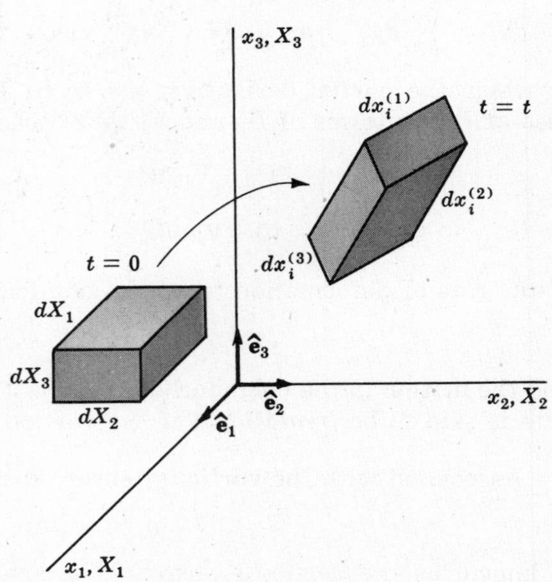

Fig. 4-2

Due to the motion, this parallelepiped is moved and distorted, but because the motion is assumed continuous the volume element does not break up. In fact, because of the relationship (*3.33*), $dx_i = (\partial x_i/\partial X_j)\,dX_j$ between the material and spatial line elements, the "line of particles" that formed dX_1 now form the differential line segment $dx_i^{(1)} = (\partial x_i/\partial X_1)\,dX_1$. Similarly dX_2 becomes $dx_i^{(2)} = (\partial x_i/\partial X_2)\,dX_2$ and dX_3 becomes $dx_i^{(3)} = (\partial x_i/\partial X_3)\,dX_3$. Therefore the differential volume element dV is a skewed parallelepiped having edges $dx_i^{(1)}$, $dx_i^{(2)}$, $dx_i^{(3)}$ and a volume given by the box product

$$dV \;=\; d\mathbf{x}^{(1)} \times d\mathbf{x}^{(2)} \cdot d\mathbf{x}^{(3)} \;=\; \epsilon_{ijk}\, dx_i^{(1)}\, dx_j^{(2)}\, dx_k^{(3)} \tag{4.37}$$

But (4.37) is seen to be equal to

$$dV \;=\; \epsilon_{ijk} \frac{\partial x_i}{\partial X_1}\frac{\partial x_j}{\partial X_2}\frac{\partial x_k}{\partial X_3}\, dX_1\, dX_2\, dX_3 \;=\; J\, dV_0 \tag{4.38}$$

where $J = |\partial x_i/\partial X_j|$ is the Jacobian defined by (4.3).

Using (4.38), it is now possible to obtain the material derivative of dV as

$$\frac{d}{dt}(dV) \;=\; \frac{d}{dt}(J\,dV_0) \;=\; \frac{dJ}{dt}\, dV_0 \tag{4.39}$$

since dV_0 is time independent, so that $\dfrac{d}{dt}(dV_0) = 0$. The material derivative of the Jacobian J may be shown to be (see Problem 4.28)

$$\frac{dJ}{dt} = J\frac{\partial v_i}{\partial x_i} \qquad \text{or} \qquad \dot{J} = J\nabla_{\mathbf{x}} \cdot \mathbf{v} \tag{4.40}$$

and hence (4.39) may be put into the form

$$\frac{d}{dt}(dV) \;=\; \frac{\partial v_i}{\partial x_i}\, dV \qquad \text{or} \qquad \frac{d}{dt}(dV) \;=\; \nabla_{\mathbf{x}} \cdot \mathbf{v}\, dV \tag{4.41}$$

For the initial configuration of a continuum, a differential element of area having the magnitude dS^0 may be represented in terms of its unit normal vector n_i by the expression $dS^0 n_i$. For the current configuration of the continuum in motion, the particles initially making up the area $dS^0 n_i$ now fill an area element represented by the vector $dS\, n_i$ or dS_i. It may be shown that

$$dS_i \;=\; J\frac{\partial X_j}{\partial x_i}\, dX_j \qquad \text{or} \qquad d\mathbf{S} \;=\; J\, d\mathbf{X} \cdot \mathbf{X}\nabla_{\mathbf{x}} \tag{4.42}$$

from which the material derivative of the element of area is

$$\frac{d}{dt}(dS_i) \;=\; \frac{d}{dt}\left(J\frac{\partial X_j}{\partial x_i}\right) dX_j \;=\; \frac{\partial v_j}{\partial x_j}\, dS_i \;-\; \frac{\partial v_j}{\partial x_i}\, dS_j \tag{4.43}$$

The material derivative of the squared length of the differential line element dx_i may be calculated as follows,

$$\frac{d}{dt}(dx^2) \;=\; \frac{d}{dt}(dx_i\, dx_i) \;=\; 2\frac{d(dx_i)}{dt}\, dx_i \tag{4.44}$$

However, since $dx_i = (\partial x_i/\partial X_j)\, dX_j$,

$$\frac{d}{dt}(dx_i) \;=\; \frac{d}{dt}\left(\frac{\partial x_i}{\partial X_j}\right) dX_j \;=\; \frac{\partial v_i}{\partial X_j}\, dX_j \;=\; \frac{\partial v_i}{\partial x_k}\frac{\partial x_k}{\partial X_j}\, dX_j \;=\; \frac{\partial v_i}{\partial x_k}\, dx_k \tag{4.45}$$

and (4.44) becomes

$$\frac{d}{dt}(dx^2) \;=\; 2\frac{\partial v_i}{\partial x_k}\, dx_k\, dx_i \qquad \text{or} \qquad \frac{d}{dt}(dx^2) \;=\; 2\, d\mathbf{x} \cdot \nabla_{\mathbf{x}}\mathbf{v} \cdot d\mathbf{x} \tag{4.46}$$

The expression on the right-hand side in the indicial form of (4.46) is symmetric in i and k, and accordingly may be written

$$\frac{\partial v_i}{\partial x_k}\, dx_k\, dx_i \;+\; \frac{\partial v_k}{\partial x_i}\, dx_i\, dx_k \;=\; \left(\frac{\partial v_i}{\partial x_k} + \frac{\partial v_k}{\partial x_i}\right) dx_i\, dx_k \tag{4.47}$$

or, from (4.20),

$$\frac{d}{dt}(dx^2) = 2D_{ij}\,dx_i\,dx_j \quad \text{or} \quad \frac{d}{dt}(dx^2) = 2\,d\mathbf{x}\cdot\mathbf{D}\cdot d\mathbf{x} \tag{4.48}$$

4.7 MATERIAL DERIVATIVES OF VOLUME, SURFACE AND LINE INTEGRALS

Not all properties of a continuum may be defined for a specific particle as functions of the coordinates such as those given by (4.7) and (4.9). Some properties are defined as integrals over a finite portion of the continuum. In particular, let any scalar, vector or tensor property be represented by the volume integral

$$P_{ij\ldots}(t) = \int_V P^*_{ij\ldots}(\mathbf{x},t)\,dV \tag{4.49}$$

where V is the volume that the considered part of the continuum occupies at time t. The material derivative of $P_{ij\ldots}(t)$ is

$$\frac{d}{dt}[P_{ij\ldots}(t)] = \frac{d}{dt}\int_V P^*_{ij\ldots}(\mathbf{x},t)\,dV \tag{4.50}$$

and since the differentiation is with respect to a definite portion of the continuum (i.e. a specific mass system), the operations of differentiation and integration may be interchanged. Therefore

$$\frac{d}{dt}\int_V P^*_{ij\ldots}(\mathbf{x},t)\,dV = \int_V \frac{d}{dt}[P^*_{ij\ldots}(\mathbf{x},t)\,dV] \tag{4.51}$$

which, upon carrying out the differentiation and using (4.41), results in

$$\frac{d}{dt}\int_V P^*_{ij\ldots}(\mathbf{x},t)\,dV = \int_V \left[\frac{dP^*_{ij\ldots}(\mathbf{x},t)}{dt} + P^*_{ij\ldots}(\mathbf{x},t)\frac{\partial v_p}{\partial x_p}\right]dV \tag{4.52}$$

Since the material derivative operator is given by (4.12) as $d/dt = \partial/\partial t + v_p\,\partial/\partial x_p$, (4.52) may be put into the form

$$\frac{d}{dt}\int_V P^*_{ij\ldots}(\mathbf{x},t)\,dV = \int_V \left[\frac{\partial P^*_{ij\ldots}(\mathbf{x},t)}{\partial t} + \frac{\partial}{\partial x_p}(v_p P^*_{ij\ldots}(\mathbf{x},t))\right]dV \tag{4.53}$$

By using Gauss' theorem (1.157), the second term of the right-hand integral of (4.53) may be converted to a surface integral, and the material derivative then given by

$$\frac{d}{dt}\int_V P^*_{ij\ldots}(\mathbf{x},t)\,dV = \int_V \frac{\partial P^*_{ij\ldots}(\mathbf{x},t)}{\partial t}\,dV + \int_S v_p[P^*_{ij\ldots}(\mathbf{x},t)]\,dS_p \tag{4.54}$$

This equation states that the rate of increase of the property $P_{ij\ldots}(t)$ in that portion of the continuum instantaneously occupying V is equal to the sum of the amount of the property created within V plus the flux $v_p[P^*_{ij\ldots}(\mathbf{x},t)]$ through the bounding surface S of V.

The procedure for determining the material derivatives of surface and line integrals is essentially the same as that used above for the volume integral. Thus for any tensorial property of a continuum represented by the surface integral

$$Q_{ij\ldots}(t) = \int_S Q^*_{ij\ldots}(\mathbf{x},t)\,dS_p \tag{4.55}$$

where S is the surface occupied by the considered part of the continuum at time t, then, as before,

$$\frac{d}{dt}\int_S Q^*_{ij\ldots}(\mathbf{x},t)\,dS_p = \int_S \frac{d}{dt}[Q^*_{ij\ldots}(\mathbf{x},t)\,dS_p] \tag{4.56}$$

and, from (4.43), the differentiation in (4.56) yields

$$\frac{dQ_{ij\ldots}(t)}{dt} = \int_S \left[\frac{dQ^*_{ij\ldots}(\mathbf{x},t)}{dt} + \frac{\partial v_q}{\partial x_q} Q^*_{ij\ldots}(\mathbf{x},t)\right] dS_p - \int_S \left[Q^*_{ij\ldots}\frac{\partial v_p}{\partial x_q} dS_p\right] \tag{4.57}$$

For properties expressed in line integral form such as

$$R_{ij\ldots}(t) = \int_C R^*_{ij\ldots}(\mathbf{x},t)\, dx_p \tag{4.58}$$

the material derivative is given by

$$\frac{d}{dt}\int_C R^*_{ij\ldots}(\mathbf{x},t)\, dx_p = \int_C \frac{d}{dt}\left[R^*_{ij\ldots}(\mathbf{x},t)\, dx_p\right] \tag{4.59}$$

Differentiating the right hand integral as indicated in (4.59), and making use of (4.45), results in the material derivative

$$\frac{d}{dt}\left[R_{ij\ldots}(t)\right] = \int_C \frac{d[R^*_{ij\ldots}(\mathbf{x},t)]}{dt}\, dx_p + \int_C \frac{\partial v_p}{\partial x_q}\left[R^*_{ij\ldots}(\mathbf{x},t)\right] dx_q \tag{4.60}$$

Solved Problems

MATERIAL DERIVATIVES. VELOCITY. ACCELERATION (Sec. 4.1-4.3)

4.1. The spatial (Eulerian) description of a continuum motion is given by $x_1 = X_1 e^t + X_3(e^t - 1)$, $x_2 = X_3(e^t - e^{-t}) + X_2$, $x_3 = X_3$. Show that the Jacobian J does not vanish for this motion and determine the material (Lagrangian) description by inverting the displacement equations.

By (4.3) the Jacobian determinant is

$$J = |\partial x_i/\partial X_j| = \begin{vmatrix} e^t & 0 & e^t - 1 \\ 0 & 1 & e^t - e^{-t} \\ 0 & 0 & 1 \end{vmatrix} = e^t$$

Inverting the motion equations, $X_1 = x_1 e^{-t} + x_3(e^{-t} - 1)$, $X_2 = x_2 - x_3(e^t - e^{-t})$, $X_3 = x_3$. Note that in each description when $t = 0$, $x_i = X_i$.

4.2. A continuum motion is expressed by $x_1 = X_1$, $x_2 = e^t(X_2 + X_3)/2 + e^{-t}(X_2 - X_3)/2$, $x_3 = e^t(X_2 + X_3)/2 - e^{-t}(X_2 - X_3)/2$. Determine the velocity components in both their material and spatial forms.

From the second and third equations, $X_2 + X_3 = e^{-t}(x_2 + x_3)$ and $X_2 - X_3 = e^t(x_2 - x_3)$. Solving these simultaneously the inverse equations become $X_1 = x_1$, $X_2 = e^{-t}(x_2 + x_3)/2 + e^t(x_2 - x_3)/2$, $X_3 = e^{-t}(x_2 + x_3)/2 - e^t(x_2 - x_3)/2$. Accordingly, the displacement components $u_i = x_i - X_i$ may be written in either the Lagrangian form $u_1 = 0$, $u_2 = e^t(X_2 + X_3)/2 + e^{-t}(X_2 - X_3)/2 - X_2$, $u_3 = e^t(X_2 + X_3)/2 - e^{-t}(X_2 - X_3)/2 - X_3$, or in the Eulerian form $u_1 = 0$, $u_2 = x_2 - e^{-t}(x_2 + x_3)/2 - e^t(x_2 - x_3)/2$, $u_3 = x_3 - e^{-t}(x_2 + x_3)/2 + e^t(x_2 - x_3)/2$.

By (4.14), $v_i = \partial u_i/\partial t = \partial X_i/\partial t$ and the velocity components in Lagrangian form are $v_1 = 0$, $v_2 = e^t(X_2 + X_3)/2 - e^{-t}(X_2 - X_3)/2$, $v_3 = e^t(X_2 + X_3)/2 + e^{-t}(X_2 - X_3)/2$. Using the relationships $X_2 + X_3 = e^{-t}(x_2 + x_3)$ and $X_2 - X_3 = e^t(x_2 - x_3)$ these components reduce to $v_1 = 0$, $v_2 = x_3$, $v_3 = x_2$.

Also, from (4.15), for the Eulerian case,

$$du_2/dt \;=\; v_2 \;=\; e^{-t}(x_2+x_3)/2 \;-\; e^{t}(x_2-x_3)/2 \;+\; v_2(2-e^{-t}-e^{t})/2 \;+\; v_3(-e^{-t}+e^{t})/2$$

$$du_3/dt \;=\; v_3 \;=\; e^{-t}(x_2+x_3)/2 \;+\; e^{t}(x_2-x_3)/2 \;+\; v_2(-e^{-t}+e^{t})/2 \;+\; v_3(2-e^{-t}-e^{t})/2$$

Solving these equations simultaneously for v_2 and v_3, the result is as before $v_2 = x_3$, $v_3 = x_2$.

4.3. **A velocity field is described by $v_1 = x_1/(1+t)$, $v_2 = 2x_2/(1+t)$, $v_3 = 3x_3/(1+t)$. Determine the acceleration components for this motion.**

By (4.18),

$$dv_1/dt \;=\; a_1 \;=\; -x_1/(1+t)^2 + x_1/(1+t)^2 \;=\; 0$$

$$dv_2/dt \;=\; a_2 \;=\; -2x_2/(1+t)^2 + 4x_2/(1+t)^2 \;=\; 2x_2/(1+t)^2$$

$$dv_3/dt \;=\; a_3 \;=\; -3x_3/(1+t)^2 + 9x_3/(1+t)^2 \;=\; 6x_3/(1+t)^2$$

4.4. **Integrate the velocity equations of Problem 4.3 to obtain the displacement relations $x_i = x_i(\mathbf{X}, t)$ and from these determine the acceleration components in Lagrangian form for the motion.**

By (4.13), $v_1 = dx_1/dt = x_1/(1+t)$; separating variables, $dx_1/x_1 = dt/(1+t)$ which upon integration gives $\ln x_1 = \ln(1+t) + \ln C$ where C is a constant of integration. Since $x_1 = X_1$ when $t = 0$, $C = X_1$ and so $x_1 = X_1(1+t)$. Similar integrations yield $x_2 = X_2(1+t)^2$ and $x_3 = X_3(1+t)^3$.

Thus from (4.14) and (4.17), $v_1 = X_1$, $v_2 = 2X_2(1+t)$, $v_3 = 3X_3(1+t)^2$ and $a_1 = 0$, $a_2 = 2X_2$, $a_3 = 6X_3(1+t)$.

4.5. **The motion of a continuum is given by $x_1 = A + (e^{-B\lambda}/\lambda)\sin\lambda(A+\omega t)$, $x_2 = -B - (e^{-B\lambda}/\lambda)\cos\lambda(A+\omega t)$, $x_3 = X_3$. Show that the particle paths are circles and that the velocity magnitude is constant. Also determine the relationship between X_1 and X_2 and the constants A and B.**

By writing $x_1 - A = (e^{-B\lambda}/\lambda)\sin\lambda(A+\omega t)$, $x_2 + B = (-e^{-B\lambda}/\lambda)\cos\lambda(A+\omega t)$, then squaring and adding, t is eliminated and the path lines are the circles $(x_1-A)^2 + (x_2+B)^2 = e^{-2B\lambda}/\lambda$. From (4.6), $v_1 = \omega e^{-B\lambda}\cos\lambda(A+\omega t)$, $v_2 = \omega e^{-B\lambda}\sin\lambda(A+\omega t)$, $v_3 = 0$ and $v^2 = v_1^2 + v_2^2 + v_3^2 = \omega^2 e^{-2B\lambda}$. Finally, when $t = 0$, $x_i = X_i$ and so $X_1 = A + (e^{-B\lambda}/\lambda)\sin\lambda A$, $X_2 = -B - (e^{-B\lambda}/\lambda)\cos\lambda A$.

4.6. **A velocity field is specified by the vector $\mathbf{v} = x_1^2 t\hat{\mathbf{e}}_1 + x_2 t^2\hat{\mathbf{e}}_2 + x_1x_3t\hat{\mathbf{e}}_3$. Determine the velocity and acceleration of the particle at $P(1,3,2)$ when $t = 1$.**

By direct substitution, $\mathbf{v}_P = \hat{\mathbf{e}}_1 + 3\,\hat{\mathbf{e}}_2 + 2\,\hat{\mathbf{e}}_3$. Using the vector form of (4.18) the acceleration field is given by

$$\mathbf{a} \;=\; x_1^2\hat{\mathbf{e}}_1 + 2x_2t\,\hat{\mathbf{e}}_2 + x_1x_3\hat{\mathbf{e}}_3 + (x_1^2 t\,\hat{\mathbf{e}}_1 + x_2t^2\,\hat{\mathbf{e}}_2 + x_1x_3t\,\hat{\mathbf{e}}_3) \cdot (2x_1t\,\hat{\mathbf{e}}_1\hat{\mathbf{e}}_1 + x_3t\,\hat{\mathbf{e}}_1\hat{\mathbf{e}}_3 + t^2\,\hat{\mathbf{e}}_2\hat{\mathbf{e}}_2 + x_1t\,\hat{\mathbf{e}}_3\hat{\mathbf{e}}_3)$$

or

$$\mathbf{a} \;=\; (x_1^2 + 2x_1^3t^2)\,\hat{\mathbf{e}}_1 + (2x_2t + x_2t^4)\,\hat{\mathbf{e}}_2 + (x_1x_3 + 2x\;x_3t^2)\,\hat{\mathbf{e}}_3$$

Thus $\mathbf{a}_P = 3\,\hat{\mathbf{e}}_1 + 9\,\hat{\mathbf{e}}_2 + 6\,\hat{\mathbf{e}}_3$.

4.7. **For the velocity field of Problem 4.3 determine the streamlines and path lines of the flow and show that they coincide.**

At every point on a streamline the tangent is in the direction of the velocity. Hence for the differential tangent vector $d\mathbf{x}$ along the streamline, $\mathbf{v} \times d\mathbf{x} = 0$ and accordingly the differential equations of the streamlines become $dx_1/v_1 = dx_2/v_2 = dx_3/v_3$. For the given flow these equations are $dx_1/x_1 = dx_2/2x_2 = dx_3/3x_3$. Integrating and using the conditions $x_i = X_i$ when $t = 0$, the equations of the streamlines are $(x_1/X_1)^2 = x_2/X_2$, $(x_1/X_1)^3 = x_3/X_3$, $(x_2/X_2)^3 = (x_3/X_3)^2$.

Integration of the velocity expressions $dx_i/dt = v_i$ as was carried out in Problem 4.4 yields the displacement equations $x_1 = X_1(1+t)$, $x_2 = X_2(1+t)^2$, $x_3 = X_3(1+t)^3$. Eliminating t from these equations gives the path lines which are identical with the streamlines presented above.

4.8. The magnetic field strength of an electromagnetic continuum is given by $\lambda = e^{-At}/r$ where $r^2 = x_1^2 + x_2^2 + x_3^2$ and A is a constant. If the velocity field of the continuum is given by $v_1 = Bx_1x_3t$, $v_2 = Bx_2^2t^2$, $v_3 = Bx_3x_2$, determine the rate of change of magnetic intensity for the particle at $P(2,-1,2)$ when $t=1$.

Since $\partial(r^{-1})/\partial x_i = -x_i/r^3$, equation (*4.11*) gives

$$\dot{\lambda} \;=\; -Ae^{-At}/r \;-\; e^{-At}(Bx_1^2x_3t + Bx_2^3t^2 + Bx_3^2x_2)/r^3$$

Thus for P at $t=1$, $\dot{\lambda}_P = -e^{-A}(3A+B)/9$.

4.9. A velocity field is given by $v_1 = 4x_3 - 3x_2$, $v_2 = 3x_1$, $v_3 = -4x_1$. Determine the acceleration components at $P(b,0,0)$ and $Q(0,4b,3b)$ and note that the velocity field corresponds to a rigid body rotation of angular velocity 5 about the axis along $\hat{\mathbf{e}} = (4\hat{\mathbf{e}}_2 + 3\hat{\mathbf{e}}_3)/5$.

From (*4.18*), $a_1 = -25x_1$, $a_2 = -9x_2 + 12x_3$, $a_3 = 12x_2 - 16x_3$. Thus at $P(b,0,0)$, $\mathbf{a} = -25b\,\hat{\mathbf{e}}_1$ which is a normal component of acceleration. Also, at $Q(0,4b,3b)$ which is on the axis of rotation, $\mathbf{a} = 0$. Note that $\mathbf{v} = \mathbf{w}\times\mathbf{x} = (4\hat{\mathbf{e}}_2 + 3\hat{\mathbf{e}}_3)\times(x_1\hat{\mathbf{e}}_1 + x_2\hat{\mathbf{e}}_2 + x_3\hat{\mathbf{e}}_3) = (4x_3 - 3x_2)\hat{\mathbf{e}}_1 + 3x_1\hat{\mathbf{e}}_2 - 4x_1\hat{\mathbf{e}}_3$.

RATE OF DEFORMATION, VORTICITY (Sec. 4.4-4.5)

4.10. A certain flow is given by $v_1 = 0$, $v_2 = A(x_1x_2 - x_3^2)e^{-Bt}$, $v_3 = A(x_2^2 - x_1x_3)e^{-Bt}$ where A and B are constants. Determine the velocity gradient $\partial v_i/\partial x_j$ for this motion and from it compute the rate of deformation tensor $\mathbf{D}$ and the spin tensor $\mathbf{V}$ for the point $P(1,0,3)$ when $t=0$.

By (*4.19*), $$\partial v_i/\partial x_j \;=\; \begin{pmatrix} 0 & 0 & 0 \\ x_2 & x_1 & -2x_3 \\ -x_3 & 2x_2 & -x_1 \end{pmatrix} Ae^{-Bt}$$ which may be evaluated at P when $t=0$

and decomposed according to (*4.20*) and (*4.21*) as

$$\mathbf{Y} \;=\; \mathbf{D} + \mathbf{V} \;=\; \begin{pmatrix} 0 & 0 & 0 \\ 0 & A & -6A \\ -3A & 0 & -A \end{pmatrix} \;=\; \begin{pmatrix} 0 & 0 & -1.5A \\ 0 & A & -3A \\ -1.5A & -3A & -A \end{pmatrix} + \begin{pmatrix} 0 & 0 & 1.5A \\ 0 & 0 & -3A \\ -1.5A & 3A & 0 \end{pmatrix}$$

4.11. For the motion $x_1 = X_1$, $x_2 = X_2 + X_1(e^{-2t}-1)$, $x_3 = X_3 + X_1(e^{-3t}-1)$ compute the rate of deformation $\mathbf{D}$ and the vorticity tensor $\mathbf{V}$. Compare $\mathbf{D}$ with $d\epsilon_{ij}/dt$, the rate of change of the Eulerian small strain tensor $\mathbf{E}$.

Here the displacement components are $u_1 = 0$, $u_2 = x_1(e^{-2t}-1)$, $u_3 = x_1(e^{-3t}-1)$ and from (*4.14*) the velocity components are $v_1 = 0$, $v_2 = -2x_1e^{-2t}$, $v_3 = -3x_1e^{-3t}$. Decomposition of the velocity gradient $\partial v_i/\partial x_j$ gives $\partial v_i/\partial x_j = D_{ij} + V_{ij}$. Thus

$$\partial v_i/\partial x_j \;=\; \begin{pmatrix} 0 & 0 & 0 \\ -2e^{-2t} & 0 & 0 \\ -3e^{-3t} & 0 & 0 \end{pmatrix} \;=\; \begin{pmatrix} 0 & -e^{-2t} & -3e^{-3t}/2 \\ -e^{-2t} & 0 & 0 \\ -3e^{-3t}/2 & 0 & 0 \end{pmatrix} + \begin{pmatrix} 0 & e^{-2t} & 3e^{-3t}/2 \\ -e^{-2t} & 0 & 0 \\ -3e^{-3t}/2 & 0 & 0 \end{pmatrix}$$

Likewise, decomposition of the displacement gradient gives $\partial u_i/\partial x_j = \epsilon_{ij} + \omega_{ij}$. Thus

$$\partial u_i/\partial x_j \;=\; \begin{pmatrix} 0 & 0 & 0 \\ e^{-2t} & 0 & 0 \\ e^{-3t} & 0 & 0 \end{pmatrix} \;=\; \frac{1}{2}\begin{pmatrix} 0 & e^{-2t} & e^{-3t} \\ e^{-2t} & 0 & 0 \\ e^{-3t} & 0 & 0 \end{pmatrix} + \frac{1}{2}\begin{pmatrix} 0 & -e^{-2t} & -e^{-3t} \\ e^{-2t} & 0 & 0 \\ e^{-3t} & 0 & 0 \end{pmatrix}$$

Comparing $\mathbf{D}$ with $d\mathbf{E}/dt$,

$$d\epsilon_{ij}/dt = \begin{pmatrix} 0 & -e^{-2t} & -3e^{-3t}/2 \\ -e^{-2t} & 0 & 0 \\ -3e^{-3t}/2 & 0 & 0 \end{pmatrix} = D_{ij}$$

The student should show that $d\omega_{ij}/dt = V_{ij}$.

4.12. A vortex line is one whose tangent at every point in a moving continuum is in the direction of the vorticity vector $\mathbf{q}$. Show that the equations for vortex lines are $dx_1/q_1 = dx_2/q_2 = dx_3/q_3$.

Let $d\mathbf{x}$ be a differential distance vector in the direction of $\mathbf{q}$. Then $\mathbf{q} \times d\mathbf{x} \equiv 0$, or

$$(q_2\,dx_3 - q_3\,dx_2)\,\hat{\mathbf{e}}_1 + (q_3\,dx_1 - q_1\,dx_3)\,\hat{\mathbf{e}}_2 + (q_1\,dx_2 - q_2\,dx_1)\,\hat{\mathbf{e}}_3 \equiv 0$$

from which $dx_1/q_1 = dx_2/q_2 = dx_3/q_3$.

4.13. Show that for the velocity field $\mathbf{v} = (Ax_3 - Bx_2)\hat{\mathbf{e}}_1 + (Bx_1 - Cx_3)\hat{\mathbf{e}}_2 + (Cx_2 - Ax_1)\hat{\mathbf{e}}_3$ the vortex lines are straight lines and determine their equations.

From (*4.29*), $\mathbf{q} = \nabla_{\mathbf{x}} \times \mathbf{v} = 2(C\hat{\mathbf{e}}_1 + A\hat{\mathbf{e}}_2 + B\hat{\mathbf{e}}_3)$, and by Problem 4.12, the d.e. for the vortex lines are $A\,dx_3 = B\,dx_2$, $B\,dx_1 = C\,dx_3$, $C\,dx_2 = A\,dx_1$. Integrating these in turn yields the equations of the vortex lines $x_3 = Bx_2/A + K_1$, $x_1 = Cx_3/B + K_2$, $x_2 = Ax_1/C + K_3$ where the K_i are constants of integration.

4.14. Show that the velocity field of Problem 4.13 represents a rigid body rotation by showing that $\mathbf{D} \equiv 0$.

Calculating the velocity gradient $\partial v_i/\partial x_j$, it is found to be antisymmetric. Thus $\partial v_i/\partial x_j =$
$$\begin{pmatrix} 0 & -B & A \\ B & 0 & -C \\ -A & C & 0 \end{pmatrix} = V_{ij} \text{ and } D_{ij} \equiv 0.$$

4.15. For the rigid body rotation $\mathbf{v} = 3x_3\hat{\mathbf{e}}_1 - 4x_3\hat{\mathbf{e}}_2 + (4x_2 - 3x_1)\hat{\mathbf{e}}_3$, determine the rate of rotation vector $\boldsymbol{\Omega}$ and show that $\mathbf{v} = \boldsymbol{\Omega} \times \mathbf{x}$.

From (*4.30*), $2\boldsymbol{\Omega} = \mathbf{q}$, or $\boldsymbol{\Omega} = 4\hat{\mathbf{e}}_1 + 3\hat{\mathbf{e}}_2$. This vector is along the axis of rotation. Thus

$$(4\hat{\mathbf{e}}_1 + 3\hat{\mathbf{e}}_2) \times (x_1\hat{\mathbf{e}}_1 + x_2\hat{\mathbf{e}}_2 + x_3\hat{\mathbf{e}}_3) = 3x_3\hat{\mathbf{e}}_1 - 4x_3\hat{\mathbf{e}}_2 + (4x_2 - 3x_1)\hat{\mathbf{e}}_3 = \mathbf{v}$$

4.16. A steady velocity field is given by $\mathbf{v} = (x_1^3 - x_1x_2^2)\hat{\mathbf{e}}_1 + (x_1^2x_2 + x_2)\hat{\mathbf{e}}_2$. Determine the unit relative velocity with respect to $P(1,1,3)$ of the particles at $Q_1(1,0,3)$, $Q_2(1,3/4,3)$, $Q_3(1,7/8,3)$ and show that these values approach the relative velocity given by (*4.26*).

By direct calculation $\mathbf{v}_P - \mathbf{v}_{Q_1} = -\hat{\mathbf{e}}_1 + 2\hat{\mathbf{e}}_2$, $4(\mathbf{v}_P - \mathbf{v}_{Q_2}) = -7\hat{\mathbf{e}}_1/4 + 2\hat{\mathbf{e}}_2$ and $8(\mathbf{v}_P - \mathbf{v}_{Q_3}) = -15\hat{\mathbf{e}}_1/8 + 2\hat{\mathbf{e}}_2$. The velocity gradient matrix is

$$[\partial v_i/\partial x_j] = \begin{bmatrix} 3x_1^2 - x_2^2 & -2x_1x_2 & 0 \\ 2x_1x_2 & x_1^2 + 1 & 0 \\ 0 & 0 & 0 \end{bmatrix}$$

and at $P(1,1,3)$ in the negative x_2 direction,

$$(dv_i/dx)_{\hat{\mathbf{e}}_2} = \begin{bmatrix} 2 & -2 & 0 \\ 2 & 2 & 0 \\ 0 & 0 & 0 \end{bmatrix} \begin{bmatrix} 0 \\ -1 \\ 0 \end{bmatrix} = \begin{bmatrix} +2 \\ -2 \\ 0 \end{bmatrix}$$

Thus $(d\mathbf{v}/dx)_{\hat{\mathbf{e}}_2} = -2\hat{\mathbf{e}}_1 + 2\hat{\mathbf{e}}_2$ which is the value approached by the relative unit velocities $\mathbf{v}_P - \mathbf{v}_{Q_i}$.

4.17. For the steady velocity field $\mathbf{v} = 3x_1^2x_2\hat{\mathbf{e}}_1 + 2x_2^2x_3\hat{\mathbf{e}}_2 + x_1x_2x_3^2\hat{\mathbf{e}}_3$, determine the rate of extension at $P(1,1,1)$ in the direction of $\hat{\boldsymbol{\nu}} = (3\hat{\mathbf{e}}_1 - 4\hat{\mathbf{e}}_3)/5$.

Here the velocity gradient is $[\partial v_i/\partial x_j] = \begin{bmatrix} 6x_1x_2 & 3x_1^2 & 0 \\ 0 & 4x_2x_3 & 2x_2^2 \\ x_2x_3^2 & x_1x_3^2 & 2x_1x_2x_3 \end{bmatrix}$ and its symmetric part at P is $[D_{ij}] = \begin{bmatrix} 6 & 1.5 & 0.5 \\ 1.5 & 4 & 1.5 \\ 0.5 & 1.5 & 2 \end{bmatrix}$.

Thus from (*4.34*) for $\hat{\boldsymbol{\nu}} = (3\hat{\mathbf{e}}_1 - 4\hat{\mathbf{e}}_3)/5$,

$$d = [3/5, 0, -4/5] \begin{bmatrix} 6 & 1.5 & 0.5 \\ 1.5 & 4 & 1.5 \\ 0.5 & 1.5 & 2 \end{bmatrix} \begin{bmatrix} 3/5 \\ 0 \\ -4/5 \end{bmatrix} = 74/25$$

4.18. For the motion of Problem 4.17 determine the rate of shear at P between the orthogonal directions $\hat{\boldsymbol{\nu}} = (3\hat{\mathbf{e}}_1 - 4\hat{\mathbf{e}}_3)/5$ and $\hat{\boldsymbol{\mu}} = (4\hat{\mathbf{e}}_1 + 3\hat{\mathbf{e}}_3)/5$.

In analogy with the results of Problem 3.20 the shear rate $\dot{\gamma}_{\mu\nu}$ is given by $\dot{\gamma}_{\mu\nu} = \hat{\boldsymbol{\mu}} \cdot 2\mathbf{D} \cdot \hat{\boldsymbol{\nu}}$, or in matrix form

$$\dot{\gamma}_{\mu\nu} = [4/5, 0, 3/5] \begin{bmatrix} 12 & 3 & 1 \\ 3 & 8 & 3 \\ 1 & 3 & 4 \end{bmatrix} \begin{bmatrix} 3/5 \\ 0 \\ -4/5 \end{bmatrix} = 89/25$$

4.19. A steady velocity field is given by $v_1 = 2x_3$, $v_2 = 2x_3$, $v_3 = 0$. Determine the principal directions and principal values (rates of extension) of the rate of deformation tensor for this motion.

Here $[\partial v_i/\partial x_j] = \begin{bmatrix} 0 & 0 & 2 \\ 0 & 0 & 2 \\ 0 & 0 & 0 \end{bmatrix} = \begin{bmatrix} 0 & 0 & 1 \\ 0 & 0 & 1 \\ 1 & 1 & 0 \end{bmatrix} + \begin{bmatrix} 0 & 0 & 1 \\ 0 & 0 & 1 \\ -1 & -1 & 0 \end{bmatrix}$ and for principal values λ of D_{ij},

$$\begin{vmatrix} -\lambda & 0 & 1 \\ 0 & -\lambda & 1 \\ 1 & 1 & -\lambda \end{vmatrix} = 0 = -\lambda^3 + 2\lambda$$

Thus $\lambda_{\text{I}} = +\sqrt{2}$, $\lambda_{\text{II}} = 0$, $\lambda_{\text{III}} = -\sqrt{2}$.

The transformation matrix to principal axis directions is

$$[a_{ij}] = \begin{bmatrix} -1/2 & -1/2 & 1/\sqrt{2} \\ 1/\sqrt{2} & -1/\sqrt{2} & 0 \\ 1/2 & 1/2 & 1/\sqrt{2} \end{bmatrix}$$

with the rate of deformation matrix in principal form $[D^*_{ij}] = \begin{bmatrix} +\sqrt{2} & 0 & 0 \\ 0 & 0 & 0 \\ 0 & 0 & -\sqrt{2} \end{bmatrix}$.

4.20. Determine the maximum shear rate $\dot{\gamma}_{\max}$ for the motion of Problem 4.19.

Analogous to principal shear strains of Chapter 3, the maximum shear rate is $\dot{\gamma}_{\max} = (\lambda_{\mathrm{I}} - \lambda_{\mathrm{III}})/2 = \sqrt{2}$.

This result is also available by observing that the motion is a simple shearing parallel to the x_1x_2 plane in the direction of the unit vector $\hat{\boldsymbol{\nu}} = (\hat{\mathbf{e}}_1 + \hat{\mathbf{e}}_2)/\sqrt{2}$. Thus, as before,

$$\dot{\gamma}_{\max} = \dot{\gamma}_{\mu\nu} = [0, 0, 1]\begin{bmatrix} 0 & 0 & 1 \\ 0 & 0 & 1 \\ 1 & 1 & 0 \end{bmatrix}\begin{bmatrix} 1/\sqrt{2} \\ 1/\sqrt{2} \\ 0 \end{bmatrix} = \sqrt{2}$$

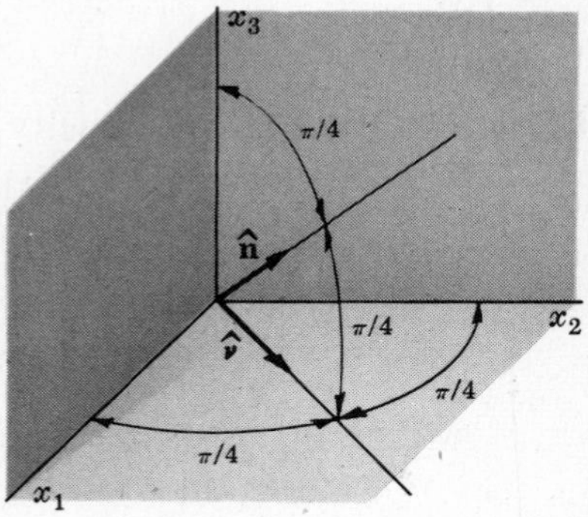

Fig. 4-3

It is also worth noting that the maximum rate of extension for this motion occurs in the direction $\hat{\mathbf{n}} = (\hat{\mathbf{e}}_1 + \hat{\mathbf{e}}_2 + \sqrt{2}\,\hat{\mathbf{e}}_3)/2$ as found in Problem 4.19. Thus

$$\lambda_{\mathrm{I}} = d^{(\hat{\mathbf{n}})} = [1/2, 1/2, \sqrt{2}/2]\begin{bmatrix} 0 & 0 & 1 \\ 0 & 0 & 1 \\ 1 & 1 & 0 \end{bmatrix}\begin{bmatrix} 1/2 \\ 1/2 \\ \sqrt{2}/2 \end{bmatrix} = \sqrt{2}$$

MATERIAL DERIVATIVES OF VOLUMES, AREAS, INTEGRALS, ETC. (Sec. 4.6-4.7)

4.21. Calculate the second material derivative of the scalar product of two line elements, i.e. determine $d^2(dx^2)/dt^2$.

From (*4.45*), $\dfrac{d(dx_i)}{dt} = \dfrac{\partial v_i}{\partial x_k}\,dx_k$; and it is shown in (*4.48*) that $\dfrac{d(dx^2)}{dt} = 2D_{ij}\,dx_i\,dx_j$. Therefore

$$\frac{d^2(dx^2)}{dt^2} = 2\left[\frac{dD_{ij}}{dt}\,dx_i\,dx_j + D_{ij}\frac{\partial v_i}{\partial x_k}\,dx_k\,dx_j + D_{ij}\,dx_i\frac{\partial v_j}{\partial x_k}\,dx_k\right]$$

and by simple manipulation of the dummy indices,

$$\frac{d^2(dx^2)}{dt^2} = 2\left[\frac{dD_{ij}}{dt} + D_{kj}\frac{\partial v_k}{\partial x_i} + D_{ik}\frac{\partial v_k}{\partial x_j}\right]dx_i\,dx_j$$

4.22. Determine the material derivative $\dfrac{d}{dt}\displaystyle\int_S p_i\,dS_i$ of the flux of the vector property p_i through the surface S.

By (*4.57*),

$$\frac{d}{dt}\int_S p_i\,dS_i = \int_S\left[\frac{dp_i}{dt} + p_i\frac{\partial v_k}{\partial x_k}\right]dS_i - \int_S p_i\frac{\partial v_k}{\partial x_i}\,dS_k = \int_S\left[\frac{dp_i}{dt} + p_i\frac{\partial v_k}{\partial x_k} - p_k\frac{\partial v_i}{\partial x_k}\right]dS_k$$

4.23. Show that the transport theorem derived in Problem 4.22 may be written in symbolic notation as

$$\frac{d}{dt}\int_S \mathbf{p}\cdot\hat{\mathbf{n}}\,dS = \int_S\left[\frac{\partial\mathbf{p}}{\partial t} + \mathbf{v}(\nabla\cdot\mathbf{p}) + \nabla\times(\mathbf{p}\times\mathbf{v})\right]\cdot\hat{\mathbf{n}}\,dS$$

By a direct transcription into symbolic notation of the result in Problem 4.22,

$$\frac{d}{dt}\int_S \mathbf{p}\cdot\hat{\mathbf{n}}\,dS = \int_S\left[\frac{dp}{dt} + \mathbf{p}(\nabla\cdot\mathbf{v}) - (\mathbf{p}\cdot\nabla)\mathbf{v}\right]\cdot\hat{\mathbf{n}}\,dS$$

$$= \int_S\left[\frac{\partial\mathbf{p}}{\partial t} + (\mathbf{v}\cdot\nabla)\mathbf{p} + \mathbf{p}(\nabla\cdot\mathbf{v}) - (\mathbf{p}\cdot\nabla)\mathbf{v}\right]\cdot\hat{\mathbf{n}}\,dS$$

Now use of the vector identity $\nabla \times (\mathbf{p} \times \mathbf{v}) = \mathbf{p}(\nabla \cdot \mathbf{v}) - \mathbf{v}(\nabla \cdot \mathbf{p}) + (\mathbf{v} \cdot \nabla)\mathbf{p} - (\mathbf{p} \cdot \nabla)\mathbf{v}$ (see Problem 1.65) gives

$$\frac{d}{dt}\int_S \mathbf{p} \cdot \hat{\mathbf{n}}\, dS = \int_S \left[\frac{\partial \mathbf{p}}{\partial t} + \mathbf{v}(\nabla \cdot \mathbf{p}) + \nabla \times (\mathbf{p} \times \mathbf{v})\right] \cdot \hat{\mathbf{n}}\, dS$$

4.24. Express *Reynold's transport theorem* as given by equations (*4.53*) and (*4.54*) in symbolic notation.

Let $P^*(\mathbf{x}, t)$ be any tensor function of the Eulerian coordinates and time. Then (*4.53*) is

$$\frac{d}{dt}\int_V P^*(\mathbf{x}, t)\, dV = \int_V \left[\frac{\partial P^*}{\partial t} + \nabla \cdot (P^* \mathbf{v})\right] dV$$

and by Gauss' divergence theorem this becomes (*4.54*),

$$\frac{d}{dt}\int_V P^*(\mathbf{x}, t)\, dV = \int_V \frac{\partial P^*}{\partial t}\, dV + \int_S P^* \mathbf{v} \cdot \hat{\mathbf{n}}\, dS$$

4.25. If the function $P^*(\mathbf{x}, t)$ in Problem 4.24 is the scalar 1, the integral on the left is simply the instantaneous volume of a portion of the continuum. Determine the material derivative of this volume.

Using the vector form of (*4.53*) as given in Problem 4.24, $\frac{d}{dt}\int_V dV = \int_V \nabla \cdot \mathbf{v}\, dV$. Here $\nabla \cdot \mathbf{v}\, dV$ represents the rate of change of dV, and so $\nabla \cdot \mathbf{v}$ is known as the cubical rate of dilatation. This relationship may also be established by a direct differentiation of (*4.38*). See Problem 4.43.

MISCELLANEOUS PROBLEMS

4.26. From the definition of the vorticity vector (*4.29*), $\mathbf{q} = \operatorname{curl} \mathbf{v}$, show that $q_i = \epsilon_{ijk} V_{kj}$ and that $2V_{ij} = \epsilon_{jik} q_k$.

By (*4.29*), $q_i = \epsilon_{ijk} v_{k,j} = \epsilon_{ijk}(v_{[k,j]} + v_{(k,j)})$ and since $\epsilon_{ijk} v_{(k,j)} = 0$ (see, for example, Problem 1.50), $q_i = \epsilon_{ijk} v_{[k,j]} = \epsilon_{ijk} V_{kj}$. From this result $\epsilon_{irs} q_i = \epsilon_{irs} \epsilon_{ijk} V_{kj} = (\delta_{rj}\delta_{sk} - \delta_{rk}\delta_{sj}) V_{kj} = 2V_{sr}$.

4.27. Show that the acceleration $\mathbf{a}$ may be written as $\mathbf{a} = \frac{\partial \mathbf{v}}{\partial t} + \mathbf{q} \times \mathbf{v} + \frac{1}{2}\nabla v^2$.

From (*4.18*), $a_i = \frac{\partial v_i}{\partial t} + v_k \frac{\partial v_i}{\partial x_k}$ and so

$$a_i = \frac{\partial v_i}{\partial t} + v_k\left(\frac{\partial v_i}{\partial x_k} - \frac{\partial v_k}{\partial x_i}\right) + v_k \frac{\partial v_k}{\partial x_i}$$

$$= \frac{\partial v_i}{\partial t} + 2v_k V_{ik} + \frac{1}{2}\frac{\partial(v_k v_k)}{\partial x_i} = \frac{\partial v_i}{\partial t} + \epsilon_{ijk} q_j v_k + \frac{1}{2}\frac{\partial(v_k v_k)}{\partial x_i}$$

which, as the student should confirm, is the indicial form of the required equation.

4.28. Show that $d(\ln J)/dt = \operatorname{div} \mathbf{v}$.

Let $\partial x_i/\partial X_p$ be written here as $x_{i,P}$ so that $J = \epsilon_{PQR} x_{1,P} x_{2,Q} x_{3,R}$ and $\dot{J}$ becomes the sum of the three determinants, $\dot{J} = \epsilon_{PQR}(\dot{x}_{1,P} x_{2,Q} x_{3,R} + x_{1,P} \dot{x}_{2,Q} x_{3,R} + x_{1,P} x_{2,Q} \dot{x}_{3,R})$. Now $\dot{x}_{1,P} = v_{1,s} x_{s,P}$, etc., and so $\dot{J} = \epsilon_{PQR}(v_{1,s} x_{s,P} x_{2,Q} x_{3,R} + x_{1,P} v_{2,s} x_{s,Q} x_{3,R} + x_{1,P} x_{2,Q} v_{3,s} x_{s,R})$. Of the nine 3×3 determinants resulting from summation on s in this expression, the three non-vanishing ones yield $\dot{J} = v_{1,1}J + v_{2,2}J + v_{3,3}J = v_{s,s}J$. Thus $\dot{J} = J\nabla \cdot \mathbf{v}$ and so $d(\ln J)/dt = \operatorname{div} \mathbf{v}$.

4.29. Show that for steady motion $(\partial v_i/\partial t = 0)$ of a continuum the streamlines and pathlines coincide.

As shown in Problem 4.7, at a given instant t streamlines are the solutions of the differential equations $dx_1/v_1 = dx_2/v_2 = dx_3/v_3$. Pathlines are solutions of the differential equations $dx_i/dt = v_i(\mathbf{x}, t)$. If $v_i = v_i(\mathbf{x})$, these equations become $dt = dx_1/v_1 = dx_2/v_2 = dx_3/v_3$ which coincide with the streamline differential equations.

4.30. For the steady velocity field $v_1 = x_1^2 x_2 + x_2^3$, $v_2 = -x_1^3 - x_1 x_2^2$, $v_3 = 0$ determine expressions for the principal values of the rate of deformation tensor $\mathbf{D}$ at an arbitrary point $P(x_1, x_2, x_3)$.

By (*4.19*) $\partial v_i/\partial x_j = D_{ij} + V_{ij}$, or

$$\begin{pmatrix} 2x_1x_2 & x_1^2 + 3x_2^2 & 0 \\ -3x_1^2 - x_2^2 & -2x_1x_2 & 0 \\ 0 & 0 & 0 \end{pmatrix} = \begin{pmatrix} 2x_1x_2 & -x_1^2 + x_2^2 & 0 \\ -x_1^2 + x_2^2 & -2x_1x_2 & 0 \\ 0 & 0 & 0 \end{pmatrix} + \begin{pmatrix} 0 & 2(x_1^2 + x_2^2) & 0 \\ -2(x_1^2 + x_2^2) & 0 & 0 \\ 0 & 0 & 0 \end{pmatrix}$$

Principal values $d_{(i)}$ are solutions of

$$\begin{vmatrix} 2x_1x_2 - d & -x_1^2 + x_2^2 & 0 \\ -x_1^2 + x_2^2 & -2x_1x_2 - d & 0 \\ 0 & 0 & -d \end{vmatrix} = 0 = -d[-4x_1^2x_2^2 + d^2 - (x_2^2 - x_1^2)^2]$$

Thus $d_{(1)} = 0$, $d_{(2)} = -(x_1^2 + x_2^2)$, $d_{(3)} = (x_1^2 + x_2^2)$. Note here that $d_{\mathrm{I}} = (x_1^2 + x_2^2)$, $d_{\mathrm{II}} = 0$, $d_{\mathrm{III}} = -(x_1^2 + x_2^2)$.

4.31. Prove equation (*4.43*) by taking the material derivative of dS_i in its cross product form $dS_i = \epsilon_{ijk}\, dx_j^{(2)}\, dx_k^{(3)}$.

Using (*3.33*), $dS_i = \epsilon_{ijk}(\partial x_j/\partial X_2)\, dX_2 (\partial x_k/\partial X_3)\, dX_3$ and $\dfrac{\partial x_i}{\partial X_1}\, dS_i = \epsilon_{ijk} \dfrac{\partial x_i}{\partial X_1}\dfrac{\partial x_j}{\partial X_2}\dfrac{\partial x_k}{\partial X_3}\, dX_2\, dX_3 = J\, dX_2\, dX_3$. Thus $\dfrac{\partial X_1}{\partial x_p}\dfrac{\partial x_i}{\partial X_1}\, dS_i = \delta_{ip}\, dS_i = dS_p = \dfrac{\partial X_1}{\partial x_p} J\, dX_2\, dX_3$ and by Problem 4.28,

$$\begin{aligned} \frac{dS_p}{dt} &= \left(\frac{\partial X_1}{\partial x_p} J \frac{\partial v_q}{\partial x_q} - J \frac{\partial X_1}{\partial x_q}\frac{\partial v_q}{\partial x_p}\right) dX_2\, dX_3 \\ &= \left(\epsilon_{pjk} \frac{\partial x_j}{\partial X_2}\, dX_2 \frac{\partial x_k}{\partial X_3}\, dX_3\right) \frac{\partial v_q}{\partial x_q} - \left(\epsilon_{qjk} \frac{\partial x_j}{\partial X_2}\, dX_2 \frac{\partial x_k}{\partial X_3}\, dX_3\right) \frac{\partial v_q}{\partial x_p} \\ &= (\partial v_q/\partial x_q)\, dS_p - (\partial v_q/\partial x_p)\, dS_q \end{aligned}$$

4.32. Use the results of Problems 4.27 and 4.23 to show that the material rate of change of the vorticity flux $\dfrac{d}{dt}\displaystyle\int_S \mathbf{q}\cdot\hat{\mathbf{n}}\, dS$ equals the flux of the curl of the acceleration $\mathbf{a}$.

Taking the curl of the acceleration as given in Problem 4.27,

$$\nabla \times \mathbf{a} = \nabla \times \frac{\partial \mathbf{v}}{\partial t} + \nabla \times (\mathbf{q} \times \mathbf{v}) + \nabla \times \nabla(v^2/2)$$

or

$$\nabla \times \mathbf{a} = \partial \mathbf{q}/\partial t + \nabla \times (\mathbf{q} \times \mathbf{v}) = d\mathbf{q}/dt + \mathbf{q}(\nabla \cdot \mathbf{v}) - (\mathbf{q} \cdot \nabla)\mathbf{v}$$

since $\mathbf{q} = \nabla \times \mathbf{v}$ and $\nabla \times \nabla(v^2/2) = 0$. Thus if $\mathbf{q}$ is substituted for $\mathbf{p}$ in Problem 4.23,

$$\frac{d}{dt}\int_S \mathbf{q}\cdot\hat{\mathbf{n}}\, dS = \int_S \left[\frac{d\mathbf{q}}{dt} + \mathbf{q}(\nabla\cdot\mathbf{v}) - (\mathbf{q}\cdot\nabla)\mathbf{v}\right]\cdot\hat{\mathbf{n}}\, dS = \int_S (\nabla\times\mathbf{a})\cdot\hat{\mathbf{n}}\, dS$$

4.33. For the vorticity q_i show that $\dfrac{\partial}{\partial t}\displaystyle\int_V q_i\,dV = \int_S [\epsilon_{ijk}a_k + q_j v_i - q_i v_j]\,dS_j$.

From Problem 4.32 the identity $\nabla \times \mathbf{a} = \partial\mathbf{q}/\partial t + \nabla \times (\mathbf{q}\times\mathbf{v})$ may be written in indicial form as $\partial q_i/\partial t = \epsilon_{ijk}a_{k,j} - \epsilon_{isp}(\epsilon_{pmr}q_m v_r)_{,s}$. Thus

$$\int_V \frac{\partial q_i}{\partial t}\,dV = \int_V [\epsilon_{ijk}a_{k,j} - (\epsilon_{isp}\epsilon_{pmr}q_m v_r)_{,s}]\,dV$$

and by the divergence theorem of Gauss (1.157),

$$\int_V \frac{\partial q_i}{\partial t}\,dV = \int_S \epsilon_{ijk}a_k\,dS_j - \int_S (\delta_{im}\delta_{sr} - \delta_{ir}\delta_{sm})(q_m v_r)\,dS_s = \int_S [\epsilon_{ijk}a_k + q_j v_i - q_i v_j]\,dS_j$$

Supplementary Problems

4.34. A continuum motion is given by $x_1 = X_1e^t + X_3(e^t - 1)$, $x_2 = X_2 + X_3(e^t - e^{-t})$, $x_3 = X_3$. Show that J does not vanish for this motion and obtain the velocity components.
Ans. $v_1 = (X_1 + X_3)e^t$, $v_2 = X_3(e^t + e^{-t})$, $v_3 = 0$ or $v_1 = x_1 - x_3$, $v_2 = x_3(e^t + e^{-t})$, $v_3 = 0$

4.35. A velocity field is specified in Lagrangian form by $v_1 = -X_2e^{-t}$, $v_2 = -X_3$, $v_3 = 2t$. Determine the acceleration components in Eulerian form. *Ans.* $a_1 = e^{-t}(x_2 + tx_3 - t^3)$, $a_2 = 0$, $a_3 = 2$

4.36. Show that the velocity field $v_i = \epsilon_{ijk}b_jx_k + c_i$ where b_i and c_i are constant vectors, represents a rigid body rotation and determine the vorticity vector for this motion.
Ans. $q_i = b_ix_{j,j} - b_i = 2b_i$

4.37. Show that for the flow $v_i = x_i/(1+t)$ the streamlines and path lines coincide.

4.38. The electrical field strength in a region containing a fluid flow is given by $\lambda = (A\cos 3t)/r$ where $r^2 = x_1^2 + x_2^2$ and A is a constant. The velocity field of the fluid is $v_1 = x_1^2x_2 + x_2^3$, $v_2 = -x_1^3 - x_1x_2^2$, $v_3 = 0$. Determine $d\lambda/dt$ at $P(x_1, x_2, x_3)$. *Ans.* $d\lambda/dt = (-3A\sin 3t)/r$

4.39. Show that for the velocity field $v_1 = x_1^2x_2 + x_2^3$, $v_2 = -x_1^3 - x_1x_2^2$, $v_3 = 0$ the streamlines are circular.

4.40. For the continuum motion $x_1 = X_1$, $x_2 = e^t(X_2 + X_3)/2 + e^{-t}(X_2 - X_3)/2$, $x_3 = e^t(X_2 + X_3)/2 - e^{-t}(X_2 - X_3)/2$, show that $D_{ij} = d\epsilon_{ij}/dt$ at $t = 0$. Compare these tensors at $t = 0.5$.

4.41. For the velocity field $v_1 = x_1^2x_2 + x_2^3$, $v_2 = -(x_1^3 + x_1x_2^2)$, $v_3 = 0$, determine the principal axes and principal values of $\mathbf{D}$ at $P(1,2,3)$.

$$Ans.\quad D^*_{ij} = \begin{pmatrix} 5 & 0 & 0 \\ 0 & 0 & 0 \\ 0 & 0 & -5 \end{pmatrix}; \quad a_{ij} = \begin{pmatrix} 3/\sqrt{10} & 1/\sqrt{10} & 0 \\ 0 & 0 & 1 \\ 1/\sqrt{10} & -3/\sqrt{10} & 0 \end{pmatrix}$$

4.42. For the velocity field of Problem 4.41 determine the rate of extension in the direction $\hat{\mathbf{v}} = (\hat{\mathbf{e}}_1 - 2\hat{\mathbf{e}}_2 + 2\hat{\mathbf{e}}_3)/3$ at $P(1,2,3)$. What is the maximum shear rate at P?
Ans. $d^{(\hat{\mathbf{v}})} = -24/9$, $\dot{\gamma}_{\max} = 5$

4.43. Show that $d(\partial x_i/\partial X_j)/dt = v_{i,k}x_{k,j}$ and use this to derive (*4.41*) of the text directly from (*4.38*).

4.44. Prove the identity $\epsilon_{pqr}(v_sv_{r,s})_{,q} = q_pv_{q,q} + v_qq_{p,q} - q_qv_{p,q}$ where v_i is the velocity and q_i the vorticity. Also show that $v_{i,j}v_{j,i} = D_{ij}D_{ij} - q_iq_i/2$.

4.45. Prove that the material derivative of the total vorticity is given by

$$\frac{d}{dt}\int_V q_i\,dV = \int_S [\epsilon_{ijk}a_k + q_jv_i]\,dS_j$$

Chapter 5

Fundamental Laws of Continuum Mechanics

5.1 CONSERVATION OF MASS. CONTINUITY EQUATION

Associated with every material continuum there is the property known as *mass*. The amount of mass in that portion of the continuum occupying the spatial volume V at time t is given by the integral

$$m = \int_V \rho(\mathbf{x}, t)\, dV \tag{5.1}$$

in which $\rho(\mathbf{x}, t)$ is a continuous function of the coordinates called the *mass density*. The law of *conservation of mass* requires that the mass of a specific portion of the continuum remain constant, and hence that the material derivative of (*5.1*) be zero. Therefore from (*4.52*) with $P^*_{ij\ldots}(\mathbf{x}, t) \equiv \rho(\mathbf{x}, t)$, the rate of change of m in (*5.1*) is

$$\frac{dm}{dt} = \frac{d}{dt}\int_V \rho(\mathbf{x}, t)\, dV = \int_V \left[\frac{d\rho}{dt} + \rho\frac{\partial v_k}{\partial x_k}\right] dV = 0 \tag{5.2}$$

Since this equation holds for an arbitrary volume V, the integrand must vanish, or

$$\frac{d\rho}{dt} + \rho v_{k,k} = 0 \quad \text{or} \quad \frac{d\rho}{dt} + \rho(\nabla\cdot\mathbf{v}) = 0 \tag{5.3}$$

This equation is called the *continuity equation*; using the material derivative operator it may be put into the alternative form

$$\frac{\partial\rho}{\partial t} + (\rho v_k)_{,k} = 0 \quad \text{or} \quad \frac{\partial\rho}{\partial t} + \nabla\cdot(\rho\mathbf{v}) = 0 \tag{5.4}$$

For an *incompressible* continuum the mass density of each particle is independent of time, so that $d\rho/dt = 0$ and (*5.3*) yields the result

$$v_{k,k} = 0 \quad \text{or} \quad \operatorname{div}\mathbf{v} = 0 \tag{5.5}$$

The velocity field $\mathbf{v}(\mathbf{x}, t)$ of an incompressible continuum can therefore be expressed by the equation

$$v_i = \epsilon_{ijk}s_{k,j} \quad \text{or} \quad \mathbf{v} = \nabla\times\mathbf{s} \tag{5.6}$$

in which $\mathbf{s}(\mathbf{x}, t)$ is called the *vector potential* of $\mathbf{v}$.

The continuity equation may also be expressed in the Lagrangian, or material form. The conservation of mass requires that

$$\int_{V_0} \rho_0(\mathbf{X}, 0)\, dV_0 = \int_V \rho(\mathbf{x}, t)\, dV \tag{5.7}$$

where the integrals are taken over the same particles, i.e. V is the volume now occupied by the material which occupied V_0 at time $t = 0$. Using (*4.1*) and (*4.38*), the right hand integral in (*5.7*) may be converted so that

$$\int_{V_0} \rho_0(\mathbf{X}, 0)\, dV_0 = \int_{V_0} \rho(\mathbf{x}(\mathbf{X}, t), t)J\, dV_0 = \int_{V_0} \rho(\mathbf{X}, t)J\, dV_0 \tag{5.8}$$

Since this relationship must hold for any volume V_0, it follows that

$$\rho_0 = \rho J \tag{5.9}$$

which implies that the product ρJ is independent of time since V is arbitrary, or that

$$\frac{d}{dt}(\rho J) = 0 \tag{5.10}$$

Equation (*5.10*) is the *Lagrangian differential form* of the continuity equation.

5.2 LINEAR MOMENTUM PRINCIPLE. EQUATIONS OF MOTION. EQUILIBRIUM EQUATIONS

A moving continuum which occupies the volume V at time t is shown in Fig. 5-1. Body forces b_i per unit mass are given. On the differential element dS of the bounding surface, the stress vector is $t_i^{(\hat{n})}$. The velocity field $v_i = du_i/dt$ is prescribed throughout the region occupied by the continuum. For this situation, the total *linear momentum* of the mass system within V is given by

$$P_i(t) = \int_V \rho v_i \, dV \tag{5.11}$$

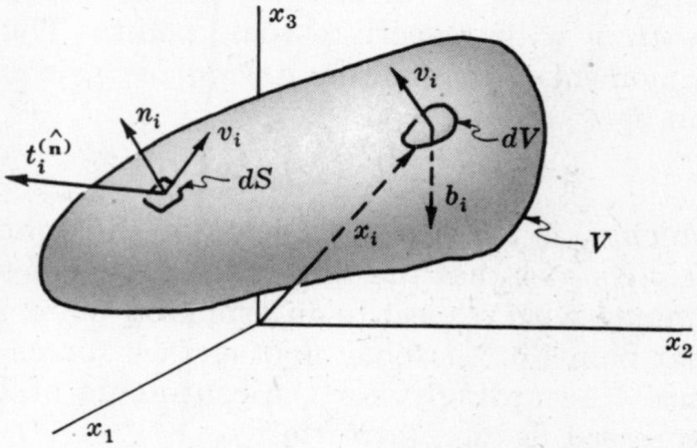

Fig. 5-1

Based upon Newton's second law, the *principle of linear momentum* states that the time rate of change of an arbitrary portion of a continuum is equal to the resultant force acting upon the considered portion. Therefore if the internal forces between particles of the continuum in Fig. 5-1 obey Newton's third law of action and reaction, the momentum principle for this mass system is expressed by

$$\int_S t_i^{(\hat{n})} \, dS + \int_V \rho b_i \, dV = \frac{d}{dt}\int_V \rho v_i \, dV$$

or

$$\int_S \mathbf{t}^{(\hat{n})} \, dS + \int_V \rho \mathbf{b} \, dV = \frac{d}{dt}\int_V \rho \mathbf{v} \, dV \tag{5.12}$$

Upon substituting $t_i^{(\hat{n})} = \sigma_{ji} n_j$ into the first integral of (*5.12*) and converting the resulting surface integral by the divergence theorem of Gauss, (*5.12*) becomes

$$\int_V (\sigma_{ji,j} + \rho b_i) \, dV = \frac{d}{dt}\int_V \rho v_i \, dV \quad \text{or} \quad \int_V (\nabla_{\mathbf{x}} \cdot \mathbf{\Sigma} + \rho \mathbf{b}) \, dV = \frac{d}{dt}\int \rho \mathbf{v} \, dV \tag{5.13}$$

In calculating the material derivative in (*5.13*), the continuity equation in the form given by (*5.10*) may be used. Thus

$$\frac{d}{dt}\int_V \rho v_i \, dV = \frac{d}{dt}\int_{V_0} \rho v_i J \, dV_0 = \int_{V_0} \left[v_i \frac{d(\rho J)}{dt} + \rho J \frac{dv_i}{dt} \right] dV_0 = \int_V \frac{dv_i}{dt} \rho \, dV \tag{5.14}$$

Replacing the right hand side of (*5.13*) by the right hand side of (*5.14*) and collecting terms results in the *linear momentum principle* in *integral form*,

$$\int_V (\sigma_{ji,j} + \rho b_i - \rho \dot{v}_i) \, dV = 0 \quad \text{or} \quad \int_V (\nabla_{\mathbf{x}} \cdot \mathbf{\Sigma} + \rho \mathbf{b} - \rho \dot{\mathbf{v}}) \, dV = 0 \tag{5.15}$$

Since the volume V is arbitrary, the integrand of (*5.15*) must vanish. The resulting equations,

$$\sigma_{ji,j} + \rho b_i = \rho \dot{v}_i \quad \text{or} \quad \nabla_{\mathbf{x}} \cdot \mathbf{\Sigma} + \rho \mathbf{b} = \rho \dot{\mathbf{v}} \tag{5.16}$$

are known as the *equations of motion.*

The important case of static equilibrium, in which the acceleration components vanish, is given at once from (*5.16*) as

$$\sigma_{ji,j} + \rho b_i = 0 \quad \text{or} \quad \nabla_{\mathbf{x}} \cdot \mathbf{\Sigma} + \rho \mathbf{b} = 0 \tag{5.17}$$

These are the *equilibrium equations,* used extensively in solid mechanics.

5.3 MOMENT OF MOMENTUM (ANGULAR MOMENTUM) PRINCIPLE

The *moment of momentum* is, as the name implies, simply the *moment of linear momentum* with respect to some point. Thus for the continuum shown in Fig. 5-1, the total moment of momentum or *angular momentum* as it is often called, with respect to the origin, is

$$N_i(t) = \int_V \epsilon_{ijk} x_j \rho v_k \, dV \quad \text{or} \quad \mathbf{N} = \int_V (\mathbf{x} \times \rho \mathbf{v}) \, dV \tag{5.18}$$

in which x_j is the position vector of the volume element dV. The *moment of momentum principle* states that the time rate of change of the angular momentum of any portion of a continuum with respect to an arbitrary point is equal to the resultant moment (with respect to that point) of the body and surface forces acting on the considered portion of the continuum. Accordingly, for the continuum of Fig. 5-1, the moment of momentum principle is expressed in integral form by

$$\int_S \epsilon_{ijk} x_j t_k^{(\hat{\mathbf{n}})} \, dS + \int_V \epsilon_{ijk} x_j \rho b_k \, dV = \frac{d}{dt} \int_V \epsilon_{ijk} x_j \rho v_k \, dV$$

or

$$\int_S (\mathbf{x} \times \mathbf{t}^{(\hat{\mathbf{n}})}) \, dS + \int_V (\mathbf{x} \times \rho \mathbf{b}) \, dV = \frac{d}{dt} \int_V (\mathbf{x} \times \rho \mathbf{v}) \, dV \tag{5.19}$$

Equation (*5.19*) is valid for those continua in which the forces between particles are equal, opposite and collinear, and in which distributed moments are absent.

The moment of momentum principle does not furnish any new differential equation of motion. If the substitution $t_k^{(\hat{\mathbf{n}})} = \sigma_{pk} n_p$ is made in (*5.19*), and the symmetry of the stress tensor assumed, the equation is satisfied identically by using the relationship given in (*5.16*). If stress symmetry is not assumed, such symmetry may be shown to follow directly from (*5.19*), which upon substitution of $t_k^{(\hat{\mathbf{n}})} = \sigma_{pk} n_p$, reduces to

$$\int_V \epsilon_{ijk} \sigma_{jk} \, dV = 0 \quad \text{or} \quad \int_V \mathbf{\Sigma}_v \, dV = 0 \tag{5.20}$$

Since the volume V is arbitrary,

$$\epsilon_{ijk} \sigma_{jk} = 0 \quad \text{or} \quad \mathbf{\Sigma}_v = 0 \tag{5.21}$$

which by expansion demonstrates that $\sigma_{jk} = \sigma_{kj}$.

5.4 CONSERVATION OF ENERGY. FIRST LAW OF THERMODYNAMICS. ENERGY EQUATION

If mechanical quantities only are considered, the *principle of conservation of energy* for the continuum of Fig. 5-1 may be derived directly from the equation of motion given by (*5.16*). To accomplish this, the scalar product between (*5.16*) and the velocity v_i is first computed, and the result integrated over the volume V. Thus

$$\int_V \rho v_i \dot{v}_i \, dV = \int_V v_i \sigma_{ji,j} \, dV + \int_V \rho v_i b_i \, dV \tag{5.22}$$

But
$$\int_V \rho v_i \dot{v}_i \, dV = \frac{d}{dt}\int_V \rho \frac{v_i v_i}{2} dV = \frac{d}{dt}\int_V \frac{\rho v^2}{2} dV = \frac{dK}{dt} \tag{5.23}$$

which represents the time rate of change of the *kinetic energy* K in the continuum. Also, $v_i \sigma_{ji,j} = (v_i \sigma_{ji})_{,j} - v_{i,j}\sigma_{ji}$ and by (*4.19*) $v_{i,j} = D_{ij} + V_{ij}$, so that (*5.22*) may be written

$$\frac{dK}{dt} + \int_V D_{ij}\sigma_{ji} \, dV = \int_V (v_i \sigma_{ji})_{,j} \, dV + \int_V \rho v_i b_i \, dV \tag{5.24}$$

since $V_{ij}\sigma_{ji} = 0$. Finally, converting the first integral on the right hand side of (*5.24*) to a surface integral by the divergence theorem of Gauss, and making use of the identity $t_i^{(\hat{n})} = \sigma_{ji} n_j$, the *energy equation* for a continuum appears in the form

$$\frac{dK}{dt} + \int_V D_{ij}\sigma_{ji} \, dV = \int_S v_i t_i^{(\hat{n})} \, dS + \int_V \rho b_i v_i \, dV \tag{5.25}$$

This equation relates the time rate of change of total mechanical energy of the continuum on the left side to the rate of work done by the surface and body forces on the right hand side of the equation. The integral on the left side is known as the time rate of change of *internal mechanical energy,* and written dU/dt. Therefore (*5.25*) may be written briefly as

$$\frac{dK}{dt} + \frac{dU}{dt} = \frac{đW}{dt} \tag{5.26}$$

where $đW/dt$ represents the rate of work, and the special symbol $đ$ is used to indicate that this quantity is *not* an *exact* differential.

If both mechanical and non-mechanical energies are to be considered, the principle of conservation of energy in its most general form must be used. In this form the conservation principle states that *the time rate of change of the kinetic plus the internal energy is equal to the sum of the rate of work plus all other energies supplied to, or removed from the continuum per unit time.* Such energies supplied may include *thermal* energy, *chemical* energy, or *electromagnetic* energy. In the following, only *mechanical* and *thermal* energies are considered, and the energy principle takes on the form of the well-known *first law of thermodynamics.*

For a thermomechanical continuum it is customary to express the time rate of change of internal energy by the integral expression

$$\frac{dU}{dt} = \frac{d}{dt}\int_V \rho u \, dV = \int_V \rho \dot{u} \, dV \tag{5.27}$$

where u is called the *specific internal energy.* (The symbol u for specific energy is so well established in the literature that it is used in the energy equations of this chapter since there appears to be only a negligible chance that it will be mistaken in what follows for the magnitude of the displacement vector u_i.) Also, if the vector c_i is defined as the *heat flux* per unit area per unit time by conduction, and z is taken as the *radiant heat constant* per unit mass per unit time, the rate of increase of total heat into the continuum is given by

$$\frac{đQ}{dt} = -\int_S c_i n_i \, dS + \int_V \rho z \, dV \tag{5.28}$$

Therefore the energy principle for a thermomechanical continuum is given by

$$\frac{dK}{dt} + \frac{dU}{dt} = \frac{đW}{dt} + \frac{đQ}{dt} \tag{5.29}$$

or, in terms of the energy integrals, as

$$\frac{d}{dt}\int_V \rho\frac{v_i v_i}{2}\,dV + \int_V \rho\dot{u}\,dV = \int_S t_i^{(\hat{n})} v_i\,dS + \int_V \rho v_i b_i\,dV + \int_V \rho z\,dV - \int_S c_i n_i\,dS \qquad (5.30)$$

Converting the surface integrals in (*5.30*) to volume integrals by the divergence theorem of Gauss, and again using the fact that V is arbitrary, leads to the local form of the energy equation:

$$\frac{d}{dt}\left(\frac{v^2}{2}+u\right) = \frac{1}{\rho}(\sigma_{ij}v_i)_{,j} + b_i v_i - \frac{1}{\rho}c_{i,i} + z$$

or

$$\frac{du}{dt} = \frac{1}{\rho}\boldsymbol{\Sigma}:\mathbf{D} - \frac{1}{\rho}\nabla\cdot\mathbf{c} + \mathbf{b}\cdot\mathbf{v} + z \qquad (5.31)$$

Within the arbitrarily small volume element for which the local energy equation (*5.31*) is valid, the balance of momentum given by (*5.16*) must also hold. Therefore by taking the scalar product between (*5.16*) and the velocity $\rho\dot{v}_i v_i = v_i\sigma_{ji,j} + \rho v_i b_i$ and, after some simple manipulations, subtracting this product from (*5.31*), the result is the reduced, but highly useful form of the local energy equation,

$$\frac{du}{dt} = \frac{1}{\rho}\sigma_{ij}D_{ij} - \frac{1}{\rho}c_{i,i} + z \qquad (5.32)$$

This equation expresses the rate of change of *internal energy* as the sum of the *stress power* plus the *heat* added to the continuum.

5.5 EQUATIONS OF STATE. ENTROPY. SECOND LAW OF THERMODYNAMICS

The complete characterization of a thermodynamic system (here, a continuum) is said to describe the *state* of the system. This description is specified, in general, by several thermodynamic and kinematic quantities called *state variables.* A change with time of the state variables characterizes a *thermodynamic process.* The state variables used to describe a given system are usually not all independent. Functional relationships exist among the state variables and these relationships are expressed by the so-called *equations of state.* Any state variable which may be expressed as a single-valued function of a set of other state variables is known as a *state function.*

As presented in the previous section, the first law of thermodynamics postulates the interconvertibility of mechanical and thermal energy. The relationship expressing conversion of heat and work into kinetic and internal energies during a thermodynamic process is set forth in the energy equation. The first law, however, leaves unanswered the question of the extent to which the conversion process is *reversible* or *irreversible.* All real processes are irreversible, but the reversible process is a very useful hypothesis since energy dissipation may be assumed negligible in many situations. The basic criterion for irreversibility is given by the *second law of thermodynamics* through its statement on the limitations of *entropy production.*

The second law of thermodynamics postulates the existence of two distinct state functions; T the *absolute temperature,* and S the *entropy,* with certain following properties. T is a positive quantity which is a function of the empirical temperature θ, only. The entropy is an extensive property, i.e. the total entropy in the system is the sum of the entropies of its parts. In continuum mechanics the *specific entropy* (per unit mass), or *entropy density* is denoted by s, so that the total entropy L is given by $L = \int_V \rho s\,dV$. The entropy of a system can change either by interactions that occur with the surroundings, or by changes that take place within the system. Thus

$$ds = ds^{(e)} + ds^{(i)} \qquad (5.33)$$

where ds is the increase in specific entropy, $ds^{(e)}$ is the increase due to interaction with the exterior, and $ds^{(i)}$ is the internal increase. The change $ds^{(i)}$ is never negative. It is zero for a reversible process, and positive for an irreversible process. Therefore

$$ds^{(i)} > 0 \quad \text{(irreversible process)} \tag{5.34}$$

$$ds^{(i)} = 0 \quad \text{(reversible process)} \tag{5.35}$$

In a reversible process, if $dq_{(R)}$ denotes the heat supplied per unit mass to the system, the change $ds^{(e)}$ is given by

$$ds^{(e)} = \frac{dq_{(R)}}{T} \quad \text{(reversible process)} \tag{5.36}$$

5.6 THE CLAUSIUS-DUHEM INEQUALITY. DISSIPATION FUNCTION

According to the second law, the time rate of change of total entropy L in a continuum occupying a volume V is never less than the sum of the entropy influx through the continuum surface plus the entropy produced internally by body sources. Mathematically, this entropy principle may be expressed in integral form as the *Clausius-Duhem inequality,*

$$\frac{d}{dt}\int_V \rho s\,dV \geqq \int_V \rho e\,dV - \int_S \frac{c_i n_i}{T}\,dS \tag{5.37}$$

where e is the local entropy source per unit mass. The equality in (*5.37*) holds for reversible processes; the inequality applies to irreversible processes.

The Clausius-Duhem inequality is valid for arbitrary choice of volume V so that transforming the surface integral in (*5.37*) by the divergence theorem of Gauss, the local form of the *internal entropy production rate* γ, per unit mass, is given by

$$\gamma \equiv \frac{ds}{dt} - e - \frac{1}{\rho}\left(\frac{c_i}{T}\right)_{,i} \geqq 0 \tag{5.38}$$

This inequality must be satisfied for every process and for any assignment of state variables. For this reason it plays an important role in imposing restrictions upon the so-called constitutive equations discussed in the following section.

In much of continuum mechanics, it is often assumed (based upon statistical mechanics of irreversible processes) that the stress tensor may be split into two parts according to the scheme,

$$\sigma_{ij} = \sigma_{ij}^{(C)} + \sigma_{ij}^{(D)} \tag{5.39}$$

where $\sigma_{ij}^{(C)}$ is a *conservative* stress tensor, and $\sigma_{ij}^{(D)}$ is a *dissipative* stress tensor. With this assumption the energy equation (*5.32*) may be written with the use of (*4.25*) as

$$\frac{du}{dt} = \frac{1}{\rho}\sigma_{ij}^{(C)}\dot{\epsilon}_{ij} + \frac{1}{\rho}\sigma_{ij}^{(D)}\dot{\epsilon}_{ij} + \frac{dq}{dt} \tag{5.40}$$

In this equation, $\frac{1}{\rho}\sigma_{ij}^{(D)}\dot{\epsilon}_{ij}$ is the rate of energy dissipated per unit mass by the stress, and dq/dt is the rate of heat influx per unit mass into the continuum. If the continuum undergoes a reversible process, there will be no energy dissipation, and furthermore, $dq/dt = dq_{(R)}/dt$, so that (*5.40*) and (*5.36*) may be combined to yield

$$\frac{du}{dt} = \frac{1}{\rho}\sigma_{ij}^{(C)}\dot{\epsilon}_{ij} + T\frac{ds}{dt} \tag{5.41}$$

Therefore in the irreversible process described by (*5.40*), the entropy production rate may be expressed by inserting (*5.41*). Thus

$$\frac{ds}{dt} = \frac{1}{T}\frac{dq}{dt} + \frac{1}{\rho T}\sigma_{ij}^{(D)}\dot{\epsilon}_{ij} \tag{5.42}$$

The scalar $\sigma_{ij}^{(D)}\dot{\epsilon}_{ij}$ is called the *dissipation function.* For an irreversible, adiabatic process ($dq = 0$), $ds/dt > 0$ by the second law, so from (*5.42*) it follows that the dissipation function is *positive definite,* since both ρ and T are always positive.

5.7 CONSTITUTIVE EQUATIONS. THERMOMECHANICAL AND MECHANICAL CONTINUA

In the preceding sections of this chapter, several equations have been developed that must hold for every process or motion that a continuum may undergo. For a thermo-mechanical continuum in which the mechanical and thermal phenomena are coupled, the basic equations are

(*a*) the *equation* of *continuity,* (*5.4*)

$$\frac{\partial\rho}{\partial t} + (\rho v_k)_{,k} = 0 \qquad \text{or} \qquad \frac{\partial\rho}{\partial t} + \nabla\cdot(\rho\mathbf{v}) = 0 \tag{5.43}$$

(*b*) the *equation* of *motion,* (*5.16*)

$$\sigma_{ji,j} + \rho b_i = \rho\dot{v}_i \qquad \text{or} \qquad \nabla_{\mathbf{x}}\cdot\mathbf{\Sigma} + \rho\mathbf{b} = \rho\dot{\mathbf{v}} \tag{5.44}$$

(*c*) the *energy equation,* (*5.32*)

$$\frac{du}{dt} = \frac{1}{\rho}\sigma_{ij}D_{ij} - \frac{1}{\rho}c_{i,i} + z \qquad \text{or} \qquad \frac{du}{dt} = \frac{1}{\rho}\mathbf{\Sigma}:\mathbf{D} - \rho\nabla\cdot\mathbf{c} + z \tag{5.45}$$

Assuming that body forces b_i and the distributed heat sources z are prescribed, (*5.43*), (*5.44*) and (*5.45*) consist of *five* independent equations involving *fourteen* unknown functions of *time* and *position.* The unknowns are the *density* ρ, the three *velocity* components v_i, (or, alternatively, the *displacement* components u_i), the six independent *stress* components σ_{ij}, the three components of the *heat flux* vector c_i, and the specific internal energy u. In addition, the Clausius-Duhem inequality (*5.38*)

$$\frac{ds}{dt} - e - \frac{1}{\rho}\left(\frac{q_i}{T}\right)_{,i} \geqq 0 \tag{5.46}$$

which governs entropy production, must hold. This introduces two additional unknowns: the *entropy density* s, and T, the *absolute temperature.* Therefore *eleven* additional equations must be supplied to make the system determinate. Of these, *six* will be in the form known as *constitutive equations,* which characterize the particular physical properties of the continuum under study. Of the remaining five, *three* will be in the form of *temperature-heat conduction* relations, and *two* will appear as *thermodynamic equations of state*; for example, perhaps as the *caloric* equation of state and the *entropic* equation of state. Specific formulation of the thermomechanical continuum problem is given in a subsequent chapter.

It should be pointed out that the function of the constitutive equations is to establish a mathematical relationship among the statical, kinematical and thermal variables, which will describe the behavior of the material when subjected to applied mechanical or thermal forces. Since real materials respond in an extremely complicated fashion under various loadings, constitutive equations do not attempt to encompass all the observed phenomena related to a particular material, but, rather, to define certain *ideal materials,* such as the ideal elastic solid or the ideal viscous fluid. Such idealizations or *material models* as they are sometimes called, are very useful in that they portray reasonably well over a definite range of loads and temperatures the behavior of real substances.

In many situations the interaction of mechanical and thermal processes may be neglected. The resulting analysis is known as the *uncoupled thermoelastic theory* of continua. Under this assumption the purely mechanical processes are governed by (*5.43*) and (*5.44*) since the energy equation (*5.45*) for this case is essentially a first integral of the equation of motion. The system of equations formed by (*5.43*) and (*5.44*) consists of *four* equations involving *ten* unknowns. *Six* constitutive equations are required to make the system determinate. In the uncoupled theory, the constitutive equations contain only the *statical* (stresses) and *kinematic* (velocities, displacements, strains) variables and are often referred to as *stress-strain relations*. Also, in the uncoupled theory, the temperature field is usually regarded as known, or at most, the heat-conduction problem must be solved separately and independently from the mechanical problem. In *isothermal* problems the temperature is assumed uniform and the problem is purely mechanical.

Solved Problems

CONTINUITY EQUATION (Sec. 5.1)

5.1. An *irrotational* motion of a continuum is described in Chapter 4 as one for which the vorticity vanishes identically. Determine the form of the continuity equation for such motions.

By (*4.29*), curl $\mathbf{v} = 0$ when $\mathbf{q} \equiv 0$, and so $\mathbf{v}$ becomes the gradient of a scalar field $\phi(x_i, t)$ (see Problem 1.50). Thus $v_i = \phi_{,i}$ and (*5.3*) is now $d\rho/dt + \rho\phi_{,kk} = 0$ or $d\rho/dt + \rho\nabla^2\phi = 0$.

5.2. If $P^{**}_{ij\ldots}(\mathbf{x}, t)$ represents any scalar, vector or tensor property per unit mass of a continuum so that $P^{*}_{ij\ldots}(\mathbf{x}, t) = \rho P^{**}_{ij\ldots}(\mathbf{x}, t)$ show that

$$\frac{d}{dt}\int_V \rho P^{**}_{ij\ldots}(\mathbf{x}, t)\, dV = \int_V \rho \frac{dP^{**}_{ij\ldots}(\mathbf{x}, t)}{dt}\, dV$$

By (*4.52*),

$$\frac{d}{dt}\int_V \rho P^{**}_{ij\ldots}\, dV = \int_V \left[\frac{d}{dt}(\rho P^{**}_{ij\ldots}) + \rho P^{**}_{ij\ldots}\frac{\partial v_k}{\partial x_k}\right] dV$$

$$= \int_V \left[\rho\frac{dP^{**}_{ij\ldots}}{dt} + P^{**}_{ij\ldots}\left(\frac{d\rho}{dt} + \rho\frac{\partial v_k}{\partial x_k}\right)\right] dV = \int_V \rho\frac{dP^{**}_{ij\ldots}}{dt}\, dV$$

since by (*5.3*), $d\rho/dt + \rho v_{k,k} = 0$.

5.3. Show that the material form $d(\rho J)/dt = 0$ of the continuity equation and the spatial form $d\rho/dt + \rho v_{k,k} = 0$ are equivalent.

Differentiating, $d(\rho J)/dt = (d\rho/dt)J + \rho\, dJ/dt = 0$ and from Problem 4.28, $dJ/dt = Jv_{k,k}$ so that $d(\rho J)/dt = J(d\rho/dt + \rho v_{k,k}) = 0$.

5.4. Show that the velocity field $v_i = Ax_i/r^3$, where $x_i x_i = r^2$ and A is an arbitrary constant, satisfies the continuity equation for an incompressible flow.

From (*5.5*) $v_{k,k} = 0$ for incompressible flow. Here

$$v_{i,k} = A(x_{i,k}/r^3 - 3x_i x_k/r^5) = A(\delta_{ik}/r^3 - 3x_i x_k/r^5)$$

and so $v_{k,k} = (3-3)/r^3 = 0$ to satisfy the continuity equation.

5.5. For the velocity field $v_i = x_i/(1+t)$, show that $\rho x_1 x_2 x_3 = \rho_0 X_1 X_2 X_3$.

Here $v_{k,k} = 3/(1+t)$ and integrating (*5.3*) yields $\ln \rho = -\ln(1+t)^3 + \ln C$ where C is a constant of integration. Since $\rho = \rho_0$ when $t = 0$, this equation becomes $\rho = \rho_0/(1+t)^3$. Next by integrating the velocity field $dx_i/x_i = dt/(1+t)$ (no sum on i), $x_i = X_i/(1+t)$ and hence $\rho x_1 x_2 x_3 = \rho_0 X_1 X_2 X_3$.

LINEAR AND ANGULAR MOMENTUM. EQUATIONS OF MOTION (Sec. 5.2-5.3)

5.6. Show by a direct expansion of each side that the identity $\epsilon_{ijk}\sigma_{jk}\hat{\mathbf{e}}_i = \mathbf{\Sigma}_v$ used in (*5.20*) and (*5.21*) is valid.

By (*1.15*) and (*2.8*),

$$\begin{aligned}\mathbf{\Sigma}_v &= \sigma_{11}\hat{\mathbf{e}}_1 \times \hat{\mathbf{e}}_1 + \sigma_{12}\hat{\mathbf{e}}_1 \times \hat{\mathbf{e}}_2 + \sigma_{13}\hat{\mathbf{e}}_1 \times \hat{\mathbf{e}}_3 + \cdots + \sigma_{33}\hat{\mathbf{e}}_3 \times \hat{\mathbf{e}}_3 \\ &= (\sigma_{23} - \sigma_{32})\,\hat{\mathbf{e}}_1 + (\sigma_{31} - \sigma_{13})\,\hat{\mathbf{e}}_2 + (\sigma_{12} - \sigma_{21})\,\hat{\mathbf{e}}_3\end{aligned}$$

Also, expanding $\epsilon_{ijk}\sigma_{jk}$ gives identical results, $(\sigma_{23} - \sigma_{32})$ for $i = 1$, $(\sigma_{31} - \sigma_{13})$ for $i = 2$, $(\sigma_{12} - \sigma_{21})$ for $i = 3$.

5.7. If distributed body moments m_i per unit volume act throughout a continuum, show that the equations of motion (*5.16*) remain valid but the stress tensor can no longer be assumed symmetric.

Since (*5.16*) is derived on the basis of *force* equilibrium, it is not affected. Now, however, (*5.19*) acquires an additional term so that

$$\frac{d}{dt}\int_V \epsilon_{ijk} x_j \rho v_k \, dV = \int_S \epsilon_{ijk} x_j t_k^{(\hat{\mathbf{n}})} \, dS + \int_V (\epsilon_{ijk} x_j \rho b_k + m_i) \, dV$$

which reduces to (see Problem 2.9) $\int_V (\epsilon_{ijk}\sigma_{jk} + m_i)\, dV = 0$, and because V is arbitrary, $\epsilon_{ijk}\sigma_{jk} + m_i = 0$ for this case.

5.8. The momentum principle in differential form (the so-called local or "in the small" form) is expressed by the equation $\partial(\rho v_i)/\partial t = \rho b_i + (\sigma_{ij} - \rho v_i v_j)_{,j}$. Show that the equation of motion (*5.16*) follows from this equation.

Carrying out the indicated differentiation and rearranging the terms in the resulting equation yields

$$v_i(\partial\rho/\partial t + \rho_{,j} v_j + \rho v_{j,j}) + \rho(\partial v_i/\partial t + v_j v_{i,j}) = \rho b_i + \sigma_{ij,j}$$

The first term on the left is zero by (*5.4*) and the second term is ρa_i. Thus $\rho a_i = \rho b_i + \sigma_{ij,j}$ which is (*5.16*).

5.9. Show that (*5.19*) reduces to (*5.20*).

Substituting $\sigma_{pk} n_p$ for $t_k^{(\hat{\mathbf{n}})}$ in (*5.19*) and applying the divergence theorem (*1.157*) to the resulting surface integral gives

$$\int_V \epsilon_{ijk}\{(x_j \sigma_{pk})_{,p} + x_j \rho b_k\}\, dV = \frac{d}{dt}\int_V \epsilon_{ijk}\rho(x_j v_k)\, dV$$

Using the results of Problem 5.2, the indicated differentiations here lead to

$$\int_V \epsilon_{ijk}\{x_{j,p}\sigma_{pk} + x_j(\sigma_{pk,p} + \rho b_k - \rho \dot{v}_k) - \rho v_j v_k\}\, dV = 0$$

The term in parentheses is zero by (*5.16*), also $x_{j,p} = \delta_{jp}$ and $\epsilon_{ijk} v_j v_k = 0$, so that finally $\int_V \epsilon_{ijk}\sigma_{jk}\, dV = 0$.

5.10. For a rigid body rotation about a point, $v_i = \epsilon_{ijk}\omega_j x_k$. Show that for this velocity (*5.19*) reduces to the well-known momentum principle of rigid body dynamics.

The left hand side of (*5.19*) is the total moment M_i of all surface and body forces relative to the origin. Thus for $v_i = \epsilon_{ijk}\omega_j x_k$,

$$M_i = \frac{d}{dt}\int_V \epsilon_{ijk}x_j\rho\epsilon_{kpq}\omega_p x_q\,dV = \frac{d}{dt}\int_V \omega_p\rho(\delta_{ip}\delta_{jq} - \delta_{iq}\delta_{jp})x_j x_q\,dV$$

$$= \frac{d}{dt}\left[\omega_p\int_V \rho(\delta_{ip}x_q x_q - x_p x_i)\,dV\right] = \frac{d}{dt}(\omega_p I_{ip})$$

where $I_{ip} = \int_V \rho(\delta_{ip}x_q x_q - x_p x_i)\,dV$ is the moment of inertia tensor.

ENERGY. ENTROPY. DISSIPATION FUNCTION (Sec. 5.4-5.6)

5.11. Show that for a rigid body rotation with $v_i = \epsilon_{ijk}\omega_j x_k$, the kinetic energy integral of (*5.23*) reduces to the familiar form given in rigid body dynamics.

From (*5.23*),

$$K = \int_V \rho\frac{v_i v_i}{2}\,dV = \frac{1}{2}\int_V \rho\epsilon_{ijk}\omega_j x_k\epsilon_{ipq}\omega_p x_q\,dV$$

$$= \frac{1}{2}\int_V \rho\omega_p\omega_j(\delta_{jp}\delta_{kq} - \delta_{jq}\delta_{kp})x_k x_q\,dV$$

$$= \frac{\omega_j\omega_p}{2}\int_V \rho(\delta_{jp}x_q x_q - x_p x_j)\,dV = \frac{\omega_j\omega_p I_{jp}}{2}$$

In symbolic notation note that $K = \dfrac{\boldsymbol{\omega}\cdot\mathbf{I}\cdot\boldsymbol{\omega}}{2}$.

5.12. At a certain point in a continuum the rate of deformation and stress tensors are given by

$$D_{ij} = \begin{pmatrix} 1 & 6 & 4 \\ 6 & 3 & 2 \\ 4 & 2 & 5 \end{pmatrix} \quad \text{and} \quad \sigma_{ij} = \begin{pmatrix} 4 & 0 & -1 \\ 0 & -2 & 7 \\ -1 & 7 & 8 \end{pmatrix}$$

Determine the value λ of the stress power $D_{ij}\sigma_{ij}$ at the point.

Multiplying each element of D_{ij} by its counterpart in σ_{ij} and adding, $\lambda = 4 + 0 - 4 + 0 - 6 + 14 - 4 + 14 + 40 = 58$.

5.13. If $\sigma_{ij} = -p\delta_{ij}$ where p is a positive constant, show that the stress power may be expressed by the equation $D_{ij}\sigma_{ij} = \dfrac{p}{\rho}\dfrac{d\rho}{dt}$.

By (*4.19*), $D_{ij} = v_{i,j} - V_{ij}$; and since $V_{ij}\sigma_{ij} = 0$, it follows that $D_{ij}\sigma_{ij} = v_{i,j}(-p\delta_{ij}) = -pv_{i,i}$. From the continuity equation (*5.3*), $v_{i,i} = -(1/\rho)(d\rho/dt)$ and so $D_{ij}\sigma_{ij} = (p/\rho)(d\rho/dt)$ for $\sigma_{ij} = -p\delta_{ij}$.

5.14. Determine the form of the energy equation if $\sigma_{ij} = (-p + \lambda^* D_{kk})\delta_{ij} + 2\mu^* D_{ij}$ and the heat conduction obeys the Fourier law $c_i = -kT_{,i}$.

From (*5.32*),

$$\rho\frac{du}{dt} = (-p + \lambda^* D_{kk})\delta_{ij}D_{ij} + 2\mu^* D_{ij}D_{ij} + kT_{,ii} + z$$

$$= \frac{p}{\rho}\frac{d\rho}{dt} + (\lambda^* + 2\mu^*)(\mathrm{I_D})^2 - 4\mu^*\mathrm{II_D} + kT_{,ii} + z$$

where $\mathrm{I_D}$ and $\mathrm{II_D}$ are the first and second invariants respectively of the rate of deformation tensor.

5.15. If $\sigma_{ij} = -p\delta_{ij}$, determine an equation for the rate of change of specific entropy during a reversible thermodynamic process.

Here $\sigma_{ij} = \sigma_{ij}^{(C)}$ and (5.41) gives $T\dfrac{ds}{dt} = \dfrac{du}{dt} + \dfrac{p}{\rho^2}\dfrac{d\rho}{dt}$ upon use of the result in Problem 5.13.

5.16. For the stress having $\sigma_{ij}^{(D)} = \beta D_{ik}D_{kj}$, determine the dissipation function in terms of the invariants of the rate of deformation tensor **D**.

Here by (4.25), $\sigma_{ij}^{(D)}\dot{\epsilon}_{ij} = \beta D_{ik}D_{kj}D_{ij}$ which is the trace of $\mathbf{D}^3$ (see page 16) and may be evaluated by using the principal axis values $D_{(1)}, D_{(2)}, D_{(3)}$. Thus by (1.138) the trace

$$\begin{aligned} D_{ij}D_{ik}D_{kj} &= D_{(1)}^3 + D_{(2)}^3 + D_{(3)}^3 \\ &= (D_{(1)} + D_{(2)} + D_{(3)})^3 - 3(D_{(1)} + D_{(2)} + D_{(3)})(D_{(1)}D_{(2)} + D_{(2)}D_{(3)} + D_{(3)}D_{(1)}) \\ &\qquad + 3D_{(1)}D_{(2)}D_{(3)} \end{aligned}$$

Therefore $\sigma_{ij}^{(D)}\dot{\epsilon}_{ij} = \beta[\mathrm{I}_\mathbf{D}^3 - 3\mathrm{I}_\mathbf{D}\mathrm{II}_\mathbf{D} + 3\mathrm{III}_\mathbf{D}]$.

CONSTITUTIVE EQUATIONS (Sec. 5.7)

5.17. For the constitutive equations $\sigma_{ij} = K_{ijpq}D_{pq}$ show that because of the symmetry of the stress and rate of deformation tensors the fourth order tensor K_{ijpq} has at most 36 distinct components. Display the components in a 6×6 array.

Since $\sigma_{ij} = \sigma_{ji}$, $K_{ijpq} = K_{jipq}$; and since $D_{ij} = D_{ji}$, $K_{ijpq} = K_{ijqp}$. If K_{ijpq} is considered as the outer product of two symmetric tensors $A_{ij}B_{pq} = K_{ijpq}$, it is clear that since both A_{ij} and B_{ij} have six independent components, K_{ijpq} will have at most 36 distinct components.

The usual arrangement followed in displaying the components of K_{ijpq} is

$$K_{ijpq} = \begin{pmatrix} K_{1111} & K_{1122} & K_{1133} & K_{1123} & K_{1131} & K_{1112} \\ K_{2211} & K_{2222} & K_{2233} & K_{2223} & K_{2231} & K_{2212} \\ K_{3311} & K_{3322} & K_{3333} & K_{3323} & K_{3331} & K_{3312} \\ K_{2311} & K_{2322} & K_{2333} & K_{2323} & K_{2331} & K_{2312} \\ K_{3111} & K_{3122} & K_{3133} & K_{3123} & K_{3131} & K_{3112} \\ K_{1211} & K_{1222} & K_{1233} & K_{1223} & K_{1231} & K_{1212} \end{pmatrix}$$

5.18. If the continuum having the constitutive relations $\sigma_{ij} = K_{ijpq}D_{pq}$ of Problem 5.17 is assumed *isotropic* so that K_{ijpq} has the same array of components in any rectangular Cartesian system of axes, show that by a cyclic labeling of the coordinate axes the 36 components may be reduced to 26.

The coordinate directions may be labeled in six different ways as shown in Fig. 5-2. Isotropy of K_{ijpq} then requires that $K_{1122} = K_{1133} = K_{2233} = K_{2211} = K_{3311} = K_{3322}$ and that $K_{1212} = K_{1313} = K_{2323} = K_{2121} = K_{3131} = K_{3232}$ which reduces the 36 components to 26. By suitable reflections and rotations of the coordinate axes these 26 components may be reduced to 2 for the case of isotropy.

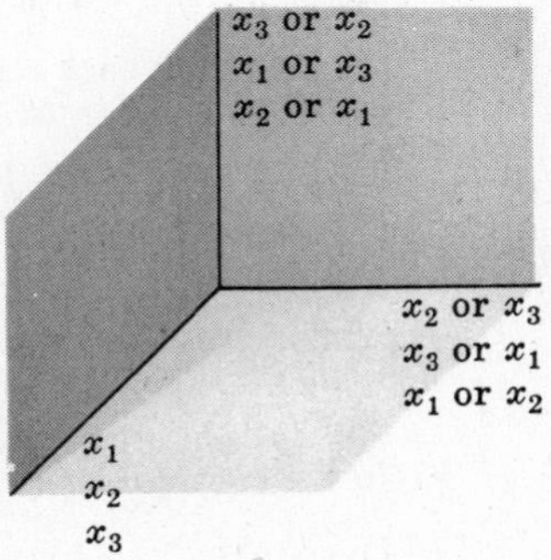

Fig. 5-2

5.19. For isotropy K_{ijpq} may be represented by $K_{ijpq} = \lambda^*\delta_{ij}\delta_{pq} + \mu^*(\delta_{ip}\delta_{jq} + \delta_{iq}\delta_{jp})$. Use this to develop the constitutive equation $\sigma_{ij} = K_{ijpq}D_{pq}$ in terms of λ^* and μ^*.

$$\begin{aligned} \sigma_{ij} &= \lambda^*\delta_{ij}\delta_{pq}D_{pq} + \mu^*(\delta_{ip}\delta_{jq} + \delta_{iq}\delta_{jp})D_{pq} \\ &= \lambda^*\delta_{ij}D_{pp} + \mu^*(D_{ij} + D_{ji}) = \lambda^*\delta_{ij}D_{pp} + 2\mu^*D_{ij} \end{aligned}$$

5.20. Show that the constitutive equation of Problem 5.19 may be split into the equivalent equations $\sigma_{ii} = (3\lambda^* + 2\mu^*)D_{ii}$ and $s_{ij} = 2\mu^* D'_{ij}$ where s_{ij} and D'_{ij} are the deviator tensors of stress and rate of deformation, respectively.

Substituting $\sigma_{ij} = s_{ij} + \delta_{ij}\sigma_{kk}/3$ and $D_{ij} = D'_{ij} + \delta_{ij}D_{kk}/3$ into $\sigma_{ij} = \lambda^*\delta_{ij}D_{kk} + 2\mu^* D_{ij}$ of Problem 5.19 results in the equation $s_{ij} + \delta_{ij}\sigma_{kk}/3 = \lambda^*\delta_{ij}D_{kk} + 2\mu^*(D'_{ij} + \delta_{ij}D_{kk}/3)$. From this when $i \neq j$, $s_{ij} = 2\mu^* D'_{ij}$ and hence $\sigma_{kk} = (3\lambda^* + 2\mu^*)D_{kk}$.

MISCELLANEOUS PROBLEMS

5.21. Show that $\dfrac{d}{dt}\left(\dfrac{q_i}{\rho}\right) = (\epsilon_{ijk}a_{k,j} + q_j v_{i,j})/\rho$ where ρ is the density, a_i the acceleration and q_i the vorticity vector.

By direct differentiation $\dfrac{d}{dt}\left(\dfrac{q_i}{\rho}\right) = \dfrac{\dot{q}_i}{\rho} - \dfrac{q_i\dot{\rho}}{\rho^2}$. But $\dot{q}_i = \epsilon_{ijk}a_{k,j} + q_j v_{i,j} - q_i v_{j,j}$ (see Problem 4.32); and by the continuity equation (*5.3*), $\dot{\rho} = -\rho v_{i,i}$. Thus

$$\frac{d}{dt}\left(\frac{q_i}{\rho}\right) = \frac{1}{\rho}(\epsilon_{ijk}a_{k,j} + q_j v_{i,j} - q_i v_{j,j} + q_i v_{j,j}) = (\epsilon_{ijk}a_{k,j} + q_j v_{i,j})/\rho$$

5.22. A two dimensional incompressible flow is given by $v_1 = A(x_1^2 - x_2^2)/r^4$, $v_2 = A(2x_1x_2)/r^4$, $v_3 = 0$, where $r^2 = x_1^2 + x_2^2$. Show that the continuity equation is satisfied by this motion.

By (*5.5*), $v_{i,i} = 0$ for incompressible flow. Here $v_{1,1} = A[-4x_1(x_1^2 - x_2^2)/r^6 + 2x_1/r^4]$ and $v_{2,2} = A[2x_1/r^4 - 8x_1x_2^2/r^6]$. Adding, $v_{1,1} + v_{2,2} = 0$.

5.23. Show that the flow of Problem 5.22 is irrotational.

By (*4.29*), $\text{curl}\,\mathbf{v} = 0$ for irrotational flow. Thus

$$\text{curl}\,\mathbf{v} = \begin{vmatrix} \hat{\mathbf{e}}_1 & \hat{\mathbf{e}}_2 & \hat{\mathbf{e}}_3 \\ \partial/\partial x_1 & \partial/\partial x_2 & \partial/\partial x_3 \\ A(x_1^2 - x_2^2)/r^4 & 2Ax_1x_2/r^4 & 0 \end{vmatrix}$$

$$= A[2x_2/r^4 - 8x_1^2x_2/r^6 + 2x_2/r^4 + 4x_2(x_1^2 - x_2^2)/r^6]\,\hat{\mathbf{e}}_3 = 0$$

5.24. In a two dimensional incompressible steady flow, $v_1 = -Ax_2/r^2$ where $r^2 = x_1^2 + x_2^2$. Determine v_2 if $v_2 = 0$ at $x_1 = 0$ for all x_2. Show that the motion is irrotational and that the streamlines are circles.

From (*5.5*), $v_{i,i} = 0$ or $v_{1,1} = -v_{2,2} = 2Ax_1x_2/r^4$ for this incompressible flow. Integrating with respect to x_2 and imposing the given conditions on v_2 yields $v_2 = Ax_1/r^2$.

For an irrotational motion, $\text{curl}\,\mathbf{v} = 0$. Here

$$\text{curl}\,\mathbf{v} = A[(x_1^2 - x_2^2)/r^4 + (-x_1^2 + x_2^2)/r^4]\,\hat{\mathbf{e}}_3 = 0$$

From Problem 4.7, page 118, the equations for streamlines are $dx_1/v_1 = dx_2/v_2$. Here these equations are $x_1\,dx_1 + x_2\,dx_2 = 0$ which integrate directly into the circles $x_1^2 + x_2^2 = \text{constant}$.

5.25. For a continuum whose constitutive equations are $\sigma_{ij} = (-p + \lambda^* D_{kk})\delta_{ij} + 2\mu^* D_{ij}$, determine the equations of motion in terms of the velocity v_i.

From (*5.16*), $\rho\dot{v}_i = \rho b_i + \sigma_{ij,j}$ or here $\rho\dot{v}_i = \rho b_i - p_{,j}\delta_{ij} + \lambda^* D_{kk,j}\delta_{ij} + 2\mu^* D_{ij,j}$. By definition, $2D_{ij} = v_{i,j} + v_{j,i}$ so that $D_{kk} = v_{k,k}$ and $2D_{ij,j} = v_{i,jj} + v_{j,ij}$. Therefore

$$\rho\dot{v}_i = \rho b_i - p_{,i} + (\lambda^* + \mu^*)v_{j,ij} + \mu^* v_{i,jj}$$

In symbolic notation this equation is

$$\rho\dot{\mathbf{v}} = \rho\mathbf{b} - \nabla p + (\lambda^* + \mu^*)\nabla(\nabla \cdot \mathbf{v}) + \mu^*\nabla^2\mathbf{v}$$

5.26. If the continuum of Problem 5.25 is considered incompressible, show that the divergence of the vorticity vanishes and give the form of the equations of motion for this case.

By (*4.29*), $q_i = \epsilon_{ijk}v_{k,j}$; and $\operatorname{div}\mathbf{q} = \epsilon_{ijk}v_{k,ji} = 0$ since ϵ_{ijk} is antisymmetric and $v_{k,ji}$ is symmetric in i and j. Thus for $\nabla \cdot \mathbf{v} = 0$ the equations of motion become $\rho\dot{v}_i = \rho b_i - p_{,i} + \mu^* v_{i,jj}$ or in Gibbs notation $\rho\dot{\mathbf{v}} = \rho\mathbf{b} - \nabla p + \mu^*\nabla^2\mathbf{v}$.

5.27. Determine the material rate of change of the kinetic energy of the continuum which occupies the volume V and give the meaning of the resulting integrals.

By (*5.23*), $dK/dt = \int_V \rho v_i \dot{v}_i \, dV$. Also the total stress power of the surface forces is $\int_S v_i t_i^{(\hat{n})} \, dS$ which may be written $\int_S v_i \sigma_{ij} n_j \, dS$ and by the divergence theorem (*1.157*) and the equations of motion (*5.16*) expressed as the volume integrals $\int_S v_i \sigma_{ij} n_j \, dS = \int_V \sigma_{ij} v_{i,j} \, dV + \int_V \rho(v_i\dot{v}_i - b_i v_i) \, dV$. Thus

$$\frac{dK}{dt} = \int_V \rho b_i v_i \, dV - \int_V \sigma_{ij} v_{i,j} \, dV + \int_S v_i t_i^{(\hat{n})} \, dS$$

This sum of integrals represents the rate of work done by the body forces, the internal stresses and the surface stresses, respectively.

5.28. A continuum for which $\sigma_{ij}^{(D)} = \lambda^* D_{kk}\delta_{ij} + 2\mu^* D_{ij}$ undergoes an incompressible irrotational flow with a velocity potential ϕ such that $\mathbf{v} = \operatorname{grad}\phi$. Determine the dissipation function $\sigma_{ij}^{(D)}\dot{\epsilon}_{ij}$.

Here $\sigma_{ij}^{(D)}\dot{\epsilon}_{ij} = \sigma_{ij}^{(D)} D_{ij} = (\lambda^* D_{kk}\delta_{ij} + 2\mu^* D_{ij})D_{ij} = 2\mu^* D_{ij}D_{ij}$ since $D_{kk} = v_{k,k} = 0$ for incompressible flow. Also since $v_i = \phi_{,i}$, the scalar $D_{ij}D_{ij} = \phi_{,ij}\phi_{,ij}$ and so $\sigma_{ij}^{(D)} D_{ij} = 2\mu^*\phi_{,ij}\phi_{,ij}$.

Because the motion is incompressible and irrotational, $\phi_{,ii} = 0$ and $2\phi_{,ij}\phi_{,ij} = (\phi_{,i}\phi_{,i})_{,jj} = \nabla^2(\nabla\phi)^2$. It is also interesting to note that

$$\nabla^4(\phi^2) = (\phi\phi)_{,iijj} = 2(\phi_{,iijj}\phi + 4\phi_{,ijj}\phi_{,i} + \phi_{,ii}\phi_{,jj} + 2\phi_{,ij}\phi_{,ij})$$

which for $\phi_{,ii} = 0$ reduces to $4\phi_{,ij}\phi_{,ij}$. Thus $\phi_{ij}^{(D)}\dot{\epsilon}_{ij} = \mu^*\nabla^2(\nabla\phi)^2 = \mu^*\nabla^4(\phi^2)/2$.

5.29. For a continuum with $\sigma_{ij} = -p\delta_{ij}$, the *specific enthalpy* $h = u + p/\rho$. Show that the energy equation may be written $\dot{h} = \dot{p}/\rho + T\dot{s}$ using this definition for the enthalpy.

From (*5.41*), $\dot{u} = -p\delta_{ij}D_{ij}/\rho + T\dot{s}$ for the given stress; and by the result of Problem 5.13 and the definition of h, $\dot{u} = \dot{h} - \dot{p}/\rho - p\dot{\rho}/\rho^2 = -p\dot{\rho}/\rho^2 + T\dot{s}$. Canceling and rearranging, $\dot{h} = \dot{p}/\rho + T\dot{s}$.

5.30. If the continuum of Problem 5.25 undergoes an incompressible flow, determine the equation of motion in terms of the vorticity $\mathbf{q}$ in the absence of body forces and assuming constant density.

For incompressible flow, $\nabla \cdot \mathbf{v} = 0$; and if $\mathbf{b} \equiv 0$, the equation of motion in Problem 5.25 reduces to $\rho\dot{v}_i = -p_{,i} + \mu^* v_{i,jj}$. Taking the cross product $\nabla \times$ with this equation for $\rho = $ constant gives $\epsilon_{pqi}\dot{v}_{i,q} = -\epsilon_{pqi}p_{,iq}/\rho + (\mu^*/\rho)\epsilon_{pqi}v_{i,jjq}$. But $\epsilon_{pqi}p_{,iq} = 0$ and by (*4.29*) the result is $\dot{q}_p = (\mu^*/\rho)q_{,jj}$. In symbolic notation, $d\mathbf{q}/dt = (\mu^*/\rho)\nabla^2\mathbf{q}$.

Supplementary Problems

5.31. Show that for the rate of rotation vector $\mathbf{\Omega}$, $\dfrac{d}{dt}\left(\dfrac{\mathbf{\Omega}}{\rho}\right) = \dfrac{\mathbf{\Omega}\cdot\nabla\mathbf{v}}{\rho}$.

5.32. Show that the flow represented by $v_1 = -2x_1x_2x_3/r^4$, $v_2 = (x_1^2 - x_2^2)x_3/r^4$, $v_3 = x_2/r^2$ where $r^2 = x_1^2 + x_2^2$, satisfies conditions for an incompressible flow. Is this motion irrotational?

5.33. In terms of Cartesian coordinates x, y, z the continuity equation is

$$\partial\rho/\partial t + \partial(\rho v_x)/\partial x + \partial(\rho v_y)/\partial y + \partial(\rho v_z)/\partial z = 0$$

Show that in terms of cylindrical coordinates r, θ, z this equation becomes

$$r(\partial\rho/\partial t) + \partial(r\rho v_r)/\partial r + \partial(\rho v_\theta)/\partial\theta + r(\partial(\rho v_z)/\partial z) = 0$$

5.34. Show that the flow $v_r = (1-r^2)\cos\theta/r^2$, $v_\theta = (1+r^2)\sin\theta/r^2$, $v_z = 0$ satisfies the continuity equation in cylindrical coordinates when the density ρ is a constant.

5.35. If $P_{ij\ldots}(\mathbf{x}, t)$ is an arbitrary scalar, vector or tensor function, show that

$$\int_S P_{ij\ldots}\sigma_{pq}n_q\,dS = \int_V [\sigma_{pq}P_{ij\ldots,q} + \rho P_{ij\ldots}(\dot{v}_p - b_p)]\,dV$$

5.36. If a continuum is subjected to a body moment per unit mass $\mathbf{h}$ in addition to body force $\mathbf{b}$, and a couple stress $\mathbf{g}^{(\hat{\mathbf{n}})}$ in addition to the stress $\mathbf{t}^{(\hat{\mathbf{n}})}$, the angular momentum balance may be written

$$\frac{d}{dt}\int_V \rho(\mathbf{m} + \mathbf{x}\times\mathbf{v})\,dV = \int_V (\mathbf{h} + \mathbf{x}\times\mathbf{b})\,dV + \int_S (\mathbf{g}^{(\hat{\mathbf{n}})} + \mathbf{x}\times\mathbf{t}^{(\hat{\mathbf{n}})})\,dS$$

where $\mathbf{m}$ is distributed angular momentum per unit mass. If $\hat{\mathbf{n}}\cdot\mathbf{G} = \mathbf{g}^{(\hat{\mathbf{n}})}$, show that the local form of this relation is $\rho\,d\mathbf{m}/dt = \mathbf{h} + \nabla\cdot\mathbf{G} + \mathbf{\Sigma}_v$.

5.37. If a continuum has the constitutive equation $\sigma_{ij} = -p\delta_{ij} + \beta D_{ij} + \alpha D_{ik}D_{kj}$, show that $\sigma_{ii} = 3(-p - 2\alpha \mathrm{II}_D/3)$. Assume incompressibility, $D_{ii} = 0$.

5.38. For a continuum having $\sigma_{ij} = -p\delta_{ij}$, show that $du = T\,ds - p\,dv$ where *in this problem* $v = 1/\rho$, the specific volume.

5.39. If $T\,ds/dt = -v_{i,i}/\rho$ and the specific free energy is defined by $\Psi = u - Ts$, show that the energy equation may be written $\rho\,d\Psi/dt + \rho s\,dT/dt = \sigma_{ij}D_{ij}$.

5.40. For a thermomechanical continuum having the constitutive equation

$$\sigma_{ij} = \lambda\epsilon_{kk}\delta_{ij} + 2\mu\epsilon_{ij} - (3\lambda + 2\mu)\alpha\,\delta_{ij}(T - T_0)$$

where T_0 is a reference temperature, show that $\epsilon_{kk} = 3\alpha(T - T_0)$ when $\sigma_{ij} = s_{ij} = \sigma_{ij} - \sigma_{kk}\delta_{ij}/3$.

Chapter 6

Linear Elasticity

6.1 GENERALIZED HOOKE'S LAW. STRAIN ENERGY FUNCTION

In classical linear elasticity theory it is assumed that *displacements* and *displacement gradients* are sufficiently small that no distinction need be made between the Lagrangian and Eulerian descriptions. Accordingly in terms of the displacement vector u_i, the linear strain tensor is given by the equivalent expressions

$$l_{ij} = \epsilon_{ij} = \frac{1}{2}\left(\frac{\partial u_i}{\partial X_j} + \frac{\partial u_j}{\partial X_i}\right) = \frac{1}{2}\left(\frac{\partial u_i}{\partial x_j} + \frac{\partial u_j}{\partial x_i}\right) = \tfrac{1}{2}(u_{i,j} + u_{j,i})$$

or $$\mathbf{L} = \mathbf{E} = \tfrac{1}{2}(\mathbf{u}\nabla_{\mathbf{X}} + \nabla_{\mathbf{X}}\mathbf{u}) = \tfrac{1}{2}(\mathbf{u}\nabla_{\mathbf{x}} + \nabla_{\mathbf{x}}\mathbf{u}) = \tfrac{1}{2}(\mathbf{u}\nabla + \nabla\mathbf{u}) \qquad (6.1)$$

In the following it is further assumed that the deformation processes are *adiabatic* (no heat loss or gain) and *isothermal* (constant temperature) unless specifically stated otherwise.

The constitutive equations for a linear elastic solid relate the stress and strain tensors through the expression

$$\sigma_{ij} = C_{ijkm}\epsilon_{km} \quad \text{or} \quad \mathbf{\Sigma} = \tilde{\mathbf{C}} : \mathbf{E} \qquad (6.2)$$

which is known as the *generalized Hooke's law*. In (*6.2*) the *tensor* of *elastic constants* C_{ijkm} has 81 components. However, due to the symmetry of both the stress and strain tensors, there are at most 36 distinct elastic constants. For the purpose of writing Hooke's law in terms of these 36 components, the double indexed system of stress and strain components is often replaced by a single indexed system having a range of 6. Thus in the notation

$$\begin{array}{ll} \sigma_{11} = \sigma_1 & \sigma_{23} = \sigma_{32} = \sigma_4 \\ \sigma_{22} = \sigma_2 & \sigma_{13} = \sigma_{31} = \sigma_5 \\ \sigma_{33} = \sigma_3 & \sigma_{12} = \sigma_{21} = \sigma_6 \end{array} \qquad (6.3)$$

and

$$\begin{array}{ll} \epsilon_{11} = \epsilon_1 & 2\epsilon_{23} = 2\epsilon_{32} = \epsilon_4 \\ \epsilon_{22} = \epsilon_2 & 2\epsilon_{13} = 2\epsilon_{31} = \epsilon_5 \\ \epsilon_{33} = \epsilon_3 & 2\epsilon_{12} = 2\epsilon_{21} = \epsilon_6 \end{array} \qquad (6.4)$$

Hooke's law may be written

$$\sigma_K = C_{KM}\epsilon_M \quad (K, M = 1,2,3,4,5,6) \qquad (6.5)$$

where C_{KM} represents the 36 elastic constants, and where upper case Latin subscripts are used to emphasize the range of 6 on these indices.

When thermal effects are neglected, the energy balance equation (*5.32*) may be written

$$\frac{du}{dt} = \frac{1}{\rho}\sigma_{ij}D_{ij} = \frac{1}{\rho}\sigma_{ij}\dot{\epsilon}_{ij} \qquad (6.6)$$

The internal energy in this case is purely mechanical and is called the *strain energy* (per unit mass). From (*6.6*),

$$du = \frac{1}{\rho}\sigma_{ij}\,d\epsilon_{ij} \tag{6.7}$$

and if u is considered a function of the nine strain components, $u = u(\epsilon_{ij})$, its differential is given by

$$du = \frac{\partial u}{\partial \epsilon_{ij}}\,d\epsilon_{ij} \tag{6.8}$$

Comparing (*6.7*) and (*6.8*), it is observed that

$$\frac{1}{\rho}\sigma_{ij} = \frac{\partial u}{\partial \epsilon_{ij}} \tag{6.9}$$

The *strain energy density* u^* (per unit volume) is defined as

$$u^* = \rho u \tag{6.10}$$

and since ρ may be considered a constant in the small strain theory, u^* has the property that

$$\sigma_{ij} = \rho\frac{\partial u}{\partial \epsilon_{ij}} = \frac{\partial u^*}{\partial \epsilon_{ij}} \tag{6.11}$$

Furthermore, the zero state of strain energy may be chosen arbitrarily; and since the stress must vanish with the strains, the simplest form of strain energy function that leads to a linear stress-strain relation is the quadratic form

$$u^* = \tfrac{1}{2}C_{ijkm}\epsilon_{ij}\epsilon_{km} \tag{6.12}$$

From (*6.2*), this equation may be written

$$u^* = \tfrac{1}{2}\sigma_{ij}\epsilon_{ij} \quad \text{or} \quad u^* = \tfrac{1}{2}\mathbf{\Sigma}:\mathbf{E} \tag{6.13}$$

In the single indexed system of symbols, (*6.12*) becomes

$$u^* = \tfrac{1}{2}C_{KM}\epsilon_K\epsilon_M \tag{6.14}$$

in which $C_{KM} = C_{MK}$. Because of this symmetry on C_{KM}, the number of independent elastic constants is at most 21 if a strain energy function exists.

6.2 ISOTROPY. ANISOTROPY. ELASTIC SYMMETRY

If the elastic properties are independent of the reference system used to describe it, a material is said to be *elastically isotropic*. A material that is not isotropic is called *anisotropic*. Since the elastic properties of a Hookean solid are expressed through the coefficients C_{KM}, a general anisotropic body will have an *elastic-constant matrix* of the form

$$[C_{KM}] = \begin{bmatrix} C_{11} & C_{12} & C_{13} & C_{14} & C_{15} & C_{16} \\ C_{21} & C_{22} & C_{23} & C_{24} & C_{25} & C_{26} \\ C_{31} & C_{32} & C_{33} & C_{34} & C_{35} & C_{36} \\ C_{41} & C_{42} & C_{43} & C_{44} & C_{45} & C_{46} \\ C_{51} & C_{52} & C_{53} & C_{54} & C_{55} & C_{56} \\ C_{61} & C_{62} & C_{63} & C_{64} & C_{65} & C_{66} \end{bmatrix} \tag{6.15}$$

When a strain energy function exists for the body, $C_{KM} = C_{MK}$, and the 36 constants in (*6.15*) are reduced to 21.

A *plane of elastic symmetry* exists at a point where the elastic constants have the same values for every pair of coordinate systems which are the reflected images of one another with respect to the plane. The axes of such coordinate systems are referred to as "equivalent elastic directions." If the x_1x_2 plane is one of elastic symmetry, the constants C_{KM} are invariant under the coordinate transformation

$$x_1' = x_1, \quad x_2' = x_2, \quad x_3' = -x_3 \qquad (6.16)$$

as shown in Fig. 6-1. The transformation matrix of (*6.16*) is given by

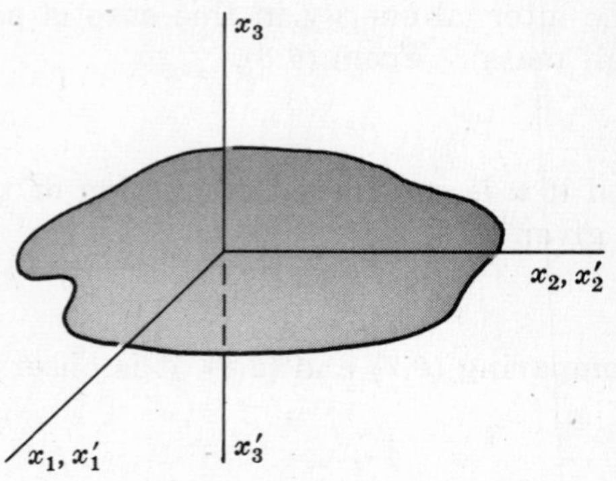

Fig. 6-1

$$[a_{ij}] = \begin{bmatrix} 1 & 0 & 0 \\ 0 & 1 & 0 \\ 0 & 0 & -1 \end{bmatrix} \qquad (6.17)$$

Inserting the values of (*6.17*) into the transformation laws for the linear stress and strain tensors, (*2.27*) and (*3.78*) respectively, the elastic matrix for a material having x_1x_2 as a plane of symmetry is

$$[C_{KM}] = \begin{bmatrix} C_{11} & C_{12} & C_{13} & 0 & 0 & C_{16} \\ C_{21} & C_{22} & C_{23} & 0 & 0 & C_{26} \\ C_{31} & C_{32} & C_{33} & 0 & 0 & C_{36} \\ 0 & 0 & 0 & C_{44} & C_{45} & 0 \\ 0 & 0 & 0 & C_{54} & C_{55} & 0 \\ C_{61} & C_{62} & C_{63} & 0 & 0 & C_{66} \end{bmatrix} \qquad (6.18)$$

The 20 constants in (*6.18*) are reduced to 13 when a strain energy function exists.

If a material possesses three mutually perpendicular planes of elastic symmetry, the material is called *orthotropic* and its elastic matrix is of the form

$$[C_{KM}] = \begin{bmatrix} C_{11} & C_{12} & C_{13} & 0 & 0 & 0 \\ C_{21} & C_{22} & C_{23} & 0 & 0 & 0 \\ C_{31} & C_{32} & C_{33} & 0 & 0 & 0 \\ 0 & 0 & 0 & C_{44} & 0 & 0 \\ 0 & 0 & 0 & 0 & C_{55} & 0 \\ 0 & 0 & 0 & 0 & 0 & C_{66} \end{bmatrix} \qquad (6.19)$$

having 12 independent constants, or 9 if $C_{KM} = C_{MK}$.

An *axis of elastic symmetry* of order N exists at a point when there are sets of equivalent elastic directions which can be superimposed by a rotation through an angle of $2\pi/N$ about the axis. Certain cases of axial and plane elastic symmetry are equivalent.

6.3 ISOTROPIC MEDIA. ELASTIC CONSTANTS

Bodies which are elastically equivalent in all directions possess complete symmetry and are termed *isotropic*. Every plane and every axis is one of elastic symmetry in this case.

For isotropy, the number of independent elastic constants reduces to 2, and the elastic matrix is symmetric regardless of the existence of a strain energy function. Choosing as the two independent constants the well-known Lamé constants, λ and μ, the matrix (*6.19*) reduces to the isotropic elastic form

$$[C_{KM}] = \begin{bmatrix} \lambda+2\mu & \lambda & \lambda & 0 & 0 & 0 \\ \lambda & \lambda+2\mu & \lambda & 0 & 0 & 0 \\ \lambda & \lambda & \lambda+2\mu & 0 & 0 & 0 \\ 0 & 0 & 0 & \mu & 0 & 0 \\ 0 & 0 & 0 & 0 & \mu & 0 \\ 0 & 0 & 0 & 0 & 0 & \mu \end{bmatrix} \tag{6.20}$$

In terms of λ and μ, Hooke's law (*6.2*) for an isotropic body is written

$$\sigma_{ij} = \lambda\delta_{ij}\epsilon_{kk} + 2\mu\epsilon_{ij} \quad \text{or} \quad \mathbf{\Sigma} = \lambda\mathbf{I}\epsilon + 2\mu\mathbf{E} \tag{6.21}$$

where $\epsilon = \epsilon_{kk} = \mathrm{I}_{\mathbf{E}}$. This equation may be readily inverted to express the strains in terms of the stresses as

$$\epsilon_{ij} = \frac{-\lambda}{2\mu(3\lambda+2\mu)}\delta_{ij}\sigma_{kk} + \frac{1}{2\mu}\sigma_{ij} \quad \text{or} \quad \mathbf{E} = \frac{-\lambda}{2\mu(3\lambda+2\mu)}\mathbf{I}\Theta + \frac{1}{2\mu}\mathbf{\Sigma} \tag{6.22}$$

where $\Theta = \sigma_{kk} = \mathrm{I}_{\mathbf{\Sigma}}$, the symbol traditionally used in elasticity for the first stress invariant.

For a simple uniaxial state of stress in the x_1 direction, engineering constants E and ν may be introduced through the relationships $\sigma_{11} = E\epsilon_{11}$ and $\epsilon_{22} = \epsilon_{33} = -\nu\epsilon_{11}$. The constant E is known as *Young's modulus*, and ν is called *Poisson's ratio*. In terms of these elastic constants Hooke's law for isotropic bodies becomes

$$\sigma_{ij} = \frac{E}{1+\nu}\left(\epsilon_{ij} + \frac{\nu}{1-2\nu}\delta_{ij}\epsilon_{kk}\right) \quad \text{or} \quad \mathbf{\Sigma} = \frac{E}{1+\nu}\left(\mathbf{E} + \frac{\nu}{1-2\nu}\mathbf{I}\epsilon\right) \tag{6.23}$$

or, when inverted,

$$\epsilon_{ij} = \frac{1+\nu}{E}\sigma_{ij} - \frac{\nu}{E}\delta_{ij}\sigma_{kk} \quad \text{or} \quad \mathbf{E} = \frac{1+\nu}{E}\mathbf{\Sigma} - \frac{\nu}{E}\mathbf{I}\Theta \tag{6.24}$$

From a consideration of a uniform hydrostatic pressure state of stress, it is possible to define the *bulk modulus*,

$$K = \frac{E}{3(1-2\nu)} \quad \text{or} \quad K = \frac{3\lambda+2\mu}{3} \tag{6.25}$$

which relates the pressure to the cubical dilatation of a body so loaded. For a so-called state of pure shear, the *shear modulus* G relates the shear components of stress and strain. G is actually equal to μ and the expression

$$\mu = G = \frac{E}{2(1+\nu)} \tag{6.26}$$

may be proven without difficulty.

6.4 ELASTOSTATIC PROBLEMS. ELASTODYNAMIC PROBLEMS

In an elastostatic problem of a homogeneous isotropic body, certain field equations, namely,

(*a*) Equilibrium equations,

$$\sigma_{ji,j} + \rho b_i = 0 \quad \text{or} \quad \nabla\cdot\mathbf{\Sigma} + \rho\mathbf{b} = 0 \tag{6.27}$$

(*b*) Hooke's law,

$$\sigma_{ij} = \lambda\delta_{ij}\epsilon_{kk} + 2\mu\epsilon_{ij} \quad \text{or} \quad \mathbf{\Sigma} = \lambda\mathbf{I}\epsilon + 2\mu\mathbf{E} \tag{6.28}$$

(*c*) Strain-displacement relations,

$$\epsilon_{ij} = \tfrac{1}{2}(u_{i,j} + u_{j,i}) \quad \text{or} \quad \mathbf{E} = \tfrac{1}{2}(\mathbf{u}\nabla + \nabla\mathbf{u}) \tag{6.29}$$

must be satisfied at all interior points of the body. Also, prescribed conditions on stress and/or displacements must be satisfied on the bounding surface of the body.

The boundary value problems of elasticity are usually classified according to boundary conditions into problems for which

(1) displacements are prescribed everywhere on the boundary,

(2) stresses (surface tractions) are prescribed everywhere on the boundary,

(3) displacements are prescribed over a portion of the boundary, stresses are prescribed over the remaining part.

For all three categories the body forces are assumed to be given throughout the continuum.

For those problems in which boundary displacement components are given everywhere by an equation of the form

$$u_i = g_i(\mathbf{X}) \quad \text{or} \quad \mathbf{u} = \mathbf{g}(\mathbf{X}) \tag{6.30}$$

the strain-displacement relations (*6.29*) may be substituted into Hooke's law (*6.28*) and the result in turn substituted into (*6.27*) to produce the governing equations,

$$\mu u_{i,jj} + (\lambda+\mu)u_{j,ji} + \rho b_i = 0 \quad \text{or} \quad \mu\nabla^2\mathbf{u} + (\lambda+\mu)\nabla\nabla\cdot\mathbf{u} + \rho\mathbf{b} = 0 \tag{6.31}$$

which are called the *Navier-Cauchy* equations. The solution of this type of problem is therefore given in the form of the displacement vector u_i, satisfying (*6.31*) throughout the continuum and fulfilling (*6.30*) on the boundary.

For those problems in which surface tractions are prescribed everywhere on the boundary by equations of the form

$$t_i^{(\hat{\mathbf{n}})} = \sigma_{ij}n_j \quad \text{or} \quad \mathbf{t}^{(\hat{\mathbf{n}})} = \mathbf{\Sigma}\cdot\hat{\mathbf{n}} \tag{6.32}$$

the equations of compatibility (*3.104*) may be combined with Hooke's law (*6.24*) and the equilibrium equation (*6.27*) to produce the governing equations,

$$\sigma_{ij,kk} + \frac{1}{1+\nu}\sigma_{kk,ij} + \rho(b_{i,j} + b_{j,i}) + \frac{\nu}{1-\nu}\delta_{ij}\rho b_{k,k} = 0$$

or

$$\nabla^2\mathbf{\Sigma} + \frac{1}{1+\nu}\nabla\nabla\Theta + \rho(\nabla\mathbf{b} + \mathbf{b}\nabla) + \frac{\nu}{1-\nu}\mathbf{I}\rho\nabla\cdot\mathbf{b} = 0 \tag{6.33}$$

which are called the *Beltrami-Michell* equations of compatibility. The solution for this type of problem is given by specifying the stress tensor which satisfies (*6.33*) throughout the continuum and fulfills (*6.32*) on the boundary.

For those problems having "mixed" boundary conditions, the system of equations (*6.27*), (*6.28*) and (*6.29*) must be solved. The solution gives the stress and displacement fields throughout the continuum. The stress components must satisfy (*6.32*) over some portion of the boundary, while the displacements satisfy (*6.30*) over the remainder of the boundary.

In the formulation of elastodynamics problems, the equilibrium equations (*6.27*) must be replaced by the equations of motion (*5.16*)

$$\sigma_{ij,j} + \rho b_i = \rho\dot{v}_i \quad \text{or} \quad \nabla\cdot\mathbf{\Sigma} + \rho\mathbf{b} = \rho\dot{\mathbf{v}} \tag{6.34}$$

and *initial conditions* as well as *boundary conditions* must be specified. In terms of the displacement field u_i, the governing equation here, analogous to (6.31) in the elastostatic case is

$$\mu u_{i,jj} + (\lambda+\mu)u_{j,ji} + \rho b_i = \rho \ddot{u}_i \quad \text{or} \quad \mu\nabla^2\mathbf{u} + (\lambda+\mu)\nabla\nabla\cdot\mathbf{u} + \rho\mathbf{b} = \rho\ddot{\mathbf{u}} \tag{6.35}$$

Solutions of (6.35) appear in the form $u_i = u_i(\mathbf{x}, t)$ and must satisfy not only initial conditions on the motion, usually expressed by equations such as

$$u_i = u_i(\mathbf{x}, 0) \quad \text{and} \quad \dot{u}_i = \dot{u}_i(\mathbf{x}, 0) \tag{6.36}$$

but also boundary conditions, either on the displacements,

$$u_i = g_i(\mathbf{x}, t) \quad \text{or} \quad \mathbf{u} = \mathbf{g}(\mathbf{x}, t) \tag{6.37}$$

or on the surface tractions,

$$t_i^{(\hat{\mathbf{n}})} = t_i^{(\hat{\mathbf{n}})}(\mathbf{x}, t) \quad \text{or} \quad \mathbf{t}^{(\hat{\mathbf{n}})} = \mathbf{t}^{(\hat{\mathbf{n}})}(\mathbf{x}, t) \tag{6.38}$$

6.5 THEOREM OF SUPERPOSITION. UNIQUENESS OF SOLUTIONS. ST. VENANT PRINCIPLE

Because the equations of linear elasticity are linear equations, the principle of superposition may be used to obtain additional solutions from those previously established. If, for example, $\sigma_{ij}^{(1)}, u_i^{(1)}$ represent a solution to the system (6.27), (6.28) and (6.29) with body forces $b_i^{(1)}$, and $\sigma_{ij}^{(2)}, u_i^{(2)}$ represent a solution for body forces $b_i^{(2)}$, then $\sigma_{ij} = \sigma_{ij}^{(1)} + \sigma_{ij}^{(2)}$, $u_i = u_i^{(1)} + u_i^{(2)}$ represent a solution to the system for body forces $b_i = b_i^{(1)} + b_i^{(2)}$.

The uniqueness of a solution to the general elastostatic problem of elasticity may be established by use of the superposition principle, together with the law of conservation of energy. A proof of uniqueness is included among the exercises that follow.

St. Venant's principle is a statement regarding the differences that occur in the stresses and strains at some interior location of an elastic body, due to two separate but statically equivalent systems of surface tractions, being applied to some portion of the boundary. The principle asserts that, for locations sufficiently remote from the area of application of the loadings, the differences are negligible. This assumption is often of great assistance in solving practical problems.

6.6 TWO-DIMENSIONAL ELASTICITY. PLANE STRESS AND PLANE STRAIN

Many problems in elasticity may be treated satisfactorily by a two-dimensional, or *plane theory of elasticity*. There are two general types of problems involved in this plane analysis. Although these two types may be defined by setting down certain restrictions and assumptions on the stress and displacement fields, they are often introduced descriptively in terms of their physical prototypes. In *plane stress* problems, the geometry of the body is essentially that of a plate with one dimension much smaller than the others. The loads are applied uniformly over the thickness of the plate and act in the plane of the plate as shown in Fig. 6-2(*a*) below. In *plane strain* problems, the geometry of the body is essentially that of a prismatic cylinder with one dimension much larger than the others. The loads are uniformly distributed with respect to the large dimension and act perpendicular to it as shown in Fig. 6-2(*b*) below.

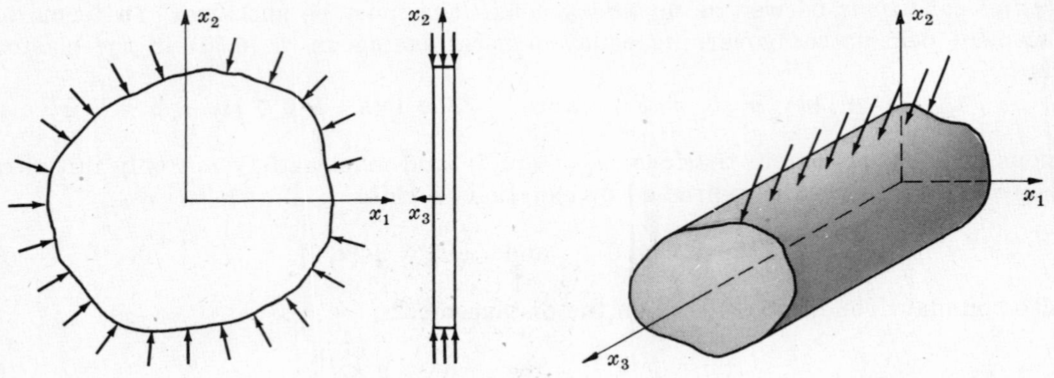

(a) (b)

Fig. 6-2

For the plane stress problem of Fig. 6-2(*a*) the stress components $\sigma_{33}, \sigma_{13}, \sigma_{23}$ are taken as zero everywhere, and the remaining components are taken as functions of x_1 and x_2 only,

$$\sigma_{\alpha\beta} = \sigma_{\alpha\beta}(x_1, x_2) \qquad (\alpha, \beta = 1, 2) \tag{6.39}$$

Accordingly, the field equations for plane stress are

(a) $$\sigma_{\alpha\beta,\beta} + \rho b_\alpha = 0 \qquad \text{or} \qquad \nabla \cdot \mathbf{\Sigma} + \rho \mathbf{b} = 0 \tag{6.40}$$

(b) $$\epsilon_{\alpha\beta} = \frac{1+\nu}{E}\sigma_{\alpha\beta} - \frac{\nu}{E}\delta_{\alpha\beta}\sigma_{\gamma\gamma} \qquad \text{or} \qquad \mathbf{E} = \frac{1+\nu}{E}\mathbf{\Sigma} - \frac{\nu}{E}\mathbf{I}\Theta$$
$$\epsilon_{33} = -\frac{\nu}{E}\sigma_{\alpha\alpha} \tag{6.41}$$

(c) $$\epsilon_{\alpha\beta} = \tfrac{1}{2}(u_{\alpha,\beta} + u_{\beta,\alpha}) \qquad \text{or} \qquad \mathbf{E} = \tfrac{1}{2}(\mathbf{u}\nabla + \nabla\mathbf{u}) \tag{6.42}$$

in which $\nabla \equiv \frac{\partial}{\partial x_1}\hat{\mathbf{e}}_1 + \frac{\partial}{\partial x_2}\hat{\mathbf{e}}_2$ and

$$\mathbf{\Sigma} = \begin{pmatrix} \sigma_{11} & \sigma_{12} & 0 \\ \sigma_{12} & \sigma_{22} & 0 \\ 0 & 0 & 0 \end{pmatrix}, \qquad \mathbf{E} = \begin{pmatrix} \epsilon_{11} & \epsilon_{12} & 0 \\ \epsilon_{12} & \epsilon_{22} & 0 \\ 0 & 0 & \epsilon_{33} \end{pmatrix} \tag{6.43}$$

Due to the particular form of the strain tensor in the plane stress case, the six compatibility equations (*3.104*) may be reduced with reasonable accuracy for very thin plates to the single equation

$$\epsilon_{11,22} + \epsilon_{22,11} = 2\epsilon_{12,12} \tag{6.44}$$

In terms of the displacement components u_α, the field equations may be combined to give the governing equation

$$\frac{E}{2(1+\nu)}\nabla^2 u_\alpha + \frac{E}{2(1-\nu)}u_{\beta,\beta\alpha} + \rho b_\alpha = 0 \qquad \text{or} \qquad \frac{E}{2(1+\nu)}\nabla^2\mathbf{u} + \frac{E}{2(1-\nu)}\nabla\nabla\cdot\mathbf{u} + \rho\mathbf{b} = 0 \tag{6.45}$$

where $\nabla^2 \equiv \frac{\partial^2}{\partial x_1^2} + \frac{\partial^2}{\partial x_2^2}$.

For the plane strain problem of Fig. 6-2(*b*) the displacement component u_3 is taken as zero, and the remaining components considered as functions of x_1 and x_2 only,

$$u_\alpha = u_\alpha(x_1, x_2) \tag{6.46}$$

In this case, the field equations may be written

$$(a)\qquad \sigma_{\alpha\beta,\beta} + \rho b_\alpha = 0 \qquad \text{or} \qquad \nabla\cdot\mathbf{\Sigma} + \rho\mathbf{b} = 0 \tag{6.47}$$

$$(b)\qquad \sigma_{\alpha\beta} = \lambda\delta_{\alpha\beta}\epsilon_{\gamma\gamma} + 2\mu\epsilon_{\alpha\beta} \qquad \text{or} \qquad \mathbf{\Sigma} = \lambda\mathbf{I}\epsilon + 2\mu\mathbf{E}$$

$$\sigma_{33} = \nu\sigma_{\alpha\alpha} = \frac{\lambda}{2(\lambda+\mu)}\sigma_{\alpha\alpha} \tag{6.48}$$

$$(c)\qquad \epsilon_{\alpha\beta} = \tfrac{1}{2}(u_{\alpha,\beta} + u_{\beta,\alpha}) \qquad \text{or} \qquad \mathbf{E} = \tfrac{1}{2}(\mathbf{u}\nabla + \nabla\mathbf{u}) \tag{6.49}$$

$$\text{in which}\qquad \mathbf{\Sigma} = \begin{pmatrix} \sigma_{11} & \sigma_{12} & 0 \\ \sigma_{12} & \sigma_{22} & 0 \\ 0 & 0 & \sigma_{33} \end{pmatrix} \qquad \text{and} \qquad \mathbf{E} = \begin{pmatrix} \epsilon_{11} & \epsilon_{12} & 0 \\ \epsilon_{12} & \epsilon_{22} & 0 \\ 0 & 0 & 0 \end{pmatrix} \tag{6.50}$$

From (*6.47*), (*6.48*), (*6.49*), the appropriate Navier equation for plane strain is

$$\mu\nabla^2 u_\alpha + (\lambda+\mu)u_{\beta,\beta\alpha} + \rho b_\alpha = 0 \qquad \text{or} \qquad \mu\nabla^2\mathbf{u} + (\lambda+\mu)\nabla\nabla\cdot\mathbf{u} + \rho\mathbf{b} = 0 \tag{6.51}$$

As in the case of plane stress, the compatibility equations for plane strain reduce to the single equation (*6.44*).

If the forces applied to the edge of the plate in Fig. 6-2(*a*) are not uniform across the thickness, but are symmetrical with respect to the middle plane of the plate, a state of *generalized plane stress* is said to exist. In formulating problems for this case, the field variables $\sigma_{\alpha\beta}$, $\epsilon_{\alpha\beta}$ and u_α must be replaced by stress, strain and displacement variables averaged across the thickness of the plate. In terms of such averaged field variables, the generalized plane stress formulation is essentially the same as the plane strain case if λ is replaced by

$$\lambda' = \frac{2\lambda\mu}{\lambda+2\mu} = \frac{\nu E}{1-\nu^2} \tag{6.52}$$

A case of *generalized plane strain* is sometimes mentioned in elasticity books when ϵ_{33} is taken as a constant other than zero in (*6.50*).

6.7 AIRY'S STRESS FUNCTION

If body forces are absent or are constant, the solution of *plane elastostatic problems* (plane strain or generalized plane stress problems) is often obtained through the use of the *Airy stress function.* Even if body forces must be taken into account, the superposition principle allows for their contribution to the solution to be introduced as a particular integral of the linear differential field equations.

For plane elastostatic problems in the absence of body forces, the equilibrium equations reduce to

$$\sigma_{\alpha\beta,\beta} = 0 \qquad \text{or} \qquad \nabla\cdot\mathbf{\Sigma} = 0 \tag{6.53}$$

and the compatibility equation (*6.44*) may be expressed in terms of stress components as

$$\nabla^2(\sigma_{11}+\sigma_{22}) = 0, \qquad \nabla^2\Theta_1 = 0 \tag{6.54}$$

The stress components are now given as partial derivatives of the Airy stress function $\phi = \phi(x_1, x_2)$ in accordance with the equations

$$\sigma_{11} = \phi_{,22}, \qquad \sigma_{12} = -\phi_{,12}, \qquad \sigma_{22} = \phi_{,11} \tag{6.55}$$

The equilibrium equations (*6.53*) are satisfied identically, and the compatibility condition (*6.54*) becomes the *biharmonic equation*

$$\nabla^2(\nabla^2\phi) = \nabla^4\phi = \phi_{,1111} + 2\phi_{,1122} + \phi_{,2222} = 0 \qquad (6.56)$$

Functions which satisfy (*6.56*) are called *biharmonic functions*. By considering biharmonic functions with single-valued second partial derivatives, numerous solutions to plane elastostatic problems may be constructed, which satisfy automatically both equilibrium and compatibility. Of course these solutions must be tailored to fit whatever boundary conditions are prescribed.

6.8 TWO-DIMENSIONAL ELASTOSTATIC PROBLEMS IN POLAR COORDINATES

Body geometry often deems it convenient to formulate two-dimensional elastostatic problems in terms of polar coordinates r and θ. Thus for transformation equations

$$x_1 = r\cos\theta, \qquad x_2 = r\sin\theta \qquad (6.57)$$

the stress components shown in Fig. 6-3 are found to lead to equilibrium equations in the form

$$\frac{\partial\sigma_{(rr)}}{\partial r} + \frac{1}{r}\frac{\partial\sigma_{(r\theta)}}{\partial\theta} + \frac{\sigma_{(rr)} - \sigma_{(\theta\theta)}}{r} + R = 0 \qquad (6.58)$$

$$\frac{1}{r}\frac{\partial\sigma_{(\theta\theta)}}{\partial\theta} + \frac{\partial\sigma_{(r\theta)}}{\partial r} + \frac{2\sigma_{(r\theta)}}{r} + Q = 0 \qquad (6.59)$$

in which R and Q represent body forces per unit volume in the directions shown.

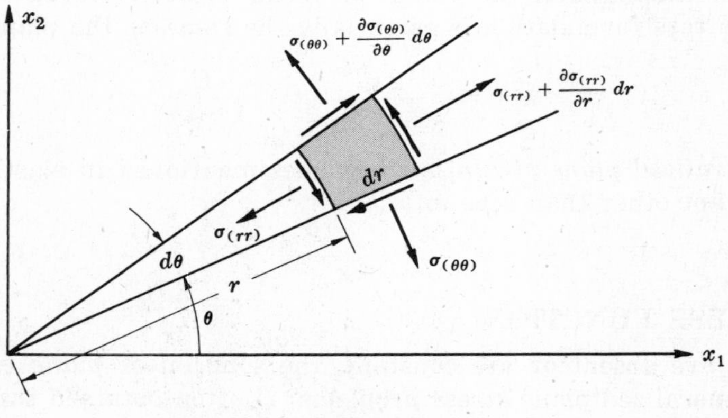

Fig. 6-3

Taking the Airy stress function now as $\Phi = \Phi(r,\theta)$, the stress components are given by

$$\sigma_{(rr)} = \frac{1}{r}\frac{\partial\Phi}{\partial r} + \frac{1}{r^2}\frac{\partial^2\Phi}{\partial\theta^2} \qquad (6.60)$$

$$\sigma_{(\theta\theta)} = \partial^2\Phi/\partial r^2 \qquad (6.61)$$

$$\sigma_{(r\theta)} = -\frac{\partial}{\partial r}\left(\frac{1}{r}\frac{\partial\Phi}{\partial\theta}\right) \qquad (6.62)$$

The compatibility condition again leads to the biharmonic equation

$$\nabla^2(\nabla^2\Phi) = \nabla^4\Phi = 0 \qquad (6.63)$$

but, in polar form, $\nabla^2 = \dfrac{\partial^2}{\partial r^2} + \dfrac{1}{r}\dfrac{\partial}{\partial r} + \dfrac{1}{r^2}\dfrac{\partial^2}{\partial\theta^2}$.

6.9 HYPERELASTICITY. HYPOELASTICITY

Modern continuum studies have led to constitutive equations which define materials that are elastic in a special sense. In this regard a material is said to be *hyperelastic* if it possesses a strain energy function U such that the material derivative of this function is equal to the stress power per unit volume. Thus the constitutive equation is of the form

$$\frac{d}{dt}(U) = \frac{1}{\rho}\sigma_{ij}D_{ij} = \frac{1}{\rho}\sigma_{ij}\dot{\epsilon}_{ij} \tag{6.64}$$

in which D_{ij} is the rate of deformation tensor. In a second classification, a material is said to be *hypoelastic* if the stress rate is a homogeneous linear function of the rate of deformation. In this case the constitutive equation is written

$$\sigma^{\nabla}_{ij} = K_{ijkm}D_{km} \tag{6.65}$$

in which the stress rate σ^{∇}_{ij} is defined as

$$\sigma^{\nabla}_{ij} = \frac{d}{dt}(\sigma_{ij}) - \sigma_{iq}V_{qj} - \sigma_{jq}V_{qi} \tag{6.66}$$

where V_{ij} is the vorticity tensor.

6.10 LINEAR THERMOELASTICITY

If thermal effects are taken into account, the components of the linear strain tensor ϵ_{ij} may be considered to be the sum

$$\epsilon_{ij} = \epsilon^{(S)}_{ij} + \epsilon^{(T)}_{ij} \tag{6.67}$$

in which $\epsilon^{(S)}_{ij}$ is the contribution from the stress field and $\epsilon^{(T)}_{ij}$ is the contribution from the temperature field. Due to a change from some reference temperature T_0 to the temperature T, the strain components of an elementary volume of an unconstrained isotropic body are given by

$$\epsilon^{(T)}_{ij} = \alpha(T - T_0)\delta_{ij} \tag{6.68}$$

where α denotes the linear coefficient of thermal expansion. Inserting (*6.68*), together with Hooke's law (*6.22*), into (*6.67*) yields

$$\epsilon_{ij} = \frac{1}{2\mu}\left(\sigma_{ij} - \frac{\lambda}{3\lambda + 2\mu}\delta_{ij}\sigma_{kk}\right) + \alpha(T - T_0)\delta_{ij} \tag{6.69}$$

which is known as the *Duhamel-Neumann* relations. Equation (*6.69*) may be inverted to give the thermoelastic constitutive equations

$$\sigma_{ij} = \lambda\delta_{ij}\epsilon_{kk} + 2\mu\epsilon_{ij} - (3\lambda + 2\mu)\alpha\delta_{ij}(T - T_0) \tag{6.70}$$

Heat conduction in an isotropic elastic solid is governed by the well-known *Fourier law* of heat conduction,

$$c_i = -kT_{,i} \tag{6.71}$$

where the scalar k, the thermal conductivity of the body, must be positive to assure a positive rate of entropy production. If now the *specific heat* at constant deformation $c^{(v)}$ is introduced through the equation

$$-c_{i,i} = \rho c^{(v)}\dot{T} \tag{6.72}$$

and the internal energy is assumed to be a function of the strain components ϵ_{ij} and the temperature T, the energy equation (*5.45*) may be expressed in the form

$$kT_{,ii} = \rho c^{(v)}\dot{T} + (3\lambda + 2\mu)\alpha T_0 \dot{\epsilon}_{ii} \tag{6.73}$$

which is known as the *coupled heat equation.*

The system of equations that formulate the general thermoelastic problem for an isotropic body consists of

(*a*) equations of motion

$$\sigma_{ij,j} + \rho b_i = \ddot{u}_i \quad \text{or} \quad \nabla \cdot \mathbf{\Sigma} + \rho \mathbf{b} = \ddot{\mathbf{u}} \tag{6.74}$$

(*b*) thermoelastic constitutive equations

$$\sigma_{ij} = \lambda \delta_{ij} \epsilon_{kk} + 2\mu \epsilon_{ij} - (3\lambda + 2\mu)\alpha \delta_{ij}(T - T_0)$$

or

$$\mathbf{\Sigma} = \lambda \mathbf{I} \epsilon + 2\mu \mathbf{E} - (3\lambda + 2\mu)\alpha \mathbf{I}(T - T_0) \tag{6.75}$$

(*c*) strain-displacement relations

$$\epsilon_{ij} = \tfrac{1}{2}(u_{i,j} + u_{j,i}) \quad \text{or} \quad \mathbf{E} = \tfrac{1}{2}(\mathbf{u}\nabla + \nabla \mathbf{u}) \tag{6.76}$$

(*d*) coupled heat equation

$$kT_{,ii} = \rho c^{(v)}\dot{T} + (3\lambda + 2\mu)\alpha T_0 \dot{e}_{kk} \quad \text{or} \quad k\nabla^2 T = \rho c^{(v)}\dot{T} + (3\lambda + 2\mu)\alpha T_0 \dot{\epsilon} \tag{6.77}$$

This system must be solved for the stress, displacement and temperature fields, subject to appropriate initial and boundary conditions. In addition, the compatibility equations must be satisfied.

There is a large collection of problems in which both the inertia and coupling effects may be neglected. For these cases the general thermoelastic problem decomposes into two separate problems which must be solved consecutively, but independently. Thus for the uncoupled, quasi-static, thermoelastic problem the basic equations are the

(*a*) heat conduction equation

$$kT_{,ii} = \rho c^{(v)}\dot{T} \quad \text{or} \quad k\nabla^2 T = \rho c^{(v)}\dot{T} \tag{6.78}$$

(*b*) equilibrium equations

$$\sigma_{ij,j} + \rho b_i = 0 \quad \text{or} \quad \nabla \cdot \mathbf{\Sigma} + \rho \mathbf{b} = 0 \tag{6.79}$$

(*c*) thermoelastic stress-strain equations

$$\sigma_{ij} = \lambda \delta_{ij} \epsilon_{kk} + 2\mu \epsilon_{ij} - (3\lambda + 2\mu)\alpha \delta_{ij}(T - T_0)$$

or

$$\mathbf{\Sigma} = \lambda \mathbf{I} \epsilon + 2\mu \mathbf{E} - (3\lambda + 2\mu)\alpha \mathbf{I}(T - T_0) \tag{6.80}$$

(*d*) strain-displacement relations

$$\epsilon_{ij} = \tfrac{1}{2}(u_{i,j} + u_{j,i}) \quad \text{or} \quad \mathbf{E} = \tfrac{1}{2}(\nabla \mathbf{u} + \mathbf{u}\nabla) \tag{6.81}$$

Solved Problems

HOOKE'S LAW. STRAIN ENERGY. ISOTROPY (Sec. 6.1-6.3)

6.1. Show that the strain energy density u^* for an isotropic Hookean solid may be expressed in terms of the strain tensor by $u^* = \lambda(\text{tr}\,\mathbf{E})^2/2 + \mu\mathbf{E}:\mathbf{E}$, and in terms of the stress tensor by $u^* = [(1+\nu)\mathbf{\Sigma}:\mathbf{\Sigma} - \nu(\text{tr}\,\mathbf{\Sigma})^2]/2E$.

Inserting (*6.21*) into (*6.13*), $u^* = (\lambda\delta_{ij}\epsilon_{kk} + 2\mu\epsilon_{ij})\epsilon_{ij}/2 = \lambda\epsilon_{ii}\epsilon_{jj}/2 + \mu\epsilon_{ij}\epsilon_{ij}$ which in symbolic notation is $u^* = \lambda(\text{tr}\,\mathbf{E})^2/2 + \mu\mathbf{E}:\mathbf{E}$.

Inserting (*6.24*) into (*6.13*), $u^* = \sigma_{ij}[(1+\nu)\sigma_{ij} - \nu\delta_{ij}\sigma_{kk}]/2E = [(1+\nu)\sigma_{ij}\sigma_{ij} - \nu\sigma_{ii}\sigma_{jj}]/2E$ which in symbolic notation is $u^* = [(1+\nu)\mathbf{\Sigma}:\mathbf{\Sigma} - \nu(\text{tr}\,\mathbf{\Sigma})^2]/2E$.

6.2. Separating the stress and strain tensors into their spherical and deviator components, express the strain energy density u^* as the sum of a dilatation energy density $u^*_{(S)}$ and distortion energy density $u^*_{(D)}$.

Inserting (*3.98*) and (*2.70*) into (*6.13*),

$$u^* = \tfrac{1}{2}(s_{ij} + \sigma_{kk}\delta_{ij}/3)(e_{ij} + \epsilon_{pp}\delta_{ij}/3) = \tfrac{1}{2}(s_{ij}e_{ij} + \sigma_{ii}e_{jj}/3 + s_{ii}\epsilon_{jj}/3 + \sigma_{ii}\epsilon_{jj}/3)$$

and since $e_{ii} = s_{ii} = 0$ this reduces to $u^* = u^*_{(S)} + u^*_{(D)} = \sigma_{ii}\epsilon_{jj}/6 + s_{ij}e_{ij}/2$.

6.3. Assuming a state of uniform compressive stress $\sigma_{ij} = -p\delta_{ij}$, develop the formulas for the bulk modulus (ratio of pressure to volume change) given in (*6.25*).

With $\sigma_{ij} = -p\delta_{ij}$, (*6.24*) becomes $\epsilon_{ij} = [(1+\nu)(-p\delta_{ij}) + \nu\delta_{ij}(3p)]/E$ and so $\epsilon_{ii} = [-3p(1+\nu) + 9p\nu]/E$. Thus $K = -p/\epsilon_{ii} = E/3(1-2\nu)$. Likewise from (*6.21*), $\sigma_{ii} = (3\lambda + 2\mu)\epsilon_{ii} = -3p$ so that $K = (3\lambda + 2\mu)/3$.

6.4. Express $u^*_{(S)}$ and $u^*_{(D)}$ of Problem 6.2 in terms of the engineering constants K and G and the strain components.

From a result in Problem 6.3, $\sigma_{ii} = 3K\epsilon_{ii}$ and so

$$u^*_{(S)} = \sigma_{ii}\epsilon_{jj}/6 = K\epsilon_{ii}\epsilon_{jj}/2 = K(\text{I}_E)^2/2$$

From (*6.21*) and (*2.70*), $\sigma_{ij} = \lambda\delta_{ij}\epsilon_{kk} + 2\mu\epsilon_{ij} = s_{ij} + \sigma_{kk}\delta_{ij}/3$ and since $\sigma_{ii} = (3\lambda + 2\mu)\epsilon_{ii}$ it follows that $s_{ij} = 2\mu(\epsilon_{ij} - \epsilon_{kk}\delta_{ij}/3)$. Thus

$$u^*_{(D)} = 2\mu(\epsilon_{ij} - \epsilon_{kk}\delta_{ij}/3)(\epsilon_{ij} - \epsilon_{pp}\delta_{ij}/3)/2 = \mu(\epsilon_{ij}\epsilon_{ij} - \epsilon_{ii}\epsilon_{jj}/3)$$

Note that the dilatation energy density $u^*_{(S)}$ appears as a function of K only, whereas the distortion energy $u^*_{(D)}$ is in terms of μ (or G), the shear modulus.

6.5. In general, u^* may be expressed in the quadratic form $u^* = C^*_{KM}\epsilon_K\epsilon_M$ in which the C^*_{KM} are not necessarily symmetrical. Show that this equation may be written in the form of (*6.14*) and that $\partial u^*/\partial\epsilon_K = \sigma_K$.

Write the quadratic form as

$$u^* = \tfrac{1}{2}C^*_{KM}\epsilon_K\epsilon_M + \tfrac{1}{2}C^*_{KM}\epsilon_K\epsilon_M = \tfrac{1}{2}C^*_{KM}\epsilon_K\epsilon_M + \tfrac{1}{2}C^*_{PN}\epsilon_N\epsilon_P = \tfrac{1}{2}(C^*_{KM} + C^*_{MK})\epsilon_K\epsilon_M = \tfrac{1}{2}C_{KM}\epsilon_K\epsilon_M$$

where $C_{KM} = C_{MK}$.

Thus the derivative $\partial u^*/\partial\epsilon_R$ is now

$$\partial u^*/\partial\epsilon_R = \tfrac{1}{2}C_{KM}(\epsilon_{K,R}\epsilon_M + \epsilon_K\epsilon_{M,R}) = \tfrac{1}{2}C_{KM}(\delta_{KR}\epsilon_M + \epsilon_K\delta_{MR}) = \tfrac{1}{2}(C_{RM}\epsilon_M + C_{KR}\epsilon_K) = C_{RM}\epsilon_M = \sigma_R$$

6.6. Show that for an orthotropic elastic continuum (three orthogonal planes of elastic symmetry) the elastic coefficient matrix is as given in (*6.19*), page 142.

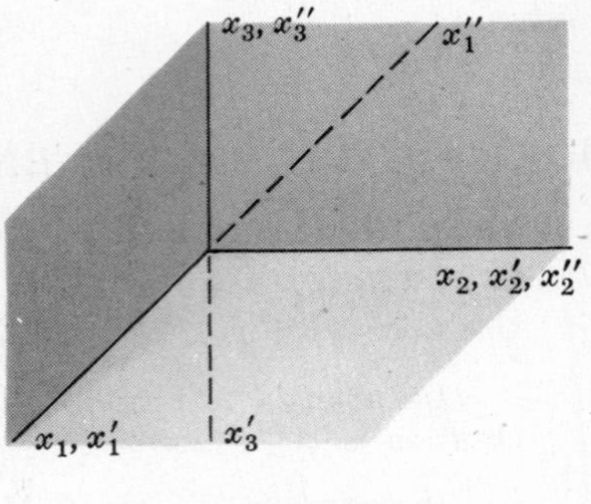

Fig. 6-4

Let the x_1x_2 (or equivalently, $x_1'x_2'$) plane be a plane of elastic symmetry (Fig. 6-4). Then $\sigma_K = C_{KM}\epsilon_M$ and also $\sigma_K' = C_{KM}\epsilon_M'$. The transformation matrix between x_i and x_i' is

$$[a_{ij}] = \begin{bmatrix} 1 & 0 & 0 \\ 0 & 1 & 0 \\ 0 & 0 & -1 \end{bmatrix}$$

and from (*2.27*) and (*3.78*), $\sigma_K' = \sigma_K$, $\epsilon_K' = \epsilon_K$ for $K = 1, 2, 3, 6$ whereas $\sigma_K' = -\sigma_K$, $\epsilon_K' = -\epsilon_K$ for $K = 4, 5$. Thus, for example, from $\sigma_1' = C_{1M}\epsilon_M'$,

$$\sigma_1' = \sigma_1 = C_{11}\epsilon_1 + C_{12}\epsilon_2 + C_{13}\epsilon_3 - C_{14}\epsilon_4 - C_{15}\epsilon_5 + C_{16}\epsilon_6$$

But from $\sigma_1 = C_{1M}\epsilon_M$,

$$\sigma_1 = C_{11}\epsilon_1 + C_{12}\epsilon_2 + C_{13}\epsilon_3 + C_{14}\epsilon_4 + C_{15}\epsilon_5 + C_{16}\epsilon_6$$

These two expressions for $\sigma_1' = \sigma_1$ are equal only if $C_{14} = C_{15} = 0$. Likewise, from $\sigma_2' = \sigma_2$, $\sigma_3' = \sigma_3$, $\sigma_4' = -\sigma_4$, $\sigma_5' = -\sigma_5$, $\sigma_6' = \sigma_6$ it is found that $C_{24} = C_{25} = C_{34} = C_{35} = C_{64} = C_{65} = C_{41} = C_{42} = C_{43} = C_{51} = C_{52} = C_{53} = C_{56} = 0$.

If x_2x_3 (or $x_2''x_3''$) is a second plane of elastic symmetry such that $\sigma_K'' = C_{KM}\epsilon_M''$, the transformation array is

$$[a_{ij}] = \begin{bmatrix} -1 & 0 & 0 \\ 0 & 1 & 0 \\ 0 & 0 & 1 \end{bmatrix}$$

and now from (*2.27*) and (*3.78*), $\sigma_K'' = \sigma_K$, $\epsilon_K'' = -\epsilon_K$ for $K = 1, 2, 3, 4$ whereas $\sigma_K'' = -\sigma_K$, $\epsilon_K'' = -\epsilon_K$ for $K = 5, 6$. Now $C_{16} = C_{26} = C_{36} = C_{45} = C_{54} = C_{61} = C_{62} = C_{63} = 0$ and the elastic coefficient matrix attains the form (*6.19*). The student should verify that elastic symmetry with respect to the (third) x_1x_3 plane is identically satisfied by this array.

6.7. Give the details of the reduction of the orthotropic elastic matrix (*6.19*) to the isotropic matrix (*6.20*).

For isotropy, elastic properties are the same with respect to all Cartesian coordinate axes. In particular, for the rotated x_i' axes shown in Fig. 6-5, the method of Problem 6.6 results in the matrix (*6.19*) being further simplified by the conditions $C_{11} = C_{22} = C_{33}$, $C_{44} = C_{55} = C_{66}$, and $C_{12} = C_{21} = C_{13} = C_{31} = C_{23} = C_{32}$.

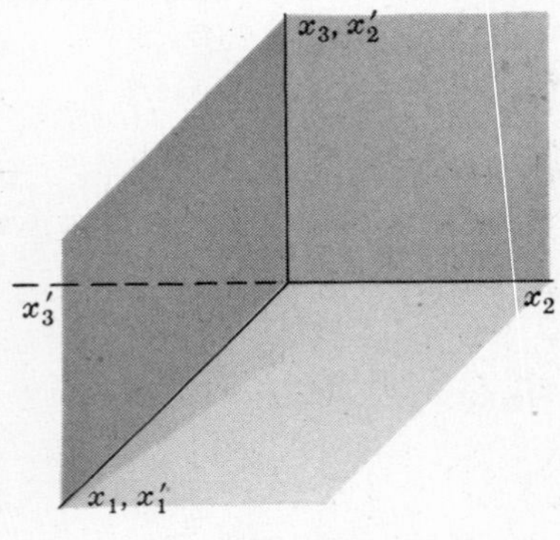

Fig. 6-5

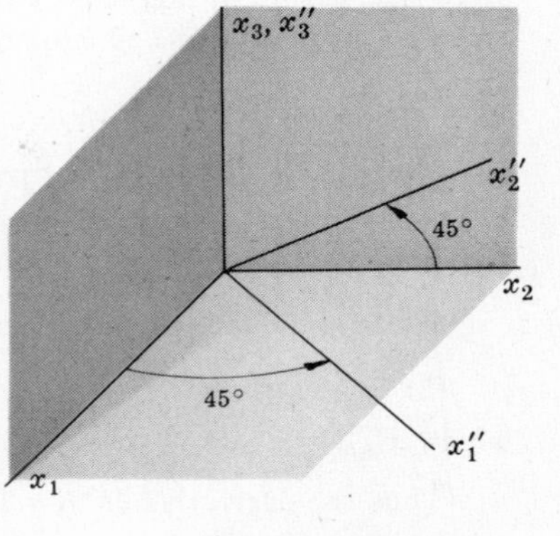

Fig. 6-6

Finally, for the axes x_i'' obtained by a 45° rotation about x_3 as in Fig. 6-6 above, the transformation matrix is

$$[a_{ij}] = \begin{bmatrix} 1/\sqrt{2} & 1/\sqrt{2} & 0 \\ -1/\sqrt{2} & 1/\sqrt{2} & 0 \\ 0 & 0 & 1 \end{bmatrix}$$

so that $\sigma_6'' = (\sigma_2 - \sigma_1)/2 = (C_{11} - C_{12})(\epsilon_2 - \epsilon_1)/2$ and $\epsilon_6'' = \epsilon_2 - \epsilon_1$. But $\sigma_6'' = C_{44}\epsilon_6''$ and so $2C_{44} = C_{11} - C_{12}$. Thus defining $\mu = C_{44}$ and $\lambda = C_{12}$, (*6.20*) is obtained.

6.8. Give the details of the inversion of (*6.21*) to obtain (*6.22*).

From (*6.21*) with $i = j$, $\sigma_{ii} = (3\lambda + 2\mu)\epsilon_{ii}$ and so $2\mu\epsilon_{ij} = \sigma_{ij} - \lambda\delta_{ij}\sigma_{kk}/(3\lambda + 2\mu)$ or $\epsilon_{ij} = \sigma_{ij}/2\mu - \lambda\delta_{ij}\sigma_{kk}/2\mu(3\lambda + 2\mu)$.

6.9. Express the engineering constants ν and E in terms of the Lamé constants λ and μ.

From (*6.25*), $E/(1-2\nu) = 3\lambda + 2\mu$; and from (*6.26*), $E/(1+\nu) = 2\mu$. Thus $(3\lambda + 2\mu)(1-2\nu) = 2\mu(1+\nu)$ from which $\nu = \lambda/2(\lambda+\mu)$. Now by (*6.26*), $E = 2\mu(1+\nu) = \mu(3\lambda + 2\mu)/(\lambda+\mu)$.

6.10. Determine the elastic coefficient matrix for a continuum having an axis of elastic symmetry of order $N = 4$. Assume $C_{KM} = C_{MK}$.

Let x_3 be the axis of elastic symmetry. A rotation $\theta = 2\pi/4 = \pi/2$ of the axes about x_3 produces equivalent elastic directions for $N = 4$. The transformation matrix is

$$[a_{ij}] = \begin{bmatrix} 0 & 1 & 0 \\ -1 & 0 & 0 \\ 0 & 0 & 1 \end{bmatrix}$$

and by (*2.27*) and (*3.78*), $\sigma_1' = \sigma_2$, $\sigma_2' = \sigma_1$, $\sigma_3' = \sigma_3$, $\sigma_4' = -\sigma_5$, $\sigma_5' = \sigma_4$, $\sigma_6' = -\sigma_6$ and $\epsilon_1' = \epsilon_2$, $\epsilon_2' = \epsilon_1$, $\epsilon_3' = \epsilon_3$, $\epsilon_4' = -\epsilon_5$, $\epsilon_5' = \epsilon_4$, $\epsilon_6' = -\epsilon_6$. Thus, for example, from $\sigma_3' = \sigma_3$, $C_{34} = C_{35} = C_{36} = 0$, $C_{31} = C_{32}$. Likewise, from the remaining five stress relations, the elastic matrix becomes

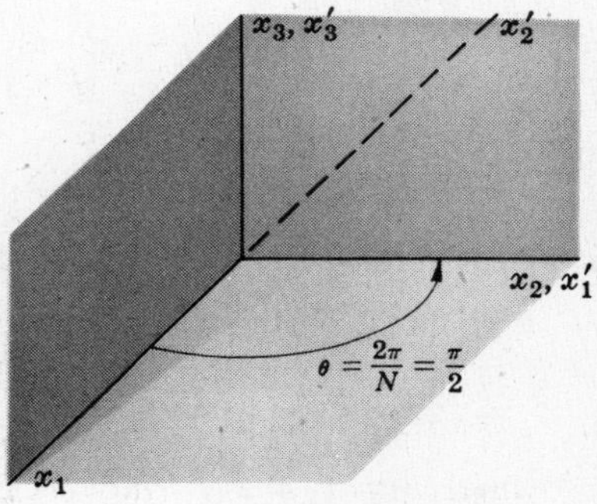

Fig. 6-7

$$[C_{KM}] = \begin{bmatrix} C_{11} & C_{12} & C_{13} & 0 & 0 & C_{16} \\ C_{12} & C_{11} & C_{13} & 0 & 0 & -C_{16} \\ C_{13} & C_{13} & C_{33} & 0 & 0 & 0 \\ 0 & 0 & 0 & C_{44} & 0 & 0 \\ 0 & 0 & 0 & 0 & C_{44} & 0 \\ C_{16} & -C_{16} & 0 & 0 & 0 & C_{66} \end{bmatrix}$$

with seven independent constants.

ELASTOSTATICS. ELASTODYNAMICS (Sec. 6.4-6.5)

6.11. Derive the Navier equations (*6.31*).

Replacing the strain components in (*6.28*) by the equivalent expressions in terms of displacements yields $\sigma_{ij} = \lambda\delta_{ij}u_{k,k} + \mu(u_{i,j} + u_{j,i})$. Thus $\sigma_{ij,j} = \lambda u_{k,ki} + \mu(u_{i,jj} + u_{j,ij})$. Substituting this into the equilibrium equations (*6.27*) and rearranging terms gives $\mu u_{i,jj} + (\lambda+\mu)u_{j,ji} + \rho b_i = 0$.

6.12. Show that if $\nabla^4 F_i = 0$, the displacement $u_i = (\lambda + 2\mu)F_{i,jj}/\mu(\lambda+\mu) - F_{j,ji}/\mu$ is a solution of the Navier equation *(6.31)* for zero body forces.

Differentiating the assumed solution, the terms $\mu u_{i,jj} = (\lambda+2\mu)F_{i,kkjj}/(\lambda+\mu) - F_{k,kijj}$ and $(\lambda+\mu)u_{j,ji} = (\lambda+2\mu)F_{j,kkji}/\mu - (\lambda+\mu)F_{k,kjji}/\mu$ are readily calculated. Inserting these into *(6.31)* gives

$$(\lambda+2\mu)F_{i,kkjj}/(\lambda+\mu) - [\mu - (\lambda+2\mu) + (\lambda+\mu)]F_{j,jkki} = 0$$

provided $F_{i,kkjj} = \nabla^4 F_i = 0$.

6.13. If body forces may be neglected, show that *(6.35)* is satisfied by $u_i = \phi_{,i} + \epsilon_{ijk}\psi_{k,j}$ provided ϕ and ψ_i each satisfies the familiar three dimensional wave equation.

Substituting the assumed u_i into *(6.35)* yields

$$\mu(\phi_{,ikk} + \epsilon_{ijk}\psi_{k,jqq}) + (\lambda+\mu)(\phi_{,jji} + \epsilon_{jpq}\psi_{q,pji}) = \rho(\ddot{\phi}_{,i} + \epsilon_{ijk}\ddot{\psi}_{k,j})$$

Since $\epsilon_{jpq}\psi_{q,pji} = 0$, this equation may be written

$$((\lambda+2\mu)\phi_{,kk} - \rho\ddot{\phi})_{,i} + \epsilon_{ijk}(\mu\psi_{k,qq} - \rho\ddot{\psi}_k)_{,j} = 0$$

which is satisfied when $\nabla^2\phi = \rho\ddot{\phi}/(\lambda+2\mu)$ and $\nabla^2\psi_k = \rho\ddot{\psi}_k/\mu$.

6.14. Writing $c^2\nabla^2\phi = \ddot{\phi}$ where $c^2 = (\lambda+2\mu)/\rho$ for the wave equation derived in Problem 6.13, show that $\phi = \dfrac{g(r+ct) + h(r-ct)}{r}$ is a solution with g and h arbitrary functions of their arguments and $r^2 = x_i x_i$.

Here it is convenient to use the spherical form $\nabla^2 \equiv \dfrac{1}{r^2}\dfrac{\partial}{\partial r}\left(r^2\dfrac{\partial}{\partial r}\right)$ since $\phi = \phi(r,t)$. Thus $r^2(\partial\phi/\partial r) = r(g'+h') - (g+h)$ where primes denote derivatives with respect to the arguments of g and h. Then $\nabla^2\phi = (g''+h'')/r$. Also $\dot{\phi} = (g'c - h'c)/r$ and $\ddot{\phi} = c^2(g''+h'')/r$. Therefore $c^2\nabla^2\phi = \ddot{\phi}$ for the given ϕ.

6.15. Derive the Beltrami-Michell equations *(6.33)* and determine the form they take when body forces are conservative, i.e. when $\rho b_i = \phi_{,i}$.

Substituting *(6.24)* into *(3.103)* yields

$$(1+\nu)(\sigma_{ij,km} + \sigma_{km,ij} - \sigma_{ik,jm} - \sigma_{jm,ik}) = \nu(\delta_{ij}\Theta_{,km} + \delta_{km}\Theta_{,ij} - \delta_{ik}\Theta_{,jm} - \delta_{jm}\Theta_{,ik})$$

where $\Theta = \mathrm{I}_\Sigma = \sigma_{ii}$. Only six of the eighty-one equations represented here are independent. Thus setting $m = k$ and using *(6.27)* gives

$$\sigma_{ij,kk} + \Theta_{,ij} + \rho(b_{i,j} + b_{j,i}) = \nu(\delta_{ij}\Theta_{,kk} + \Theta_{,ij})/(1+\nu)$$

from which $\Theta_{,kk} = -(1+\nu)\rho b_{k,k}/(1-\nu)$. Inserting this expression for $\Theta_{,kk}$ into the previous equation leads to *(6.33)*.

If $\rho b_i = \phi_{,i}$, then $\rho(b_{i,j} + b_{j,i}) = 2\phi_{,ij}$ and $\rho b_{k,k} = \phi_{,kk} = \nabla^2\phi$ so that *(6.33)* becomes

$$\nabla^2\sigma_{ij} + \Theta_{,ij}/(1+\nu) + 2\phi_{,ij} + \nu\delta_{ij}\nabla^2\phi/(1-\nu) = 0$$

TWO-DIMENSIONAL ELASTICITY (Sec. 6.6-6.8)

6.16. For plane stress parallel to the x_1x_2 plane, develop the stress-strain relations in terms of λ and μ. Show that these equations correspond to those given as *(6.41)*.

Here $\sigma_{33} = \sigma_{13} = \sigma_{23} = 0$ so that *(6.21)* yields $\epsilon_{13} = \epsilon_{23} = 0$ and $\epsilon_{33} = -\lambda(\epsilon_{11}+\epsilon_{22})/(\lambda+2\mu)$. Thus *(6.21)* reduces to $\sigma_{\alpha\beta} = 2\lambda\mu\delta_{\alpha\beta}\epsilon_{\gamma\gamma}/(\lambda+2\mu) + 2\mu\epsilon_{\alpha\beta}$ with $\alpha, \beta, \gamma = 1,2$ from which $\sigma_{\alpha\alpha} = 2\mu(3\lambda+2\mu)\epsilon_{\gamma\gamma}/(\lambda+2\mu)$ so that the equation may be inverted to give

$$\epsilon_{\alpha\beta} = -\lambda\delta_{\alpha\beta}\sigma_{\gamma\gamma}/2\mu(3\lambda+2\mu) + \sigma_{\alpha\beta}/2\mu = -\nu\delta_{\alpha\beta}\sigma_{\gamma\gamma}/E + (1+\nu)\sigma_{\alpha\beta}/E$$

Also,
$$\epsilon_{33} = -\lambda\epsilon_{\gamma\gamma}/(\lambda+2\mu) = -\lambda\sigma_{\gamma\gamma}/2\mu(3\lambda+2\mu) = -\nu\sigma_{\gamma\gamma}/E$$

6.17. For plane strain parallel to x_1x_2, develop the stress-strain relations in terms of ν and E. Show that these equations correspond to those given as (*6.48*).

Here $u_3 \equiv 0$ so that $\epsilon_{33} = 0$ and (*6.24*) gives $\sigma_{33} = \nu(\sigma_{11} + \sigma_{22}) = \lambda\sigma_{\alpha\alpha}/2(\lambda+\mu)$. Thus (*6.24*) becomes $\epsilon_{\alpha\beta} = (1+\nu)\sigma_{\alpha\beta}/E - \nu(1+\nu)\delta_{\alpha\beta}\sigma_{\gamma\gamma}/E$ from which $\epsilon_{\alpha\alpha} = (1+\nu)(1-2\nu)\sigma_{\alpha\alpha}/E$. Finally, inverting,

$$\sigma_{\alpha\beta} = \nu E\delta_{\alpha\beta}\epsilon_{\gamma\gamma}/(1+\nu)(1-2\nu) + E\epsilon_{\alpha\beta}/(1+\nu) = \lambda\delta_{\alpha\beta}\epsilon_{\gamma\gamma} + 2\mu\epsilon_{\alpha\beta}$$

6.18. Develop the Navier equation for plane stress (*6.45*) and show that it is equivalent to the corresponding equation for plane strain (*6.51*) if $\lambda' = 2\lambda\mu/(\lambda+2\mu)$ is substituted for λ.

Inverting (*6.41*) and using (*6.42*) leads to $\sigma_{\alpha\beta} = E(u_{\alpha,\beta} + u_{\beta,\alpha})/2(1+\nu) + 2\nu E\delta_{\alpha\beta}u_{\gamma,\gamma}/2(1-\nu^2)$. Differentiating with respect to x_β and substituting into (*6.40*) gives

$$Eu_{\alpha,\beta\beta}/2(1+\nu) + Eu_{\beta,\beta\alpha}/2(1-\nu) + \rho b_\alpha = \mu\nabla^2 u_\alpha + \mu(3\lambda+2\mu)u_{\beta,\beta\alpha}/(\lambda+2\mu) + \rho b_\alpha = 0$$

Thus since $\mu(3\lambda+2\mu)/(\lambda+2\mu) = (2\lambda\mu/(\lambda+2\mu)+\mu) = (\lambda'+\mu)$, (*6.45*) and (*6.51*) have the same form for the given substitution.

6.19. Determine the necessary relationship between the constants A and B if $\phi = Ax_1^2x_2^3 + Bx_2^5$ is to serve as an Airy stress function.

By (*6.56*), ϕ must be biharmonic or $\phi_{,1111} + 2\phi_{,1122} + \phi_{,2222} = 0 + 24Ax_2 + 120Bx_2 = 0$, which is satisfied when $A = -5B$.

6.20. Show that $\phi = \dfrac{3F}{4c}\left[x_1x_2 - \dfrac{x_1x_2^3}{3c^2}\right] + \dfrac{P}{4c}x_2^2$ is suitable for use as an Airy stress function and determine the stress components in the region $x_1 > 0$, $-c < x_2 < c$.

Since $\nabla^4\phi$ is identically zero, ϕ is a valid stress function. The stress components as given by (*6.55*) are $\sigma_{11} = \phi_{,22} = -3Fx_1x_2/2c^3 + P/2c$, $\sigma_{12} = -\phi_{,12} = -3F(c^2 - x_2^2)/4c^3$, $\sigma_{22} = \phi_{,11} = 0$. These stresses are those of a cantilever beam subjected to a transverse end load F and an axial pull P (Fig. 6-8).

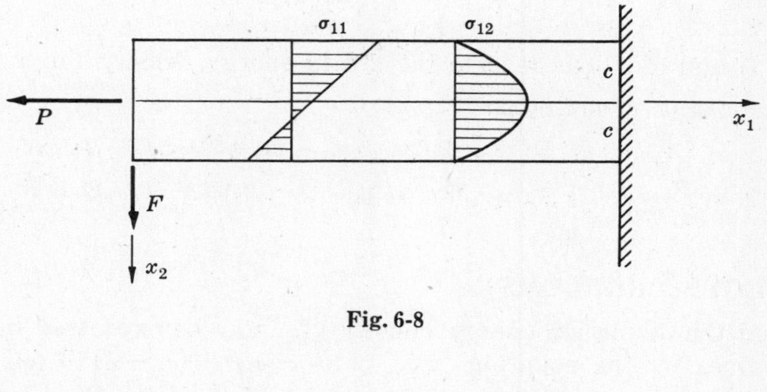

Fig. 6-8

6.21. In Problem 2.36 it was shown that the equilibrium equations were satisfied in the absence of body forces by $\sigma_{ij} = \epsilon_{ipq}\epsilon_{jmn}\phi_{qn,pm}$. Show that Airy's stress function is represented by the case $\phi_{33} = \phi(x_1, x_2)$ with $\phi_{11} = \phi_{22} = \phi_{12} = \phi_{13} = \phi_{23} \equiv 0$.

Since ϕ_{33} is the only non-vanishing component, $\sigma_{ij} = \epsilon_{ipq}\epsilon_{jmn}\phi_{qn,pm}$ becomes $\sigma_{ij} = \epsilon_{ip3}\epsilon_{jm3}\phi_{33,pm}$ which may be written $\sigma_{\alpha\beta} = \epsilon_{\alpha\gamma3}\epsilon_{\beta\zeta3}\phi_{33,\gamma\zeta}$. Thus since $\phi_{33} = \phi$, $\sigma_{\alpha\beta} = (\delta_{\alpha\beta}\delta_{\gamma\zeta} - \delta_{\alpha\zeta}\delta_{\gamma\beta})\phi_{,\gamma\zeta} = \delta_{\alpha\beta}\phi_{,\gamma\gamma} - \phi_{,\alpha\beta}$. The stress components are therefore $\sigma_{11} = \phi_{,11} + \phi_{,22} - \phi_{,11} = \phi_{,22}$, $\sigma_{12} = -\phi_{,12}$, $\sigma_{22} = \phi_{,11} + \phi_{,22} - \phi_{,22} = \phi_{,11}$.

6.22. In polar coordinates (r, θ) the Airy stress function $\Phi = B\theta$ is used in the solution of a disk of radius a subjected to a central moment M. Determine the stress components and the value of the constant B.

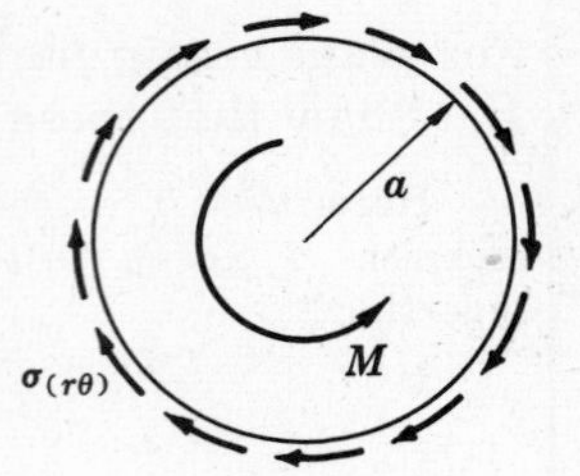

Fig. 6-9

From (6.60) and (6.61), $\sigma_{(rr)} = \sigma_{(\theta\theta)} = 0$. From (6.62) $\sigma_{(r\theta)} = B/r^2$. Equilibrium of moments about the center of the disk requires $M = \int_0^{2\pi} \sigma_{(r\theta)} a^2\, d\theta = \int_0^{2\pi} B\, d\theta = 2\pi B$. Thus $B = M/2\pi$.

LINEAR THERMOELASTICITY (Sec. 6.10)

6.23. Carry out the inversion of (6.69) to obtain the thermoelastic constitutive equations (6.70).

From (6.69) with $i = j$, $\sigma_{ii} = (3\lambda + 2\mu)(\epsilon_{ii} - 3\alpha(T - T_0))$. Solving (6.69) for σ_{ij} gives

$$\begin{aligned}\sigma_{ij} &= 2\mu\epsilon_{ij} + \lambda\delta_{ij}\sigma_{kk}/(3\lambda + 2\mu) - 2\mu\alpha\delta_{ij}(T - T_0)\\ &= 2\mu\epsilon_{ij} + \lambda\delta_{ij}(\epsilon_{kk} - 3\alpha(T - T_0)) - 2\mu\alpha\delta_{ij}(T - T_0)\\ &= 2\mu\epsilon_{ij} + \lambda\delta_{ij}\epsilon_{kk} - (3\lambda + 2\mu)\alpha\delta_{ij}(T - T_0)\end{aligned}$$

6.24. Develop the thermoelastic energy equation (6.73) by use of the free energy $f = u - Ts$.

Assuming the free energy to be a function of the strains and temperature, $f = f(\epsilon_{ij}, T)$ and substituting into (5.41) $\rho\dot{u} = \sigma_{ij}\dot{\epsilon} + \rho T\dot{s}$ where dots indicate time derivatives, the result is $(\sigma_{ij} - \rho\partial f/\partial\epsilon_{ij})\dot{\epsilon}_{ij} - \rho(s + \partial f/\partial T)\dot{T} = 0$. Since the terms in parentheses are independent of strain and temperature rates, it follows that $\sigma_{ij} = \rho\partial f/\partial\epsilon_{ij}$ and $s = -\partial f/\partial T$. From (5.38) for a reversible isothermal process, $-c_{i,i} = \rho T\dot{s} = \rho T\left(\frac{\partial s}{\partial\epsilon_{ij}}\dot{\epsilon}_{ij} + \frac{\partial s}{\partial T}\dot{T}\right)$. At constant deformation, $\dot{\epsilon}_{ij} = 0$ and comparing this equation with (6.72) gives $c^{(v)} = T(\partial s/\partial T)$ or from above, since $\partial s/\partial T = -\partial^2 f/\partial T^2$, $c^{(v)} = -\partial^2 f/\partial T^2$. Also, from above, $\rho(\partial^2 f/\partial\epsilon_{ij}\partial T) = \partial\sigma_{ij}/\partial T$ and so combining (5.38) with (6.71), $-c_{i,i} = kT_{,ii} = \rho T\left(\frac{\partial\sigma_{ij}}{\partial T}\dot{\epsilon}_{ij} + \frac{c^{(v)}}{T}\dot{T}\right)$. Finally from (6.70), $\partial\sigma_{ij}/\partial T = (3\lambda + 2\mu)\alpha\delta_{ij}T_0$ so that $kT_{,ii} = \rho c^{(v)}\dot{T} + (3\lambda + 2\mu)\alpha T_0\dot{\epsilon}_{ii}$ which is (6.73).

6.25. Use (6.13) and (6.70) to develop the strain energy density for a thermoelastic solid.

Substituting (6.70) directly into (6.13),

$$\begin{aligned}u^* &= \lambda\delta_{ij}\epsilon_{kk}\epsilon_{ij}/2 + \mu\epsilon_{ij}\epsilon_{ij} - (3\lambda + 2\mu)\alpha\delta_{ij}(T - T_0)\epsilon_{ij}/2\\ &= \lambda\epsilon_{ii}\epsilon_{jj}/2 + \mu\epsilon_{ij}\epsilon_{ij} - (3\lambda + 2\mu)\alpha(T - T_0)\epsilon_{ii}/2\end{aligned}$$

MISCELLANEOUS PROBLEMS

6.26. Show that the distortion energy density $u^*_{(D)}$ may be expressed in terms of principal stress values by the equation $u^*_{(D)} = [(\sigma_1 - \sigma_2)^2 + (\sigma_2 - \sigma_3)^2 + (\sigma_3 - \sigma_1)^2]/12G$.

From Problem 6.2, $u^*_{(D)} = s_{ij}e_{ij}/2 = s_{ij}s_{ij}/4G$ which in terms of stress components becomes

$$u^*_{(D)} = (\sigma_{ij} - \delta_{ij}\sigma_{kk}/3)(\sigma_{ij} - \delta_{ij}\sigma_{pp}/3)/4G = (\sigma_{ij}\sigma_{ij} - \sigma_{ii}\sigma_{jj}/3)/4G$$

In terms of principal stresses this is

$$\begin{aligned}u^*_{(D)} &= [\sigma_1^2 + \sigma_2^2 + \sigma_3^2 - (\sigma_1 + \sigma_2 + \sigma_3)(\sigma_1 + \sigma_2 + \sigma_3)/3]/4G\\ &= [2(\sigma_1^2 + \sigma_2^2 + \sigma_3^2 - \sigma_1\sigma_2 - \sigma_2\sigma_3 - \sigma_3\sigma_1)/3]/4G\\ &= [(\sigma_1 - \sigma_2)^2 + (\sigma_2 - \sigma_3)^2 + (\sigma_3 - \sigma_1)^2]/12G\end{aligned}$$

6.27. Use the results of Problem 6.1 to show that for an elastic material $\partial u^*/\partial \epsilon_{ij} = \sigma_{ij}$ and $\partial u^*/\partial \sigma_{ij} = \epsilon_{ij}$.

From Problem 6.1, $u^* = \lambda\epsilon_{ii}\epsilon_{jj}/2 + \mu\epsilon_{ij}\epsilon_{ij}$ and so

$$\begin{aligned}\partial u^*/\partial\epsilon_{pq} &= \lambda/2[\epsilon_{ii}(\partial\epsilon_{jj}/\partial\epsilon_{pq}) + \epsilon_{jj}(\partial\epsilon_{ii}/\partial\epsilon_{pq})] + 2\mu\epsilon_{ij}(\partial\epsilon_{ij}/\partial\epsilon_{pq})\\ &= \lambda/2[\epsilon_{ii}\delta_{jp}\delta_{jq} + \epsilon_{jj}\delta_{ip}\delta_{iq}] + 2\mu\epsilon_{ij}\delta_{ip}\delta_{jq} = \lambda/2[\epsilon_{ii}\delta_{pq} + \epsilon_{jj}\delta_{pq}] + 2\mu\epsilon_{pq}\\ &= \lambda\epsilon_{ii}\delta_{pq} + 2\mu\epsilon_{pq} = \sigma_{pq}\end{aligned}$$

Likewise from Problem 6.1, $u^* = [(1+\nu)\sigma_{ij}\sigma_{ij} - \nu\sigma_{ii}\sigma_{jj}]/2E$ and so

$$\partial u^*/\partial\sigma_{pq} = [2(1+\nu)\sigma_{ij}\delta_{ip}\delta_{jq} - \nu(\sigma_{ii}\delta_{pq} + \sigma_{jj}\delta_{pq}]/2E = [(1+\nu)\sigma_{pq} - \nu\delta_{pq}\sigma_{ii}]/E = \epsilon_{pq}$$

6.28. Express the strain energy density u^* as a function of the strain invariants.

From Problem 6.1, $u^* = \lambda\epsilon_{ii}\epsilon_{jj}/2 + \mu\epsilon_{ij}\epsilon_{ij}$; and since by comparison with (*3.91*), $\text{I}_E = \epsilon_{ii}$ and $\text{II}_E = (\epsilon_{ii}\epsilon_{jj} - \epsilon_{ij}\epsilon_{ij})/2$, it follows that

$$u^* = \lambda(\text{I}_E)^2/2 + \mu(-2\text{II}_E + (\text{I}_E)^2) = (\lambda/2 + \mu)(\text{I}_E)^2 - 2\mu\text{II}_E$$

6.29. When a circular shaft of length L and radius a is subjected to end couples as shown in Fig. 6-10, the nonzero stress components are $\sigma_{13} = -G\alpha x_2$, $\sigma_{23} = G\alpha x_1$ where α is the angle of twist per unit length. Determine expressions for the strain energy density and the total strain energy in the shaft.

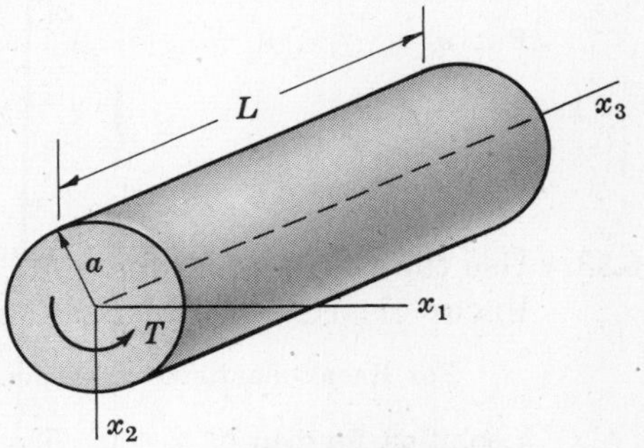

Fig. 6-10

From Problem 6.1, $u^* = [(1+\nu)\boldsymbol{\Sigma}:\boldsymbol{\Sigma} - \nu(\text{tr}\,\boldsymbol{\Sigma})^2]/2E$. Here $\text{tr}\,\boldsymbol{\Sigma} \equiv 0$ and $\boldsymbol{\Sigma}:\boldsymbol{\Sigma} = 2G^2\alpha^2r^2$ where $r^2 = x_1^2 + x_2^2$. Thus $u^* = G\alpha^2r^2/2$. The total strain energy is given by

$$U = \int_V u^*\,dV = \frac{G\alpha^2}{2}\int_0^a\int_0^{2\pi}\int_0^L r^3\,dr\,d\theta\,dx_3 = G\alpha^2a^4\pi L/4$$

Note that since $T = \int_0^{2\pi}\int_0^a G\alpha(x_1^2 + x_2^2)r\,dr\,d\theta = G\alpha a^4\pi/2$, $U = T\alpha L/2$, the external work.

6.30. Show that for a continuum having an axis of elastic symmetry of order $N = 2$, the elastic properties (Hooke's law and strain energy density) are of the same form as a continuum having one plane of elastic symmetry.

Here a rotation of axes $\theta = 2\pi/N = 2\pi/2 = \pi$ produces equivalent elastic directions. But this is precisely the same situation as the reflection about a plane of elastic symmetry.

6.31. Show that (*6.19*) with $C_{11} = C_{22} = C_{33}$, $C_{44} = C_{55} = C_{66}$ and $C_{12} = C_{13} = C_{23}$ may be reduced to (*6.20*) by an arbitrary rotation θ of axes about x_3 (Fig. 6-11).

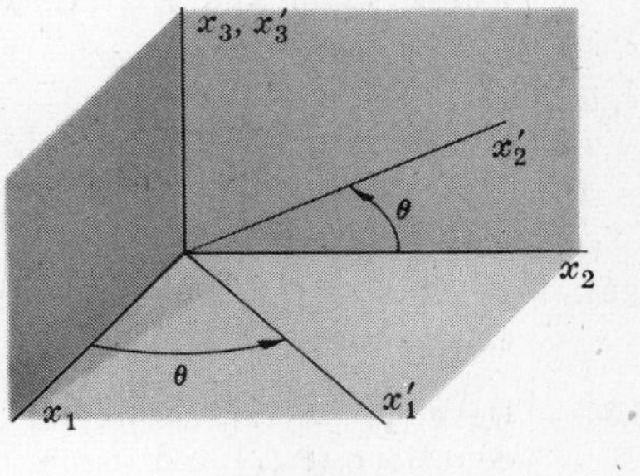

Fig. 6-11

The transformation between x_i and x_i' axes is

$$a_{ij} = \begin{pmatrix}\cos\theta & \sin\theta & 0\\ -\sin\theta & \cos\theta & 0\\ 0 & 0 & 1\end{pmatrix}$$

and from (*2.27*),

$$\sigma'_{12} \;=\; (-\sin\theta\cos\theta)\sigma_{11} + (\cos^2\theta - \sin^2\theta)\sigma_{12} + (\sin\theta\cos\theta)\sigma_{22}$$

or in single index notation,

$$\sigma'_6 \;=\; (-\sin\theta\cos\theta)\sigma_1 + (\cos^2\theta - \sin^2\theta)\sigma_6 + (\sin\theta\cos\theta)\sigma_2$$

Likewise from (*3.78*) and (*6.4*),

$$\epsilon'_6 \;=\; (-2\sin\theta\cos\theta)\epsilon_1 + (\cos^2\theta - \sin^2\theta)\epsilon_6 + (2\sin\theta\cos\theta)\epsilon_2$$

But $\sigma'_6 = C_{44}\epsilon'_6$ for an isotropic body and so here $\sigma_2 - \sigma_1 = 2C_{44}(\epsilon_2 - \epsilon_1)$. Finally from (*6.19*) with the given conditions, $\sigma_1 = C_{11}\epsilon_1 + C_{12}(\epsilon_2 + \epsilon_1)$ and $\sigma_2 = C_{11}\epsilon_2 + C_{12}(\epsilon_1 + \epsilon_2)$ and so $\sigma_2 - \sigma_1 = (C_{11} - C_{12})(\epsilon_2 - \epsilon_1)$. Therefore $(C_{11} - C_{12}) = 2C_{44}$ and with $C_{44} = \mu$, $C_{12} = \lambda$, $C_{11} = \lambda + 2\mu$ as given in (*6.20*).

6.32. For an elastic body in equilibrium under body forces b_i and surface forces $t_i^{(\hat{n})}$, show that the total strain energy is equal to one-half the work done by the external forces acting through their displacements u_i.

It is required to show that $\int_V \rho b_i u_i\,dV + \int_S t_i^{(\hat{n})} u_i\,dS = 2\int_V u^*\,dV$. Consider first the surface integral with $t_i^{(\hat{n})} = \sigma_{ji}u_j$ and convert by Gauss' theorem. Thus

$$\int_S \sigma_{ij}u_i n_j\,dS \;=\; \int_V (\sigma_{ij}u_i)_{,j}\,dV \;=\; \int_V (\sigma_{ij,j}u_i + \sigma_{ij}u_{i,j})\,dV$$

But $\sigma_{ij}u_{i,j} = \sigma_{ij}(\epsilon_{ij} + \omega_{ij}) = \sigma_{ij}\epsilon_{ij}$, and from equilibrium $\sigma_{ij,j} = -\rho b_i$. Thus

$$\int_S t_i^{(\hat{n})} u_i\,dS \;=\; -\int_V \rho b_i u_i\,dV + 2\int \sigma_{ij}\epsilon_{ij}/2\,dV$$

and the theorem is proved.

6.33. Use the result of Problem 6.32 to establish uniqueness of the elastostatic solution of a linear elastic body by assuming two solutions $\sigma_{ij}^{(1)}, u_i^{(1)}$ and $\sigma_{ij}^{(2)}, u_i^{(2)}$.

For linear elasticity superposition holds, so $\sigma_{ij} = \sigma_{ij}^{(1)} - \sigma_{ij}^{(2)}$, $u_i = u_i^{(1)} - u_i^{(2)}$ would also be a solution for which $b_i = 0$. Thus for this "difference" solution $\int_S t_i^{(\hat{n})} u_i\,dS = 2\int_V u^*\,dV$ from Problem 6.32. Since the two assumed solutions satisfy boundary conditions, the left hand integral is zero here since $t_i^{(\hat{n})} = t_i^{(1)} - t_i^{(2)}$ on the boundary for equation (*6.32*) and $u_i = u_i^{(1)} - u_i^{(2)}$ on the boundary for equation (*6.30*). Thus $\int_V u^*\,dV = 0$ and since u^* is positive definite this occurs only if $\epsilon_{ij} = \epsilon_{ij}^{(1)} - \epsilon_{ij}^{(2)} \equiv 0$, or $\epsilon_{ij}^{(1)} = \epsilon_{ij}^{(2)}$. If the strains are equal for the two assumed solutions, the stresses are also equal by Hooke's law and the displacements are equal to within a rigid body displacement. Thus uniqueness is established.

6.34. The Navier equations (*6.31*) may be put in the form $\mu u_{i,jj} + \dfrac{\mu}{1-2\nu}u_{j,ji} + \rho b_i = 0$ which for the incompressible case $(\nu = \frac{1}{2})$ are clearly indeterminate. Use the equilibrium equations for this situation to show that $\mu u_{i,jj} + \Theta_{,i}/3 + \rho b_i = 0$.

From equation (*6.24*), $\epsilon_{ii} = (1-2\nu)\sigma_{ii}/E$; and for $\nu = \frac{1}{2}$, $\epsilon_{ii} \equiv u_{i,i} = 0$. Thus from (*6.24*),

$$2\epsilon_{ij,j} \;=\; u_{i,jj} + u_{j,ij} \;=\; 2(1+\nu)\sigma_{ij,j}/E - 2\nu\delta_{ij}\sigma_{kk,j}/E$$

But $u_{j,ji} = 0$ and $E = 3G$ when $\nu = \frac{1}{2}$, so that $u_{i,jj} = -\rho b_i/G - \sigma_{kk,i}/3G$ or $\mu\nabla^2 u_i + \Theta_{,i}/3 + \rho b_i = 0$.

Supplementary Problems

6.35. Prove that the principal axes of the stress and strain tensors coincide for a homogeneous isotropic elastic body.

6.36. Develop the expression for the strain energy density u^* for an orthotropic elastic medium. Use equations (*6.14*) and (*6.19*).

Ans. $u^* = (C_{11}\epsilon_1 + 2C_{12}\epsilon_2 + 2C_{13}\epsilon_5)\epsilon_1/2 + (C_{22}\epsilon_2 + 2C_{23}\epsilon_4)\epsilon_2/2 + C_{33}\epsilon_3^2 + C_{44}\epsilon_4^2 + C_{55}\epsilon_5^2 + C_{66}\epsilon_6^2$.

6.37. Determine the form of the strain energy density for the case of (a) plane stress, (b) plane strain.
Ans. (a) $u^* = [\sigma_{11}^2 + \sigma_{22}^2 - 2\nu\sigma_{11}\sigma_{22} + 2(1+\nu)\sigma_{12}^2]/2E$
(b) $u^* = (\mu + \lambda/2)(\epsilon_{11}^2 + \epsilon_{22}^2) + \lambda\epsilon_{11}\epsilon_{22} + 2\mu\epsilon_{12}^2$

6.38. Determine the value of c for which $u_1 = A \sin\frac{2\pi}{l}(x_1 \pm ct)$, $u_2 = u_3 = 0$ is a solution of equation (6.35) when body forces are zero. *Ans.* $c = \sqrt{(\lambda + 2\mu)/\rho}$

6.39. Show that the distortion energy density $u^*_{(D)} = (\sigma_{ij}\sigma_{ij} - \sigma_{ii}\sigma_{jj}/3)/4G$ and the dilatation energy density $u^*_{(S)} = \sigma_{ii}\sigma_{jj}/18K$.

6.40. Show that $1/(1+\nu) = 2(\lambda + \mu)/(3\lambda + 2\mu)$ and $\nu/(1-\nu) = \lambda/(\lambda + 2\mu)$.

6.41. For plane strain parallel to x_1x_2, show that $b_3 \equiv 0$ and that b_1 and b_2 are functions of x_1 and x_2 only.

6.42. Use the transformation laws for stress and strain to show that the elastic constants C_{ijkm} are the components of a fourth order Cartesian tensor so that $C'_{ijkm} = a_{ip}a_{jq}a_{kr}a_{ms}C_{pqrs}$.

6.43. Show that the Airy stress function $\phi = 2x_1^4 + 12x_1^2x_2^2 - 6x_2^4$ satisfies the biharmonic equation $\nabla^4\phi = 0$ and determine the stress components assuming plane strain.

Ans. $$\sigma_{ij} = 24\begin{pmatrix} x_1^2 - 3x_2^2 & -2x_1x_2 & 0 \\ -2x_1x_2 & x_1^2 + x_2^2 & 0 \\ 0 & 0 & 2\nu(x_1^2 - x_2^2) \end{pmatrix}$$

6.44. Determine the strains associated with the stresses of Problem 6.43 and show that the compatibility equation (6.44) is satisfied.

Ans. $$\epsilon_{ij} = 24\left(\frac{1+\nu}{E}\right)\begin{pmatrix} x_1^2 - 3x_2^2 - 2\nu(x_1^2 - x_2^2) & -2x_1x_2 & 0 \\ -2x_1x_2 & x_1^2 + x_2^2 - 2\nu(x_1^2 - x_2^2) & 0 \\ 0 & 0 & 0 \end{pmatrix}$$

6.45. For an elastic body having an axis of elastic symmetry of order $N = 6$, show that $C_{22} = C_{11}$, $C_{55} = C_{44}$, $C_{66} = 2(C_{11} - C_{12})$ and that C_{13} and C_{33} are the only remaining nonzero coefficients.

6.46. Show that for an elastic continuum with conservative body forces such that $\rho b_\alpha = \nabla\psi = \psi_{,\alpha}$, the compatibility condition (6.44) may be written $\nabla^2\sigma_{\alpha\alpha} = \nabla^2\psi/(1-\nu)$ for plane strain, or $\nabla^2\sigma_{\alpha\alpha} = (1+\nu)\nabla^2\psi$ for plane stress.

6.47. If $\nabla^4 F_i = 0$, show that $u_i = 2(1-\nu)\nabla^2 F_i/G - F_{j,ji}/G$ is a solution of the Navier equation (6.31) when $b_i \equiv 0$ (see Problem 6.12). If $\mathbf{F} = B(x_2\hat{\mathbf{e}}_1 - x_1\hat{\mathbf{e}}_2)/r$ where $r^2 = x_ix_i$, determine the stress components.

Ans. $\sigma_{11} = -\sigma_{22} = 6QGx_1x_2/r^5$, $\sigma_{33} = 0$, $\sigma_{12} = 3QG(x_2^2 - x_1^2)/r^5$, $\sigma_{13} = -\sigma_{23} = 3QGx_2x_3/r^5$, where $Q = 4B(1-\nu)/G$.

6.48. In polar coordinates an Airy stress function is given by $\Phi = Cr^2(\cos 2\theta - \cos 2\alpha)$ where C and α are constants. Determine C if $\sigma_{\theta\theta} = 0$, $\sigma_{r\theta} = \tau$ when $\theta = \alpha$, and $\sigma_{\theta\theta} = 0$, $\sigma_{r\theta} = -\tau$ when $\theta = -\alpha$. *Ans.* $C = \tau/(2\sin 2\alpha)$

6.49. Show that in plane strain thermoelastic problems $\sigma_{33} = \nu(\sigma_{11} + \sigma_{22}) - \alpha E(T - T_0)$ and that $\sigma_{\alpha\beta} = \lambda\delta_{\alpha\beta}\epsilon_{\alpha\alpha} + 2\mu\epsilon_{\alpha\beta} - \delta_{\alpha\beta}(3\lambda + 2\mu)\alpha(T - T_0)$. In plane stress thermoelasticity show that

$$\epsilon_{33} = -\nu(\sigma_{11} + \sigma_{22})/E + \alpha(T - T_0) \quad \text{and} \quad \epsilon_{\alpha\beta} = (1+\nu)\sigma_{\alpha\beta}/E - \nu\delta_{\alpha\beta}\sigma_{\alpha\alpha}/E + \delta_{\alpha\beta}(T - T_0)\alpha$$

6.50. In terms of the Airy stress function $\phi = \phi(x_1, x_2)$, show that for plane strain thermoelasticity the compatibility equation (6.44) may be expressed as $\nabla^4\phi = -\alpha E\nabla^2(T - T_0)/(1-\nu)$ and that for plane stress as $\nabla^4\phi = -\alpha E\nabla^2(T - T_0)$.

Chapter 7

Fluids

7.1 FLUID PRESSURE. VISCOUS STRESS TENSOR. BAROTROPIC FLOW

In any fluid at rest the stress vector $t_i^{(\hat{n})}$ on an arbitrary surface element is collinear with the normal $\hat{\mathbf{n}}$ of the surface and equal in magnitude for every direction at a given point. Thus

$$t_i^{(\hat{n})} = \sigma_{ij} n_j = -p_0 n_i \quad \text{or} \quad \mathbf{t}^{(\hat{n})} = \mathbf{\Sigma} \cdot \hat{\mathbf{n}} = -p_0 \hat{\mathbf{n}} \tag{7.1}$$

in which p_0 is the stress magnitude, or *hydrostatic pressure*. The negative sign indicates a compressive stress for a positive value of the pressure. Here every direction is a principal direction, and from (*7.1*)

$$\sigma_{ij} = -p_0 \delta_{ij} \quad \text{or} \quad \mathbf{\Sigma} = -p_0 \mathbf{I} \tag{7.2}$$

which represents a spherical state of stress often referred to as hydrostatic pressure. From (*7.2*), the shear stress components are observed to be zero in a fluid at rest.

For a fluid in motion, the shear stress components are usually not zero, and it is customary in this case to resolve the stress tensor according to the equation

$$\sigma_{ij} = -p\delta_{ij} + \tau_{ij} \quad \text{or} \quad \mathbf{\Sigma} = -p\mathbf{I} + \mathbf{\Gamma} \tag{7.3}$$

where τ_{ij} is called the *viscous stress tensor* and p is the *pressure*.

All real fluids are both compressible and viscous. However, these characteristics vary widely in different fluids so that it is often possible to neglect their effects in certain situations without significant loss of accuracy in calculations based upon such assumptions. Accordingly, an *inviscid*, or so-called *perfect fluid* is one for which τ_{ij} is taken identically zero even when motion is present. *Viscous fluids* on the other hand are those for which τ_{ij} must be considered. For a compressible fluid, the pressure p is essentially the same as the pressure associated with classical thermodynamics. From (*7.3*), the mean normal stress is given by

$$\tfrac{1}{3}\sigma_{ii} = -p + \tfrac{1}{3}\tau_{ii} \quad \text{or} \quad \tfrac{1}{3}\Theta = -p + \tfrac{1}{3}\Gamma \tag{7.4}$$

For a fluid at rest, τ_{ij} vanishes and p reduces to p_0 which in this case is equal to the negative of the mean normal stress. For an incompressible fluid, the thermodynamic pressure is not defined separately from the mechanical conditions so that p must be considered as an independent mechanical variable in such fluids.

In a compressible fluid, the pressure p, the density ρ and the absolute temperature T are related through a kinetic equation of state having the form

$$p = p(\rho, T) \tag{7.5}$$

An example of such an equation of state is the well-known ideal gas law

$$p = \rho R T \tag{7.6}$$

where R is the gas constant. If the changes of state of a fluid obey an equation of state that does not contain the temperature, i.e. $p = p(\rho)$, such changes are termed *barotropic.* An isothermal process for a perfect gas is an example of a special case which obeys the barotropic assumption.

7.2 CONSTITUTIVE EQUATIONS. STOKESIAN FLUIDS. NEWTONIAN FLUIDS

The viscous stress components of the stress tensor for a fluid are associated with the dissipation of energy. In developing constitutive relations for fluids, it is generally assumed that the viscous stress tensor τ_{ij} is a function of the rate of deformation tensor D_{ij}. If the functional relationship is a nonlinear one, as expressed symbolically by

$$\tau_{ij} = f_{ij}(D_{pq}) \quad \text{or} \quad \mathbf{T} = \mathbf{f}(\mathbf{D}) \tag{7.7}$$

the fluid is called a *Stokesian fluid.* When the function is a linear one of the form

$$\tau_{ij} = K_{ijpq}D_{pq} \quad \text{or} \quad \mathbf{T} = \tilde{\mathbf{K}} : \mathbf{D} \tag{7.8}$$

where the constants K_{ijpq} are called *viscosity coefficients,* the fluid is known as a *Newtonian fluid.* Some authors classify fluids simply as *Newtonian* and *non-Newtonian.*

Following a procedure very much the same as that carried out for the generalized Hooke's law of an elastic media in Chapter 6, the constitutive equations for an isotropic homogeneous Newtonian fluid may be determined from (*7.7*) and (*7.8*). The final form is

$$\sigma_{ij} = -p\delta_{ij} + \lambda^*\delta_{ij}D_{kk} + 2\mu^*D_{ij} \quad \text{or} \quad \mathbf{\Sigma} = -p\mathbf{I} + \lambda^*\mathbf{I}(\text{tr}\,\mathbf{D}) + 2\mu^*\mathbf{D} \tag{7.9}$$

where λ^* and μ^* are viscosity coefficients of the fluid. From (*7.9*), the mean normal stress is given by

$$\tfrac{1}{3}\sigma_{ii} = -p + \tfrac{1}{3}(3\lambda^* + 2\mu^*)D_{ii} = -p + \kappa^*D_{ii}$$

or

$$\tfrac{1}{3}(\text{tr}\,\mathbf{\Sigma}) = -p + \tfrac{1}{3}(3\lambda^* + 2\mu^*)(\text{tr}\,\mathbf{D}) = -p + \kappa^*(\text{tr}\,\mathbf{D}) \tag{7.10}$$

where $\kappa^* = \frac{1}{3}(3\lambda^* + 2\mu^*)$ is called the *coefficient of bulk viscosity.* The condition that

$$\kappa^* = \lambda^* + \tfrac{2}{3}\mu^* = 0 \tag{7.11}$$

is known as *Stokes' condition,* and guarantees that the pressure p is defined as the average of the normal stresses for a compressible fluid at rest. In this way the thermodynamic pressure is defined in terms of the mechanical stresses.

In terms of the deviator components $s_{ij} = \sigma_{ij} - \delta_{ij}\sigma_{kk}/3$ and $D'_{ij} = D_{ij} - \delta_{ij}D_{kk}/3$, equation (*7.9*) above may be rewritten in the form

$$s_{ij} + \tfrac{1}{3}\delta_{ij}\sigma_{kk} = -p\delta_{ij} + \delta_{ij}(\lambda^* + \tfrac{2}{3}\mu^*)D_{ii} + 2\mu^*D'_{ij}$$

or

$$\mathbf{S} + \tfrac{1}{3}\mathbf{I}(\text{tr}\,\mathbf{\Sigma}) = -p\mathbf{I} + \mathbf{I}(\lambda^* + \tfrac{2}{3}\mu^*)(\text{tr}\,\mathbf{D}) + 2\mu^*\mathbf{D}' \tag{7.12}$$

Therefore in view of the relationship (*7.10*), equation (*7.12*) may be expressed by the pair of equations

$$s_{ij} = 2\mu^*D'_{ij} \quad \text{or} \quad \mathbf{S} = 2\mu^*\mathbf{D}' \tag{7.13}$$

$$\sigma_{ii} = -3p + 3\kappa^*D_{ii} \quad \text{or} \quad \text{tr}\,\mathbf{\Sigma} = -3p + 3\kappa^*(\text{tr}\,\mathbf{D}) \tag{7.14}$$

the first of which relates the shear effects in the fluid and the second gives the volumetric relationship.

7.3 BASIC EQUATIONS FOR NEWTONIAN FLUIDS. NAVIER-STOKES-DUHEM EQUATIONS

In Eulerian form, the basic equations required to formulate the problem of motion for a Newtonian fluid are

(*a*) the continuity equation (*5.3*),

$$\dot{\rho} + \rho v_{i,i} = 0 \quad \text{or} \quad \dot{\rho} + \rho(\nabla_{\mathbf{x}} \cdot \mathbf{v}) = 0 \tag{7.15}$$

(*b*) the equations of motion (*5.16*),

$$\sigma_{ij,j} + \rho b_i = \rho \dot{v}_i \quad \text{or} \quad \nabla_{\mathbf{x}} \cdot \mathbf{\Sigma} + \rho \mathbf{b} = \rho \dot{\mathbf{v}} \tag{7.16}$$

(*c*) the energy equation (*5.32*),

$$\dot{u} = \frac{1}{\rho}\sigma_{ij}D_{ij} - \frac{1}{\rho}c_{i,i} + z \quad \text{or} \quad \dot{u} = \frac{1}{\rho}\mathbf{\Sigma}:\mathbf{D} - \frac{1}{\rho}\nabla_{\mathbf{x}} \cdot c + z \tag{7.17}$$

(*d*) the constitutive equations (*7.9*),

$$\sigma_{ij} = -p\delta_{ij} + \lambda^* \delta_{ij} D_{kk} + 2\mu^* D_{ij} \quad \text{or} \quad \mathbf{\Sigma} = -p\mathbf{I} + \lambda^*\mathbf{I}(\text{tr}\,\mathbf{D}) + 2\mu^*\mathbf{D} \tag{7.18}$$

(*e*) the kinetic equation of state (*7.5*),

$$p = p(\rho, T) \tag{7.19}$$

If thermal effects are considered, as they very often must be in fluids problems, the additional equations

(*f*) the Fourier law of heat conduction (*6.71*),

$$c_i = -kT_{,i} \quad \text{or} \quad \mathbf{c} = -k\nabla T \tag{7.20}$$

(*g*) the caloric equation of state,

$$u = u(\rho, T) \tag{7.21}$$

are required. The system of equations (*7.15*) through (*7.21*) represents sixteen equations in sixteen unknowns and is therefore determinate.

If (*7.18*) above is substituted into (*7.16*) and the definition $2D_{ij} = (v_{i,j} + v_{j,i})$ is used, the equations that result from the combination are the *Navier-Stokes-Duhem equations of motion,*

$$\rho\dot{v}_i = \rho b_i - p_{,i} + (\lambda^* + \mu^*)v_{j,ji} + \mu^* v_{i,jj}$$

or

$$\rho\dot{\mathbf{v}} = \rho\mathbf{b} - \nabla p + (\lambda^* + \mu^*)\nabla(\nabla \cdot \mathbf{v}) + \mu^*\nabla^2\mathbf{v} \tag{7.22}$$

When the flow is incompressible ($v_{j,j} = 0$), (*7.22*) reduce to the *Navier-Stokes equations* for *incompressible* flow,

$$\rho\dot{v}_i = \rho b_i - p_{,i} + \mu^* v_{i,jj} \quad \text{or} \quad \rho\dot{\mathbf{v}} = \rho\mathbf{b} - \nabla p + \mu^*\nabla^2\mathbf{v} \tag{7.23}$$

If Stokes condition is assumed ($\lambda^* = -\frac{2}{3}\mu^*$), (*7.22*) reduce to the *Navier-Stokes* equations for *compressible* flow

$$\rho\dot{\mathbf{v}}_i = \rho b_i - p_{,i} + \tfrac{1}{3}\mu^* v_{j,ji} + \mu^* v_{i,jj}$$

or

$$\rho\dot{\mathbf{v}} = \rho\mathbf{b} - \nabla p + \tfrac{1}{3}\mu^*\nabla(\nabla \cdot \mathbf{v}) + \mu^*\nabla^2\mathbf{v} \tag{7.24}$$

The Navier-Stokes equations (*7.23*), together with the continuity equation (*7.15*) form a complete set of four equations in four unknowns: the pressure p and the three velocity components v_i. In any given problem, the solutions of this set of equations must satisfy boundary and initial conditions on traction and velocity components. For a viscous fluid, the appropriate boundary conditions at a fixed surface require both the normal and tangential components of velocity to vanish. This condition results from the experimentally established fact that a fluid adheres to and obtains the velocity of the boundary. For an inviscid fluid, only the normal velocity component is required to vanish on a fixed surface.

If the Navier-Stokes equations are put into dimensionless form, several ratios of the normalizing parameters appear. One of the most significant and commonly used ratios is the Reynolds number $N_{(R)}$ which expresses the ratio of inertia to viscous forces. Thus if a flow is characterized by a certain length L, velocity V and density ρ, the Reynolds number is

$$N_{(R)} = VL/\nu \tag{7.25}$$

where $\nu = \mu^*/\rho$ is called the *kinematic viscosity*. For very large Reynolds numbers, the viscous contribution to the shear stress terms of the momentum equations may be neglected. In *turbulent flow*, the apparent stresses act on the time mean flow in a manner similar to the viscous stress effects in a *laminar flow*. If turbulence is not present, inertia effects outweigh viscous effects and the fluid behaves as though it were inviscid. The ability of a flow to support turbulent motions is related to the Reynolds number. It is only in the case of laminar flow that the constitutive relations (*7.18*) apply to real fluids.

7.4 STEADY FLOW. HYDROSTATICS. IRROTATIONAL FLOW

The motion of a fluid is referred to as a *steady flow* if the velocity components are independent of time. For this situation, the derivative $\partial v_i/\partial t$ is zero, and hence the material derivative of the velocity

$$\frac{dv_i}{dt} \equiv \dot{v}_i = \frac{\partial v_i}{\partial t} + v_j v_{i,j} \quad \text{or} \quad \frac{d\mathbf{v}}{dt} \equiv \dot{\mathbf{v}} = \frac{\partial \mathbf{v}}{\partial t} + \mathbf{v}\cdot\nabla_{\mathbf{x}}\mathbf{v} \tag{7.26}$$

reduces to the simple form

$$\dot{v}_i = v_j v_{i,j} \quad \text{or} \quad \dot{\mathbf{v}} = \mathbf{v}\cdot\nabla_{\mathbf{x}}\mathbf{v} \tag{7.27}$$

A steady flow in which the velocity is zero everywhere, causes the Navier-Stokes equations (*7.22*) to reduce to

$$\rho b_i = p_{,i} \quad \text{or} \quad \rho\mathbf{b} = \nabla_{\mathbf{x}} p \tag{7.28}$$

which describes the *hydrostatic equilibrium* situation. If the barotropic condition $\rho = \rho(p)$ is assumed, a *pressure function*

$$P(p) = \int_{p_0}^{p} \frac{dp}{\rho} \tag{7.29}$$

may be defined. Furthermore, if the body force may be prescribed by a potential function

$$b_i = -\Omega_{,i} \quad \text{or} \quad \mathbf{b} = -\nabla\Omega \tag{7.30}$$

equations (*7.28*) take on the form

$$(\Omega + P)_{,i} = 0 \quad \text{or} \quad \nabla(\Omega + P) = 0 \tag{7.31}$$

A flow in which the spin, or vorticity tensor (*4.21*),

$$V_{ij} = \frac{1}{2}\left(\frac{\partial v_i}{\partial x_j} - \frac{\partial v_j}{\partial x_i}\right) \quad \text{or} \quad \mathbf{V} = \tfrac{1}{2}(\mathbf{v}\nabla - \nabla\mathbf{v}) \tag{7.32}$$

vanishes everywhere is called an *irrotational flow*. The vorticity vector q_i is related to the vorticity tensor by the equation

$$q_i = \epsilon_{ijk} V_{kj} \quad \text{or} \quad \mathbf{q} = \mathbf{V}_v \tag{7.33}$$

and therefore also vanishes for irrotational flow. Furthermore,

$$q_i = \epsilon_{ijk} v_{k,j} \quad \text{or} \quad \mathbf{q} = \nabla \times \mathbf{v} \tag{7.34}$$

and since $\nabla \times \mathbf{v} = 0$ is necessary and sufficient for a *velocity potential* ϕ to exist, the velocity vector for irrotational flow may be expressed by

$$v_i = -\phi_{,i} \quad \text{or} \quad \mathbf{v} = -\nabla\phi \tag{7.35}$$

7.5 PERFECT FLUIDS. BERNOULLI EQUATION. CIRCULATION

If the viscosity coefficients λ^* and μ^* are zero, the resulting fluid is called an *inviscid* or *perfect* (frictionless) *fluid* and the Navier-Stokes-Duhem equations (*7.22*) reduce to the form

$$\rho\dot{v}_i = \rho b_i - p_{,i} \quad \text{or} \quad \rho\dot{\mathbf{v}} = \rho\mathbf{b} - \nabla p \tag{7.36}$$

which is known as the *Euler equation of motion*. For a barotropic fluid with conservative body forces, (*7.29*) and (*7.30*) may be introduced so that (*7.36*) becomes

$$\dot{v}_i = -(\Omega + P)_{,i} \quad \text{or} \quad \dot{\mathbf{v}} = -\nabla(\Omega + P) \tag{7.37}$$

For steady flow (*7.37*) may be written

$$v_j v_{i,j} = -(\Omega + P)_{,i} \quad \text{or} \quad \mathbf{v}\cdot\nabla\mathbf{v} = -\nabla(\Omega + P) \tag{7.38}$$

If the Euler equation (*7.37*) is integrated along a streamline, the result is the well-known Bernoulli equation in the form (see Problem 7.17)

$$\Omega + P + v^2/2 + \int \frac{\partial v_i}{\partial t}\, dx_i = C(t) \tag{7.39}$$

For steady motion, $\partial v_i/\partial t = 0$ and $C(t)$ becomes the Bernoulli constant C which is, in general, different along different streamlines. If the flow is irrotational as well, a single constant C holds everywhere in the field of flow.

When the only body force present is gravity, the potential $\Omega = gh$ where g is the gravitational constant and h is the elevation above some reference level. Thus with $h_p = P/g$ defined as the *pressure* head, and $v^2/2g = h_v$ defined as the *velocity* head, Bernoulli's equation requires the total head along any streamline to be constant. For incompressible fluids (liquids), the equation takes the form

$$h + h_p + h_v = h + p/\rho g + v^2/2g = \text{constant} \tag{7.40}$$

By definition, the *velocity circulation* around a closed path of fluid particles is given by the line integral

$$\Gamma_c = \oint v_i\, dx_i \quad \text{or} \quad \Gamma_c = \oint \mathbf{v}\cdot d\mathbf{x} \tag{7.41}$$

From Stokes theorem (*1.153*) or (*1.154*), page 23, the line integral (*7.41*) may be converted to the surface integral

$$\Gamma_c = \int_S n_i \epsilon_{ijk} v_{k,j}\, dS \quad \text{or} \quad \Gamma_c = \int_S \hat{\mathbf{n}}\cdot(\nabla \times \mathbf{v})\, dS \tag{7.42}$$

where $\hat{\mathbf{n}}$ is the unit normal to the surface S enclosed by the path. If the flow is irrotational, $\nabla \times \mathbf{v} = 0$ and the circulation is zero. In this case the integrand of (*7.41*) is the perfect differential $d\phi = -\mathbf{v}\cdot d\mathbf{x}$ with ϕ the velocity potential.

The material derivative $d\Gamma_c/dt$ of the circulation may be determined by using (*4.60*) which when applied to (*7.41*) gives

$$\dot{\Gamma}_c = \oint (\dot{v}_i\,dx_i + v_i\,dv_i) \quad \text{or} \quad \dot{\Gamma}_c = \oint (\dot{\mathbf{v}}\cdot d\mathbf{x} + \mathbf{v}\cdot d\mathbf{v}) \tag{7.43}$$

For a barotropic, inviscid fluid with conservative body forces the circulation may be shown to be a constant. This is known as *Kelvin's theorem* of constant circulation.

7.6 POTENTIAL FLOW. PLANE POTENTIAL FLOW

The term *potential flow* is often used to denote an irrotational flow since the condition of irrotationality, $\nabla \times \mathbf{v} = 0$, is necessary and sufficient for the existence of the velocity potential ϕ of (*7.35*). For a compressible irrotational flow, the Euler equation and the continuity equation may be linearized and combined as is done in acoustics to yield the governing wave equation

$$\ddot{\phi} = c^2\phi_{,ii} \quad \text{or} \quad \ddot{\phi} = c^2\nabla^2\phi \tag{7.44}$$

where c is the velocity of sound in the fluid. For a steady irrotational flow of a compressible barotropic fluid, the Euler equation and continuity equation may be combined to give

$$(c^2\delta_{ij} - v_iv_j)v_{j,i} = 0 \quad \text{or} \quad c^2\nabla\cdot\mathbf{v} - \mathbf{v}\cdot(\mathbf{v}\cdot\nabla\mathbf{v}) = 0 \tag{7.45}$$

which is the so-called *gas dynamical equation.*

For incompressible potential flow the continuity equation attains the form

$$\phi_{,ii} = 0 \quad \text{or} \quad \nabla^2\phi = 0 \tag{7.46}$$

and solutions of this *Laplace equation* provide the velocity components through the definition (*7.35*). Boundary conditions on velocity must also be satisfied. On a fixed boundary, for example, $\partial\phi/\partial n = 0$. An important feature of this formulation rests in the fact that the Laplace equation is linear so that superposition of solutions is possible.

In a two-dimensional incompressible flow parallel to the x_1x_2 plane, $v_3 = 0$, and the continuity equation becomes

$$v_{\alpha,\alpha} = 0 \quad \text{or} \quad \nabla\cdot\mathbf{v} = 0 \tag{7.47}$$

where, as usual in this book, Greek subscripts have a range of two. By (*7.47*), regardless of whether the flow is irrotational or not, it is possible to introduce the *stream function* $\psi = \psi(x_1, x_2)$ such that

$$v_\alpha = -\epsilon_{\alpha\beta 3}\psi_{,\beta} \tag{7.48}$$

If the plane flow is, indeed, irrotational so that

$$v_\alpha = -\phi_{,\alpha} \quad \text{or} \quad \mathbf{v} = -\nabla\phi \tag{7.49}$$

then from (*7.48*) and (*7.49*) the stream function and velocity potential are seen to satisfy the *Cauchy-Riemann* conditions

$$\phi_{,1} = \psi_{,2} \quad \text{and} \quad \phi_{,2} = -\psi_{,1} \tag{7.50}$$

By eliminating ϕ and ψ in turn from (*7.50*) it is easily shown that

$$\phi_{,\alpha\alpha} = 0 \quad \text{or} \quad \nabla^2\phi = 0 \tag{7.51}$$

$$\psi_{,\alpha\alpha} = 0 \quad \text{or} \quad \nabla^2\psi = 0 \tag{7.52}$$

Thus both ϕ and ψ are harmonic functions when the flow is irrotational. Furthermore, the complex potential

$$\Phi(z) = \phi(x_1, x_2) + i\psi(x_1, x_2) \tag{7.53}$$

is an analytic function of the complex variable, $z = x_1 + ix_2$ so that its derivative $d\Psi/dz$ defines the *complex velocity*

$$d\Psi/dz = -v_1 + iv_2 \tag{7.54}$$

Solved Problems

FUNDAMENTALS OF FLUIDS. NEWTONIAN FLUIDS (Sec. 7.1-7.3)

7.1. Show that the deviator s_{ij} for the stress tensor σ_{ij} of (*7.3*) is equal to t_{ij}, the deviator of τ_{ij} of (*7.3*).

From (*7.3*), $\sigma_{ii} = -3p + \tau_{ii}$ and so here

$$s_{ij} = \sigma_{ij} - \delta_{ij}\sigma_{kk}/3 = -p\delta_{ij} + \tau_{ij} - \delta_{ij}(-3p + \tau_{kk})/3 = \tau_{ij} - \delta_{ij}\tau_{kk}/3 = t_{ij}$$

7.2. Determine the mean normal stress $\sigma_{ii}/3$ for an incompressible Stokesian (nonlinear) fluid for which $\tau_{ij} = \alpha D_{ij} + \beta D_{ik} D_{kj}$ where α and β are constants.

From (*7.3*), $\sigma_{ij} = -p\delta_{ij} + \alpha D_{ij} + \beta D_{ik} D_{kj}$ and so $\sigma_{ii} = -3p + \alpha D_{ii} + \beta D_{ik} D_{ki}$. But $D_{ik} = D_{ki}$ and $D_{ii} = v_{i,i} = 0$ for an incompressible fluid so that

$$\sigma_{ii}/3 = -p + \beta D_{ij} D_{ij}/3 = -p - 2\beta \mathrm{II}_D/3$$

where II_D is the second invariant of the rate of deformation tensor.

7.3. Frictionless adiabatic, or isentropic flow of an ideal gas, is a barotropic flow for which $p = c\rho^k$ where C and k are constants with $k = c^{(p)}/c^{(v)}$, the ratio of specific heat at constant pressure to that at constant volume. Determine the temperature-density and temperature-pressure relationships for such a flow.

Inserting $p = C\rho^k$ into equation (*7.6*), the temperature-density relationship is $\rho^{k-1}/T = R/C$, a constant. Also, since $\rho = (p/C)^{1/k}$ here, (*7.6*) yields the temperature-pressure relationship as $p^{(k-1)/k}/T = R/C^{1/k}$, a constant.

7.4. Determine the constitutive equation for a Newtonian fluid with zero bulk viscosity, i.e. with $\kappa^* \equiv 0$.

If $\kappa^* \equiv 0$, $\lambda^* = -2\mu^*/3$ by (*7.11*) and so (*7.9*) becomes $\sigma_{ij} = -p\delta_{ij} - (2\mu^*/3)\delta_{ij} D_{kk} + 2\mu^* D_{ij}$ which is expressed in terms of the rate of deformation deviator by

$$\sigma_{ij} = -p\delta_{ij} + 2\mu^*(D_{ij} - \delta_{ij} D_{kk}/3) = -p\delta_{ij} + 2\mu^* D'_{ij}$$

If the deviator stress s_{ij} is introduced, this constitutive relation is given by the two equations $s_{ij} = 2\mu^* D'_{ij}$ and $\sigma_{ii} = -3p$.

7.5. **Determine an expression for the "stress power" $\sigma_{ij}D_{ij}$ of a Newtonian fluid having equation (7.9) as its constitutive relation.**

From (7.9) and the stress power definition,

$$\sigma_{ij}D_{ij} = -p\delta_{ij}D_{ij} + \lambda^*\delta_{ij}D_{kk}D_{ij} + 2\mu^*D_{ij}D_{ij} = -pD_{ii} + \lambda^*D_{ii}D_{jj} + 2\mu^*D_{ij}D_{ij}$$

In symbolic notation, this expression is written

$$\mathbf{\Sigma} : \mathbf{D} = -p(\text{tr}\,\mathbf{D}) + \lambda^*(\text{tr}\,\mathbf{D})^2 + 2\mu^*\mathbf{D}:\mathbf{D}$$

In terms of D'_{ij} the expression is

$$\sigma_{ij}D_{ij} = -pD_{ii} + \lambda^*D_{ii}D_{jj} + 2\mu^*(D'_{ij} + \delta_{ij}D_{kk}/3)(D'_{ij} + \delta_{ij}D_{qq}/3) = -pD_{ii} + \kappa^*D_{ii}D_{jj} + 2\mu^*D'_{ij}D'_{ij}$$

In symbolic notation,

$$\mathbf{\Sigma} : \mathbf{D} = -p(\text{tr}\,\mathbf{D}) + \kappa^*(\text{tr}\,\mathbf{D})^2 + 2\mu^*\mathbf{D}':\mathbf{D}'$$

7.6. **Determine the conditions under which the mean normal pressure $p_{(m)} = -\sigma_{ii}/3$ is equal to the thermodynamic pressure p for a Newtonian fluid.**

With the constitutive equations in the form (7.13) and (7.14), the latter equation gives $p_{(m)} - p = -\kappa^*D_{ii}$. Thus $p_{(m)} = p$ when $\kappa^* = 0$ (by (7.11) when $\lambda^* = -\frac{2}{3}\mu^*$) or when $D_{ii} = 0$.

7.7. **Verify the Navier-Stokes-Duhem equations of motion (7.22) for a Newtonian fluid and determine the form of the energy equation (7.17) for this fluid if the heat conduction follows the Fourier law (7.20).**

Since $D_{ii} = v_{i,i}$, equation (7.18) may be written $\sigma_{ij} = -p\delta_{ij} + \lambda^*\delta_{ij}v_{k,k} + \mu^*(v_{i,j} + v_{j,i})$. Thus

$$\sigma_{ij,j} = -p_{,j}\delta_{ij} + \lambda^*\delta_{ij}v_{k,kj} + \mu^*(v_{i,jj} + v_{j,ij}) = -p_{,i} + (\lambda^* + \mu^*)v_{j,ji} + \mu^*v_{i,jj}$$

and with this expression inserted into (7.16) a direct verification of (7.22) is complete.

Substituting the above equation for σ_{ij} together with (7.20) into the energy equation (7.17), the result is

$$\rho\dot{u} = [-p\delta_{ij} + \lambda^*\delta_{ij}v_{k,k} + \mu^*(v_{i,j} + v_{j,i})](v_{i,j} + v_{j,i})/2 - kT_{,ii} + \rho z$$

which reduces to

$$\rho\dot{u} = -pv_{i,i} + \lambda^*v_{i,i}v_{j,j} + \mu^*(v_{i,j} + v_{j,i})(v_{i,j} + v_{j,i})/2 - kT_{,ii} + \rho z$$

7.8. **Determine the traction force T_i acting on the closed surface S which surrounds the volume V of a Newtonian fluid for which the bulk viscosity is zero.**

The element of traction is $dT_i = t_i^{(\hat{n})}\,dS$ and the total traction force is $T_i = \int_S t_i^{(\hat{n})}\,dS$ which because of the stress principle is $T_i = \int_S \sigma_{ji}n_j\,dS$. From Problem 7.4, this becomes

$$T_i = \int_S (-p\delta_{ij} + 2\mu^*D'_{ij})n_j\,dS$$

for a zero bulk modulus fluid; and upon application of Gauss' theorem,

$$T_i = \int_V (2\mu^*D'_{ij,j} - p_{,i})\,dV$$

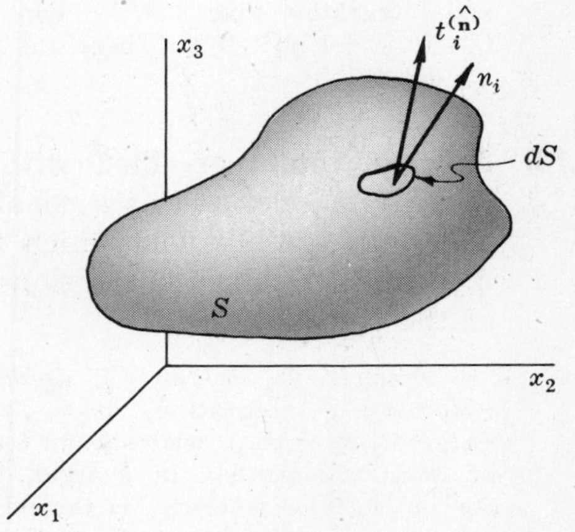

Fig. 7-1

7.9. In an axisymmetric flow along the x_3 axis the velocity is taken as a function of x_3 and r where $r^2 = x_1^2 + x_2^2$. If the velocity is expressed by $\mathbf{v} = q\hat{\mathbf{e}}_r + v_3\hat{\mathbf{e}}_3$ where $\hat{\mathbf{e}}_r$ is the unit radial vector, determine the form of the continuity equation.

Equation (*5.4*) gives the continuity equation in symbolic notation as $\partial\rho/\partial t + \nabla \cdot (\rho\mathbf{v}) = 0$. Here the cylindrical form of the operator ∇ may be used to give $\nabla \cdot (\rho\mathbf{v}) = \dfrac{1}{r}\dfrac{\partial(r\rho q)}{\partial r} + \dfrac{\partial(\rho v_3)}{\partial x_3}$. Inserting this into (*5.4*) and simplifying, the continuity equation becomes

$$r(\partial\rho/\partial t) + \partial(r\rho q)/\partial r + \partial(r\rho v_3)/\partial x_3 = 0$$

7.10. In a two-dimensional flow parallel to the x_1x_2 plane, v_3 and $\partial/\partial x_3$ are zero. Determine the Navier-Stokes equations for an incompressible fluid and the form of the continuity equation for this case.

From (*7.23*) with $i = 3$, $\rho b_3 = p_{,3}$ and when $i = 1, 2$, $\rho\dot{v}_\alpha = \rho b_\alpha - p_{,\alpha} + \mu^* v_{\alpha,\beta\beta}$. The continuity equation (*7.15*) reduces to $v_{\alpha,\alpha} = 0$.

If body forces were zero and $v_1 = v_1(x_1, x_2, t)$, $v_2 = 0$, $p = p(x_1, x_2, t)$ the necessary equations would be $\rho\dot{v}_1 = -\partial p/\partial x_1 + \mu^*(\partial^2 v_1/\partial x_1^2 + \partial^2 v_1/\partial x_2^2)$ and $\partial v_1/\partial x_1 = 0$.

HYDROSTATICS. STEADY AND IRROTATIONAL FLOW (Sec. 7.4)

7.11. Assuming air is an ideal gas whose temperature varies linearly with altitude as $T = T_0 - \alpha x_3$ where T_0 is ground level temperature and x_3 measures height above the earth, determine the air pressure in the atmosphere as a function of x_3 under hydrostatic conditions.

From (*7.6*) in this case, $p = \rho R(T_0 - \alpha x_3)$; and from (*7.28*) with the body force $b_3 = -g$, the gravitational constant, $dp/dx_3 = -\rho g = -pg/R(T_0 - \alpha x_3)$. Separating variables and integrating yields $\ln p = (g/R\alpha)\ln(T_0 - \alpha x_3) + \ln C$ where C is a constant of integration. Thus $p = C(T_0 - \alpha x_3)^{g/R\alpha}$ and if $p = p_0$ when $x_3 = 0$, $C = p_0 T_0^{-g/R\alpha}$ and so $p = p_0(1 - \alpha x_3/T_0)^{g/R\alpha}$.

7.12. A barotropic fluid having the equation of state $p = \lambda\rho^k$ where λ and k are constants is at rest in a gravity field in the x_3 direction. Determine the pressure in the fluid with respect to x_3 and p_0, the pressure at $x_3 = 0$.

From (*7.28*), $dp/dx_3 = -\rho g$, $dp/dx_1 = dp/dx_2 = 0$. Note that pressure in x_1 and x_2 directions is constant in the absence of body forces b_1 and b_2. Since here $\rho = (p/\lambda)^{1/k}$, $p^{-1/k}dp = -g\lambda^{-1/k}dx_3$ and integration gives $(k/(k-1))p^{(k-1)/k} = -g\lambda^{-1/k}x_3 + C$. But $p = p_0$ when $x_3 = 0$ so that $C = (k/(k-1))p_0^{(k-1)/k}$. Therefore $x_3 = (kp_0/(k-1)g\rho_0)(1 - (p/p_0)^{(k-1)/k})$ where $\rho_0 = (p_0/\lambda)^{1/k}$.

7.13. A large container filled with an incompressible liquid is accelerated at a constant rate $\mathbf{a} = a_2\hat{\mathbf{e}}_2 + a_3\hat{\mathbf{e}}_3$ in a gravity field which is parallel to the x_3 direction. Determine the slope of the free surface of the liquid.

From (*7.28*), $dp/dx_1 = 0$, $dp/dx_2 = \rho a_2$ and $dp/dx_3 = -\rho(g - a_3)$. Integrating, $p = \rho a_2 x_2 + f(x_3)$ and $p = -\rho(g - a_3)x_3 + h(x_2)$ where f and h are arbitrary functions of their arguments. In general, therefore, $p = \rho a_2 x_2 - \rho(g - a_3)x_3 + p_0$ where p_0 is the pressure at the origin of coordinates on the free surface. Since $p = p_0$ everywhere on the free surface, the equation of that surface is $x_2/x_3 = (g - a_3)/a_2$.

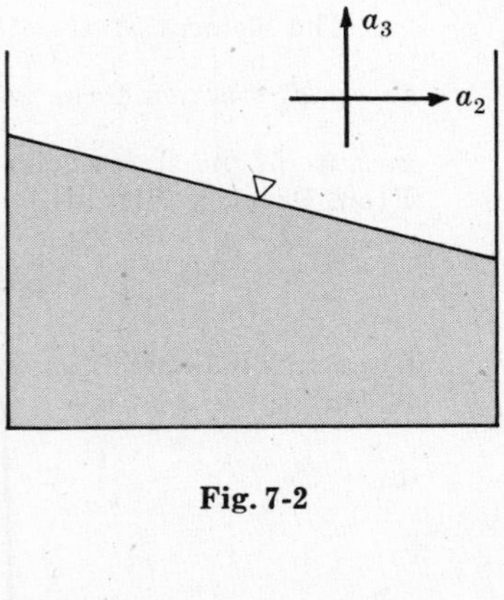

Fig. 7-2

7.14. **If a fluid motion is very slow so that higher order terms in the velocity are negligible, a limiting case known as *creeping flow* results. For this case show that in a steady incompressible flow with zero body forces the pressure is a harmonic function, i.e. $\nabla^2 p = 0$.**

For incompressible flow the Navier-Stokes equations (*7.23*) are

$$\rho(\partial v_i/\partial t + v_j v_{i,j}) \;=\; \rho b_i - p_{,i} + \mu^* v_{i,jj}$$

and for creeping flow these linearize to the form

$$\rho(\partial v_i/\partial t) \;=\; \rho b_i - p_{,i} + \mu^* v_{i,jj}$$

Hence for steady flow with zero body forces, $p_{,i} = \mu^* v_{i,jj}$. Taking the divergence of this equation yields $p_{,ii} = \mu^* v_{i,ijj}$; and since the continuity equation for incompressible flow is $v_{i,i} = 0$, it follows that here $p_{,ii} = \nabla^2 p = 0$.

7.15. **Express the continuity equation and the Navier-Stokes-Duhem equations in terms of the velocity potential ϕ for an irrotational motion.**

By (*7.35*), $v_i = -\phi_{,i}$ so that from (*7.15*) the continuity equation becomes $\dot{\rho} - \rho\nabla^2\phi = 0$. Also with $v_i = -\phi_{,i}$, (*7.22*) becomes

$$-\rho\dot{\phi}_{,i} \;=\; \rho b_i - p_{,i} - (\lambda^* + \mu^*)\phi_{,jji} - \mu^*\phi_{,ijj}$$

or

$$-\rho(\partial\phi_{,i}/\partial t + \phi_{,k}\phi_{,ik}) \;=\; \rho b_i - p_{,i} - (\lambda^* + 2\mu^*)\phi_{,jji}$$

In symbolic notation this equation is written

$$-\rho\nabla(\partial\phi/\partial t + (\nabla\phi)^2/2) \;=\; \rho\mathbf{b} - \nabla p - (\lambda^* + 2\mu^*)\nabla(\nabla^2\phi)$$

7.16. **Determine the pressure function $P(p)$ for a barotropic fluid having the equation of state $p = \lambda\rho^k$ where λ and k are constants.**

From the definition (*7.29*),

$$P(p) \;=\; \int_{p_0}^{p} \frac{dp}{\rho} \;=\; \int_{p_0}^{p} (p/\lambda)^{-1/k}\,dp \;=\; \frac{k\lambda^{1/k}}{k-1}\Big[p^{(k-1)/k}\Big]_{p_0}^{p} \;=\; \frac{k}{k-1}\left(\frac{p}{\rho} - \frac{p_0}{\rho_0}\right)$$

Also since $dp = \lambda k\rho^{k-1}\,d\rho$, the same result may be obtained from

$$P(p) \;=\; \int_{\rho_0}^{\rho} \lambda k\rho^{k-2}\,d\rho \;=\; \frac{\lambda k}{k-1}\Big[\rho^{k-1}\Big]_{\rho_0}^{\rho} \;=\; \frac{k}{k-1}\left(\frac{p}{\rho} - \frac{p_0}{\rho_0}\right)$$

PERFECT FLUIDS. BERNOULLI EQUATION. CIRCULATION (Sec. 7.5)

7.17. **Derive equation (*7.39*) by integrating Euler's equation (*7.37*) along a streamline.**

Let dx_i be an increment of displacement along a streamline. Taking the scalar product of this increment with (*7.37*) and integrating gives

$$\int \frac{\partial v_i}{\partial t}\,dx_i + \int v_j v_{i,j}\,dx_i + \int \Omega_{,i}\,dx_i + \int P_{,i}\,dx_i \;=\; C(t)$$

Since $\Omega_{,i}\,dx_i = d\Omega$ and $P_{,i}\,dx_i = dP$ the last two terms integrate at once. Also, along a streamline, $dx_i = (v_i/v)\,ds$ where ds is the increment of distance. Thus in the second integral,

$$v_j v_{i,j}\,dx_i \;=\; v_j v_{i,j}(v_i/v)\,ds \;=\; v_i v_{i,j}(v_j/v)\,ds \;=\; v_i v_{i,j}\,dx_j \;=\; v_i\,dv_i$$

Therefore $\displaystyle\int v_j v_{i,j}\,dx_i = \int v_i\,dv_i = \tfrac{1}{2}v_i v_i = \tfrac{1}{2}v^2$, and (*7.39*) is achieved.

7.18. The barotropic fluid of Problem 7.16 flows from a large closed tank through a thin smooth pipe. If the pressure in the tank is N times the atmospheric pressure, determine the speed of the emerging fluid.

Applying Bernoulli's equation for steady flow between point A, at rest in fluid of the tank and point B, in the emerging free stream, (*7.39*) assumes the form $\Omega_A + P_A + \frac{1}{2}v_A^2 = \Omega_B + P_B + \frac{1}{2}v_B^2$. But $v_A = 0$, and if gravity is assumed negligible this equation becomes (see Problem 7.16),

$$\frac{k}{k-1}\left(\frac{p_A}{\rho_A} - \frac{p_B}{\rho_B}\right) = \tfrac{1}{2}v_B^2 \qquad \text{or} \qquad v_B^2 = \frac{2k}{k-1}\frac{p_B}{\rho_B}\left(\frac{N\rho_B}{\rho_A} - 1\right)$$

Since $\rho_B/\rho_A = (p_B/p_A)^{-1/k} = N^{-1/k}$ the result may be written

$$v_B^2 = \frac{2k}{k-1}\frac{p_B}{\rho_B}(N^{(k-1)/k} - 1)$$

7.19. Show that for a barotropic, inviscid fluid with conservative body forces the rate of change of the circulation is zero (Kelvin's theorem).

From (*7.43*) $\dot{\Gamma}_c = \oint (\dot{v}_i\, dx_i + v_i\, dv_i)$ and by (*7.37*), $\dot{v}_i = -(\Omega + P)_{,i}$ for the case at hand. Thus $\dot{\Gamma}_c = \oint (-\Omega_{,i}\, dx_i - P_{,i}\, dx_i + v_i\, dv_i) = -\oint (d\Omega + dP - d(v^2/2)) = -\oint d(\Omega + P - v^2/2) = 0$, the integrand being a perfect differential.

7.20. Determine the circulation around the square $x_1 = \pm 1$, $x_2 = \pm 1$, $x_3 = 0$ (see Fig. 7-3) for the two-dimensional flow $\mathbf{v} = (x_1 + x_2)\hat{\mathbf{e}}_1 + (x_1^2 - x_2)\hat{\mathbf{e}}_2$.

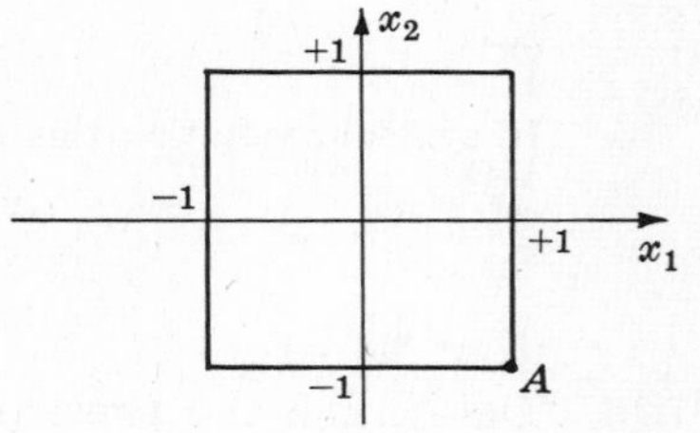

Fig. 7-3

Using the symbolic form of (*7.42*) with $\hat{\mathbf{n}} = \hat{\mathbf{e}}_3$ and $\nabla \times \mathbf{v} = (2x_1 - 1)\,\hat{\mathbf{e}}_3$,

$$\Gamma_c = \int_{-1}^{1}\int_{-1}^{1} (2x_1 - 1)\, dx_1\, dx_2 = -4$$

The same result is obtained from (*7.41*) where

$$\Gamma_c = \oint \mathbf{v} \cdot d\mathbf{x}$$

$$= \int_{-1}^{1} (1 - x_2)\, dx_2 + \int_{1}^{-1} (x_1 + 1)\, dx_1 + \int_{1}^{-1} (1 - x_2)\, dx_2 + \int_{-1}^{1} (x_1 - 1)\, dx_1 = -4$$

with the integration proceeding counterclockwise from A.

POTENTIAL FLOW. PLANE POTENTIAL FLOW (Sec. 7.6)

7.21. Give the derivation of the gas dynamical equation (*7.45*) and express this equation in terms of the velocity potential ϕ.

For a steady flow the continuity equation (*5.4*) becomes $\rho_{,i}v_i + \rho v_{i,i} = 0$ and the Euler equation (*7.36*) becomes $\rho v_j v_{i,j} + p_{,i} = 0$ if body forces are neglected. For a barotropic fluid, $p = p(\rho)$ and so $dp = (\partial p/\partial x_i)(\partial x_i/\partial \rho)d\rho$; or rearranging, $p_{,i} = (dp/d\rho)\rho_{,i} = c^2\rho_{,i}$ where c is the local velocity of sound. Inserting this into the Euler equation and multiplying by v_i gives $\rho v_i v_j v_{i,j} + c^2 v_i \rho_{,i} = 0$. From the continuity equation $c^2 v_i \rho_{,i} = -c^2 v_{i,i} = -c^2\delta_{ij}v_{i,j}$ and so $(c^2\delta_{ij} - v_i v_j)v_{i,j} = 0$. In terms of $\phi_{,i} = -v_i$ this becomes $(c^2\delta_{ij} - \phi_{,i}\phi_{,j})\phi_{,ij} = 0$.

7.22. Show that the function $\phi = A(-x_1^2 - x_2^2 + 2x_3^2)$ satisfies the Laplace equation and determine the resulting velocity components.

Substituting ϕ into (*7.46*) gives $-2A - 2A + 4A \equiv 0$. From (*7.35*), $v_1 = 2Ax_1$, $v_2 = 2Ax_2$, $v_3 = -4Ax_3$. Also, by the analysis of Problem 4.7 the streamlines in the x_1 plane are represented by $x_2^2 x_3 = \text{constant}$; in the x_2 plane by $x_1^2 x_3 = \text{constant}$. Thus the flow is in along the x_3 axis against the x_1x_2 plane (fixed wall).

7.23. Show that the stream function $\psi(x_1, x_2)$ is constant along any streamline.

From (7.48) and the differential equation of a streamline, $dx_1/v_1 = dx_2/v_2$ (see Problem 4.7), $-dx_1/\psi_{,2} = dx_2/\psi_{,1}$ or $\psi_{,1}dx_1 + \psi_{,2}dx_2 = d\psi = 0$. Thus ψ = a constant along any streamline.

7.24. Verify that $\phi = A(x_1^2 - x_2^2)$ is a valid velocity potential and describe the flow field.

For the given ϕ, (7.46) is satisfied identically by $2A - 2A = 0$; and from (7.49), $v_1 = -2Ax_1$, $v_2 = +2Ax_2$. The streamlines are determined by integrating $dx_1/x_1 = -dx_2/x_2$ to give the rectangular hyperbolas $x_1x_2 = C$ (Fig. 7-4). The equipotential lines $A(x_1^2 - x_2^2) = C_1$ form an orthogonal set of rectangular hyperbolas with the streamlines. Finally from (7.50), $\psi = -2Ax_1x_2 + C_0$ and is seen to be constant along the streamlines as was asserted in Problem 7.23.

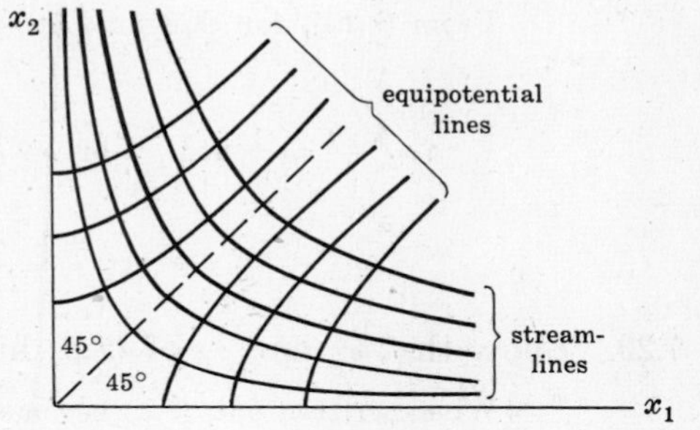

Fig. 7-4

7.25. A velocity potential is given by $\phi = Ax_1 + Bx_1/r^2$ where $r^2 = x_1^2 + x_2^2$. Determine the stream function ψ for this flow.

From (7.50), $\psi_{,1} = -\phi_{,2} = 2Bx_1x_2/r^4$ so that by integrating, $\psi = -Bx_2/r^2 + f(x_2)$ where $f(x_2)$ is an arbitrary function of x_2. Differentiating, $\psi_{,2} = -B(x_1^2 - x_2^2)/r^4 + f'(x_2)$. But from (7.50), $\psi_{,2} = \phi_{,1} = A + B(-x_1^2 + x_2^2)/r^4$. Thus $f'(x_2) = A$ and $f(x_2) = Ax_2 + C$. Finally then $\psi = Ax_2 - Bx_2/r^2 + C$.

7.26. Differentiate the complex potential $\Phi(z) = A/z$ to obtain the velocity components.

Here $d\Phi/dz = -A/z^2 = -A/(x_1 + ix_2)^2$ which after some algebra becomes $d\Phi/dz = -A(x_1^2 - x_2^2)/r^4 + i2Ax_1x_2/r^4$. Thus

$$v_1 = A(x_1^2 - x_2^2)r^4 \quad \text{and} \quad v_2 = 2Ax_1x_2/r^4$$

Note that since $\Phi = A/z = A(x_1 - ix_2)/r^2$, $\phi = Ax_1/r^2$ and $\psi = -Ax_2/r^2$. Also note that

$$v_1 = -\phi_{,1} = A(x_1^2 - x_2^2)/r^4 \quad \text{and} \quad v_2 = -\phi_{,2} = 2Ax_1x_2/r^4$$

MISCELLANEOUS PROBLEMS

7.27. Derive the one-dimensional continuity equation for the flow of an inviscid incompressible fluid through a stream tube.

Let V be the volume between arbitrary cross sections A and B of the stream tube shown in Fig. 7-5. In integral form, for this volume (5.2) becomes $\int_V \nabla \cdot \mathbf{v}\, dV = 0$ since ρ is constant here. Converting by Gauss' theorem, $\int_S \hat{\mathbf{n}} \cdot \mathbf{v}\, dS = 0$ where $\hat{\mathbf{n}}$ is the outward unit normal to the surface S enclosing V. Since $\hat{\mathbf{n}} \perp \mathbf{v}$ on the lateral surface, the integration reduces to

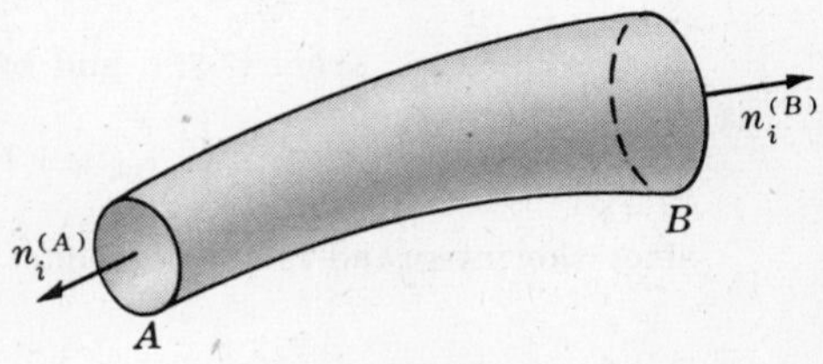

Fig. 7-5

$$\int_{S_A} \hat{\mathbf{n}}_A \cdot \mathbf{v}_A\, dS + \int_{S_B} \hat{\mathbf{n}}_B \cdot \mathbf{v}_B\, dS = 0$$

The velocity is assumed uniform and perpendicular over S_A and S_B; and since $\mathbf{v}_B = -v_B\hat{\mathbf{n}}_B$, $v_A \int_{S_A} dS - v_B \int_{S_B} dS = 0$ or $v_AS_A = v_BS_B$ = a constant.

7.28. The stress tensor at a given point for a Newtonian fluid with zero bulk viscosity is $\sigma_{ij} = \begin{pmatrix} -6 & 2 & -1 \\ 2 & -9 & 4 \\ -1 & 4 & -3 \end{pmatrix}$. Determine τ_{ij}.

From (*7.14*), for this fluid $p = -\sigma_{ii}/3 = 6$. Then from (*7.3*),

$$\tau_{ij} = \sigma_{ij} + 6\delta_{ij} \quad \text{or} \quad \begin{pmatrix} -6 & 2 & -1 \\ 2 & -9 & 4 \\ -1 & 4 & -3 \end{pmatrix} + \begin{pmatrix} 6 & 0 & 0 \\ 0 & 6 & 0 \\ 0 & 0 & 6 \end{pmatrix} = \begin{pmatrix} 0 & 2 & -1 \\ 2 & -3 & 4 \\ -1 & 4 & 3 \end{pmatrix}$$

7.29. Show that σ_{ij} and τ_{ij} of (*7.3*) have the same principal axes.

When written out, (*7.3*) becomes $\sigma_{11} = -p + \tau_{11}$, $\sigma_{22} = -p + \tau_{22}$, $\sigma_{33} = -p + \tau_{33}$, $\sigma_{12} = \tau_{12}$, $\sigma_{23} = \tau_{23}$, $\sigma_{13} = \tau_{13}$. For principal directions x_i^* of σ_{ij}, $\sigma_{12}^* = \sigma_{23}^* = \sigma_{13}^* = 0$ and by the last three equations of (*7.3*), $\sigma_{ij}^* = \tau_{ij}^* = 0$ for $i \neq j$. Thus x_i^* are principal axes for τ_{ij} also.

7.30. A dissipation potential Φ_D is often defined for a Newtonian fluid by the relationship $\Phi_D = (\kappa/2)D_{ii}D_{jj} + \mu^* D'_{ij}D'_{ij}$. Show that $\partial\Phi_D/\partial D_{ij} = \tau_{ij}$.

Here $\partial\Phi_D/\partial D_{pq} = (\kappa/2)[D_{ii}(\partial D_{jj}/\partial D_{pq}) + (\partial D_{ii}/\partial D_{pq})D_{jj}] + 2\mu^*[D'_{ij}(\partial D'_{ij}/\partial D_{pq})]$. But $\partial D_{ii}/\partial D_{pq} = \delta_{ip}\delta_{iq} = \delta_{pq}$ and $\partial D'_{ij}/\partial D_{pq} = \delta_{ip}\delta_{jq} - \delta_{ij}\delta_{pq}/3$ so that

$$\partial\Phi_D/\partial D_{pq} = \kappa D_{ii}\delta_{pq} + 2\mu^*(D_{ij} - \delta_{ij}D_{kk}/3)(\delta_{ip}\delta_{jq} - \delta_{ij}\delta_{pq}/3) = \kappa D_{ii}\delta_{pq} + 2\mu^*(D_{pq} - \delta_{pq}D_{ii}/3)$$

Finally since $\kappa = \lambda^* + 2\mu^*/3$,

$$\partial\Phi_D/\partial D_{pq} = \lambda^*\delta_{pq}D_{ii} + 2\mu^* D_{pq} = \tau_{pq}$$

7.31. Determine the pressure-density relationship for the ideal gas discussed in Problem 7.11.

At $x_3 = 0$, $\rho = \rho_0$ and $p = p_0$. The ideal gas law (*7.6*) is here $p = \rho R(T_0 - \alpha x_3)$ so that $p_0 = \rho_0 R T_0$; and from the pressure elevation relationship $p = p_0(1 - \alpha x_3/T_0)^{g/R\alpha}$ of Problem 7.11, $\rho/\rho_0 = (T/T_0)^{(g/R\alpha - 1)}$. Thus writing $p = p_0(1 - \alpha x_3/T_0)^{g/R\alpha}$ in the form $p/p_0 = (T/T_0)^{g/R\alpha}$, it is seen that $T/T_0 = (p/p_0)^{+R\alpha/g}$ and so $\rho/\rho_0 = (p/p_0)^{(1 - R\alpha/g)}$.

7.32. For a barotropic inviscid fluid with conservative body forces show that the material derivative of the total vorticity, $\frac{d}{dt}\int_V q_i\,dV = \int_S v_i q_j\,dS_j$.

From (*4.54*) and the results of Problem 4.33, $\frac{d}{dt}\int_V q_i\,dV = \int_S (\epsilon_{ijk}a_k + q_j v_i)\,dS_j$. But here $a_k = -(\Omega + P)_{,k}$ from (*7.37*); and by the divergence theorem (*1.157*),

$$\int_S \epsilon_{ijk}(\Omega + P)_{,k}\,dS_j = \int_V \epsilon_{ijk}(\Omega + P)_{,kj}\,dV = 0$$

since the integrand is zero (product of a symmetric and antisymmetric tensor). Hence

$$\frac{d}{dt}\int_V q_i\,dV = \int_S q_j v_i\,dS_j$$

7.33. For an incompressible Newtonian fluid moving inside a closed rigid container at rest, show that the time rate of change of kinetic energy of the fluid is $-\mu^*\int_V q^2\,dV$ assuming zero body forces. q is the magnitude of the vorticity vector.

From Problem 5.27, the time rate of change of kinetic energy of a continuum is

$$\frac{dK}{dt} = \int_V \rho b_i v_i \, dV - \int_V \sigma_{ij} v_{i,j} \, dV + \int_S v_i t_i^{(\hat{n})} \, dS$$

In this problem the first and third integrals are zero; and for a Newtonian fluid by (7.18),

$$\frac{dK}{dt} = -\int_V \sigma_{ij} v_{i,j} \, dV = -\int_V (-p\delta_{ij} + \lambda^*\delta_{ij}D_{kk} + 2\mu^* D_{ij}) v_{i,j} \, dV$$

But incompressibility means $v_{i,i} = D_{ii} = 0$ and so

$$\frac{dK}{dt} = -2\mu^* \int_V D_{ij} v_{i,j} \, dV = -\mu^* \int_V (v_{i,j} + v_{j,i}) v_{i,j} \, dV$$

$$= -\mu^* \int_V (\epsilon_{kji} q_k) v_{i,j} \, dV = -\mu^* \int_V q_k (\epsilon_{kji} v_{i,j}) \, dV = -\mu^* \int_V q_k q_k \, dV$$

7.34. Show that for a perfect fluid with negligible body forces the rate of change of circulation $\dot{\Gamma}_c$ may be given by $-\int_S \epsilon_{ijk}(1/\rho)_{,j} p_{,k} \, dS_i$.

From (7.43), $\dot{\Gamma}_c = \oint \dot{v}_i \, dx_i + \oint v_i \, dv_i$; and since $\oint d(\frac{1}{2}v^2) = 0$, the second integral is zero. From (7.36) with $b_i = 0$, $\dot{v}_i = -p_{,i}/\rho$ and so now

$$\dot{\Gamma}_c = -\oint (p_{,k}/\rho) \, dx_k = -\int_S \epsilon_{ijk}(p_{,k}/\rho)_{,j} n_i \, dS$$

where (7.42) has been used in converting to the surface integral. Differentiating as indicated,

$$\dot{\Gamma}_c = -\int_S \epsilon_{ijk}[(1/\rho)_{,j} p_{,k} + p_{,kj}/\rho] \, dS_i = -\int_S \epsilon_{ijk}(1/\rho)_{,j} \, p_{,k} \, dS_i$$

Supplementary Problems

7.35. The constitutive equation for an isotropic fluid is given by $\sigma_{ij} = -p\delta_{ij} + K_{ijpq} D_{pq}$ with K_{ijpq} constants independent of the coordinates. Show that the principal axes of stress and rate of deformation coincide.

7.36. Show that $(1/\rho)(d\rho/dt) = 0$ is a condition for $-\sigma_{ii}/3 = p$ for a Newtonian fluid.

7.37. Show that the constitutive relations for a Newtonian fluid with zero bulk viscosity may be expressed by the pair of equations $s_{ij} = 2\mu^* D_{ij}$ and $-\sigma_{ii} = 3p$.

7.38. Show that in terms of the vorticity vector $\mathbf{q}$ the Navier-Stokes equations may be written $\dot{\mathbf{v}} = \mathbf{b} - \nabla p/\rho - \nu^* \nabla \times \mathbf{q}$ where $\nu^* = \mu^*/\rho$ is the kinematic viscosity. Show that for irrotational motion this equation reduces to (7.36).

7.39. If a fluid moves radially with the velocity $\mathbf{v} = \mathbf{v}(r, t)$ where $r^2 = x_i x_i$, show that the equation of continuity is $\frac{\partial \rho}{\partial t} + v \frac{\partial \rho}{\partial r} + \frac{\rho}{r^2} \frac{\partial}{\partial r}(r^2 v) = 0$.

7.40. A liquid rotates as a rigid body with constant angular velocity ω about the vertical x_3 axis. If gravity is the only body force, show that $p/\rho - \omega^2 r^2/2 + g x_3 = \text{constant}$.

7.41. For an ideal gas under isothermal conditions (constant temperature $= T_0$), show that $\rho/\rho_0 = p/p_0 = e^{-(g/RT_0 x_3)}$ where ρ_0 and p_0 are the density and pressure at $x_3 = 0$.

7.42. Show that if body forces are conservative so that $b_i = -\Omega_{,i}$, the Navier-Stokes-Duhem equations for the irrotational motion of a barotropic fluid may be integrated to yield $-\rho(\partial\phi/\partial t + (\nabla\phi)^2/2) + \rho\Omega + P + (\lambda^* + 2\mu^*)\nabla^2\phi = f(t)$. (See Problem 7.15.)

7.43. Show that the velocity and vorticity for an inviscid flow having conservative body forces and constant density satisfy the relation $\dot{q}_i - q_j v_{i,j} = 0$. For steady flow of the same fluid, show that $v_j q_{i,j} = q_j v_{i,j}$.

7.44. For a barotropic fluid having $\rho = \rho(p)$ and $P(p)$ defined by (*7.29*), show that $\text{grad}\, P = \text{grad}\, p/\rho$.

7.45. Show that the Bernoulli equation (*7.39*) for steady motion of an ideal gas takes the form (*a*) $\Omega + p \ln(p/\rho) + v^2/2 =$ constant, for isothermal flow, (*b*) $\Omega + (k/k-1)(p/\rho) + v^2/2 =$ constant, for isentropic flow.

7.46. Show that the velocity field $v_1 = -2x_1x_2x_3/r^4$, $v_2 = (x_1^2 - x_2^2)x_3/r^4$, $v_3 = x_2/r^2$ where $r^2 = x_1^2 + x_2^2 + x_3^2$ is a possible flow for an incompressible fluid. Is the motion irrotational? *Ans.* Yes

7.47. If the velocity potential $\Phi(z) = \phi + i\psi$ is an analytic function of the complex variable $z = x_1 + ix_2 = re^{i\theta}$ show that in polar coordinates $\dfrac{\partial\phi}{\partial r} = \dfrac{1}{r}\dfrac{\partial\psi}{\partial\theta}$ and $\dfrac{1}{r}\dfrac{\partial\phi}{\partial\theta} = -\dfrac{\partial\psi}{\partial r}$.

7.48. If body forces are zero, show that for irrotational potential flow $\psi_{,ii} = \nu^* = \mu^*/\rho$ is the kinematic viscosity.

Chapter 8

Plasticity

8.1 BASIC CONCEPTS AND DEFINITIONS

Elastic deformations, which were considered in Chapter 6, are characterized by complete recovery to the undeformed configuration upon removal of the applied loads. Also, elastic deformations depend solely upon the stress magnitude and not upon the straining or loading history. Any deformational response of a continuum to applied loads, or to environmental changes, that does not obey the constitutive laws of classical elasticity may be spoken of as an *inelastic deformation*. In particular, irreversible deformations which result from the mechanism of *slip*, or from dislocations at the atomic level, and which thereby lead to permanent dimensional changes are known as *plastic deformations*. Such deformations occur only at stress intensities above a certain threshold value known as the *elastic limit*, or *yield stress*, which is denoted here by σ_Y.

In the *theory of plasticity*, the primary concerns are with the mathematical formulation of stress-strain relationships suitable for the phenomenological description of plastic deformations, and with the establishment of appropriate yield criteria for predicting the onset of plastic behavior. By contrast, the study of plastic deformation from the microscopic point of view resides in the realm of solid state physics.

The phrase *plastic flow* is used extensively in plasticity to designate an on-going plastic deformation. However, unlike a fluid flow, such a continuing plastic flow may be related to the amount of deformation as well as the rate of deformation. Indeed, a solid in the "plastic" state can sustain shear stresses even when at rest.

Many of the basic concepts of plasticity may be introduced in an elementary way by consideration of the stress-strain diagram for a simple one-dimensional tension (or compression) test of some hypothetical material as shown by Fig. 8-1. In this plot, σ is the nominal stress (force/original area), whereas the strain ϵ may represent either the *conventional (engineering) strain* defined here by

$$e = (L - L_0)/L_0 \tag{8.1}$$

where L is the current specimen length and L_0 the original length, or the *natural (logarithmic) strain* defined by

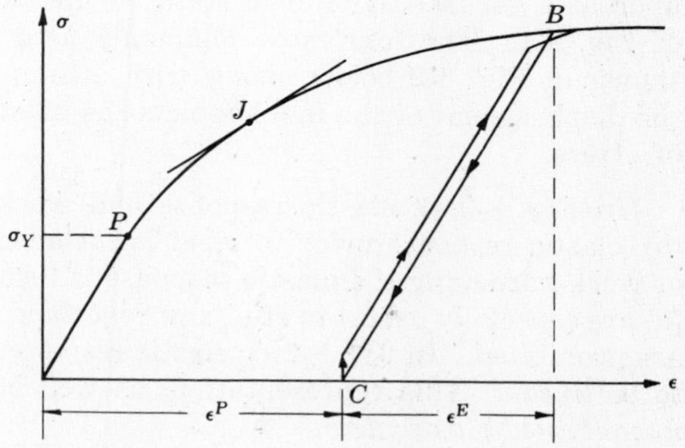

Fig. 8-1

$$\epsilon = \ln(L/L_0) = \ln(1+e) = e - e^2/2 + O(e^3) \tag{8.2}$$

For small strains, these two measures of strain are very nearly equal as seen by (*8.2*) and it is often permissible to neglect the difference.

The yield point P, corresponding to the yield stress σ_Y, separates the stress-strain curve of Fig. 8-1 into an *elastic range* and a *plastic range*. Unfortunately, the yield point is not always well-defined. It is sometimes taken at the *proportional limit*, which lies at the upper end of the linear portion of the curve. It may also be chosen as the point J, known as *Johnson's apparent elastic limit*, and defined as that point where the slope of the curve attains 50% of its initial value. Various offset methods are also used to define the yield point, one such being the stress value at 0.2 per cent permanent strain.

In the initial elastic range, which may be linear or nonlinear, an increase in load causes the stress-strain-state-point to move upward along the curve, and a decrease in load, or unloading causes the point to move downward along the same path. Thus a one-to-one stress-strain relationship exists in the elastic range.

In the plastic range, however, unloading from a point such as B in Fig. 8-1 results in the state point following the path BC which is essentially parallel with the linear elastic portion of the curve. At C, where the stress reaches zero, the permanent plastic strain ϵ^P remains. The recoverable elastic strain from B is labeled ϵ^E in Fig. 8-1. A reloading from C back to B would follow very closely the path BC but with a rounding at B, and with a small *hysteresis loop* resulting from the energy loss in the unloading-reloading cycle. Upon a return to B a load increase is required to cause further deformation, a condition referred to as *work hardening*, or *strain hardening*. It is clear therefore that in the plastic range the stress depends upon the entire loading, or strain history of the material.

Although it is recognized that temperature will have a definite influence upon the plastic behavior of a real material, it is customary in much of plasticity to assume isothermal conditions and consider temperature as a parameter. Likewise, it is common practice in traditional plasticity to neglect any effect that rate of loading would have upon the stress-strain curve. Accordingly, plastic deformations are assumed to be time-independent and separate from such phenomena as creep and relaxation.

8.2 IDEALIZED PLASTIC BEHAVIOR

Much of the three-dimensional theory for analyzing plastic behavior may be looked upon as a generalization of certain idealizations of the one-dimensional stress-strain curve of Fig. 8-1. The four most commonly used of these idealized stress-strain diagrams are shown in Fig. 8-2 below, along with a simple mechanical model of each. In the models the displacement of the mass depicts the plastic deformation, and the force F plays the role of stress.

In Fig. 8-2(a), elastic response and work-hardening are missing entirely, whereas in (b), elastic response prior to yield is included but work-hardening is not. In the absence of work-hardening the plastic response is termed *perfectly plastic*. Representations (a) and (b) are especially useful in studying *contained plastic deformation*, where large deformations are prohibited. In Fig. 8-2(c), elastic response is omitted and the work-hardening is assumed to be *linear*. This representation, as well as (a), has been used extensively in analyzing *uncontained plastic flow*.

The stress-strain curves of Fig. 8-2 appear in the context of tension curves. The compression curve for a previously unworked specimen (no history of plastic deformation) is taken as the reflection with respect to the origin of the tension curve. However, if a stress reversal (tension to compression, or vice versa) is carried out with a real material that has been work-hardened, a definite lowering of the yield stress is observed in the second type of loading. This phenomenon is known as the *Bauschinger effect*, and will be neglected in this book.

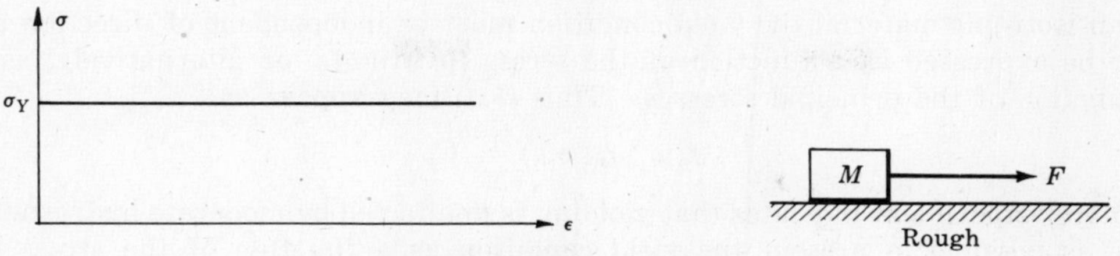

(a) Rigid — Perfectly Plastic

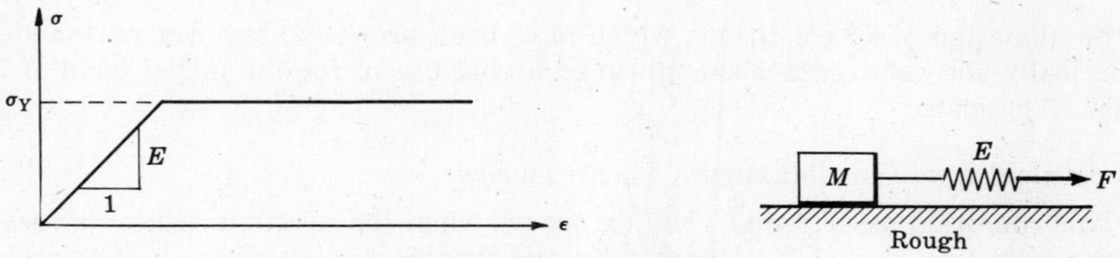

(b) Elastic — Perfectly Plastic

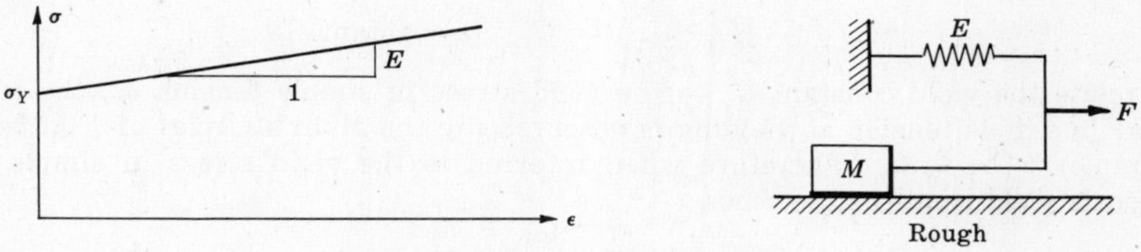

(c) Rigid — Linear Work Hardening

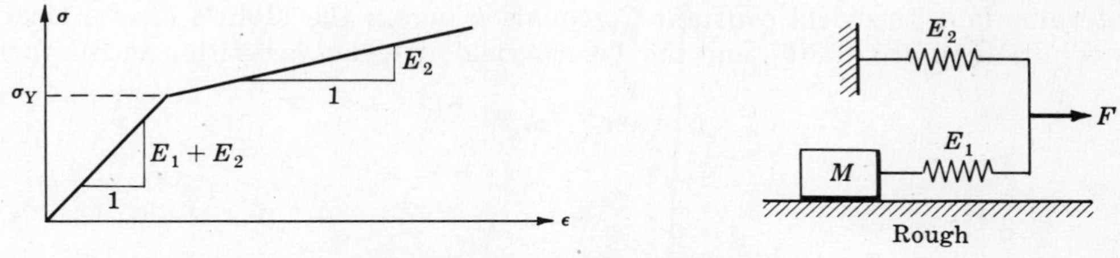

(d) Elastic — Linear Work Hardening

Fig. 8-2

8.3 YIELD CONDITIONS. TRESCA AND VON MISES CRITERIA

A *yield condition* is essentially a generalization to a three-dimensional state of stress of the yield stress concept in one dimensional loading. Briefly, the yield condition is a mathematical relationship among the stress components at a point that must be satisfied for the onset of plastic behavior at the point. In general, the yield condition may be expressed by the equation

$$f(\sigma_{ij}) = C_Y \tag{8.3}$$

where C_Y is known as the *yield constant*, or as is sometimes done by the equation

$$f_1(\sigma_{ij}) = 0 \tag{8.4}$$

in which $f_1(\sigma_{ij})$ is called the *yield function*.

For an isotropic material the yield condition must be independent of direction and may therefore be expressed as a function of the stress invariants, or alternatively, as a symmetric function of the principal stresses. Thus (*8.3*) may appear as

$$f_2(\sigma_{\rm I}, \sigma_{\rm II}, \sigma_{\rm III}) = C_Y \tag{8.5}$$

Furthermore, experiment indicates that yielding is unaffected by moderate hydrostatic stress so that it is possible to present the yield condition as a function of the stress deviator invariants in the form

$$f_3(\mathrm{II}_{\Sigma_D}, \mathrm{III}_{\Sigma_D}) = 0 \tag{8.6}$$

Of the numerous yield conditions which have been proposed, two are reasonably simple mathematically and yet accurate enough to be highly useful for the initial yield of isotropic materials. These are:

(1) *Tresca yield condition* (Maximum Shear Theory)

This condition asserts that yielding occurs when the maximum shear stress reaches the prescribed value C_Y. Mathematically, the condition is expressed in its simplest form when given in terms of principal stresses. Thus for $\sigma_{\rm I} > \sigma_{\rm II} > \sigma_{\rm III}$, the Tresca yield condition is given from (*2.54b*) as

$$\tfrac{1}{2}(\sigma_{\rm I} - \sigma_{\rm III}) = C_Y \quad \text{(a constant)} \tag{8.7}$$

To relate the yield constant C_Y to the yield stress in simple tension σ_Y, the maximum shear in simple tension at yielding is observed (by the Mohr's circles of Fig. 8-3(*a*), for example) to be $\sigma_Y/2$. Therefore when referred to the yield stress in simple tension, Tresca's yield condition becomes

$$\sigma_{\rm I} - \sigma_{\rm III} = \sigma_Y \tag{8.8}$$

The yield point for a state of stress that is so-called *pure shear* may also be used as a reference stress in establishing the yield constant C_Y. Thus if the pure shear yield point value is k, the yield constant C_Y equals k (again the Mohr's circles clearly show this result, as in Fig. 8-3(*b*), and the Tresca yield criterion is written in the form

$$\sigma_{\rm I} - \sigma_{\rm III} = 2k \tag{8.9}$$

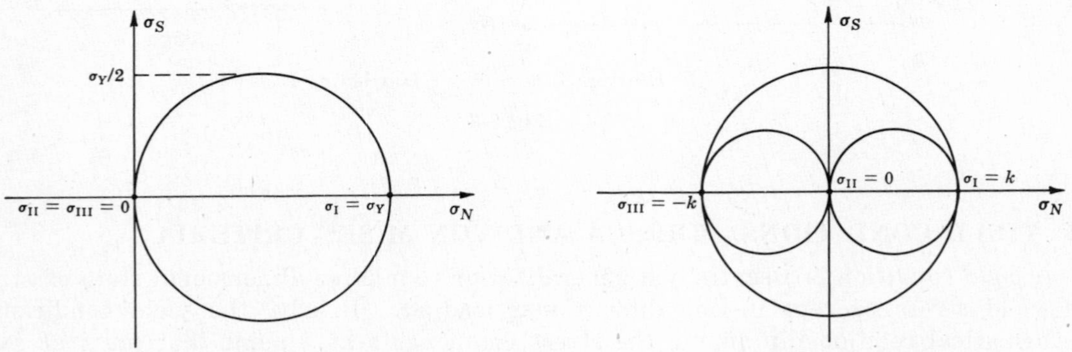

(*a*) Simple Tension (*b*) Pure Shear

Fig. 8-3

(2) *von Mises yield condition* (Distortion Energy Theory)

This condition asserts that yielding occurs when the second deviator stress invariant attains a specified value. Mathematically, the von Mises yield condition states

$$-\mathrm{II}_{\Sigma_D} = C_Y \tag{8.10}$$

which is usually written in terms of the principal stresses as

$$(\sigma_{\mathrm{I}} - \sigma_{\mathrm{II}})^2 + (\sigma_{\mathrm{II}} - \sigma_{\mathrm{III}})^2 + (\sigma_{\mathrm{III}} - \sigma_{\mathrm{I}})^2 = 6C_Y \tag{8.11}$$

With reference to the yield stress in simple tension, it is easily shown that (*8.11*) becomes

$$(\sigma_{\mathrm{I}} - \sigma_{\mathrm{II}})^2 + (\sigma_{\mathrm{II}} - \sigma_{\mathrm{III}})^2 + (\sigma_{\mathrm{III}} - \sigma_{\mathrm{I}})^2 = 2\sigma_Y^2 \tag{8.12}$$

Also, with respect to the pure shear yield value k, von Mises condition (*8.11*) appears in the form

$$(\sigma_{\mathrm{I}} - \sigma_{\mathrm{II}})^2 + (\sigma_{\mathrm{II}} - \sigma_{\mathrm{III}})^2 + (\sigma_{\mathrm{III}} - \sigma_{\mathrm{I}})^2 = 6k^2 \tag{8.13}$$

There are several variations for presenting (*8.12*) and (*8.13*) when stress components other than the principal stresses are employed.

8.4 STRESS SPACE. THE Π-PLANE. YIELD SURFACE

A stress space is established by using stress magnitude as the measure of distance along the coordinate axes. In the Haigh-Westergaard stress space of Fig. 8-4 the coordinate axes are associated with the principal stresses. Every point in this space corresponds to a state of stress, and the position vector of any such point $P(\sigma_{\mathrm{I}}, \sigma_{\mathrm{II}}, \sigma_{\mathrm{III}})$ may be resolved into a component OA along the line OZ, which makes equal angles with the coordinate axes, and a component OB in the plane (known as the Π-plane) which is perpendicular to OZ and passes through the origin. The component along OZ, for which $\sigma_{\mathrm{I}} = \sigma_{\mathrm{II}} = \sigma_{\mathrm{III}}$, represents hydrostatic stress, so that the component in the Π-plane represents the deviator portion of the stress state. It is easily shown that the equation of the Π-plane is given by

$$\sigma_{\mathrm{I}} + \sigma_{\mathrm{II}} + \sigma_{\mathrm{III}} = 0 \tag{8.14}$$

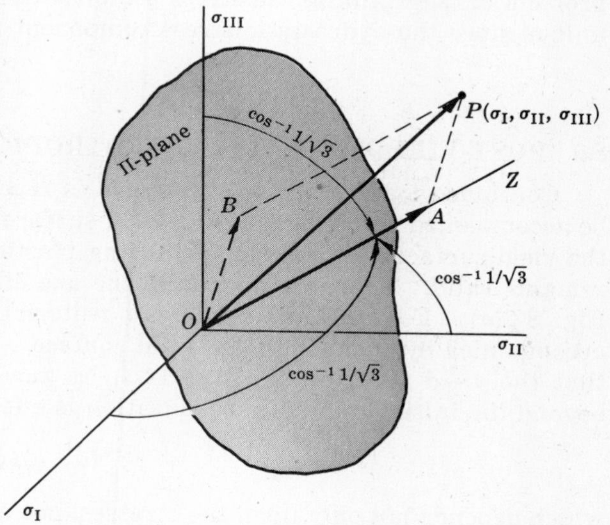

Fig. 8-4

In stress space, the yield condition (*8.5*), $f_2(\sigma_{\mathrm{I}}, \sigma_{\mathrm{II}}, \sigma_{\mathrm{III}}) = C_Y$, defines a surface, the so-called *yield surface*. Since the yield conditions are independent of hydrostatic stress, such yield surfaces are general cylinders having their generators parallel to OZ. Stress points that lie inside the cylindrical yield surface represent elastic stress states, those which lie on the yield surface represent incipient plastic stress states. The intersection of the yield surface with the Π-plane is called the *yield curve*.

In a true view of the Π-plane, looking along OZ toward the origin O, the principal stress axes appear symmetrically placed 120° apart as shown in Fig. 8-5(*a*) below. The yield curves for the Tresca and von Mises yield conditions appear in the Π-plane as shown in Fig. 8-5(*b*) and (*c*) below. In Fig. 8-5(*b*), these curves are drawn with reference to (*8.7*) and (*8.11*), using the yield stress in simple tension as the basis. For this situation, the von Mises circle of radius $\sqrt{2/3}\,\sigma_Y$ is seen to circumscribe the regular Tresca hexagon. In Fig. 8-5(*c*), the two yield curves are based upon the yield stress k in pure shear. Here the von Mises circle is inscribed in the Tresca hexagon.

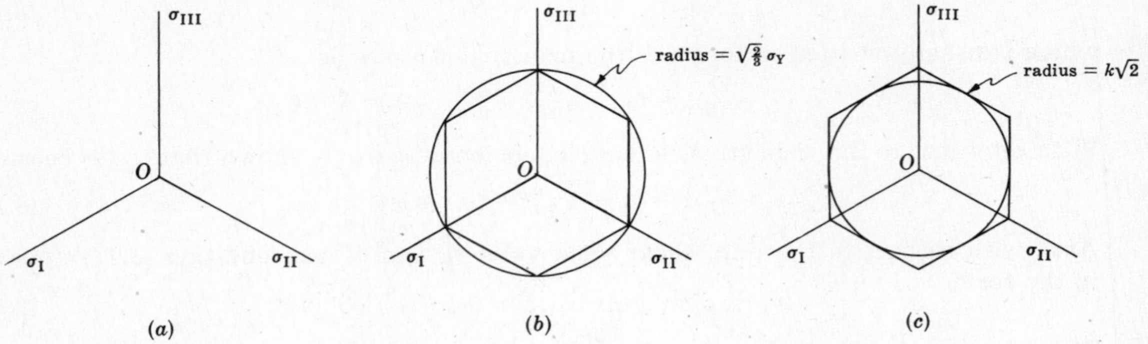

Fig. 8-5

The location in the Π-plane of the projection of an arbitrary stress point $P(\sigma_I, \sigma_{II}, \sigma_{III})$ is straightforward since each of the stress space axes makes $\cos^{-1}\sqrt{2/3}$ with the Π-plane. Thus the projected deviatoric components are $(\sqrt{2/3}\,\sigma_I, \sqrt{2/3}\,\sigma_{II}, \sqrt{2/3}\,\sigma_{III})$. The inverse problem of determining the stress components for an arbitrary point in the Π-plane is not unique since the hydrostatic stress component may have any value.

8.5 POST-YIELD BEHAVIOR. ISOTROPIC AND KINEMATIC HARDENING

Continued loading after initial yield is reached leads to plastic deformation which may be accompanied by changes in the yield surface. For an assumed *perfectly plastic* material the yield surface does not change during plastic deformation and the initial yield condition remains valid. This corresponds to the one-dimensional perfectly plastic case depicted by Fig. 8-2(*a*). For a *strain hardening* material, however, plastic deformation is generally accompanied by changes in the yield surface. To account for such changes it is necessary that the yield function $f_1(\sigma_{ij})$ of (*8.4*) be generalized to define subsequent yield surfaces beyond the initial one. A generalization is effected by introduction of the *loading function*

$$f_1^*(\sigma_{ij}, \epsilon_{ij}^P, K) = 0 \tag{8.15}$$

which depends not only upon the stresses, but also upon the plastic strains ϵ_{ij}^P and the work-hardening characteristics represented by the parameter K. Equation (*8.15*) defines a loading surface in the sense that $f_1^* = 0$ is the yield surface, $f_1^* < 0$ is a surface in the (elastic) region inside the yield surface and $f_1^* > 0$, being outside the yield surface, has no meaning.

Differentiating (*8.15*) by the chain rule of calculus,

$$df_1^* = \frac{\partial f_1^*}{\partial \sigma_{ij}}\,d\sigma_{ij} + \frac{\partial f_1^*}{\partial \epsilon_{ij}^P}\,d\epsilon_{ij}^P + \frac{\partial f_1^*}{\partial K}\,dK \tag{8.16}$$

Thus with $f_1^* = 0$ and $(\partial f_1^*/\partial \sigma_{ij})\,d\sigma_{ij} < 0$, *unloading* is said to occur; with $f_1^* = 0$ and $(\partial f_1^*/\partial \sigma_{ij})\,d\sigma_{ij} = 0$, *neutral loading* occurs; and with $f_1^* = 0$ and $(\partial f_1^*/\partial \sigma_{ij})\,d\sigma_{ij} > 0$, *loading* occurs. The manner in which the plastic strains ϵ_{ij}^P enter into the function (*8.15*) when loading occurs is defined by the *hardening rules,* two especially simple cases of which are described in what follows.

The assumption of *isotropic hardening* under loading conditions postulates that the yield surface simply increases in size and maintains its original shape. Thus in the Π-plane the yield curves for von Mises and Tresca conditions are the concentric circles and regular hexagons shown in Fig. 8-6 below.

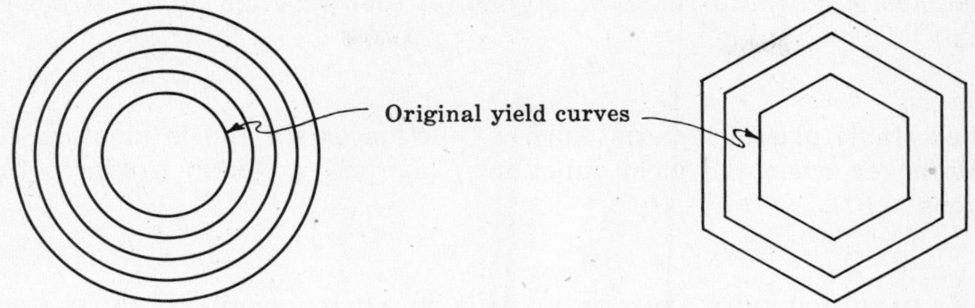

(a) Mises Circles (b) Tresca Hexagons

Fig. 8-6

In *kinematic hardening*, the initial yield surface is translated to a new location in stress space without change in size or shape. Thus (8.4) defining an initial yield surface is replaced by

$$f_1(\sigma_{ij} - \alpha_{ij}) = 0 \qquad (8.17)$$

where the α_{ij} are coordinates of the center of the new yield surface. If *linear hardening* is assumed,

$$\dot{\alpha}_{ij} = c\dot{\epsilon}^P_{ij} \qquad (8.18)$$

where c is a constant. In a one-dimensional case, the Tresca yield curve would be translated as shown in Fig. 8-7.

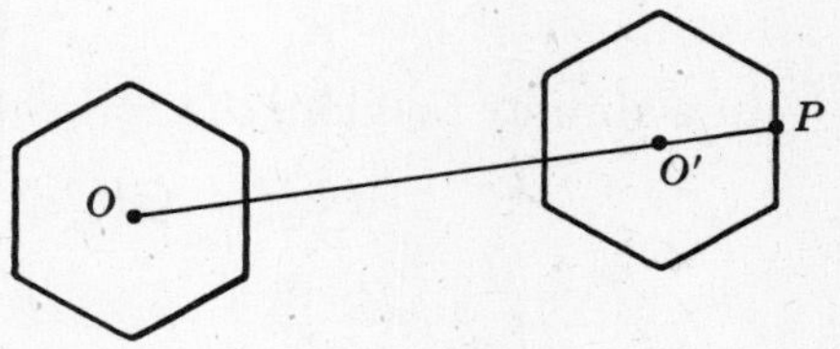

Fig. 8-7

8.6 PLASTIC STRESS-STRAIN EQUATIONS. PLASTIC POTENTIAL THEORY

Once plastic deformation is initiated, the constitutive equations of elasticity are no longer valid. Because plastic strains depend upon the entire loading history of the material, plastic stress-strain relations very often are given in terms of strain increments – the so-called *incremental theories*. By neglecting the elastic portion and by assuming that the principal axes of strain increment coincide with the principal stress axes, the *Levy-Mises* equations relate the total strain increments to the deviatoric stress components through the equations

$$d\epsilon_{ij} = s_{ij}\, d\lambda \qquad (8.19)$$

Here the proportionality factor $d\lambda$ appears in differential form to emphasize that incremental strains are being related to finite stress components. The factor $d\lambda$ may change during loading and is therefore a scalar multiplier and not a fixed constant. Equations (8.19) represent the *flow rule* for a rigid-perfectly plastic material.

If the strain increment is split into elastic and plastic portions according to

$$d\epsilon_{ij} = d\epsilon^E_{ij} + d\epsilon^P_{ij} \qquad (8.20)$$

and the plastic strain increments related to the stress deviator components by

$$d\epsilon^P_{ij} = s_{ij}\, d\lambda \qquad (8.21)$$

the resulting equations are known as the *Prandtl-Reuss* equations. Equations (8.21) represent the flow rule for an elastic-perfectly plastic material. They provide a relationship between the plastic strain increments and the current stress deviators but do not specify the strain increment magnitudes.

The name *plastic potential function* is given to that function of the stress components $g(\sigma_{ij})$ for which

$$d\epsilon_{ij}^P = \frac{\partial g}{\partial \sigma_{ij}} d\lambda \tag{8.22}$$

For a so-called stable plastic material such a function exists and is identical to the yield function. Moreover when the yield function $f_1(\sigma_{ij}) = \text{II}_{\Sigma_D}$, (*8.22*) produces the Prandtl-Reuss equations (*8.21*).

8.7 EQUIVALENT STRESS. EQUIVALENT PLASTIC STRAIN INCREMENT

With regard to the mathematical formulation of strain hardening rules, it is useful to define the *equivalent* or *effective stress* σ_{EQ} as

$$\sigma_{EQ} = \frac{1}{\sqrt{2}}\{[(\sigma_{11}-\sigma_{22})^2 + (\sigma_{22}-\sigma_{33})^2 + (\sigma_{33}-\sigma_{11})^2] + 6(\sigma_{12}^2 + \sigma_{23}^2 + \sigma_{31}^2)\}^{1/2} \tag{8.23}$$

This expression may be written in compact form as

$$\sigma_{EQ} = \sqrt{3s_{ij}s_{ij}/2} = \sqrt{3\text{II}_{\Sigma_D}} \tag{8.24}$$

In a similar fashion, the *equivalent* or *effective plastic strain increment* $d\epsilon_{EQ}^P$ is defined by

$$d\epsilon_{EQ}^P = \{\tfrac{2}{9}[(d\epsilon_{11}^P - d\epsilon_{22}^P)^2 + (d\epsilon_{22}^P - d\epsilon_{33}^P)^2 + (d\epsilon_{33}^P - d\epsilon_{11}^P)^2] + \tfrac{4}{3}[(d\epsilon_{12}^P)^2 + (d\epsilon_{23}^P)^2 + (d\epsilon_{31}^P)^2]\}^{1/2} \tag{8.25}$$

which may be written compactly in the form

$$d\epsilon_{EQ}^P = \sqrt{\tfrac{2}{3}\, d\epsilon_{ij}^P\, d\epsilon_{ij}^P} \tag{8.26}$$

In terms of the equivalent stress and strain increments defined by (*8.24*) and (*8.25*) respectively, $d\lambda$ of (*8.21*) becomes

$$d\lambda = \frac{3}{2}\frac{d\epsilon_{EQ}^P}{\sigma_{EQ}} \tag{8.27}$$

8.8 PLASTIC WORK. STRAIN-HARDENING HYPOTHESES

The rate at which the stresses do work, or the *stress power* as it is called, has been given in (*5.32*) as $\sigma_{ij}D_{ij}$ per unit volume. From (*4.25*), $d\epsilon_{ij} = D_{ij}\,dt$, so that the work increment per unit volume may be written

$$dW = \sigma_{ij}\,d\epsilon_{ij} \tag{8.28}$$

and using (*8.20*) this may be split into

$$dW = \sigma_{ij}(d\epsilon_{ij}^E + d\epsilon_{ij}^P) = dW^E + dW^P \tag{8.29}$$

For a plastically incompressible material, the *plastic work increment* becomes

$$dW^P = \sigma_{ij}\,d\epsilon_{ij}^P = s_{ij}\,d\epsilon_{ij}^P \tag{8.30}$$

Furthermore, if the same material obeys the Prandtl-Reuss equations (*8.21*), the plastic work increment may be expressed as

$$dW^P = \sigma_{EQ}\,d\epsilon_{EQ}^P \tag{8.31}$$

and (*8.21*) rewritten in the form

$$d\epsilon_{ij}^P = \frac{3}{2}\frac{dW^P}{\sigma_{EQ}^2}s_{ij} \tag{8.32}$$

There are two widely considered hypotheses proposed for computing the current yield stress under *isotropic strain hardening* plastic flow. One, known as the *work-hardening hypothesis,* assumes that the current yield surface depends only upon the total plastic work done. Thus with the total plastic work given as the integral

$$W^P = \int \sigma_{ij}\, d\epsilon_{ij}^P \tag{8.33}$$

the yield criterion may be expressed symbolically by the equation

$$f_1(\sigma_{ij}) = F(W^P) \tag{8.34}$$

for which the precise functional form must be determined experimentally. A second hardening hypothesis, known as the *strain-hardening hypothesis,* assumes that the hardening is a function of the amount of plastic strain. In terms of the total equivalent strain

$$\epsilon_{EQ}^P = \int d\epsilon_{EQ}^P \tag{8.35}$$

this hardening rule is expressed symbolically by the equation

$$f_1(\sigma_{ij}) = H(\epsilon_{EQ}^P) \tag{8.36}$$

for which the functional form is determined from a uniaxial stress-strain test of the material. For the Mises yield criterion, the hardening rules (*8.34*) and (*8.36*) may be shown to be equivalent.

8.9 TOTAL DEFORMATION THEORY

In contrast to the *incremental theory* of plastic strain as embodied in the stress-strain increment equations (*8.19*) and (*8.21*), the so-called *total deformation theory* of Hencky relates stress and total strain. The equations take the form

$$e_{ij} = (\phi + \tfrac{1}{2}G)s_{ij} \tag{8.37}$$

$$\epsilon_{ii} = (1-2\nu)\sigma_{ii}/E \tag{8.38}$$

In terms of equivalent stress and strain, the parameter ϕ may be expressed as

$$\phi = \frac{3}{2}\frac{\epsilon_{EQ}^P}{\sigma_{EQ}} \tag{8.39}$$

where here $\epsilon_{EQ}^P = \sqrt{2\epsilon_{ij}^P\epsilon_{ij}^P/3}$ so that

$$\epsilon_{ij}^P = \frac{3}{2}\frac{\epsilon_{EQ}^P}{\sigma_{EQ}}s_{ij} \tag{8.40}$$

8.10 ELASTOPLASTIC PROBLEMS

Situations in which both elastic and plastic strains of approximately the same order exist in a body under load are usually referred to as *elasto-plastic problems.* A number of well-known examples of such problems occur in beam theory, torsion of shafts and thick-walled tubes and spheres subjected to pressure. In general, the governing equations for the elastic region, the plastic region and the elastic-plastic interface are these:

(*a*) Elastic region

1. Equilibrium equations (*2.23*), page 49
2. Stress-strain relations (*6.23*) or (*6.24*), page 143
3. Boundary conditions on stress or displacement
4. Compatibility conditions

(b) Plastic region

1. Equilibrium equations (*2.23*), page 49
2. Stress-strain increment relations (*8.21*)
3. Yield condition (*8.8*) or (*8.11*)
4. Boundary conditions on plastic boundary when such exists

(c) Elastic-plastic interface

1. Continuity conditions on stress and displacement

8.11 ELEMENTARY SLIP LINE THEORY FOR PLANE PLASTIC STRAIN

In unrestricted plastic flow such as occurs in metal-forming processes, it is often possible to neglect elastic strains and consider the material to be rigid-perfectly plastic. If the flow may be further assumed to be a case of plane strain, the resulting velocity field may be studied using *slip line theory.*

Taking the x_1x_2 plane as the plane of flow, the stress tensor is given in the form

$$\sigma_{ij} = \begin{pmatrix} \sigma_{11} & \sigma_{12} & 0 \\ \sigma_{12} & \sigma_{22} & 0 \\ 0 & 0 & \sigma_{33} \end{pmatrix} \tag{8.41}$$

and since elastic strains are neglected, the plastic strain-rate tensor applicable to the situation is

$$\dot{\epsilon}_{ij} = \begin{pmatrix} \dot{\epsilon}_{11} & \dot{\epsilon}_{12} & 0 \\ \dot{\epsilon}_{12} & \dot{\epsilon}_{22} & 0 \\ 0 & 0 & 0 \end{pmatrix} \tag{8.42}$$

In (*8.41*) and (*8.42*) the variables are functions of x_1 and x_2 only, and also

$$\dot{\epsilon}_{ij} = \tfrac{1}{2}(v_{i,j} + v_{i,j}) \tag{8.43}$$

where v_i are the velocity components.

For the assumed plane strain condition, $d\epsilon_{33} = 0$; and so from the Prandtl-Reuss equations (*8.21*), the stress σ_{33} is given by

$$\sigma_{33} = \tfrac{1}{2}(\sigma_{11} + \sigma_{22}) \tag{8.44}$$

Adopting the standard slip-line notation $\sigma_{33} = -p$, and $\sqrt{(\sigma_{11} - \sigma_{22})^2/4 + (\sigma_{12})^2} = k$, the principal stress values of (*8.41*) are found to be

$$\begin{aligned} \sigma_{(1)} &= -p + k \\ \sigma_{(2)} &= -p \\ \sigma_{(3)} &= -p - k \end{aligned} \tag{8.45}$$

The principal stress directions are given with respect to the x_1x_2 axes as shown in Fig. 8-8, where $\tan 2\theta = 2\sigma_{12}/(\sigma_{11} - \sigma_{22})$.

As was shown in Section 2.11, the maximum shear directions are at 45° with respect to the principal stress directions. In Fig. 8-8, the maximum shear directions are designated as the α and β directions. From the

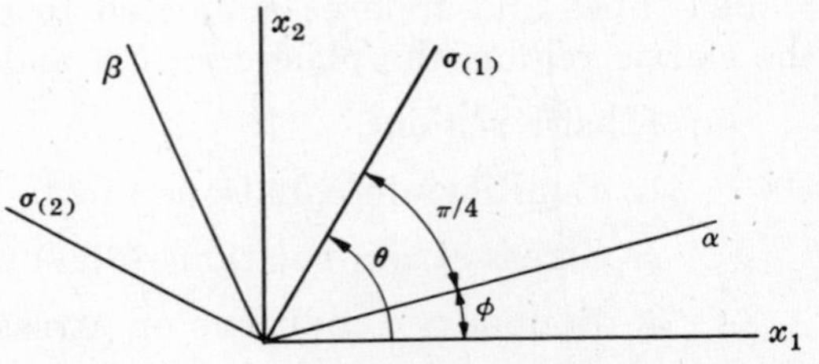

Fig. 8-8

geometry of this diagram, $\theta = \pi/4 + \phi$ so that

$$\tan 2\phi = -\frac{1}{\tan 2\theta} \tag{8.46}$$

and for a given stress field in a plastic flow, two families of curves along the directions of maximum shear at every point may be established. These curves are called *shear lines,* or *slip lines.*

For a small curvilinear element bounded by the two pairs of slip lines shown in Fig. 8-9,

$$\begin{aligned} \sigma_{11} &= -p - k \sin 2\phi \\ \sigma_{22} &= -p + k \sin 2\phi \\ \sigma_{12} &= k \cos 2\phi \end{aligned} \tag{8.47}$$

and from the equilibrium equations it may be shown that

$$\begin{aligned} p + 2k\phi &= C_1 \quad \text{a constant along an } \alpha \text{ line} \\ p - 2k\phi &= C_2 \quad \text{a constant along a } \beta \text{ line} \end{aligned} \tag{8.48}$$

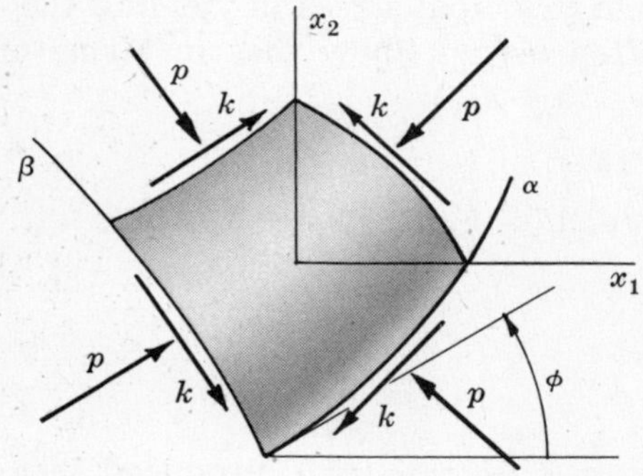

Fig. 8-9

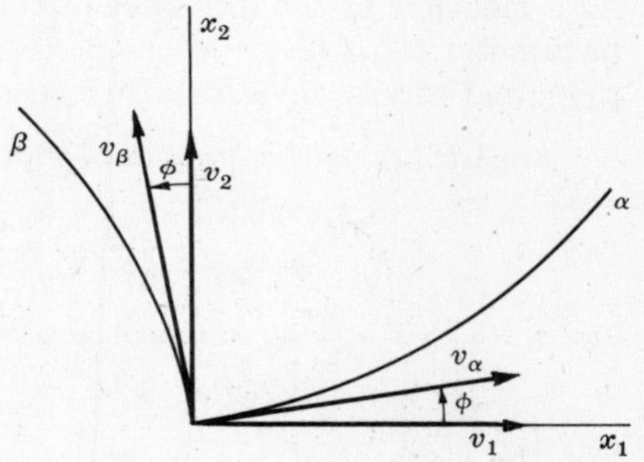

Fig. 8-10

With respect to the velocity components, Fig. 8-10 shows that relative to the α and β lines,

$$\begin{aligned} v_1 &= v_\alpha \cos\phi - v_\beta \sin\phi \\ v_2 &= v_\alpha \sin\phi + v_\beta \cos\phi \end{aligned} \tag{8.49}$$

For an isotropic material, the principal axes of stress and plastic strain-rate coincide. Therefore if x_1 and x_2 are slip-line directions, $\dot{\epsilon}_{11}$ and $\dot{\epsilon}_{22}$ are zero along the slip-lines so that

$$\left\{\frac{\partial}{\partial x_1}(v_\alpha \cos\phi - v_\beta \sin\phi)\right\}_{\phi=0} = 0 \tag{8.50}$$

$$\left\{\frac{\partial}{\partial x_2}(v_\alpha \sin\beta + v_\beta \cos\phi)\right\}_{\phi=0} = 0 \tag{8.51}$$

These equations lead to the relationships

$$dv_1 - v_2\, d\phi = 0 \quad \text{on } \alpha \text{ lines} \tag{8.52}$$

$$dv_2 + v_1\, d\phi = 0 \quad \text{on } \beta \text{ lines} \tag{8.53}$$

Finally, for statically determinate problems, the slip line field may be found from (*8.48*), and using this slip line field, the velocity field may be determined from (*8.52*) and (*8.53*).

Solved Problems

BASIC CONCEPTS. YIELD PHENOMENA (Sec. 8.1-8.4)

8.1. Making use of the definitions (*8.1*) and (*8.2*), derive the relationship between natural and engineering strain. How are the strain increments of these quantities related?

From (*8.1*), $L/L_0 = e+1$ and so (*8.2*) becomes $\epsilon = \ln(e+1)$. Differentiating this equation, $d\epsilon/de = 1/(e+1) = L_0/L$ since $dL = L\,d\epsilon = L_0\,de$.

8.2. Under a load P in a one-dimensional test the true stress is $\sigma = P/A$ while the engineering stress is $S = P/A_0$ where A_0 is the original area and A is the current area. For a constant volume plastic deformation ($A_0L_0 = AL$), determine the condition for maximum load.

Here $S = P/A_0 = (P/A)(A/A_0) = \sigma(L_0/L) = \sigma/(1+e)$, and on an S-e plot the maximum load occurs where the slope $dS/de = 0$. Differentiation gives $dS/de = (d\sigma/d\epsilon - \sigma)/(1+e)^2$ and this is zero when $d\sigma/d\epsilon = \sigma$. From Problem 8.1, this condition may be expressed by $d\sigma/de = \sigma/(1+e)$.

8.3. As a measure of the influence of the intermediate principal stress in yielding the Lode parameter, $\mu = (2\sigma_{II} - \sigma_I - \sigma_{III})/(\sigma_I - \sigma_{III})$ is often used. Show that in terms of the principal stress deviators this becomes $\mu = 3s_{II}/(s_I - s_{III})$.

From (*2.71*), $\sigma_I = s_I + \sigma_M$, etc., with $\sigma_M = \sigma_{ii}/3$. Thus

$$\begin{aligned}\mu &= [2(s_{II}+\sigma_M) - (s_I+\sigma_M) - (s_{III}+\sigma_M)]/[(s_I+\sigma_M) - (s_{III}+\sigma_M)]\\ &= [3s_{II} - (s_I + s_{II} + s_{III})]/(s_I - s_{III})\end{aligned}$$

But $s_I + s_{II} + s_{III} = \mathrm{I}_{\Sigma_D} \equiv 0$ and so $\mu = 3s_{II}/(s_I - s_{III})$.

8.4. For the state of stress $\sigma_{11} = \sigma$, $\sigma_{22} = \sigma_{33} = 0$, $\sigma_{12} = \tau$, $\sigma_{23} = \sigma_{13} = 0$ produced in a tension-torsion test of a thin-walled tube, derive the yield curves in the σ-τ plane for the Tresca and von Mises conditions if the yield stress in simple tension is σ_Y.

For the given state of stress the principal stress values are $\sigma_I = (\sigma + \sqrt{4\tau^2+\sigma^2})/2$, $\sigma_{II} = 0$, $\sigma_{III} = (\sigma - \sqrt{4\tau^2+\sigma^2})/2$ as shown by the Mohr's diagram in Fig. 8-11. Thus from (*8.8*) the Tresca yield curve is $\sqrt{4\tau^2+\sigma^2} = \sigma_Y$, or $\sigma^2 + 4\tau^2 = \sigma_Y^2$, an ellipse in the σ-τ plane. Likewise from (*8.12*) the Mises yield curve is the ellipse $\sigma^2 + 3\tau^2 = \sigma_Y^2$. The Tresca and Mises yield ellipses for this case are compared in the plot shown in Fig. 8-12.

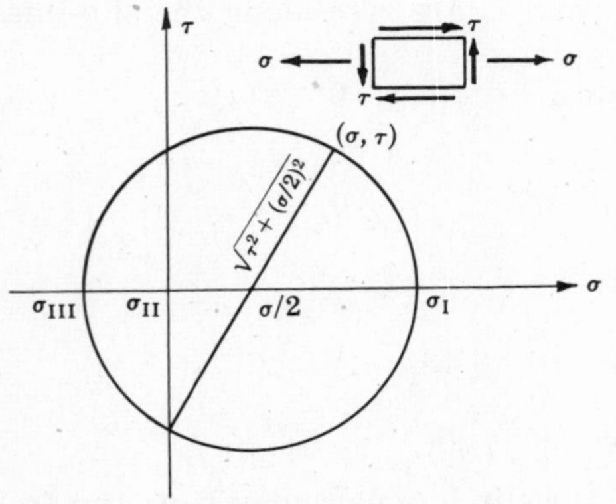

Fig. 8-11

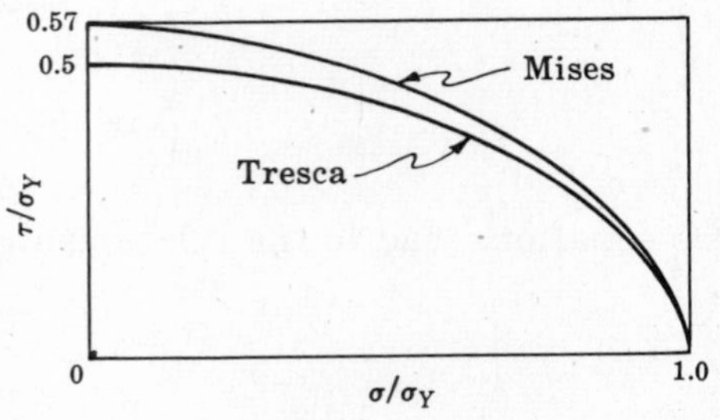

Fig. 8-12

8.5. Convert the von Mises yield condition (*8.10*) to its principal stress form as given in (*8.11*).

From (*2.72*), $-\text{II}_{\Sigma_D} = -(s_\text{I}s_\text{II} + s_\text{II}s_\text{III} + s_\text{III}s_\text{I})$; and by (*2.71*), $s_\text{I} = \sigma_\text{I} - \sigma_M$, etc., where $\sigma_M = (\sigma_\text{I} + \sigma_\text{II} + \sigma_\text{III})/3$. Hence

$$-\text{II}_{\Sigma_D} = -(\sigma_\text{I}\sigma_\text{II} + \sigma_\text{II}\sigma_\text{III} + \sigma_\text{III}\sigma_\text{I}) + (\sigma_\text{I} + \sigma_\text{II} + \sigma_\text{III})^2/3$$
$$= 2(\sigma_\text{I}^2 + \sigma_\text{II}^2 + \sigma_\text{III}^2 - \sigma_\text{I}\sigma_\text{II} - \sigma_\text{II}\sigma_\text{III} - \sigma_\text{III}\sigma_\text{I})/6$$

Thus
$$(\sigma_\text{I} - \sigma_\text{II})^2 + (\sigma_\text{II} - \sigma_\text{III})^2 + (\sigma_\text{III} - \sigma_\text{I})^2 = 6C_Y$$

8.6. With the rectangular coordinate system $OXYZ$ oriented so that the XY plane coincides with the Π-plane and the σ_III axis lies in the YOZ plane (see Fig. 8-13 and 8-4), show that the Mises yield surface intersects the Π-plane in the Mises circle of Fig. 8-5(*b*).

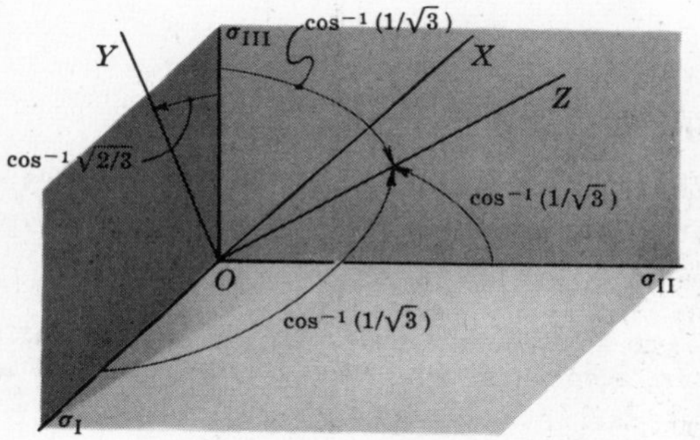

	σ_I	σ_II	σ_III
X	$-1/\sqrt{2}$	$1/\sqrt{2}$	0
Y	$-1/\sqrt{6}$	$-1/\sqrt{6}$	$2/\sqrt{6}$
Z	$1/\sqrt{3}$	$1/\sqrt{3}$	$1/\sqrt{3}$

Fig. 8-13

The table of transformation coefficients between the two sets of axes is readily determined to be as shown above. Therefore

$$\sigma_\text{I} = -X/\sqrt{2} - Y/\sqrt{6} + Z/\sqrt{3}, \qquad \sigma_\text{II} = X/\sqrt{2} - Y/\sqrt{6} + Z/\sqrt{3}, \qquad \sigma_\text{III} = 2Y/\sqrt{6} + Z/\sqrt{3}$$

and (*8.12*) becomes
$$(-\sqrt{2}\,X)^2 + (X/\sqrt{2} - 3Y/\sqrt{6})^2 + (X/\sqrt{2} + 3Y/\sqrt{6})^2 = 2\sigma_Y^2$$

which simplifies to the Mises yield circle $3X^2 + 3Y^2 = 2\sigma_Y^2$ of Fig. 8-5(*b*).

8.7. Using the transformation equations of Problem 8.6, show that (*8.14*), $\sigma_\text{I} + \sigma_\text{II} + \sigma_\text{III} = 0$, is the equation of the Π-plane.

Substituting into (*8.14*) the σ's of Problem 8.6, $\sigma_\text{I} + \sigma_\text{II} + \sigma_\text{III} = \sqrt{3}\,Z = 0$, or $Z = 0$ which is the XY plane (Π-plane).

8.8. For a biaxial state of stress with $\sigma_\text{II} = 0$, determine the yield loci for the Mises and Tresca conditions and compare them by a plot in the two-dimensional σ_I/σ_Y vs. $\sigma_\text{III}/\sigma_Y$ space.

From (*8.12*) with $\sigma_\text{II} = 0$, the Mises yield condition becomes
$$\sigma_\text{I}^2 - \sigma_\text{I}\sigma_\text{III} + \sigma_\text{III}^2 = \sigma_Y^2$$

which is the ellipse
$$(\sigma_\text{I}/\sigma_Y)^2 - (\sigma_\text{I}\sigma_\text{III}/\sigma_Y^2) + (\sigma_\text{III}/\sigma_Y)^2 = 1$$

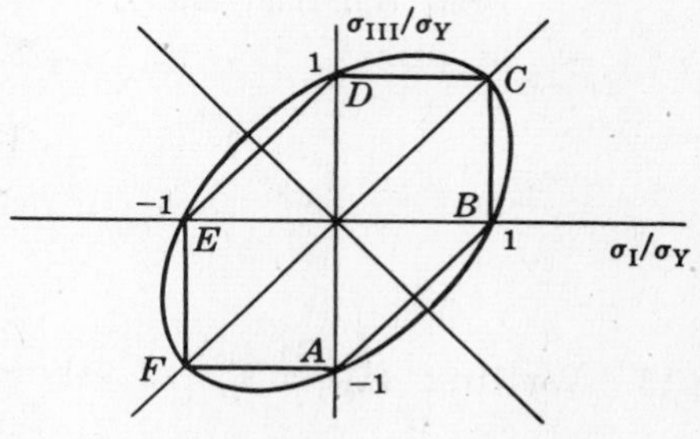

Fig. 8-14

with axes at 45° in the plot. Likewise, from (*8.8*) and the companion equations $\sigma_{III} - \sigma_{II} = \sigma_Y$, $\sigma_{II} - \sigma_I = \sigma_Y$, the Tresca yield condition results in the line segments *AB* and *ED* with equations $(\sigma_I/\sigma_Y) - (\sigma_{III}/\sigma_Y) = \pm 1$, *DC* and *FA* with equations $\sigma_{III}/\sigma_Y = \pm 1$, and *BC* and *EF* with equations $\sigma_I/\sigma_Y = \mp 1$, respectively.

8.9. The von Mises yield condition is referred to in Section 8.3 as the Distortion Energy Theory. Show that if the distortion energy per unit volume $u^*_{(D)}$ is set equal to the yield constant C_Y the result is the Mises criterion as given by (*8.12*).

From Problem 6.26, $u^*_{(D)}$ is given in terms of the principal stresses by

$$u^*_{(D)} = [(\sigma_1 - \sigma_2)^2 + (\sigma_2 - \sigma_3)^2 + (\sigma_3 - \sigma_1)^2]/12G$$

and for a uniaxial yield situation where $\sigma_1 = \sigma_Y$, $\sigma_{II} = \sigma_{III} = 0$, $u^*_{(D)} = \sigma_Y^2/6G$. Thus $C_Y = \sigma_Y^2/6G$ and, as before, the Mises yield condition is expressed by (*8.12*).

PLASTIC DEFORMATION. STRAIN-HARDENING (Sec. 8.4-8.8)

8.10. Show that the Prandtl-Reuss equations (*8.21*) imply that principal axes of plastic strain increments coincide with principal stress axes and express the equations in terms of the principal stresses.

From the form of (*8.21*), when referred to a coordinate system in which the shear stresses are zero, the plastic shear strain increments are seen to be zero also. In the principal axes system, (*8.21*) becomes $d\epsilon_I^P/s_I = d\epsilon_{II}^P/s_{II} = d\epsilon_{III}^P/s_{III} = d\lambda$. Thus $d\epsilon_I^P = (\sigma_I - \sigma_M)\,d\lambda$, $d\epsilon_{II}^P = (\sigma_{II} - \sigma_M)\,d\lambda$, etc., and by subtracting,

$$\frac{d\epsilon_I^P - d\epsilon_{II}^P}{\sigma_I - \sigma_{II}} = \frac{d\epsilon_{II}^P - d\epsilon_{III}^P}{\sigma_{II} - \sigma_{III}} = \frac{d\epsilon_{III}^P - d\epsilon_I^P}{\sigma_{III} - \sigma_I} = d\lambda$$

8.11. For the case of plastic plane strain with $\epsilon_{33} = 0$, $d\epsilon_{33} = 0$ and $\sigma_{22} = 0$, show that the Levy-Mises equations (*8.19*) lead to the conclusion that the Tresca and Mises yield conditions (when related to pure shear yield stress k) are identical.

Here (*8.19*) becomes $d\epsilon_{11} = (2\sigma_{11} - \sigma_{33})\,d\lambda/3$, $d\epsilon_{22} = -(\sigma_{11} + \sigma_{33})\,d\lambda/3$, $0 = 2\sigma_{33} - \sigma_{11}$. Thus in the absence of shear stresses, $\sigma_I = \sigma_{11}$, $\sigma_{II} = \sigma_{33} = \sigma_{11}/2$, $\sigma_{III} = 0 = \sigma_{22}$. Then from (*8.9*) the Tresca yield condition is $\sigma_I - \sigma_{III} = \sigma_{11} = 2k$. Also, from (*8.13*) for this case, Mises condition becomes $(\sigma_{11}/2)^2 + (-\sigma_{11}/2)^2 + (-\sigma_{11})^2 = 6k^2$ or $\sigma_{11}^2 = 4k^2$ and $\sigma_{11} = 2k$.

8.12. Show that the Prandtl-Reuss equations imply equality of the Lode variable μ (see Problem 8.3) and $\nu = (2d\epsilon_{II}^P - d\epsilon_I^P - d\epsilon_{III}^P)/(d\epsilon_I^P - d\epsilon_{III}^P)$.

From equations (*8.21*),

$$\begin{aligned} \nu &= (2s_{II} - s_I - s_{III})\,d\lambda/(s_I - s_{III})\,d\lambda \\ &= (2(\sigma_{II} - \sigma_M) - (\sigma_I - \sigma_M) - (\sigma_{III} - \sigma_M))/((\sigma_I - \sigma_M) - (\sigma_{III} - \sigma_M)) \\ &= (2\sigma_{II} - \sigma_I - \sigma_{III})/(\sigma_I - \sigma_{III}) = \mu \end{aligned}$$

8.13. Writing $II_{\Sigma_D} = s_{ij}s_{ij}/2$, show that $\partial II_{\Sigma_D}/\partial\sigma_{ij} = s_{ij}$.

Here $\partial II_{\Sigma_D}/\partial\sigma_{pq} = (\partial s_{ij}/\partial\sigma_{pq})s_{ij}$ where $\partial s_{ij}/\partial\sigma_{pq} = \partial(\sigma_{ij} - \delta_{ij}\sigma_{kk}/3)/\partial\sigma_{pq} = \delta_{ip}\delta_{jq} - \delta_{ij}\delta_{pq}/3$. Thus $\partial II_{\Sigma_D}/\partial\sigma_{pq} = (\delta_{ip}\delta_{jq} - \delta_{ij}\delta_{pq}/3)s_{ij} = s_{pq}$ since $s_{ii} = I_{\Sigma_D} = 0$.

8.14. Show that when the plastic potential function $g(\sigma_{ij}) = \text{II}_{\Sigma_D}$, the plastic potential equations (*8.22*) become the Prandtl-Reuss equations.

The proof follows directly from the result of Problem 8.13, since $\partial g/\partial\sigma_{ij} = s_{ij}$ in this case and (*8.22*) reduce to (*8.21*).

8.15. Expand (*8.24*) to show that the equivalent stress σ_{EQ} may be written in the form of (*8.23*).

From equation (*8.24*),

$$\sigma_{EQ}^2 = 3s_{ij}s_{ij}/2 = 3(\sigma_{ij} - \delta_{ij}\sigma_{pp}/3)(\sigma_{ij} - \delta_{ij}\sigma_{qq}/3)/2 = (3\sigma_{ij}\sigma_{ij} - \sigma_{ii}\sigma_{jj})/2$$

Expanding this gives

$$\begin{aligned}[3(\sigma_{11}^2 + \sigma_{22}^2 + \sigma_{33}^2) &+ 6(\sigma_{12}^2 + \sigma_{23}^2 + \sigma_{31}^2) - (\sigma_{11} + \sigma_{22} + \sigma_{33})^2]/2 \\ &= [2(\sigma_{11}^2 + \sigma_{22}^2 + \sigma_{33}^2 - \sigma_{11}\sigma_{22} - \sigma_{22}\sigma_{33} - \sigma_{33}\sigma_{11}) + 6(\sigma_{12}^2 + \sigma_{23}^2 + \sigma_{31}^2)]/2 \\ &= [(\sigma_{11} - \sigma_{22})^2 + (\sigma_{22} - \sigma_{33})^2 + (\sigma_{33} - \sigma_{11})^2 + 6(\sigma_{12}^2 + \sigma_{23}^2 + \sigma_{31}^2)]/2\end{aligned}$$

which confirms (*8.23*).

8.16. In plastic potential theory the plastic strain increment vector is normal to the loading (yield) surface at a regular point. If $[N_1, N_2, N_3]$ are direction numbers of the normal to the yield surface $f_1(\sigma_{ij})$, show that $d\epsilon_{\text{I}}^P/s_{\text{I}} = d\epsilon_{\text{II}}^P/s_{\text{II}} = d\epsilon_{\text{III}}^P/s_{\text{III}}$ under the Mises yield condition and flow law.

The condition of normality is expressed by $\mathbf{N} = \text{grad}\, f_1$ which requires $N_1/(\partial f_1/\partial\sigma_{\text{I}}) = N_2/(\partial f_1/\partial\sigma_{\text{II}}) = N_3/(\partial f_1/\partial\sigma_{\text{III}})$ for the Mises case where $f_1 = (\sigma_{\text{I}} - \sigma_{\text{II}})^2 + (\sigma_{\text{II}} - \sigma_{\text{III}})^2 + (\sigma_{\text{III}} - \sigma_{\text{I}})^2 - 2\sigma_Y^2 = 0$. Here $\partial f_1/\partial\sigma_{\text{I}} = 2(2\sigma_{\text{I}} - \sigma_{\text{II}} - \sigma_{\text{III}}) = 6s_{\text{I}}$, etc., and since the plastic strain increment vector is along the normal it follows that $d\epsilon_{\text{I}}^P/s_{\text{I}} = d\epsilon_{\text{II}}^P/s_{\text{II}} = d\epsilon_{\text{III}}^P/s_{\text{III}}$.

8.17. Determine the plastic strain increment ratios for (*a*) simple tension with $\sigma_{11} = \sigma_Y$, (*b*) biaxial stress with $\sigma_{11} = -\sigma_Y/\sqrt{3}$, $\sigma_{22} = \sigma_Y/\sqrt{3}$, $\sigma_{33} = \sigma_{12} = \sigma_{23} = \sigma_{13} = 0$, (*c*) pure shear with $\sigma_{12} = \sigma_Y/\sqrt{3}$.

(*a*) Here $\sigma_{11} = \sigma_{\text{I}} = \sigma_Y$, $\sigma_{\text{II}} = \sigma_{\text{III}} = 0$ and $s_{\text{I}} = 2\sigma_Y/3$, $s_{\text{II}} = s_{\text{III}} = -\sigma_Y/3$. Thus from Problem 8.16, $d\epsilon_{\text{I}}^P/2 = -d\epsilon_{\text{II}}^P/1 = -d\epsilon_{\text{III}}^P/1$.

(*b*) Here $\sigma_{\text{I}} = \sigma_Y/\sqrt{3}$, $\sigma_{\text{II}} = 0$, $\sigma_{\text{III}} = -\sigma_Y/\sqrt{3}$ and $s_{\text{I}} = \sigma_Y/\sqrt{3}$, $s_{\text{II}} = 0$, $s_{\text{III}} = -\sigma_Y/\sqrt{3}$. Thus $d\epsilon_{\text{I}}^P/1 = -d\epsilon_{\text{III}}^P/1$ and the third term is omitted since it is usually understood in the theory that if the denominator is zero the numerator will be zero too.

(*c*) Here $\sigma_{\text{I}} = \sigma_Y/\sqrt{3}$, $\sigma_{\text{II}} = 0$, $\sigma_{\text{III}} = -\sigma_Y/\sqrt{3}$ and again $d\epsilon_{\text{I}}^P/1 = -d\epsilon_{\text{III}}^P/1$.

8.18. Determine the plastic work increment dW^P and the equivalent plastic strain increment $d\epsilon_{EQ}^P$ for the biaxial stress state $\sigma_{11} = -\sigma_Y/\sqrt{3}$, $\sigma_{22} = \sigma_Y/\sqrt{3}$, $\sigma_{33} = \sigma_{12} = \sigma_{23} = \sigma_{31} = 0$ if plastic deformation is controlled so that $d\epsilon_{\text{I}}^P = C$, a constant.

In principal-axis form, (*8.30*) becomes $dW^P = \sigma_{\text{I}}\, d\epsilon_{\text{I}}^P + \sigma_{\text{II}}\, d\epsilon_{\text{II}}^P + \sigma_{\text{III}}\, d\epsilon_{\text{III}}^P$; and for the stress state given, Problem 8.17 shows that $d\epsilon_{\text{I}}^P = -d\epsilon_{\text{III}}^P$, $d\epsilon_{\text{II}}^P = 0$; hence

$$dW^P = -\sigma_Y C/\sqrt{3} + (\sigma_Y/\sqrt{3})(-C) = -2C\sigma_Y/\sqrt{3}$$

From (*8.25*),

$$\begin{aligned} d\epsilon_{EQ}^P &= \{2[(d\epsilon_{\text{I}}^P - d\epsilon_{\text{II}}^P)^2 + (d\epsilon_{\text{II}}^P - d\epsilon_{\text{III}}^P)^2 + (d\epsilon_{\text{III}}^P - d\epsilon_{\text{I}}^P)^2]\}^{1/2}/3 \\ &= \{2[C^2 + C^2 + 4C^2]\}^{1/2}/3 = 2C/\sqrt{3}\end{aligned}$$

8.19. Verify (*8.32*) by showing that for a Prandtl-Reuss material the plastic work increment is $dW^P = \sigma_{EQ}\, d\epsilon^P_{EQ}$ as given in (*8.31*).

From (*8.30*), $dW^P = s_{ij}s_{ij}\,d\lambda$ for a Prandtl-Reuss material satisfying (*8.21*). But from (*8.27*), $d\lambda = 3\,d\epsilon^P_{EQ}/2\sigma_{EQ}$ for such a material and so $dW^P = (3s_{ij}s_{ij}/2)(d\epsilon^P_{EQ}/\sigma_{EQ})$ which because of the definition (*8.24*) gives $dW^P = \sigma_{EQ}\,d\epsilon^P_{EQ}$. Thus $d\epsilon^P_{EQ} = dW^P/\sigma_{EQ}$ here and (*8.32*) follows directly from (*8.21*).

8.20. For a material obeying the Mises yield condition, the equivalent stress σ_{EQ} may be taken as the yield function in the hardening rules (*8.34*) and (*8.36*). Show that in this case $\sigma_{EQ}F' = H'$ where F' and H' are the derivatives of the hardening functions with respect to their respective arguments.

Here (*8.34*) becomes $\sigma_{EQ} = F(W^P)$ and so $d\sigma_{EQ} = F'\,dW^P$. Likewise (*8.36*) is given here by $\sigma_{EQ} = H(\epsilon^P_{EQ})$ and so $d\sigma_{EQ} = H'\,d\epsilon^P_{EQ}$. Thus $F'\,dW^P = H'\,d\epsilon^P_{EQ}$; and since from (*8.31*) (or Problem 8.19) $dW^P = \sigma_{EQ}\,d\epsilon^P_{EQ}$, it follows at once that $\sigma_{EQ}F' = H'$.

TOTAL DEFORMATION THEORY (Sec. 8.9)

8.21. The Hencky total deformation theory may be represented through the equations $\epsilon_{ij} = \epsilon^E_{ij} + \epsilon^P_{ij}$ with $\epsilon^E_{ij} = e^E_{ij} + \delta_{ij}\epsilon^E_{kk}/3 = (s_{ij}/2)G + \delta_{ij}(1-2\nu)\sigma_{kk}/3E$ and $\epsilon^P_{ij} = \phi s_{ij}$. Show that these equations are equivalent to (*8.37*) and (*8.38*).

The equation $\epsilon^P_{ij} = \phi s_{ij}$ implies $\epsilon^P_{ii} = 0$ so that $\epsilon^P_{ij} = e^P_{ij} = \phi s_{ij}$, and from $\epsilon_{ij} = \epsilon^E_{ij} + \epsilon^P_{ij}$ it follows that here $\epsilon_{jj} = \epsilon^E_{jj}$. From the same equation, $e_{ij} + \delta_{ij}\epsilon_{kk}/3 = e^E_{ij} + \delta_{ij}\epsilon^E_{kk}/3 + e^P_{ij}$ which reduces to $e_{ij} = e^E_{ij} + e^P_{ij} = (\phi + \frac{1}{2}G)s_{ij}$, (*8.37*). Also from $\epsilon_{ii} = \epsilon^E_{ii}$, $\epsilon_{ii} = (1-2\nu)\sigma_{kk}/E$, (*8.38*).

8.22. Verify that the Hencky parameter ϕ may be expressed as given in (*8.39*).

Squaring and adding the components in the equation $\epsilon^P_{ij} = \phi s_{ij}$ of Problem 8.21 gives $\epsilon^P_{ij}\epsilon^P_{ij} = \phi^2 s_{ij}s_{ij}$ or $\phi = \sqrt{3\epsilon^P_{ij}\epsilon^P_{ij}/2}\,/\sigma_{EQ}$ which when multiplied on each side by 2/3 becomes

$$\phi = 3\sqrt{2\epsilon^P_{ij}\epsilon^P_{ij}/3}\,/2\sigma_{EQ} = 3\epsilon^P_{EQ}/2\sigma_{EQ}$$

ELASTOPLASTIC PROBLEMS (Sec. 8.10)

8.23. An elastic-perfectly plastic rectangular beam is loaded steadily in pure bending. Using simple beam theory, determine the end moments M for which the remaining elastic core extends from $-a$ to a as shown in Fig. 8-15.

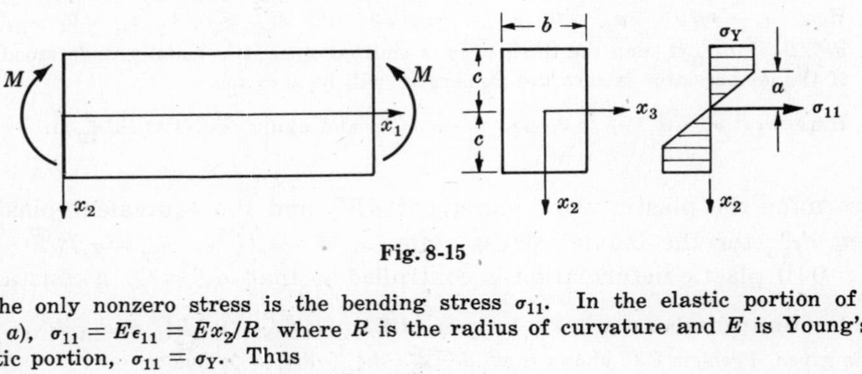

Fig. 8-15

Here the only nonzero stress is the bending stress σ_{11}. In the elastic portion of the beam $(-a < x_2 < a)$, $\sigma_{11} = E\epsilon_{11} = Ex_2/R$ where R is the radius of curvature and E is Young's modulus. In the plastic portion, $\sigma_{11} = \sigma_Y$. Thus

$$M = 2\int_0^a \frac{E}{R}(x_2)^2 b\,dx_2 + 2\int_a^c x_2\sigma_Y b\,dx_2 = b\sigma_Y(c^2 - a^2/3)$$

where $\sigma_Y = Ea/R$, the stress condition at the elastic-plastic interface, has been used. From the result obtained, $M = 2bc^2\sigma_Y/3$ at first yield (when $a = c$), and $M = bc^2\sigma_Y$ for the fully-plastic beam (when $a = 0$).

8.24. Determine the moment for a beam loaded as in Problem 8.23 if the material is a piecewise linear hardening material for which $\sigma_{11} = \sigma_Y + A(\epsilon_{11} - \sigma_Y/E)$ after yield.

The stress distribution for this beam is shown in Fig. 8-16. Again $\epsilon_{11} = x_2/R$ and so

$$M = 2\int_0^a \frac{E(x_2)^2 b}{R}\,dx_2 + 2\int_a^c \left[\sigma_Y + A\left(\frac{x_2}{R} - \frac{\sigma_Y}{E}\right)\right] x_2 b\,dx_2$$

$$= \frac{2Eba^3}{3R} + 2b\left\{\frac{\sigma_Y}{2}\left(1 - \frac{A}{E}\right)(c^2 - a^2) + \frac{A(c^3 - a^3)}{3R}\right\}$$

or using $\sigma_Y = Ea/R$ as in Problem 8.23,

$$M = c^2 b\sigma_Y(1 - A/E) + 2c^3 bA/3R + b\sigma_Y^3 R^2(A/E - 1)/3E^2$$

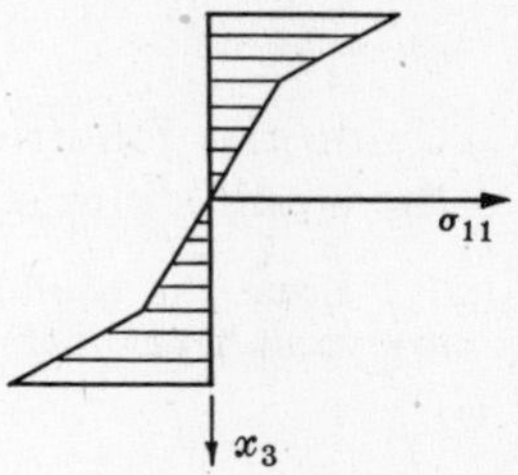

Fig. 8-16

8.25. An elastic-perfectly plastic circular shaft of radius c is twisted by end torques T as shown in Fig. 8-17. Determine the torque for which an inner elastic core of radius a remains.

The shear stress σ_{12} is given here by $\sigma_{12} = kr/a$ for $0 \leqq r \leqq a$, and by $\sigma_{12} = k$ for $a \leqq r \leqq c$ where k is the yield stress of the material in shear. Thus

$$T = 2\pi\int_0^a (kr^3/a)\,dr + 2\pi\int_a^c kr^2\,dr = \frac{2\pi k}{3}(c^3 - a^3/4)$$

Therefore the torque at first yield is $T_1 = \pi kc^3/2$ when $a = c$; and for the fully plastic condition, $T_2 = 2\pi kc^3/3 = 4T_1/3$ when $a = 0$.

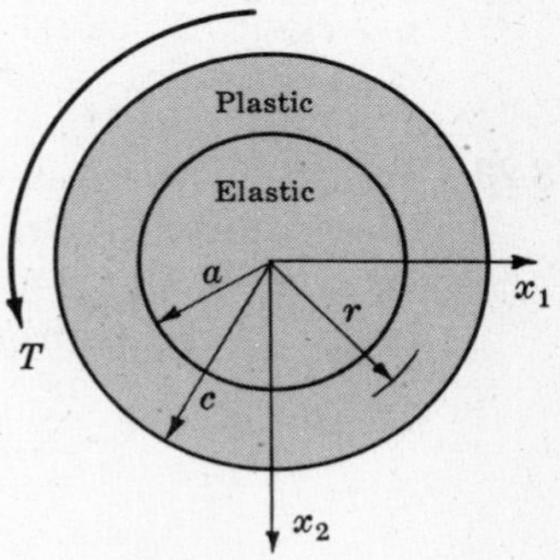

Fig. 8-17

8.26. A thick spherical shell of the dimensions shown in Fig. 8-18 is subjected to an increasing pressure p_0. Using the Mises yield condition, determine the pressure at which first yield occurs.

Because of symmetry of loading the principal stresses are the spherical components $\sigma_{(\theta\theta)} = \sigma_I = \sigma_{II}$, $\sigma_{(rr)} = \sigma_{III}$. Thus the Mises yield condition (*8.12*) becomes $\sigma_{(\theta\theta)} - \sigma_{(rr)} = \sigma_Y$. The elastic stress components may be shown to be

$$\sigma_{(rr)} = -p_0(b^3/r^3 - 1)/(b^3/a^3 - 1)$$

$$\sigma_{(\theta\theta)} = \sigma_{(\phi\phi)} = p_0(b^3/2r^3 + 1)/(b^3/a^3 - 1)$$

Therefore $\sigma_Y = 3b^3 p_0/2r^3(b^3/a^3 - 1)$ and $p_0 = 2\sigma_Y(1 - a^3/b^3)/3$ at first yield which occurs at the inner radius a.

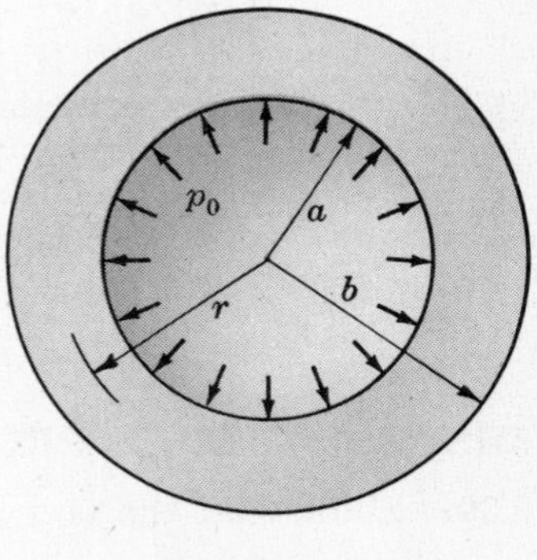

Fig. 8-18

SLIP LINE THEORY (Sec. 8.11)

8.27. Verify directly the principal stress values (*8.45*) for the stress tensor (*8.41*) with $\sigma_{33} = (\sigma_{11} + \sigma_{22})/2$ as given in (*8.44*).

The principal stress values are found from the determinant equation (*2.37*) which here becomes

$$\begin{vmatrix} \sigma_{11} - \sigma & \sigma_{12} & 0 \\ \sigma_{12} & \sigma_{22} - \sigma & 0 \\ 0 & 0 & -p - \sigma \end{vmatrix} = 0$$

Expanding by the third column,

$$(-p-\sigma)[(\sigma_{11}-\sigma)(\sigma_{22}-\sigma)-\sigma_{12}^2] = (-p-\sigma)[\sigma^2-(\sigma_{11}+\sigma_{22})\sigma-\sigma_{12}^2] = 0$$

The roots of this equation are clearly $\sigma = -p$ and $\sigma = \frac{1}{2}(\sigma_{11}+\sigma_{22}) \pm \sqrt{\frac{1}{4}(\sigma_{11}+\sigma_{22})^2+\sigma_{12}^2} = -p \pm k$.

8.28. Using the condition that the yield stress in shear k is constant, combine (*8.47*) with the equilibrium equations and integrate to prove (*8.48*).

From the equilibrium equations $\partial\sigma_{11}/\partial x_1 + \partial\sigma_{12}/\partial x_2 = 0$ and $\partial\sigma_{12}/\partial x_1 + \partial\sigma_{22}/\partial x_2 = 0$ which are valid here, (*8.47*) yields

$$-\partial p/\partial x_1 - k(2\cos 2\phi)(\partial\phi/\partial x_1) + k(-2\sin 2\phi)(\partial\phi/\partial x_2) = 0$$

and

$$-k(2\sin 2\phi)(\partial\phi/\partial x_1) - \partial p/\partial x_2 + k(2\cos 2\phi)(\partial\phi/\partial x_2) = 0$$

If x_1 is along an α line and x_2 along a β line, $\phi = 0$ and these equations become $-\partial p/\partial x_1 - 2k(\partial\phi/\partial x_1) = 0$ along the α line, $-\partial p/\partial x_2 + 2k(\partial\phi/\partial x_2) = 0$ along the β line. Integrating directly, $p + 2k\phi = C_1$ on the α line, $p - 2k\phi = C_2$ on the β line.

8.29. In the frictionless extrusion through a square die causing a fifty per cent reduction, the centered fan region is composed of straight radial β lines and circular α lines as shown in Fig. 8-19. Determine the velocity components along these slip lines in terms of the approach velocity U and the polar coordinates r and θ.

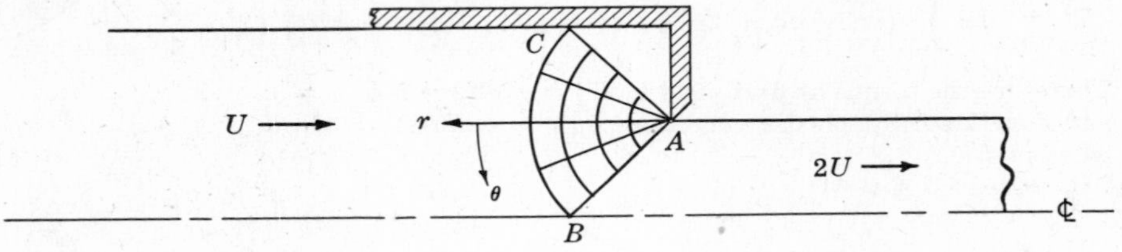

Fig. 8-19

Along the straight β lines, $d\phi = 0$; and from (*8.53*), $dv_2 = 0$ or $v_2 = \text{constant}$. From the normal velocity continuity along BC, the constant here must be $U\cos\theta$ and so $v_2 = U\cos\theta$. Along the circular α lines, $d\phi = d\theta$; and from (*8.52*),

$$v_1 = \int_{-\pi/4}^{\theta} U\cos\theta\, d\theta = U(\sin\theta + 1/\sqrt{2})$$

MISCELLANEOUS PROBLEMS

8.30. Show that the von Mises yield condition may be expressed in terms of the octahedral shear stress σ_{oct} (see Problem 2.22) by $\sigma_{oct} = \sqrt{2}\sigma_Y/3$.

In terms of principal stresses $3\sigma_{oct} = \sqrt{(\sigma_I-\sigma_{II})^2+(\sigma_{II}-\sigma_{III})^2+(\sigma_{III}-\sigma_I)^2}$ (Problem 2.22) and so $9\sigma_{oct}^2 = (\sigma_I-\sigma_{II})^2+(\sigma_{II}-\sigma_{III})^2+(\sigma_{III}-\sigma_I)^2 = 2\sigma_Y^2$ in agreement with (*8.12*).

8.31. Show that equation (*8.13*) for Mises yield condition may be written as

$$s_I^2 + s_{II}^2 + s_{III}^2 = 2k^2$$

From (*2.71*), $\sigma_I = s_I + \sigma_M$, etc., and so (*8.13*) at once becomes

$$(s_I-s_{II})^2 + (s_{II}-s_{III})^2 + (s_{III}-s_I)^2 = 6k^2$$

Expanding and rearranging, this may be written $s_I^2+s_{II}^2+s_{III}^2-(s_I+s_{II}+s_{III})^2/3 = 2k^2$. But $s_I+s_{II}+s_{III} = I_{\Sigma_D} \equiv 0$ and the required equation follows.

8.32. At what value of the Lode parameter $\mu = (2\sigma_{II} - \sigma_I - \sigma_{III})/(\sigma_I - \sigma_{III})$ are the Tresca and Mises yield conditions identical?

From the definition of μ, $\sigma_{II} = (\sigma_I + \sigma_{III})/2 + \mu(\sigma_I - \sigma_{III})/2$ which when substituted into the Mises yield condition (*8.12*) gives after some algebra (see Problem 8.42) $\sigma_I - \sigma_{III} = 2\sigma_Y/\sqrt{3+\mu^2}$. Tresca's yield condition, equation (*8.8*), is $\sigma_I - \sigma_{III} = \sigma_Y$. Thus when $\mu = 1$ the two are identical. When $\sigma_{II} = \sigma_I$, $\mu = 1$ which is sometimes called a cylindrical state of stress.

8.33. For the state of stress $\sigma_{ij} = \begin{pmatrix} \sigma & \tau & 0 \\ \tau & \sigma & 0 \\ 0 & 0 & \sigma \end{pmatrix}$ where σ and τ are constants, determine the yield condition according to Tresca and von Mises criteria.

The principal stresses here are readily shown to be $\sigma_I = \sigma + \tau$, $\sigma_{II} = \sigma$, $\sigma_{III} = \sigma - \tau$. Thus from (*8.8*), the Tresca condition $\sigma_I - \sigma_{III} = \sigma_Y$ gives $2\tau = \sigma_Y$. From (*8.12*) the Mises condition gives $\tau = \sigma_Y/\sqrt{3}$. Note that in each case yielding depends on τ, not on σ, i.e. yielding is independent of hydrostatic stress.

8.34. Show that the Prandtl-Reuss equations imply incompressible plastic deformation and write the equations in terms of actual stresses.

From (*8.21*), $d\epsilon_{ii}^P = s_{ii}\,d\lambda = 0$ since $s_{ii} = I_{\Sigma_D} \equiv 0$ and the incompressibility condition $d\epsilon_{ii}^P = 0$ is attained. In terms of stresses, $d\epsilon_{ij}^P = (\sigma_{ij} - \delta_{ij}\sigma_{kk}/3)\,d\lambda$. Thus $d\epsilon_{11}^P = (2/3)[\sigma_{11} - (\sigma_{22} + \sigma_{33})/2]\,d\lambda$, etc., for the normal components and $d\epsilon_{12}^P = \sigma_{12}\,d\lambda$, etc., for the shear components.

8.35. Using the von Mises yield condition, show that in the Π-plane the deviator stress components at yield are

$$s_I = [-2\sigma_Y \cos(\theta - \pi/6)]/3, \qquad s_{II} = [2\sigma_Y \cos(\theta + \pi/6)]/3, \qquad s_{III} = (2\sigma_Y \sin\theta)/3$$

where $\theta = \tan^{-1} Y/X$ in the notation of Problem 8.6.

The radius of the Mises yield circle is $\sqrt{2/3}\,\sigma_Y$ so that by definition $X = \sqrt{2/3}\,\sigma_Y \cos\theta$, $Y = \sqrt{2/3}\,\sigma_Y \sin\theta$ at yield. From the transformation table given in Problem 8.6 together with $\sigma_I = s_I + \sigma_M$, etc., the equations $s_I - s_{II} = -\sqrt{2}\,X = -(2/\sqrt{3})\sigma_Y\cos\theta$ and $s_I + s_{II} - 2s_{III} = -\sqrt{6}\,Y = -2\sigma_Y \sin\theta$ are obtained. Also, in the Π-plane, $s_I + s_{II} + s_{III} = 0$. Solving these three equations simultaneously yields the desired expressions, as the student should verify.

8.36. An elastic-perfectly plastic, incompressible material is loaded in plane strain between rigid plates so that $\sigma_{22} = 0$ and $\epsilon_{33} = 0$ (Fig. 8-20). Use Mises yield condition to determine the loading stress σ_{11} at first yield, and the accompanying strain ϵ_{11}.

The elastic stress-strain equation

$$E\epsilon_{33} = \sigma_{33} - \nu(\sigma_{11} + \sigma_{22})$$

reduces here to $\sigma_{33} = \nu\sigma_{11}$. Thus the principal stresses are $\sigma_I = 0$, $\sigma_{II} = -\nu\sigma_{11}$, $\sigma_{III} = -\sigma_{11}$; and by (*8.12*) we have

$$(\nu\sigma_{11})^2 + (\sigma_{11}(1-\nu))^2 + (-\sigma_{11})^2 = 2\sigma_Y^2$$

from which $\sigma_{11} = -\sigma_Y/\sqrt{1-\nu-\nu^2}$ (compressive) at yield. Likewise, from $E\epsilon_{11} = \sigma_{11} - \nu(\sigma_{22} + \sigma_{33})$ we see that here $\epsilon_{11} = -\sigma_Y(1-\nu^2)/E\sqrt{1-\nu-\nu^2}$ at yield.

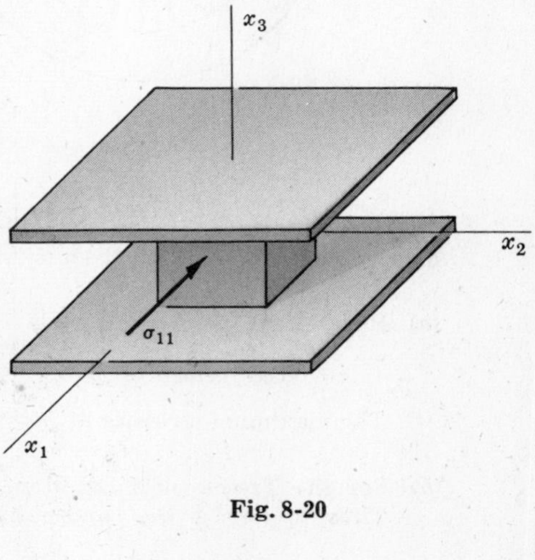

Fig. 8-20

8.37. An elastic-perfectly plastic rectangular beam is loaded in pure bending until fully plastic. Determine the residual stress in the beam upon removal of the bending moment M.

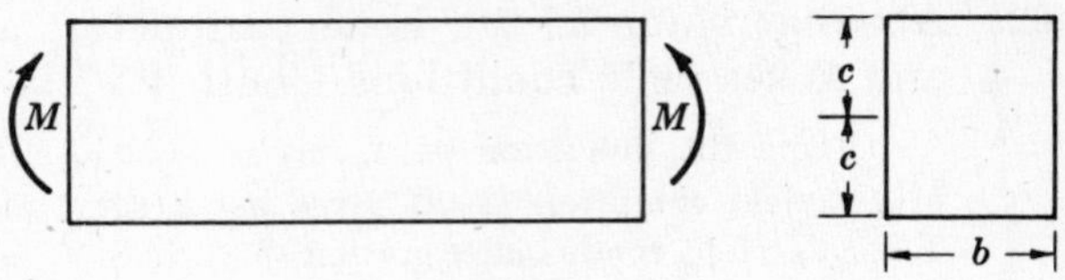

Fig. 8-21

For the fully plastic condition, the moment is (see Problem 8.23) $M = bc^2\sigma_Y$. This moment would cause an elastic stress having $\sigma = Mc/I = 3\sigma_Y/2$ at the extreme fibers, since $I = 2bc^3/3$. Thus removal of M is equivalent to applying a corresponding negative elastic stress which results in the residual stress shown in Fig. 8-22.

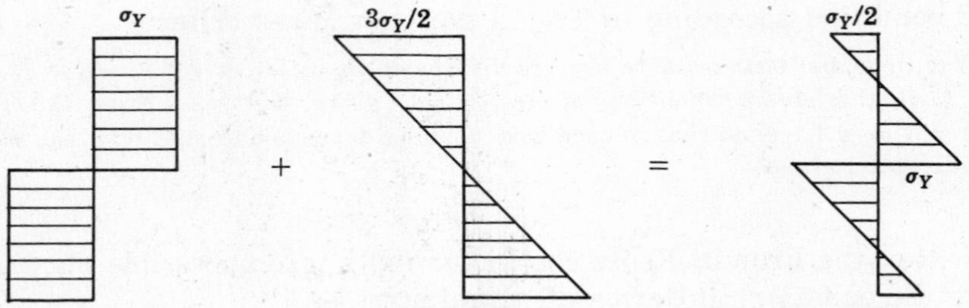

Fig. 8-22

8.38. A thick-walled cylindrical tube of the dimensions shown in Fig. 8-23 is subjected to an internal pressure p_i. Determine the value of p_i at first yield if the ends of the tube are closed. Assume (*a*) von Mises and (*b*) Tresca's yield conditions.

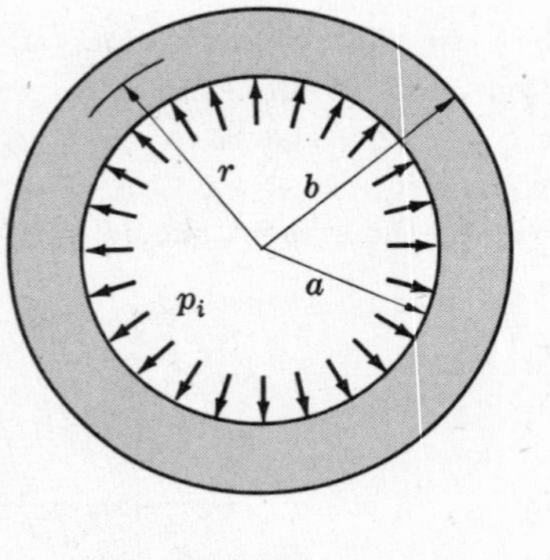

Fig. 8-23

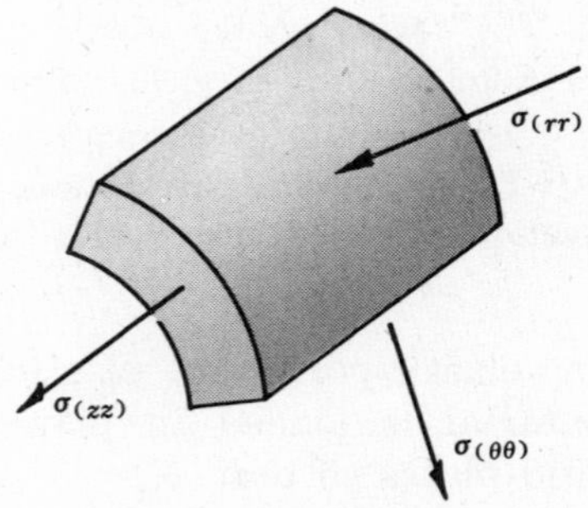

Fig. 8-24

The cylindrical stress components (Fig. 8-24) are principal stresses and for the elastic analysis may be shown to be $\sigma_{(rr)} = -p_i(b^2/r^2 - 1)/Q$, $\sigma_{(\theta\theta)} = p_i(b^2/r^2 + 1)/Q$, $\sigma_{(zz)} = p_i/Q$ where $Q = (b^2/a^2 - 1)$.

(*a*) Here Mises yield condition is

$$(\sigma_{(rr)} - \sigma_{(\theta\theta)})^2 + (\sigma_{(\theta\theta)} - \sigma_{(zz)})^2 + (\sigma_{(zz)} - \sigma_{(rr)})^2 = 2\sigma_Y^2 \quad \text{or} \quad p_i^2 b^4/r^4 = Q^2\sigma_Y^2/3$$

The maximum stress is at $r = a$, and at first yield $p_i = (\sigma_Y/\sqrt{3})(1 - a^2/b^2)$.

(*b*) For the Tresca yield condition, $\sigma_{(\theta\theta)} - \sigma_{(rr)} = \sigma_Y$ since σ_{zz} is the intermediate principal stress. Thus $2p_i b^2/r^2 = Q\sigma_Y$ and now at $r = a$, $p_i = (\sigma_Y/2)(1 - a^2/b^2)$ at first yield.

Supplementary Problems

8.39. A one-dimensional stress-strain law is given by $\sigma = K\epsilon^n$ where K and n are constants and ϵ is true strain. Show that the maximum load occurs at $\epsilon = n$.

8.40. Rework Problem 8.4 using the yield stress in shear k in place of σ_Y in the Mises and Tresca yield conditions. *Ans.* Mises: $(\sigma/\sqrt{3}\,k)^2 + (\tau/k)^2 = 1$; Tresca: $(\sigma/2k)^2 + (\tau/k)^2 = 1$

8.41. Making use of the material presented in Problem 8.6, verify the geometry of Fig. 8-5(*c*).

8.42. From the definition of Lode's parameter μ (see Problem 8.3) and the Mises yield condition, show that $\sigma_{\rm I} - \sigma_{\rm III} = 2\sigma_Y/\sqrt{3+\mu^2}$.

8.43. In the Π-plane where $\theta = \tan^{-1} Y/X$ with X and Y defined in Problem 8.6, show that $\mu = -\sqrt{3}\tan\theta$.

8.44. Show that the invariants of the deviator stress $\text{II}_{\Sigma_D} = s_{ij}s_{ij}/2$ and $\text{III}_{\Sigma_D} = s_{ij}s_{jk}s_{ki}/3$ may be written $\text{II}_{\Sigma_D} = (s_{\rm I}^2 + s_{\rm II}^2 + s_{\rm III}^2)/2$ and $\text{III}_{\Sigma_D} = (s_{\rm I}^3 + s_{\rm II}^3 + s_{\rm III}^3)/3$ respectively.

8.45. Show that von Mises yield condition may be written in the form

$$(\sigma_{11} - \sigma_{22})^2 + (\sigma_{22} - \sigma_{33})^2 + (\sigma_{33} - \sigma_{11})^2 + 6(\sigma_{12}^2 + \sigma_{23}^2 + \sigma_{31}^2) = 6k^2$$

8.46. Following the procedure of Problem 8.17, determine the plastic strain increment ratios for (*a*) biaxial tension with $\sigma_{11} = \sigma_{22} = \sigma_Y$, (*b*) tension-torsion with $\sigma_{11} = \sigma_Y/2$, $\sigma_{12} = \sigma_Y/2$.
Ans. (*a*) $d\epsilon_{11}^P = d\epsilon_{22}^P = -d\epsilon_{33}^P/2$ (*b*) $d\epsilon_{11}^P/2 = -d\epsilon_{22}^P = -d\epsilon_{33}^P = d\epsilon_{12}^P/3$

8.47. Verify the following equivalent expressions for the effective plastic strain increment $d\epsilon_{EQ}^P$ and note that in each case $d\epsilon_{EQ}^P = d\epsilon_{11}^P$ for uniaxial tension σ_{11}.

(*a*) $d\epsilon_{EQ}^P = \sqrt{2/3}\,[(d\epsilon_{11}^P)^2 + (d\epsilon_{22}^P)^2 + (d\epsilon_{33}^P)^2 + 2(d\epsilon_{12}^P)^2 + 2(d\epsilon_{23}^P)^2 + 2(d\epsilon_{31}^P)^2]^{1/2}$

(*b*) $d\epsilon_{EQ}^P = (\sqrt{2}/3)[(d\epsilon_{11}^P - d\epsilon_{22}^P)^2 + (d\epsilon_{22}^P - d\epsilon_{33}^P)^2 + (d\epsilon_{33}^P - d\epsilon_{11}^P)^2 + 6(d\epsilon_{12}^P)^2 + 6(d\epsilon_{23}^P)^2 + 6(d\epsilon_{31}^P)^2]^{1/2}$

8.48. A thin-walled elastic-perfectly plastic tube is loaded in combined tension-torsion. An axial stress $\sigma = \sigma_Y/2$ is developed first and maintained constant while the shear stress τ is steadily increased from zero. At what value of τ will yielding first occur according to the Mises condition?
Ans. $\tau = \sigma_Y/2$

8.49. The beam of triangular cross section shown in Fig. 8-25 is subjected to pure bending. Determine the location of the neutral axis (a distance b from top) of the beam when fully plastic. *Ans.* $b = h/\sqrt{2}$

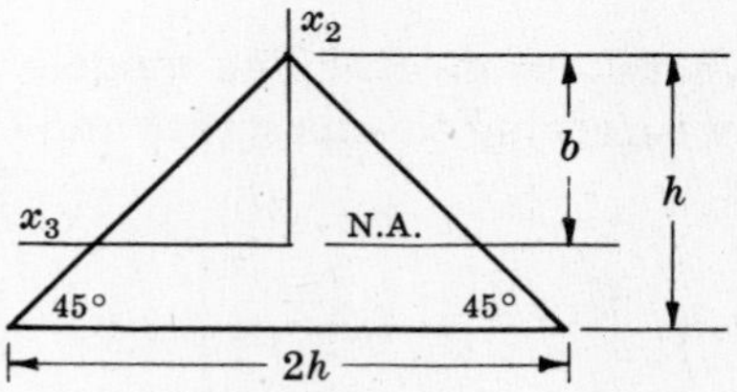

Fig. 8-25

8.50. Show that the stress tensor (*8.41*) becomes

$$\sigma_{ij} = \begin{pmatrix} -p & k & 0 \\ k & -p & 0 \\ 0 & 0 & -p \end{pmatrix}$$

when referred to the axes rotated about x_3 by an angle θ in Fig. 8-8.

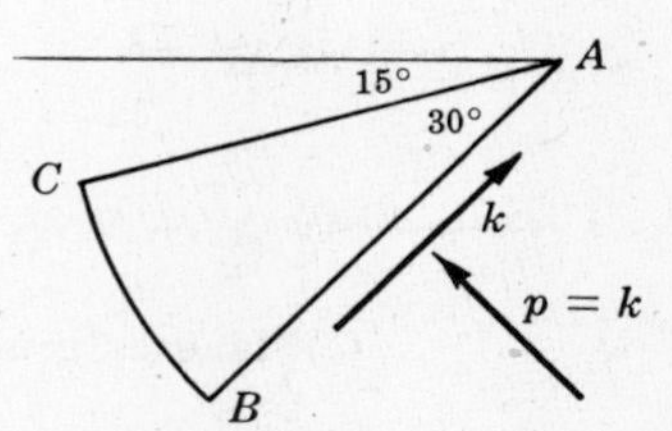

Fig. 8-26

8.51. A centered fan of α circle arcs and β radii includes an angle of 30° as shown in Fig. 8-26. The pressure on AB is k. Determine the pressure on AC.
Ans. $p = k(1 + \pi/3)$

Chapter 9

Linear Viscoelasticity

9.1 LINEAR VISCOELASTIC BEHAVIOR

Elastic solids and viscous fluids differ widely in their deformational characteristics. Elastically deformed bodies return to a natural or undeformed state upon removal of applied loads. Viscous fluids, however, possess no tendency at all for deformational recovery. Also, elastic stress is related directly to deformation whereas stress in a viscous fluid depends (except for the hydrostatic component) upon rate of deformation.

Material behavior which incorporates a blend of both elastic and viscous characteristics is referred to as *viscoelastic behavior*. The elastic (Hookean) solid and viscous (Newtonian) fluid represent opposite endpoints of a wide spectrum of viscoelastic behavior. Although viscoelastic materials are temperature sensitive, the discussion which follows is restricted to isothermal conditions and temperature enters the equations only as a parameter.

9.2 SIMPLE VISCOELASTIC MODELS

Linear viscoelasticity may be introduced conveniently from a one-dimensional viewpoint through a discussion of mechanical models which portray the deformational response of various viscoelastic materials. The mechanical elements of such models are the massless linear spring with spring constant G, and the viscous dashpot having a viscosity constant η. As shown in Fig. 9-1, the force across the spring σ is related to its elongation ϵ by

$$\sigma = G\epsilon \tag{9.1}$$

and the analogous equation for the dashpot is given by

$$\sigma = \eta\dot{\epsilon} \tag{9.2}$$

where $\dot{\epsilon} = d\epsilon/dt$. The models are given more generality and dimensional effects removed by referring to σ as *stress* and ϵ as *strain*, thereby putting these quantities on a per unit basis.

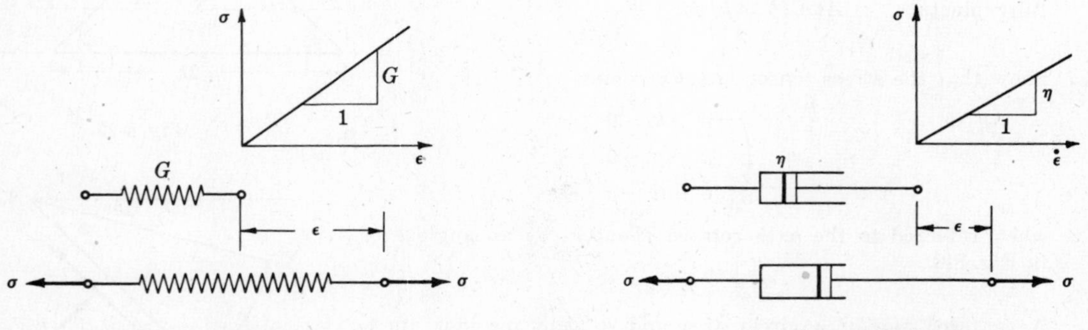

(*a*) Linear Spring (*b*) Viscous Dashpot

Fig. 9-1

The *Maxwell model* in viscoelasticity is the combination of a spring and dashpot in series as shown by Fig. 9-2(*a*). The *Kelvin* or *Voigt* model is the parallel arrangement shown in Fig. 9-2(*b*). The stress-strain relation (actually involving rates also) for the Maxwell model is

$$\frac{\dot{\sigma}}{G} + \frac{\sigma}{\eta} = \dot{\epsilon} \tag{9.3}$$

and for the Kelvin model is

$$\sigma = G\epsilon + \eta\dot{\epsilon} \tag{9.4}$$

These equations are essentially one-dimensional viscoelastic constitutive equations. It is helpful to write them in operator form by use of the *linear differential time operator* $\partial_t \equiv \partial/\partial t$. Thus (*9.3*) becomes

$$\{\partial_t/G + 1/\eta\}\sigma = \{\partial_t\}\epsilon \tag{9.5}$$

and (*9.4*) becomes

$$\sigma = \{G + \eta\partial_t\}\epsilon \tag{9.6}$$

with the appropriate operators enclosed by parentheses.

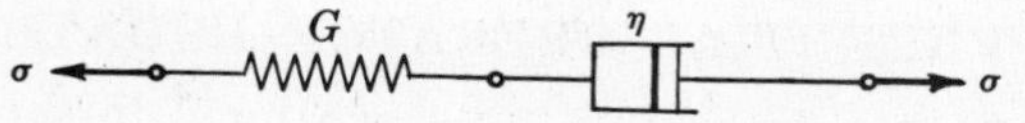

(*a*) Maxwell

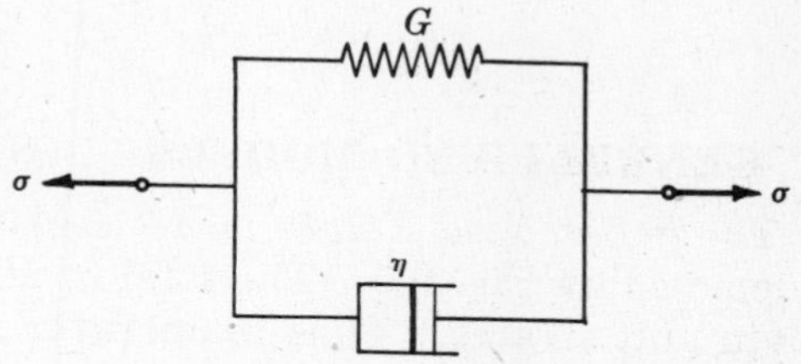

(*b*) Kelvin

Fig. 9-2

The simple Maxwell and Kelvin models are not adequate to completely represent the behavior of real materials. More complicated models afford a greater flexibility in portraying the response of actual materials. A three-parameter model constructed from two springs and one dashpot, and known as the *standard linear solid* is shown in Fig. 9-3(*a*). A three-parameter viscous model consisting of two dashpots and one spring is shown in Fig. 9-3(*b*). It should be remarked that from the point of view of the form of their constitutive equations a Maxwell unit in parallel with a spring is analogous to the standard linear solid of Fig. 9-3(*a*), and a Maxwell unit in parallel with a dashpot is analogous to the viscous model of Fig. 9-3(*b*).

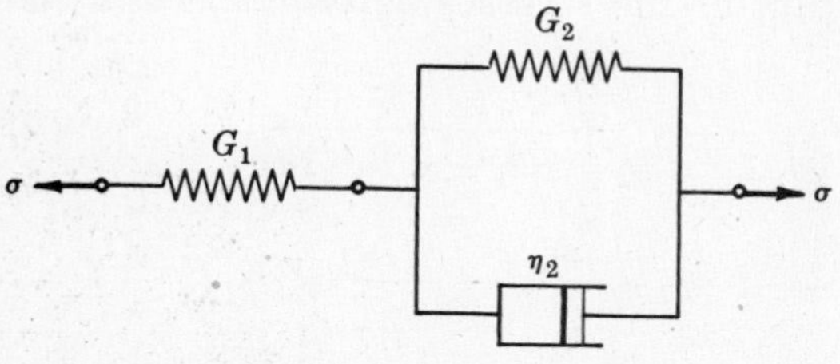

(*a*) Standard Linear Solid

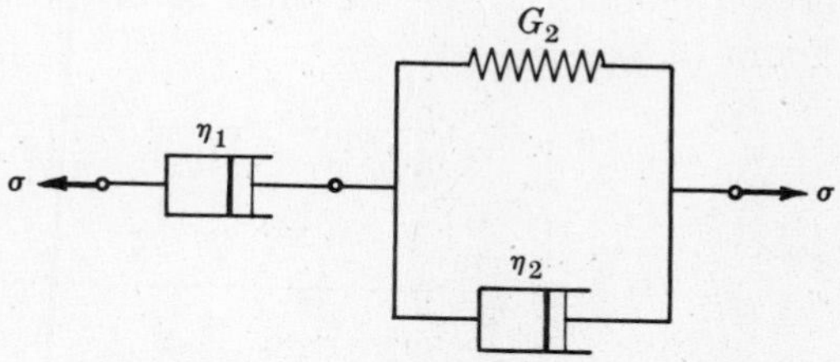

(*b*) Three-parameter Viscous Model

Fig. 9-3

A four-parameter model consisting of two springs and two dashpots may be regarded as a Maxwell unit in series with a Kelvin unit as illustrated in Fig. 9-4 below. Several equivalent forms of this model exist. The four-parameter model is capable of all three of the basic viscoelastic response patterns. Thus it incorporates "instantaneous elastic response" because of the free spring G_1, "viscous flow" because of the free dashpot η_1, and, finally, "delayed elastic response" from the Kelvin unit.

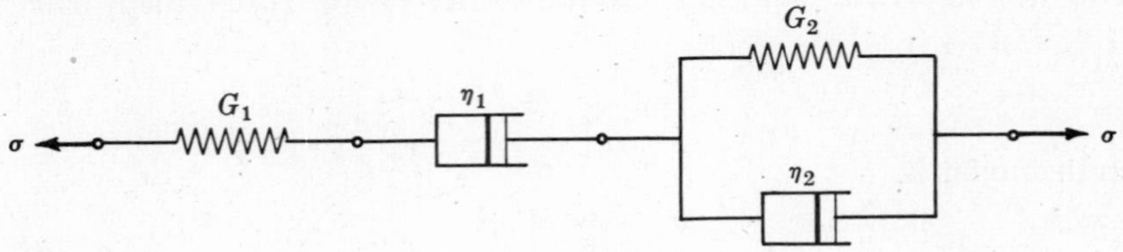

Fig. 9-4

The stress-strain equation for any of the three or four parameter models is of the general form

$$p_2\ddot{\sigma} + p_1\dot{\sigma} + p_0\sigma = q_2\ddot{\epsilon} + q_1\dot{\epsilon} + q_0\epsilon \tag{9.7}$$

where the p_i's and q_i's are coefficients made up of combinations of the G's and η's, and depend upon the specific arrangement of the elements in the model. In operator form, (*9.7*) is written

$$\{p_2\partial_t^2 + p_1\partial_t + p_0\}\sigma = \{q_2\partial_t^2 + q_1\partial_t + q_0\}\epsilon \tag{9.8}$$

9.3 GENERALIZED MODELS. LINEAR DIFFERENTIAL OPERATOR EQUATION

The *generalized Kelvin model* consists of a sequence of Kelvin units arranged in series as depicted by Fig. 9-5. The total strain of this model is equal to the sum of the individual Kelvin unit strains. Thus in operator form the constitutive equation is, by (*9.6*),

$$\epsilon = \frac{\sigma}{\{G_1 + \eta_1\partial_t\}} + \frac{\sigma}{\{G_2 + \eta_2\partial_t\}} + \cdots + \frac{\sigma}{\{G_N + \eta_N\partial_t\}} \tag{9.9}$$

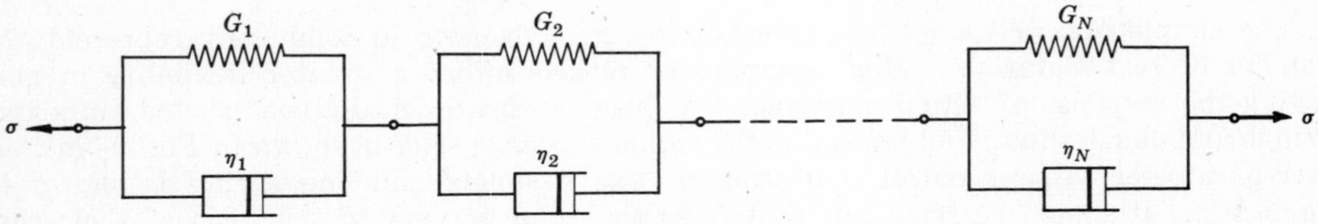

Fig. 9-5

Similarly, a sequence of Maxwell units in parallel as shown in Fig. 9-6 is called a *generalized Maxwell model*. Here the total stress is the resultant of the stresses across each unit; and so from (*9.5*),

$$\sigma = \frac{\dot{\epsilon}}{\{\partial_t/G_1 + 1/\eta_1\}} + \frac{\dot{\epsilon}}{\{\partial_t/G_2 + 1/\eta_2\}} + \cdots + \frac{\dot{\epsilon}}{\{\partial_t/G_N + 1/\eta_N\}} \tag{9.10}$$

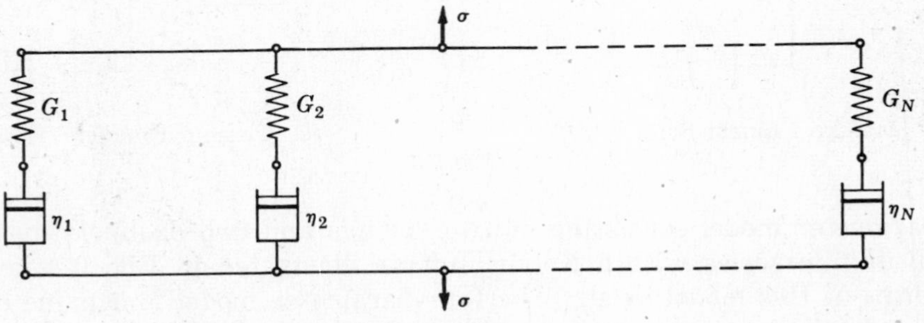

Fig. 9-6

For specific models, (*9.9*) and (*9.10*) result in equations of the form

$$p_0\sigma + p_1\dot{\sigma} + p_2\ddot{\sigma} + \cdots = q_0\epsilon + q_1\dot{\epsilon} + q_2\ddot{\epsilon} + \cdots \tag{9.11}$$

which may be expressed compactly by

$$\sum_{i=0}^{m} p_i \frac{\partial^i \sigma}{\partial t^i} = \sum_{i=0}^{n} q_i \frac{\partial^i \epsilon}{\partial t^i} \tag{9.12}$$

This *linear differential operator equation* may be written symbolically as

$$\{P\}\sigma = \{Q\}\epsilon \tag{9.13}$$

where the operators $\{P\}$ and $\{Q\}$ are defined by

$$\{P\} = \sum_{i=0}^{m} p_i \frac{\partial^i}{\partial t^i}, \qquad \{Q\} = \sum_{i=0}^{n} q_i \frac{\partial^i}{\partial t^i} \tag{9.14}$$

9.4 CREEP AND RELAXATION

The two basic experiments of viscoelasticity are the *creep* and *relaxation* tests. These tests may be performed as one-dimensional tension (compression) tests or as simple shear tests. The *creep experiment* consists of instantaneously subjecting a viscoelastic specimen to a stress σ_0 and maintaining the stress constant thereafter while measuring the strain (creep response) as a function of time. In the *relaxation experiment* an instantaneous strain ϵ_0 is imposed and maintained on the specimen while measuring the stress (relaxation) as a function of time. Mathematically, the creep and relaxation loadings are expressed in terms of the *unit step function* $[U(t-t_1)]$, defined by

$$[U(t-t_1)] = \begin{cases} 1 & t < t_1 \\ 0 & t > t_1 \end{cases} \tag{9.15}$$

and shown in Fig. 9-7.

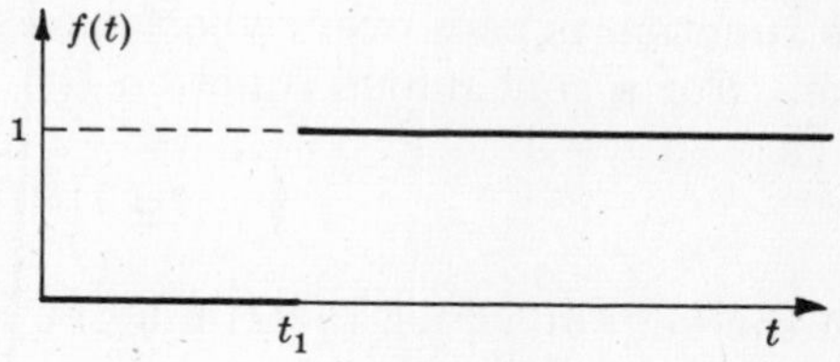

Fig. 9-7

For the creep loading,

$$\sigma = \sigma_0[U(t)] \tag{9.16}$$

where $[U(t)]$ represents the unit step function applied at time $t_1 = 0$. The creep response of a Kelvin material is determined by solving the differential equation

$$\dot{\epsilon} + \frac{\epsilon}{\tau} = \frac{\sigma_0[U(t)]}{\eta} \tag{9.17}$$

which results from the introduction of (*9.16*) into (*9.4*). Here $\tau = \eta/G$ is called the *retardation time*. For any continuous function of time $f(t)$, it may be shown that with t' as the variable of integration,

$$\int_{-\infty}^{t} f(t')[U(t'-t_1)]\,dt' = [U(t-t_1)]\int_{t_1}^{t} f(t')\,dt' \tag{9.18}$$

by means of which (*9.17*) may be integrated to yield the Kelvin creep response

$$\epsilon(t) = \frac{\sigma_0}{G}(1-e^{-t/\tau})[U(t)] \tag{9.19}$$

The creep loading, together with the creep response for the Kelvin and Maxwell models (materials) is shown in Fig. 9-8 below.

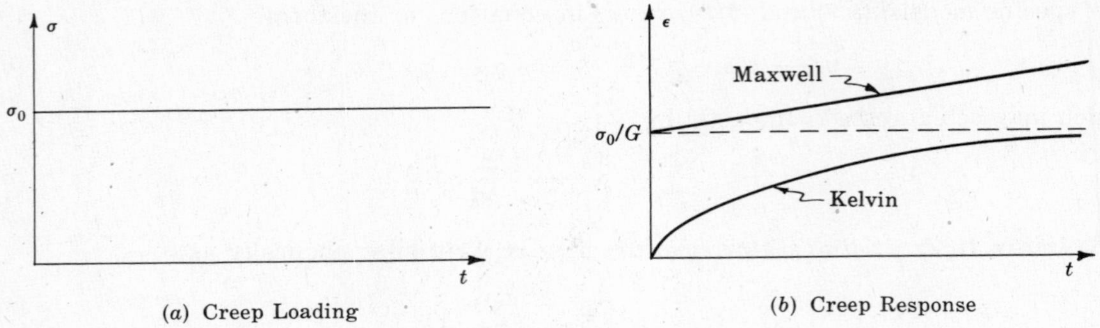

Fig. 9-8

The stress relaxation which occurs in a Maxwell material upon application of the strain

$$\epsilon = \epsilon_0[U(t)] \tag{9.20}$$

is given by the solution of the differential equation

$$\dot{\sigma} + \sigma/\tau = G\epsilon_0[\delta(t)] \tag{9.21}$$

obtained by inserting the time derivative of (*9.20*) into (*9.3*). Here $[\delta(t)] = d[U(t)]/dt$ is a singularity function called the *unit impulse function*, or *Dirac delta function*. By definition,

$$[\delta(t-t_1)] = 0, \quad t \neq t_1 \tag{9.22a}$$

$$\int_{-\infty}^{\infty} [\delta(t-t_1)]\,dt = 1 \tag{9.22b}$$

This function is zero everywhere except at $t = t_1$ where it is said to have an indeterminate spike. For a continuous function $f(t)$, it may be shown that when $t > t_1$,

$$\int_{-\infty}^{t} f(t')\,[\delta(t'-t_1)]\,dt' = f(t_1)\,[U(t-t_1)] \tag{9.23}$$

with the help of which (*9.21*) may be integrated to give the Maxwell stress relaxation

$$\sigma(t) = G\epsilon_0 e^{-t/\tau}[U(t)] \tag{9.24}$$

The stress relaxation for a Kelvin material is given directly by inserting $\dot{\epsilon} = \epsilon_0[\delta(t)]$ into (*9.4*) to yield

$$\sigma(t) = G\epsilon_0[U(t)] + \eta\epsilon_0[\delta(t)] \tag{9.25}$$

The delta function in (*9.25*) indicates that it would require an infinite stress to produce an instantaneous finite strain in a Kelvin body.

9.5 CREEP FUNCTION. RELAXATION FUNCTION. HEREDITARY INTEGRALS

The creep response of any material (model) to the creep loading $\sigma = \sigma_0[U(t)]$ may be written in the form

$$\epsilon(t) = \Psi(t)\sigma_0 \tag{9.26}$$

where $\Psi(t)$ is known as the *creep function*. For example, the creep function for the generalized Kelvin model of Fig. 9-5 is determined from (*9.19*) to be

$$\Psi(t) = \sum_{i=1}^{N} J_i(1 - e^{-t/\tau_i})[U(t)] \tag{9.27}$$

where $J_i = 1/G_i$ is called the *compliance*. If the number of Kelvin units increases indefinitely so that $N \to \infty$ in such a way that the finite set of constants (τ_i, J_i) may be replaced by the continuous compliance function $J(\tau)$, the Kelvin creep function becomes

$$\Psi(t) = \int_0^\infty J(\tau)(1 - e^{-t/\tau})\, d\tau \tag{9.28}$$

The function $J(\tau)$ is referred to as the *"distribution of retardation times"*, or *retardation spectrum.*

In analogy with the creep response, the stress relaxation for any model subjected to the strain $\epsilon = \epsilon_0[U(t)]$ may be written in the form

$$\sigma(t) = \phi(t)\epsilon_0 \tag{9.29}$$

where $\phi(t)$ is called the *relaxation function.* For the generalized Maxwell model of Fig. 9-6, the relaxation function is determined from (*9.24*) as

$$\phi(t) = \sum_{i=1}^{N} G_i e^{-t/\tau_i}[U(t)] \tag{9.30}$$

Here, as $N \to \infty$ the function $G(\tau)$ replaces the set of constants (G_i, τ_i) and the relaxation function is defined by

$$\phi(t) = \int_0^\infty G(\tau)e^{-t/\tau}\, d\tau \tag{9.31}$$

The function $G(\tau)$ is known as the *"distribution of relaxation times"*, or *relaxation spectrum.*

In linear viscoelasticity, the superposition principle is valid. Thus the total "effect" of a sum of "causes" is equal to the sum of the "effects" of each of the "causes". Accordingly, if the stepped stress history of Fig. 9-9(*a*) is applied to a material for which the creep function is $\Psi(t)$, the creep response will be

$$\epsilon(t) = \sigma_0\Psi(t) + \sigma_1\Psi(t-t_1) + \sigma_2\Psi(t-t_2) + \sigma_3\Psi(t-t_3) = \sum_{i=0}^{3} \sigma_i\Psi(t-t_i) \tag{9.32}$$

Therefore the arbitrary stress history $\sigma = \sigma(t)$ of Fig. 9-9(*b*) may be analyzed as an infinity of step loadings, each of magnitude $d\sigma$ and the creep response given by the superposition integral

$$\epsilon(t) = \int_{-\infty}^{t} \frac{d\sigma(t')}{dt'}\Psi(t-t')\, dt' \tag{9.33}$$

Such integrals are often referred to as *hereditary integrals* since the strain at any time is seen to depend upon the entire stress history.

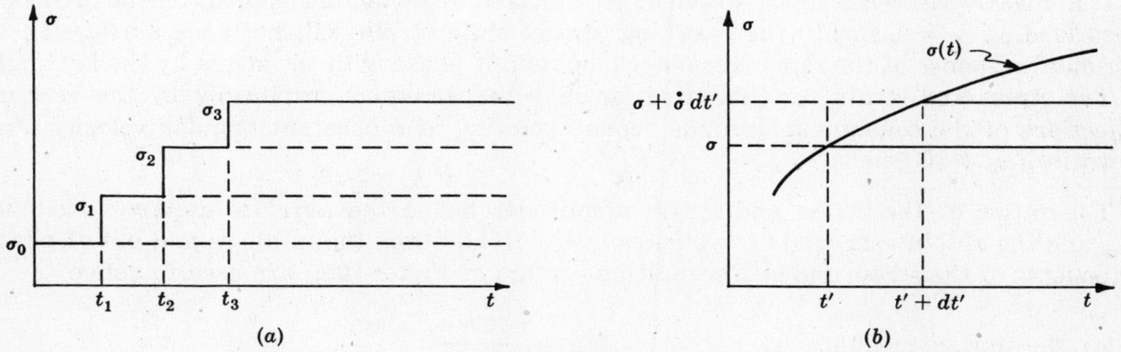

Fig. 9-9

For a material initially "dead", i.e. completely free of stress and strain at time zero, the lower limit in (*9.33*) may be replaced by zero and the creep response expressed as

$$\epsilon(t) = \int_0^t \frac{d\sigma(t')}{dt'} \Psi(t-t')\, dt' \qquad (9.34)$$

Furthermore, if the stress loading involves a step discontinuity of magnitude σ_0 at $t = 0$, (*9.34*) is usually written in the form

$$\epsilon(t) = \sigma_0 \Psi(t) + \int_0^t \frac{d\sigma(t')}{dt'} \Psi(t-t')\, dt' \qquad (9.35)$$

Following similar arguments as above, the stress as a function of time may be represented through a superposition integral involving the strain history $\epsilon(t)$ and the relaxation function $\phi(t)$. In analogy with (*9.33*) the stress is given by

$$\sigma(t) = \int_{-\infty}^t \frac{d\epsilon(t')}{dt'} \phi(t-t')\, dt' \qquad (9.36)$$

and with regard to a material that is "dead" at $t = 0$, the integrals comparable to (*9.34*) and (*9.35*) are respectively

$$\sigma(t) = \int_0^t \frac{d\epsilon(t')}{dt'} \phi(t-t')\, dt' \qquad (9.37)$$

and

$$\sigma(t) = \epsilon_0 \phi(t) + \int_0^t \frac{d\epsilon(t')}{dt'} \phi(t-t')\, dt' \qquad (9.38)$$

Since either the creep integral (*9.34*) or the relaxation integral (*9.37*) may be used to specify the viscoelastic characteristics of a given material, it follows that some relationship must exist between the creep function $\Psi(t)$ and the relaxation function $\phi(t)$. Such a relationship is not easily determined in general, but using the Laplace transform definition

$$\bar{f}(s) = \int_0^\infty f(t) e^{-st}\, dt \qquad (9.39)$$

it is possible to show that the transforms $\bar{\Psi}(s)$ and $\bar{\phi}(t)$ are related by the equation

$$\bar{\Psi}(s)\bar{\phi}(s) = 1/s^2 \qquad (9.40)$$

where s is the transform parameter.

9.6 COMPLEX MODULI AND COMPLIANCES

If a linearly viscoelastic test specimen is subjected to a one-dimensional (tensile or shear) stress loading $\sigma = \sigma_0 \sin \omega t$, the resulting steady state strain will be $\epsilon = \epsilon_0 \sin(\omega t - \delta)$, a sinusoidal response of the same frequency ω but out of phase with the stress by the lag angle δ. The stress and strain for this situation may be presented graphically by the vertical projections of the constant magnitude vectors rotating at a constant angular velocity ω as shown in Fig. 9-10 below.

The ratios of the stress and strain amplitudes define the *absolute dynamic modulus* σ_0/ϵ_0, and the *absolute dynamic compliance* ϵ_0/σ_0. In addition, the in-phase and out-of-phase components of the stress and strain rotating vectors of Fig. 9-10(*a*) are used to define

(*a*) the storage modulus $$G_1 = \frac{\sigma_0 \cos\delta}{\epsilon_0}$$

(*b*) the loss modulus $$G_2 = \frac{\sigma_0 \sin \delta}{\epsilon_0}$$

(*c*) the storage compliance $$J_1 = \frac{\epsilon_0 \cos \delta}{\sigma_0}$$

(*d*) the loss compliance $$J_2 = \frac{\epsilon_0 \sin \delta}{\sigma_0}$$

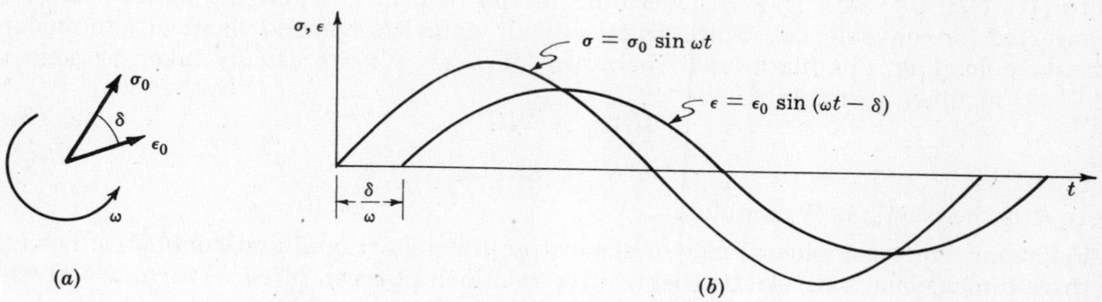

Fig. 9-10

A generalization of the above description of viscoelastic behavior is achieved by expressing the stress in complex form as

$$\sigma^* = \sigma_0 e^{i\omega t} \tag{9.41}$$

and the resulting strain also in complex form as

$$\epsilon^* = \epsilon_0 e^{i(\omega t - \delta)} \tag{9.42}$$

From (*9.41*) and (*9.42*) the *complex modulus* $G^*(i\omega)$ is defined as the complex quantity

$$\sigma^*/\epsilon^* = G^*(i\omega) = (\sigma_0/\epsilon_0)e^{i\delta} = G_1 + iG_2 \tag{9.43}$$

whose real part is the storage modulus and whose imaginary part is the loss modulus. Similarly, the *complex compliance* is defined as

$$\begin{aligned} \epsilon^*/\sigma^* &= J^*(i\omega) = (\epsilon_0/\sigma_0)e^{-i\delta} \\ &= J_1 - iJ_2 \end{aligned} \tag{9.44}$$

where the real part is the storage compliance and the imaginary part the negative of the loss compliance. In Fig. 9-11 the vector diagrams of G^* and J^* are shown. Note that $G^* = 1/J^*$.

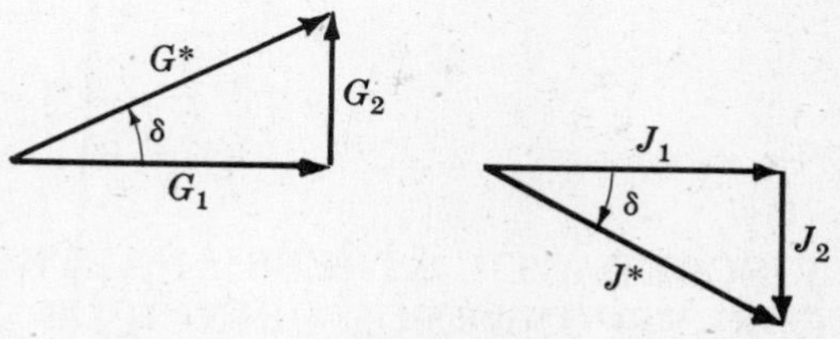

Fig. 9-11

9.7 THREE DIMENSIONAL THEORY

In developing the three dimensional theory of linear viscoelasticity, it is customary to consider separately viscoelastic behavior under conditions of so-called pure shear and pure dilatation. Thus distortional and volumetric effects are prescribed independently, and subsequently combined to provide a general theory. Mathematically, this is handled by resolving the stress and strain tensors into their deviatoric and spherical parts, for each of which viscoelastic constitutive relations are then written. The stress tensor decomposition is given by (*2.70*) as

$$\sigma_{ij} = s_{ij} + \delta_{ij}\sigma_{kk}/3 \tag{9.45}$$

and the small strain tensor by (*3.98*) as

$$\epsilon_{ij} = e_{ij} + \delta_{ij}\epsilon_{kk}/3 \tag{9.46}$$

Using the notation of these equations, the three dimensional generalization of the viscoelastic constitutive equation (*9.13*) in differential operator form is written by the combination

$$\{P\}s_{ij} = 2\{Q\}e_{ij} \tag{9.47a}$$

and

$$\{M\}\sigma_{ii} = 3\{N\}\epsilon_{ii} \tag{9.47b}$$

where $\{P\}$, $\{Q\}$, $\{M\}$ and $\{N\}$ are operators of the form (*9.14*) and the numerical factors are inserted for convenience. Since practically all materials respond elastically to moderate hydrostatic loading, the dilatational operators $\{M\}$ and $\{N\}$ are usually taken as constants and (*9.47*) modified to read

$$\{P\}s_{ij} = 2\{Q\}e_{ij} \tag{9.48a}$$

$$\sigma_{ii} = 3K\epsilon_{ii} \tag{9.48b}$$

where K is the elastic bulk modulus.

Following the same general rule of separation for distortional and volumetric behavior, the three-dimensional viscoelastic constitutive relations in *creep integral* form are given by

$$e_{ij} = \int_0^t \Psi_s(t-t')\frac{\partial s_{ij}}{\partial t'}\,dt' \tag{9.49a}$$

$$\epsilon_{ii} = \int_0^t \Psi_v(t-t')\frac{\partial \sigma_{ii}}{\partial t'}\,dt' \tag{9.49b}$$

and in the *relaxation integral* form by

$$s_{ij} = \int_0^t \phi_s(t-t')\frac{\partial e_{ij}}{\partial t'}\,dt' \tag{9.50a}$$

$$\sigma_{ii} = \int_0^t \phi_v(t-t')\frac{\partial \epsilon_{ii}}{\partial t'}\,dt' \tag{9.50b}$$

The extension to three-dimensions of the complex modulus formulation of viscoelastic behavior requires the introduction of the complex bulk modulus K^*. Again, writing shear and dilatation equations separately, the appropriate equations are of the form

$$s^*_{ij} = 2G^*(i\omega)e^*_{ij} = 2(G_1 + iG_2)e^*_{ij} \tag{9.51a}$$

$$\sigma^*_{ii} = 3K^*(i\omega)\epsilon^*_{ii} = 3(K_1 + iK_2)\epsilon^*_{ii} \tag{9.51b}$$

9.8 VISCOELASTIC STRESS ANALYSIS. CORRESPONDENCE PRINCIPLE

The stress analysis problem for an isotropic viscoelastic continuum body which occupies a volume V and has the bounding surface S as shown in Fig. 9-12, is formulated as follows: Let body forces b_i be given throughout V and let the surface tractions $t_i^{(\hat{n})}(x_k, t)$ be prescribed over the portion S_1 of S, and the surface displacements $g_i(x_k, t)$ be prescribed over the portion S_2 of S. Then the governing field equations take the form of:

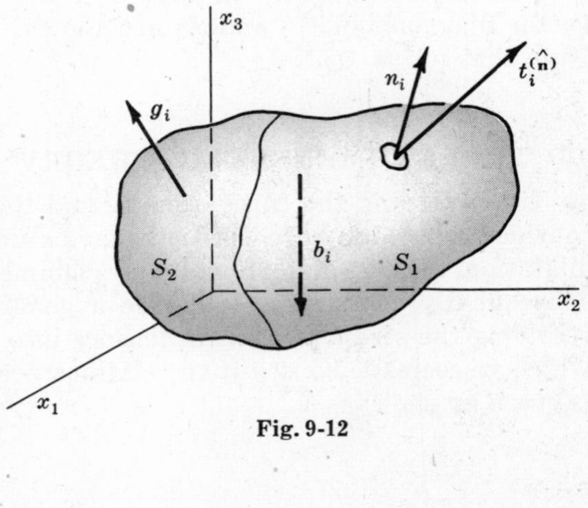

Fig. 9-12

1. Equations of motion (or of equilibrium)

$$\sigma_{ij,j} + b_i = \rho \ddot{u}_i \tag{9.52}$$

2. Strain-displacement equations

$$2\epsilon_{ij} = (u_{i,j} + u_{j,i}) \tag{9.53}$$

or strain-rate-velocity equations

$$2\dot{\epsilon}_{ij} = (v_{i,j} + v_{j,i}) \tag{9.54}$$

3. Boundary conditions

$$\sigma_{ij}(x_k, t)\, n_i(x_k) = t_i^{(\hat{n})}(x_k, t) \quad \text{on } S_1 \tag{9.55}$$

$$u_i(x_k, t) = g_i(x_k, t) \quad \text{on } S_2 \tag{9.56}$$

4. Initial conditions

$$u_i(x_k, 0) = u_0 \tag{9.57}$$

$$v_i(x_k, 0) = v_0 \tag{9.58}$$

5. Constitutive equations

   (*a*) Linear differential operator form (*9.48*)

   or

   (*b*) Hereditary integral form (*9.49*) or (*9.50*)

   or

   (*c*) Complex modulus form (*9.51*)

If the body geometry and loading conditions are sufficiently simple, and if the material behavior may be represented by one of the simpler models, the field equations above may be integrated directly (see Problem 9.22). For more general conditions, however, it is conventional to seek a solution through the use of the *correspondence principle*. This principle emerges from the analogous form between the governing field equations of elasticity and the Laplace transforms with respect to time of the basic viscoelastic field equations given above. A comparison of the pertinent equations for quasi-static isothermal problems is afforded by the following table in which barred quantities indicate Laplace transforms in accordance with the definition

$$\bar{f}(x_k, s) = \int_0^\infty f(x_k, t) e^{-st}\, dt \tag{9.59}$$

Elastic	**Transformed Viscoelastic**
1. $\sigma_{ij,j} + b_i = 0$	1. $\bar{\sigma}_{ij,j} + \bar{b}_i = 0$
2. $2\epsilon_{ij} = (u_{i,j} + u_{j,i})$	2. $2\bar{\epsilon}_{ij} = (\bar{u}_{i,j} + \bar{u}_{j,i})$
3. $\sigma_{ij} n_j = t_i^{(\hat{n})}$ on S_1	3. $\bar{\sigma}_{ij}\bar{n}_j = \bar{t}_i^{(\hat{n})}$ on S_1
$u_i = g_i$ on S_2	$\bar{u}_i = \bar{g}_i$ on S_2
4. $s_{ij} = 2Ge_{ij}$	4. $\bar{P}(s)\bar{s}_{ij} = 2\bar{Q}(s)\bar{e}_{ij}$
$\sigma_{ii} = 3K\epsilon_{ii}$	$\bar{\sigma}_{ii} = 3K\bar{\epsilon}_{ii}$

From this table it is observed that when G in the elastic equations is replaced by $\bar{Q}/\bar{P}$, the two sets of equations have the same form. Accordingly, if in the solution of the "corresponding elastic problem" G is replaced by $\bar{Q}/\bar{P}$ for the viscoelastic material involved, the result is the Laplace transform of the viscoelastic solution. Inversion of the transformed solution yields the viscoelastic solution.

The correspondence principle may also be stated for problems other than quasi-static problems. Furthermore, the form of the constitutive equations need not be the linear differential operator form but may appear as in (*9.49*), (*9.50*) or (*9.51*). The particular problem under study will dictate the appropriate form in which the principle should be used.

Solved Problems

VISCOELASTIC MODELS (Sec. 9.1-9.3)

9.1. Verify the stress-strain relations for the Maxwell and Kelvin models given by (*9.3*) and (*9.4*) respectively.

In the Maxwell model of Fig. 9-2(*a*) the total strain is the sum of the strain in the spring plus the strain of the dashpot. Thus $\epsilon = \epsilon_S + \epsilon_D$ and also $\dot{\epsilon} = \dot{\epsilon}_S + \dot{\epsilon}_D$. Since the stress across each element is σ, (*9.1*) and (*9.2*) may be used to give $\dot{\epsilon} = \dot{\sigma}/G + \sigma/\eta$.

In the Kelvin model of Fig. 9-2(*b*), $\sigma = \sigma_S + \sigma_D$ and directly from (*9.1*) and (*9.2*), $\sigma = \eta\dot{\epsilon} + G\epsilon$.

9.2. Use the operator form of the Kelvin model stress-strain relation to obtain the stress-strain law for the standard linear solid of Fig. 9-3(*a*).

Here the total strain is the sum of the strain in the spring plus the strain in the Kelvin unit. Thus $\epsilon = \epsilon_S + \epsilon_K$ or in operator form $\epsilon = \sigma/G_1 + \sigma/\{G_2 + \eta_2\partial_t\}$. From this

$$G_1\{G_2 + \eta_2\partial_t\}\epsilon = \{G_2 + \eta_2\partial_t\}\sigma + G_1\sigma$$

and so $G_1G_2\epsilon + G_1\eta_2\dot{\epsilon} = (G_1 + G_2)\sigma + \eta_2\dot{\sigma}$.

9.3. Determine the stress-strain equation for the four parameter model of Fig. 9-4. Let $\eta_1 \to \infty$ and compare with the result of Problem 9.2.

Here the total strain $\epsilon = \epsilon_K + \epsilon_M$ which in operator form is

$$\epsilon = \sigma/\{G_2 + \eta_2\partial_t\} + \{\partial_t + 1/\tau_1\}\sigma/G_1\{\partial_t\}$$

Expanding the operators and collecting terms gives

$$\ddot{\sigma} + (G_1/\eta_1 + (G_1 + G_2)/\eta_2)\dot{\sigma} + G_1G_2\sigma/\eta_1\eta_2 = G_1\ddot{\epsilon} + G_1G_2\dot{\epsilon}/\eta_2$$

As $\eta_1 \to \infty$ this becomes $\ddot{\sigma} + (G_1 + G_2)\dot{\sigma}/\eta_2 = G_1\ddot{\epsilon} + G_1G_2\dot{\epsilon}/\eta_2$ which is equivalent to the result of Problem 9.2.

9.4. Treating the model in Fig. 9-13 as a special case of the generalized Maxwell model, determine its stress-strain equation.

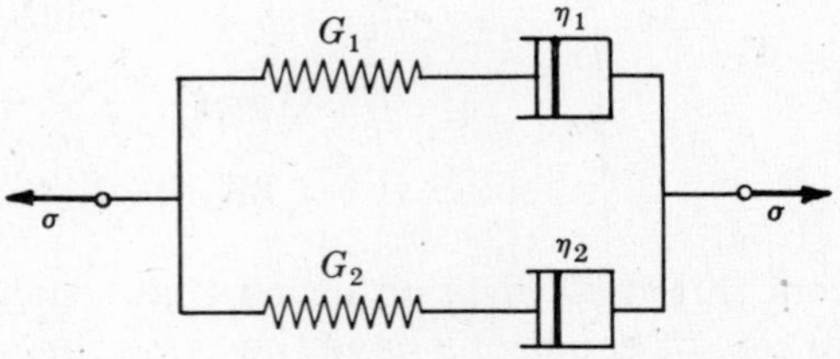

Fig. 9-13

Writing (*9.10*) for $N = 2$ in the form

$$\sigma = G_1\dot{\epsilon}/\{\partial_t + 1/\tau_1\} + G_2\dot{\epsilon}/\{\partial_t + 1/\tau_2\}$$

and operating as indicated gives

$$\{\partial_t + 1/\tau_2\}(\dot\sigma + \sigma/\tau_1) = G_1\{\partial_t + 1/\tau_2\}\dot\epsilon + G_2\{\partial_t + 1/\tau_1\}\dot\epsilon$$

which when expanded and rearranged becomes

$$\ddot\sigma + (\tau_1 + \tau_2)\dot\sigma/\tau_1\tau_2 + \sigma/\tau_1\tau_2 = (G_1 + G_2)\ddot\epsilon + (G_1/\tau_2 + G_2/\tau_1)\dot\epsilon$$

9.5. The model shown in Fig. 9-14 may be considered as a degenerate form of the generalized Maxwell model with $G_1 = \eta_2 = \infty$ for the case $N = 3$. Using these values in (*9.10*) develop the stress-strain equation for this model.

Here (*9.10*) becomes $\sigma = \eta_1\dot\epsilon + G_2\dot\epsilon/\{\partial_t\} + \dot\epsilon/\{\partial_t/G_3 + 1/\eta_3\}$ or

$$\{\partial_t/G_3 + 1/\eta_3\}\dot\sigma = \{\partial_t/G_3 + 1/\eta_3\}(\eta_1\ddot\epsilon + G_2\dot\epsilon) + \dot\epsilon$$

Application of the operators gives

$$\ddot\sigma/G_3 + \dot\sigma/\eta_3 = \eta_1\dddot\epsilon/G_3 + (1 + G_2/G_3 + \eta_1/\eta_3)\ddot\epsilon + G_2/\eta_3\dot\epsilon$$

which may also be written

$$\eta_3\ddot\sigma + G_3\dot\sigma = \eta_1\eta_3\dddot\epsilon + (G_2\eta_3 + G_3\eta_1 + G_3\eta_3)\ddot\epsilon + G_2G_3\dot\epsilon$$

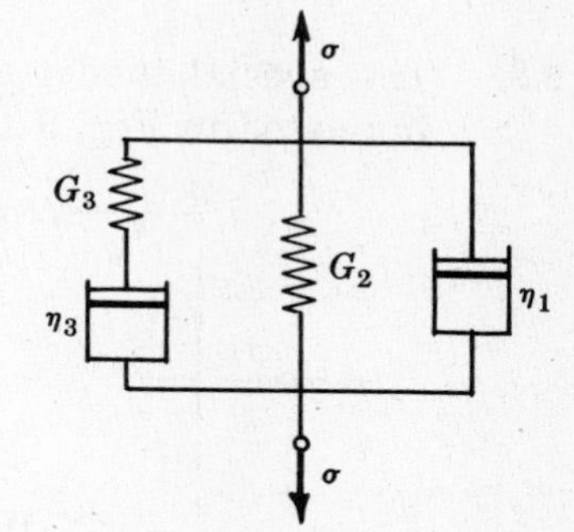

Fig. 9-14

CREEP AND RELAXATION (Sec. 9.4)

9.6. Determine the Kelvin and Maxwell creep response equations by direct integration of (*9.17*) and (*9.21*) respectively.

Using the integrating factor $e^{t/\tau}$, (*9.17*) becomes $\epsilon e^{t/\tau} = \dfrac{\sigma_0}{\eta}\displaystyle\int_0^t e^{t'/\tau}[U(t')]\,dt'$ which by formula (*9.18*) yields

$$\epsilon e^{t/\tau} = (\sigma_0[U(t)]/\eta)[\tau e^{t'/\tau}]_0^t = (\sigma_0/G)(e^{t/\tau} - 1)[U(t)] \quad \text{or} \quad \epsilon = (\sigma_0/G)(1 - e^{-t/\tau})[U(t)]$$

Use of $e^{t/\tau}$ as the integrating factor in (*9.21*) gives $\sigma e^{t/\tau} = G\epsilon_0\displaystyle\int_0^t e^{t'/\tau}[\delta(t')]\,dt'$; and by formula (*9.23*),

$$\sigma e^{t/\tau} = G\epsilon_0[U(t)] \quad \text{or} \quad \sigma = G\epsilon_0 e^{-t/\tau}[U(t)]$$

9.7. Determine the creep response of the standard linear solid of Fig. 9-3(*a*).

Since $\epsilon = \epsilon_S + \epsilon_K$ for this model the creep response from (*9.1*) and (*9.19*) is simply

$$\epsilon(t) = [1/G_1 + (1/G_2)(1 - e^{-t/\tau_2})]\sigma_0[U(t)]$$

The same result may be obtained by setting $\eta_2 = \infty$ in the generalized Kelvin $(N = 2)$ response $\epsilon = \sum_{i=1}^{2} J_i(1 - e^{-t/\tau_i})\sigma_0[U(t)]$ or by integrating directly the standard solid stress-strain law. The student should carry out the details.

9.8. The creep-recovery experiment consists of a creep loading which is maintained for a period of time and then instantaneously removed. Determine the creep-recovery response of the standard solid (Fig. 9-3(*a*)) for the loading shown in Fig. 9-15.

From Problem 9.7 the response while the load is on $(t < 2\tau_2)$ is

$$\epsilon = \sigma_0[1/G_1 + (1/G_2)(1 - e^{-t/\tau_2})]$$

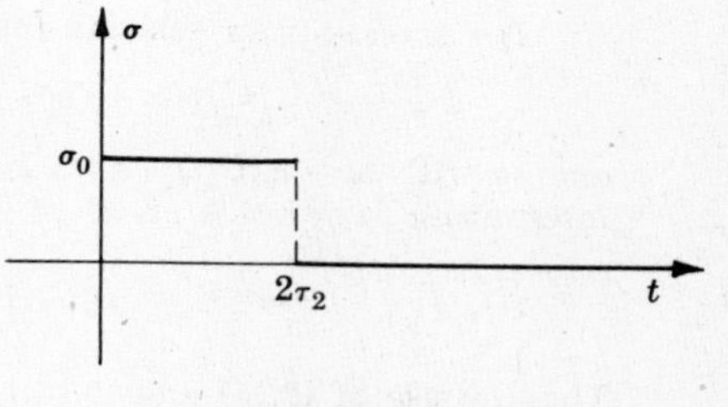

Fig. 9-15

At $t = 2\tau_2$ the load is removed and σ becomes zero at the same time that the "elastic" deformation σ_0/G_1 is recovered. For $t > 2\tau_2$ the response is governed by the equation $\dot{\epsilon} + \epsilon/\tau_2 = 0$ which is the stress-strain law for the model with $\sigma = 0$ (see Problem 9.2). The solution of this differential equation is $\epsilon = Ce^{-T/\tau_2}$ where C is a constant and $T = t - 2\tau_2$. At $T = 0$, $\epsilon = C = \sigma_0(1 - e^{-2})/G_2$ and so

$$\epsilon = \sigma_0(1 - e^{-2})e^{-T/\tau_2}/G_2 = \sigma_0(e^2 - 1)e^{-t/\tau_2}/G_2 \qquad \text{for } t > 2\tau_2$$

9.9. The special model shown in Fig. 9-16 is elongated at a constant rate $\dot{\epsilon} = \epsilon_0/t_1$ as indicated in Fig. 9-17. Determine the stress in the model under this straining.

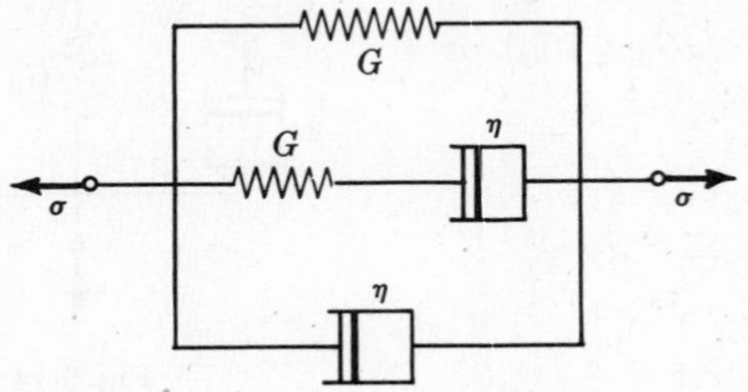

Fig. 9-16

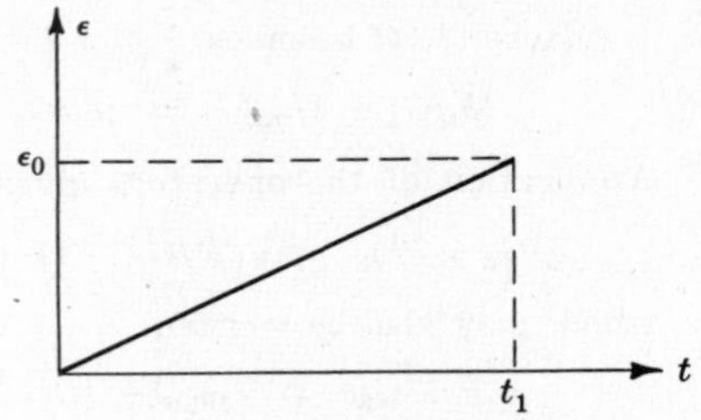

Fig. 9-17

From Problem 9.5 the stress-strain law for the model is $\dot{\sigma} + \sigma/\tau = \eta\ddot{\epsilon} + 3G\dot{\epsilon} + G\epsilon/\tau$ and so here $\dot{\sigma} + \sigma/\tau = 3G\epsilon_0/t_1 + G\epsilon_0 t/\tau t_1$. Integrating this yields $\sigma = \epsilon_0(3\eta + Gt - \eta) + Ce^{-t/\tau}$ where C is the constant of integration. When $t = 0$, $\sigma = \eta\epsilon_0/t_1$ and so $C = -\eta\epsilon_0/t_1$. Thus $\sigma = \epsilon_0(2\eta + Gt - \eta e^{-t/\tau})/t_1$. Note that the same result is obtained by integrating

$$\sigma e^{t/\tau} = \frac{\epsilon_0}{t_1}\int_0^t 3Ge^{t'/\tau}\,dt' + \frac{\epsilon_0}{t_1}\int_0^t \frac{Gt'e^{t'/\tau}}{\tau}\,dt'$$

9.10. Determine by a direct integration of the stress-strain law for the standard linear solid its stress relaxation under the strain $\epsilon = \epsilon_0[U(t)]$.

Writing the stress-strain law (see Problem 9.2) as $\dot{\sigma} + (G_1 + G_2)\sigma/\eta_2 = \epsilon_0 G_1([\delta(t)] + G_1G_2[U(t)]/\eta_2)$ for the case at hand and employing the integrating factor $e^{(G_1+G_2)t/\eta_2}$ it is seen that

$$\sigma e^{(G_1+G_2)t/\eta_2} = \epsilon_0 G_1 \int_0^t [\delta(t')]e^{(G_1+G_2)t'/\eta_2}\,dt' + \frac{\epsilon_0 G_1 G_2}{\eta_2}\int_0^t [U(t')]e^{(G_1+G_2)t'/\eta_2}\,dt'$$

Integrating this equation with the help of (*9.18*) and (*9.23*),

$$\sigma = \epsilon_0 G_1(G_2 + G_1 e^{-(G_1+G_2)t/\eta_2})[U(t)]/(G_1 + G_2)$$

CREEP AND RELAXATION FUNCTIONS. HEREDITARY INTEGRALS (Sec. 9.5)

9.11. Determine the relaxation function $\phi(t)$ for the three parameter model shown in Fig. 9-18.

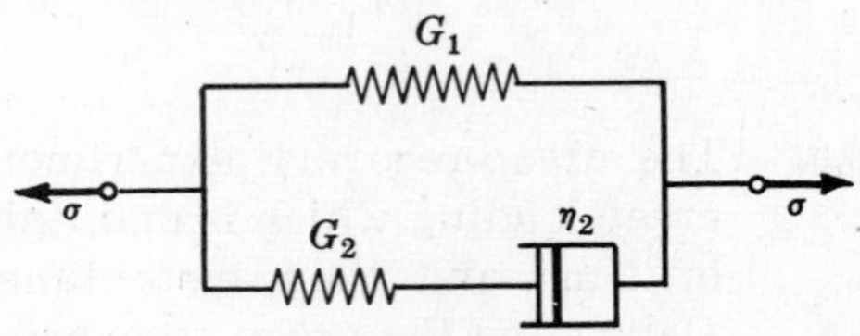

Fig. 9-18

The stress-strain relation for this model is

$$\dot{\sigma} + \sigma/\tau_2 = (G_1 + G_2)\dot{\epsilon} + G_1G_2\epsilon/\eta_2$$

and so with $\epsilon = \epsilon_0[U(t)]$ and $\dot{\epsilon} = \epsilon_0[\delta(t)]$ use of the integrating factor e^{t/τ_2} gives

$$\sigma e^{t/\tau_2} = \epsilon_0(G_1 + G_2)\int_0^t e^{t'/\tau_2}[\delta(t')]\,dt' + \frac{\epsilon_0 G_1 G_2}{\eta_2}\int_0^t e^{t'/\tau_2}[U(t')]\,dt'$$

Thus by use of (*9.18*) and (*9.23*), $\sigma = \epsilon_0(G_1 + G_2 e^{-t/\tau_2}) = \epsilon_0\phi(t)$. Note that this result may also be obtained by putting $\eta_1 \to \infty$ in (*9.30*) for the generalized Maxwell model.

9.12. Using the relaxation function $\phi(t)$ for the model of Problem 9.11, determine the creep function by means of (9.40).

The Laplace transform of $\phi(t) = G_1 + G_2e^{-t/\tau_2}$ is $\bar{\phi}(s) = G_1/s + G_2/(s+1/\tau_2)$ (see any standard table of Laplace transforms). Thus from (9.40),

$$\bar{\psi}(s) = (s+1/\tau_2)/[G_1s(s+1/\tau_2) + G_2s^2] = 1/G_1s - [G_2/G_1(G_1+G_2)]/(s+G_1/(G_1+G_2)\tau_2)$$

which may be inverted easily by a Laplace transform table to give

$$\psi = 1/G_1 - [G_2/G_1(G_1+G_2)]e^{-G_1t/(G_1+G_2)\tau_2}$$

This result may be readily verified by integration of the model's stress-strain equation under creep loading.

9.13. If a ramp type stress followed by a sustained constant stress σ_1 (Fig. 9-19) is applied to a Kelvin material, determine the resulting strain. Assume $\sigma_1/t_1 = \lambda$.

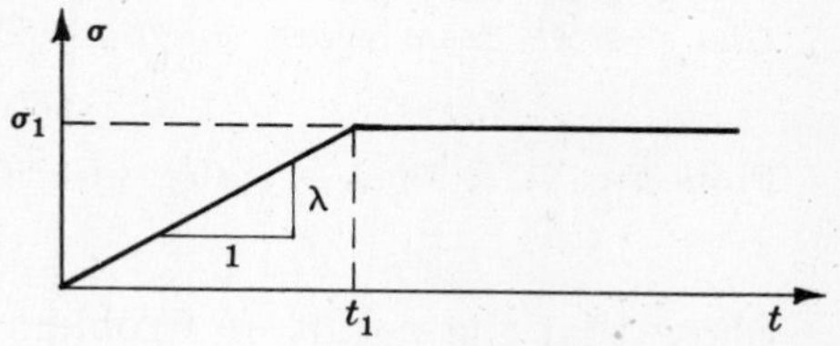

Fig. 9-19

The stress may be expressed as

$$\sigma = \lambda t[U(t)] - \lambda(t-t_1)[U(t-t_1)]$$

which when introduced into (9.4) leads to

$$\epsilon e^{t/\tau} = \frac{\lambda}{\eta}\left[\int_0^t t'e^{t'/\tau}[U(t')]\,dt' - \int_{t_1}^t (t'-t_1)e^{t'/\tau}[U(t'-t_1)]\,dt'\right]$$

Integrating with the aid of (9.18) gives

$$\epsilon = (\lambda/G)\{(t+\tau(e^{-t/\tau}-1))[U(t)] - ((t-t_1) + \tau(e^{(t_1-t)/\tau}-1))[U(t-t_1)]\}$$

which reduces as $t \to \infty$ to $\epsilon = \lambda t_1/G = \sigma_1/G$.

9.14. Using the creep integral (9.34) together with the Kelvin creep function, verify the result of Problem 9.13.

For the Kelvin body, $\psi(t) = (1-e^{-t/\tau})/G$ and (9.34) becomes

$$\epsilon(t) = \int_{-\infty}^t \frac{\lambda}{G}([U(t')] + t'[\delta(t')] - [U(t'-t_1)] - (t'-t_1)[\delta(t'-t_1)])(1-e^{-(t-t')/\tau})\,dt'$$

which by (9.18) and (9.23) reduces to

$$\epsilon = \frac{\lambda}{G}\left[[U(t)]\int_0^t (1-e^{-(t-t')/\tau})\,dt' - [U(t-t_1)]\int_{t_1}^t (1-e^{-(t-t')/\tau})\,dt'\right]$$

A straightforward evaluation of these integrals confirms the result presented in Problem 9.13.

9.15. By a direct application of the superposition principle, determine the response of a Kelvin material to the stress loading shown in Fig. 9-20.

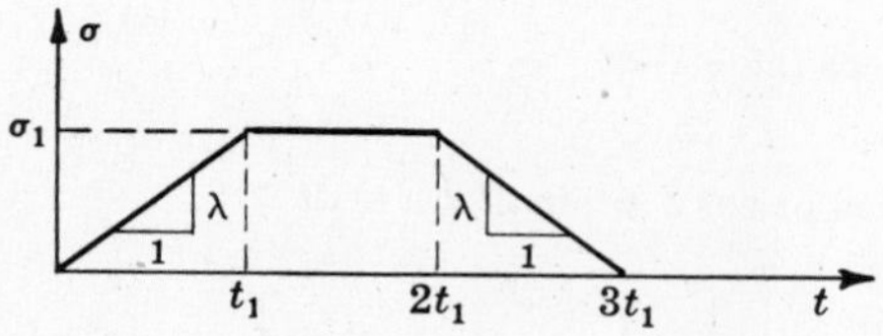

Fig. 9-20

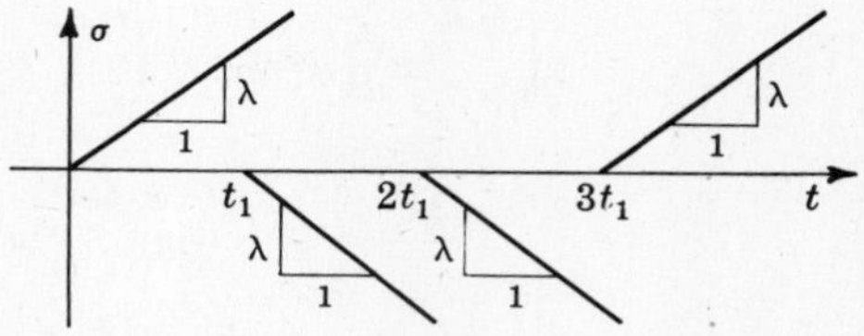

Fig. 9-21

The stress may be represented as a sequence of ramp loadings as shown in Fig. 9-21 above. From Problem 9.13, $\epsilon = (\lambda/G)[t + \tau(e^{-t/\tau} - 1)][U(t)]$ for this stress loading. In the present case therefore

$$\epsilon(t) = (\lambda/G)[(t + \tau(e^{-t/\tau} - 1))[U(t)] - ((t - t_1) + \tau(e^{-(t-t_1)/\tau} - 1))[U(t - t_1)]$$
$$-((t - 2t_1) + \tau(e^{-(t-2t_1)/\tau} - 1))[U(t - 2t_1)] + ((t - 3t_1) + \tau(e^{-(t-3t_1)/\tau} - 1))[U(t - 3t_1)]]$$

Note that as $t \to \infty$, $\epsilon \to 0$.

COMPLEX MODULI AND COMPLIANCES (Sec. 9.6)

9.16. Determine the complex modulus G^* and the lag angle δ for the Maxwell material of Fig. 9-2.

Writing (*9.3*) as $\dot{\sigma} + \sigma/\tau = G\dot{\epsilon}$ and inserting (*9.41*) and (*9.42*) gives $i\omega\sigma_0 e^{i\omega t} + \sigma_0 e^{i\omega t}/\tau = Gi\omega\epsilon_0 e^{i(\omega t - \delta)}$ from which $\sigma_0 e^{i\delta}/\epsilon_0 = G^* = Gi\omega\tau/(1 + i\omega\tau)$, or in standard form

$$G^* = G(\omega^2\tau^2 + i\omega\tau)/(1 + \omega^2\tau^2)$$

From Fig. 9-11, $\tan\delta = G_2/G_1 = G\omega\tau/G\omega^2\tau^2 = 1/\omega\tau$.

9.17. Show that the result of Problem 9.16 may also be obtained by simply replacing the operator ∂_t by $i\omega$ in equation (*9.5*) and defining $\sigma/\epsilon = G^*$.

After the suggested substitution (*9.5*) becomes $(i\omega/G + 1/\eta)\sigma = i\omega\epsilon$ from which

$$\sigma/\epsilon = Gi\omega/(i\omega + 1/\tau) = Gi\omega\tau/(1 + i\omega\tau)$$

as before.

9.18. Use equation (*9.10*) for the generalized Maxwell model to illustrate the rule that "for models in parallel, the complex moduli add".

From Problem 9.17 the complex modulus for the Maxwell model may be written $G^* = \sigma/\epsilon = Gi\omega\tau/(1 + i\omega\tau)$. Thus writing (*9.10*) as

$$\sigma = G_1\{\partial_t\}\epsilon/\{\partial_t + 1/\tau_1\} + G_2\{\partial_t\}\epsilon/\{\partial_t + 1/\tau_2\} + \cdots + G_N\{\partial_t\}\epsilon/\{\partial_t + 1/\tau_N\}$$

the generalized Maxwell complex modulus becomes

$$G^* = G_1 i\omega\tau_1/(1 + i\omega\tau_1) + G_2 i\omega\tau_2/(1 + i\omega\tau_2) + \cdots + G_N i\omega\tau_N/(1 + i\omega\tau_N) = G_1^* + G_2^* + \cdots + G_N^*$$

9.19. Verify the relationship $J_1 = 1/G_1(1 + \tan^2\delta)$ between the storage modulus and compliance.

From (*9.43*) and (*9.44*), $J^* = 1/G^*$ and so $J_1 - iJ_2 = 1/(G_1 + iG_2) = (G_1 - iG_2)/(G_1^2 + G_2^2)$. Thus

$$J_1 = G_1/(G_1^2 + G_2^2) = 1/G_1(1 + (G_2/G_1)^2) = 1/G_1(1 + \tan^2\delta)$$

9.20. Show that the energy dissipated per cycle is related directly to the loss compliance J_2 by evaluating the integral $\int \sigma\, d\epsilon$ over one cycle.

For the stress and strain vectors of Fig. 9-10, the integral $\int \sigma\, d\epsilon$ evaluated over one cycle is

$$\int_0^{2\pi/\omega} \sigma \frac{d\epsilon}{dt}\, dt = \int_0^{2\pi/\omega} (\sigma_0 \sin\omega t)\epsilon_0\omega \cos(\omega t - \delta)\, dt$$
$$= \sigma_0\epsilon_0\omega \int_0^{2\pi/\omega} \sin\omega t\,(\cos\omega t \cos\delta + \sin\omega t \sin\delta)\, dt$$
$$= \sigma_0^2\omega\left[J_1 \int_0^{2\pi/\omega} \frac{\sin 2\omega t}{2}\, dt + J_2 \int_0^{2\pi/\omega} (\sin^2\omega t)\, dt\right] = \sigma_0^2\pi J_2$$

THREE DIMENSIONAL THEORY. VISCOELASTIC STRESS ANALYSIS (Sec. 9.7-9.8)

9.21. Combine (*9.48a*) and (*9.48b*) to obtain the viscoelastic constitutive relation $\sigma_{ij} = \delta_{ij}\{R\}\epsilon_{kk} + \{S\}\epsilon_{ij}$ and determine the form of the operators $\{R\}$ and $\{S\}$.

Writing (*9.48a*) as $\{P\}(\sigma_{ij} - \delta_{ij}\sigma_{kk}/3) = 2\{Q\}(\epsilon_{ij} - \delta_{ij}\epsilon_{kk}/3)$ and replacing σ_{kk} here by the right hand side of (*9.48b*), the result after some simple manipulations is

$$\sigma_{ij} = \delta_{ij}\{(3KP - 2Q)/3P\}\epsilon_{kk} + \{2Q/P\}\epsilon_{ij}$$

9.22. A bar made of Kelvin material is pulled in tension so that $\sigma_{11} = \sigma_0[U(t)]$, $\sigma_{22} = \sigma_{33} = \sigma_{12} = \sigma_{23} = \sigma_{31} = 0$ where σ_0 is constant. Determine the strain ϵ_{11} for this loading.

From (*9.48b*), $3\epsilon_{ii} = \sigma_0[U(t)]/K$ for this case; and from (*9.48a*) with $i = j = 1$, $\{P\}(\sigma_{11} - \sigma_{11}/3) = \{2Q\}(\epsilon_{11} - \epsilon_{ii}/3)$. But from (*9.6*), $\{P\} = 1$ and $\{Q\} = \{G + \eta\partial_t\}$ for a Kelvin material; so that now

$$2\sigma_0[U(t)]/3 = 2\{G + \eta\partial_t\}(\epsilon_{11} - \sigma_0[U(t)]/9K)$$

or

$$\dot{\epsilon}_{11} + \epsilon_{11}/\tau = \sigma_0[U(t)](3K + G)/9\eta K + \sigma_0[\delta(t)]/9K$$

Solving this differential equation yields

$$\epsilon_{11} = \sigma_0(3K + G)(1 - e^{-t/\tau})[U(t)]/9KG + \sigma_0 e^{-t/\tau}[U(t)]/9K$$

As $t \to \infty$, $\epsilon_{11} \to (3K + G)\sigma_0/9KG = \sigma_0/E$.

9.23. A block of Kelvin material is held in a container with rigid walls so that $\epsilon_{22} = \epsilon_{33} = 0$ when the stress $\sigma_{11} = -\sigma_0[U(t)]$ is applied. Determine ϵ_{11} and the retaining stress components σ_{22} and σ_{33} for this situation.

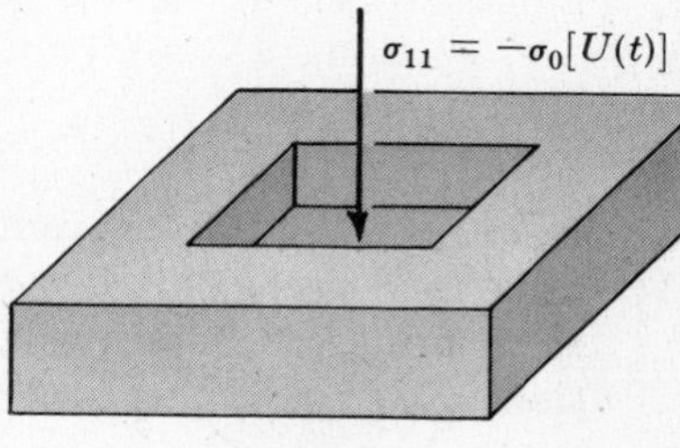

Fig. 9-22

Here $\epsilon_{ii} = \epsilon_{11}$ and $\sigma_{22} = \sigma_{33}$ so that (*9.48b*) becomes $\sigma_{11} + 2\sigma_{22} = 3K\epsilon_{11}$ and (*9.48a*) gives $2(\sigma_{11} - \sigma_{22})/3 = 2G\{1 + \tau\partial_t\}(2\epsilon_{11}/3)$ for a Kelvin body. Combining these relations yields the differential equation

$$\dot{\epsilon}_{11} + (4G + 3K)\epsilon_{11}/4G\tau = -3\sigma_0[U(t)]/4G\tau$$

which upon integration gives

$$\epsilon_{11} = -3\sigma_0[U(t)](1 - e^{-(4G+3K)t/4G\tau})/(4G + 3K)$$

Inserting this result into (*9.48a*) for $i = j = 2$ gives

$$\sigma_{22} = (+\sigma_0/2 - 9K\sigma_0(1 - e^{-(4G+3K)t/4G\tau})/(8G + 6K))[U(t)]$$

9.24. The radial stress component in an elastic half-space under a concentrated load at the origin may be expressed as

$$\sigma_{(rr)} = (P/2\pi)[(1 - 2\nu)\alpha(r, z) - \beta(r, z)]$$

where α and β are known functions. Determine the radial stress for a Kelvin viscoelastic half-space by means of the correspondence principle when $P = P_0[U(t)]$.

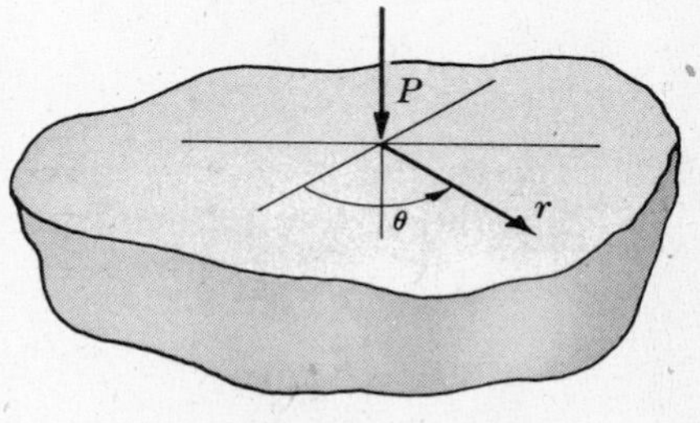

Fig. 9-23

The viscoelastic operator for the term $(1-2\nu)$ is $\{3Q\}/\{3KP+Q\}$ so that for a Kelvin body the transformed viscoelastic solution becomes

$$\bar{\sigma}_{(rr)} = \frac{3P_0}{2\pi s}\left[\frac{G+\eta s}{3K+G+\eta s}\,\alpha(r,z) - \beta(r,z)\right]$$

which may be inverted with help of partial fractions and transform tables to give the viscoelastic stress

$$\sigma_{(rr)} = \frac{3P_0}{2\pi}\left[\left(\frac{G}{3K+G} + \frac{3K}{3K+G}\,e^{-(3K+G)t/\eta}\right)\alpha(r,z) + \beta(r,z)\right]$$

9.25. **The correspondence principle may be used to obtain displacements as well as stresses. The z displacement of the surface of the elastic half space in Problem 9.24 is given by $w_{(z=0)} = P(1-\nu^2)/E\pi r$. Determine the viscoelastic displacement of the surface for the viscoelastic material of that problem.**

The viscoelastic operator corresponding to $(1-\nu^2)/E$ is $\{3K+4Q\}/4Q(3K+Q)$ which for the Kelvin body causes the transformed displacement to be

$$\bar{w}_{(z=0)} = P_0(3K+4(G+\eta s))/4\pi r s(3K+G+\eta s)(G+\eta s)$$

After considerable manipulation and inverting, the result is

$$w_{(z=0)} = \frac{P_0(3K+4G)}{4\pi r^2(3K+G)}\left[\frac{1}{G} - \frac{3e^{-(3K+G)t/\eta}}{3K+4G} - \frac{3K+G}{G(3K+4G)}\,e^{-t/\tau}\right]$$

Note that when $t=0$, $w_{(z=0)}=0$ and when $t\to\infty$, $w_{(z=0)} \to P_0(1-\nu^2)/E\pi r$, the elastic deflection.

9.26. **A simply supported uniformly loaded beam is assumed to be made of a Maxwell material. Determine the bending stress σ_{11} and the deflection $w(x_1, t)$ if the load is $p = p_0[U(t)]$.**

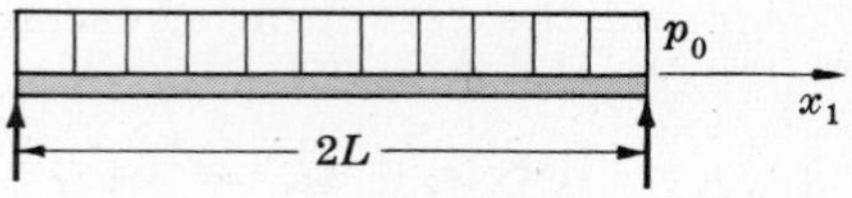

Fig. 9-24

The bending stress for a simply supported elastic beam does not depend upon material properties, so the elastic and viscoelastic bending stress here are the same. The elastic deflection of the beam is $w(x_1) = p_0\alpha(x_1)/24EI$ where $\alpha(x_1)$ is a known function. For a Maxwell body, $\{P\} = \{\partial_t + 1/\tau\}$ and $\{Q\} = \{G\partial_t\}$, so that the transformed deflection is

$$\bar{w} = \frac{p_0\alpha(x_1)}{24I}\left(\frac{3K/\tau + (3K+G)s}{9KGs^2}\right)$$

which when inverted gives

$$w(x_1,t) = \frac{p_0\alpha(x_1)}{24I}\left(\frac{t}{3\eta} + \frac{3K+G}{9KG}\right)$$

When $t=0$, $w(x_1,0) = p_0\alpha(x_1)/24EI$, the elastic deflection.

9.27. **Show that as $t\to\infty$ the stress σ_{22} in Problem 9.23 approaches σ_0 (material behaves as a fluid) if the material is considered incompressible ($\nu = 1/2$).**

From Problem 9.23,

$$\sigma_{22}|_{t\to\infty} = -\sigma_0(9K-(4G+3K))/2(4G+3K) = -\sigma_0(3K-2G)/(3K+4G)$$

which may be written in terms of ν as $\sigma_{22}|_{t\to\infty} = -\nu\sigma_0/(1-\nu)$. Thus for $\nu = 1/2$, $\sigma_{22}|_{t\to\infty} = -\sigma_0$.

MISCELLANEOUS PROBLEMS

9.28. Determine the constitutive relation for the Kelvin-Maxwell type model shown in Fig. 9-25 and deduce from the result the Kelvin and Maxwell stress-strain laws.

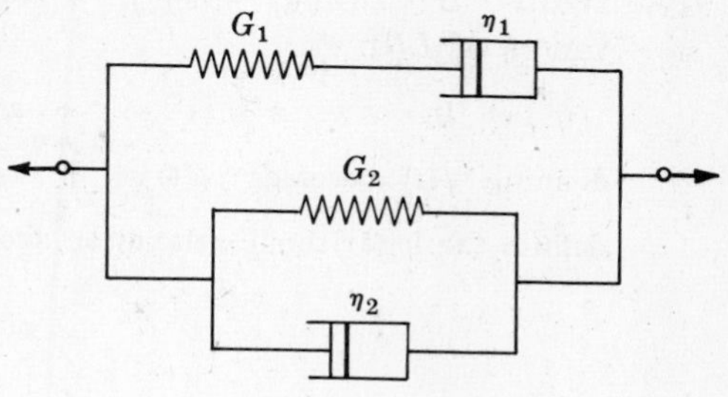

Fig. 9-25

Here

$$\sigma = \sigma_M + \sigma_K = \dot{\epsilon}/\{\partial_t/G_1 + 1/\eta_1\} + \{G_2 + \eta_2\partial_t\}\epsilon$$

which upon application of the time operators becomes

$$\dot{\sigma} + \sigma/\tau_1 = \eta_2\ddot{\epsilon} + (G_1 + G_2 + \eta_2/\tau_1)\dot{\epsilon} + (G_2/\tau_1)\epsilon$$

In this equation if $\eta_2 = 0$ (spring in parallel with Maxwell), $\dot{\sigma} + \sigma/\tau_1 = (G_1 + G_2)\dot{\epsilon} + (G_2/\tau_1)\epsilon$. Further, if $G_2 = 0$, the Maxwell law $\dot{\sigma} + \sigma/\tau_1 = G_1\dot{\epsilon}$ results. Likewise, if G_2 is taken zero first (dashpot in parallel with Maxwell), $\dot{\sigma} + \sigma/\tau_1 = \eta_2\ddot{\epsilon} + (G_1 + \eta_2/\tau_1)\dot{\epsilon}$; and when $\eta_2 = 0$, this also reduces to the Maxwell law.

If the four-parameter constitutive relation is rewritten

$$\eta_1\dot{\sigma} + G_1\sigma = \eta_1\eta_2\ddot{\epsilon} + (G_1\eta_1 + G_2\eta_1 + G_1\eta_2)\dot{\epsilon} + G_1G_2\epsilon$$

and η_1 set equal to zero, the result is the Kelvin law $\sigma = \eta_2\dot{\epsilon} + G_2\epsilon$. Likewise, if $G_1 = 0$ the reduced equation is $\dot{\sigma} = \eta_2\ddot{\epsilon} + G_2\dot{\epsilon}$, again representing the Kelvin model.

9.29. Use the superposition principle to obtain the creep recovery response for the standard linear solid of Fig. 9-3(*a*) and compare the result with that obtained in Problem 9.8.

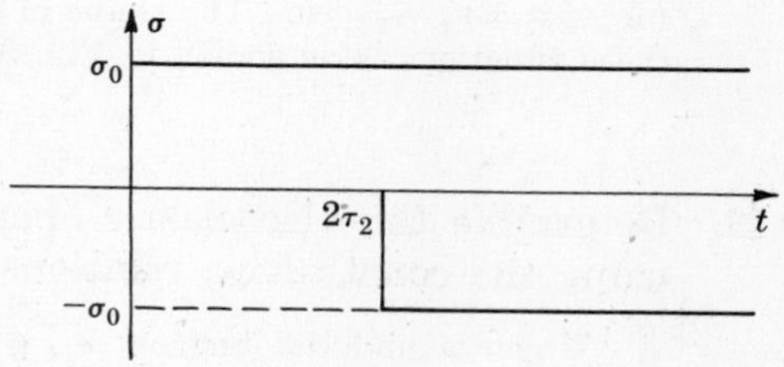

Fig. 9-26

With the stress loading expressed by

$$\sigma = \sigma_0[U(t)] - \sigma_0[U(t - 2\tau_2)]$$

(see Fig. 9-26), the strain may be written at once from the result of Problem 9.7 as

$$\epsilon = \sigma_0(1/G_1 + (1 - e^{-t/\tau_2})/G_2)[U(t)] - \sigma_0(1/G_1 + (1 - e^{-(t-2\tau_2)\tau_2})/G_2)[U(t - 2\tau_2)]$$

At times $t > 2\tau_2$ both step functions equal unity and

$$\epsilon = \sigma_0(-e^{-t/\tau_2} + e^{-(t-2\tau_2)/\tau_2})/G_2 = \sigma_0(e^2 - 1)e^{-t/\tau_2}/G_2$$

which agrees with the result in Problem 9.8.

9.30. Determine the stress in the model of Problem 9.9 when subjected to the strain history shown in Fig. 9-27. Show that eventually the "free" spring in the model carries the entire stress.

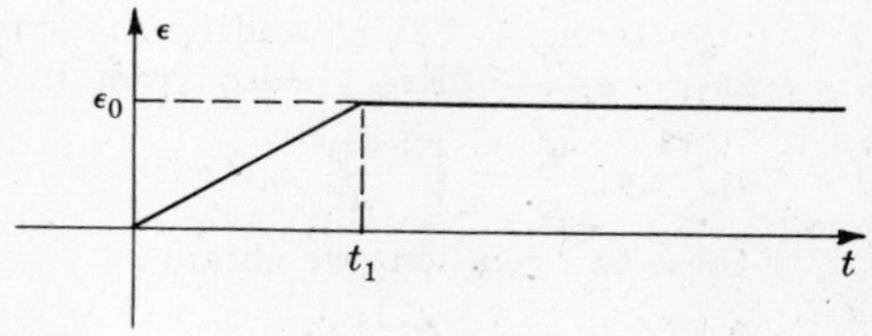

Fig. 9-27

From Problem 9.9 and the superposition principle the stress is

$$\sigma = \epsilon_0(\eta(2 - e^{-t/\tau}) + Gt)[U(t)]/t_1 - \epsilon_0(\eta(2 - e^{-(t-t_1)/\tau}) + G(t - t_1))[U(t - t_1)]/t_1$$

For times $t > t_1$ the stress is $\sigma = \epsilon_0\eta(e^{t_1/\tau} - 1)e^{-t/\tau}/t_1 + G\epsilon_0$, and as $t \to \infty$ this reduces to $\sigma = G\epsilon_0$.

9.31. The "logarithmic retardation spectrum" L is defined in terms of the retardation spectrum J by $L(\ln\tau) = \tau J(\tau)$. From this definition determine the creep function $\psi(t)$ in terms of $L(\ln\tau)$.

Let $\ln\tau = \lambda$ so that $e^\lambda = \tau$ and thus $d\tau/d\lambda = e^\lambda = \tau$, or $d\tau = \tau d(\ln\tau)$. From this, (*9.28*) defining $\psi(t)$ becomes $\psi(t) = \int_0^\infty L(\ln\tau)(1-e^{-t/\tau})\,d(\ln\tau)$. In the same way, if $H(\ln\tau) = \tau G(\tau)$ defines the logarithmic relaxation spectrum, $\phi(t)$ in (*9.31*) may be written

$$\phi(t) = \int_0^\infty H(\ln\tau)e^{-t/\tau}d(\ln\tau)$$

9.32. For the Maxwell model of Fig. 9-2(*a*) determine the storage and loss moduli, G_1 and G_2, as functions of $\ln\omega\tau$ and sketch the shape of these functions.

From Problem 9.16,

$$G^* = G(\omega^2\tau^2 + i\omega\tau)/(1+\omega^2\tau^2)$$

for a Maxwell material. Thus

$$G_1 = G\omega^2\tau^2/(1+\omega^2\tau^2) = Ge^{2\lambda}/(1+e^{2\lambda})$$

where $\lambda = \ln\omega\tau$. For $\lambda = 0$, $G_1 = G/2$; for $\lambda = \infty$, $G_1 = G$; and for $\lambda = -\infty$, $G_1 = 0$. Likewise $G_2 = Ge^\lambda/(1+e^{2\lambda})$ and for $\lambda = 0$, $G_2 = G/2$; for $\lambda = \pm\infty$, $G_2 = 0$. The shape of the curves for these functions is as shown in Fig. 9-28.

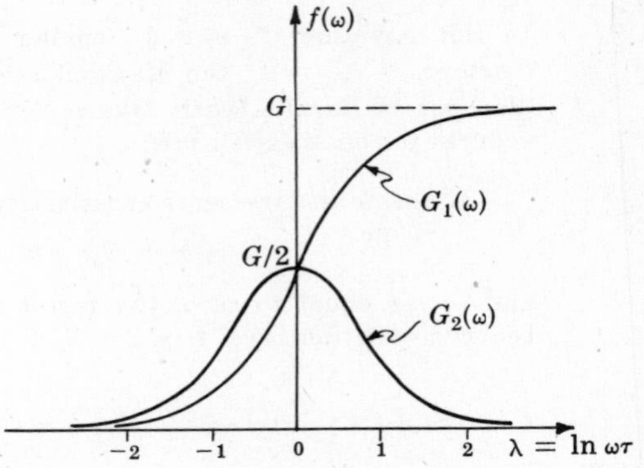

Fig. 9-28

9.33. Determine the viscoelastic operator form of the elastic constant ν (Poisson's ratio) using the constitutive relations (*9.48*).

Under a uniaxial tension $\sigma_{11} = \sigma_0$, (*9.48b*) gives $\epsilon_{ii}/3 = \sigma_0/9K$ so that (*9.48a*) for $i = j = 1$ yields $\epsilon_{11} = \{3KP + Q\}\sigma_0/\{9KQ\}$. In the same way (*9.48a*) for $i = j = 2$ yields $\epsilon_{22} = \{2Q - 3PK\}\sigma_0/\{18KQ\}$. Thus in operator form, $\nu = -\epsilon_{22}/\epsilon_{11} = \{3PK - 2Q\}/\{6KP + 2Q\}$.

9.34. A cylindrical viscoelastic body is inserted into a rigid snug-fitting container (Fig. 9-29) so that $\epsilon_{(rr)} = 0$ (no radial strain). The body is elastic in dilatation and has the creep function $\psi_s = A + Bt + Ce^{\lambda t}$ where A, B, C, λ are constants. If $\epsilon_{33} = \epsilon_0[U(t)]$, determine $\sigma_{33}(t)$.

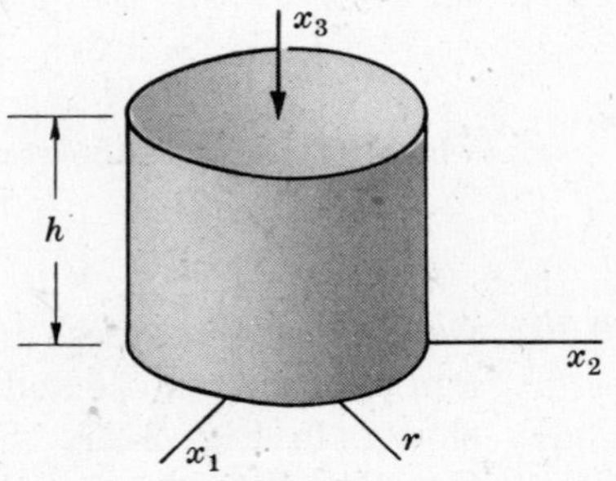

Fig. 9-29

Here $\sigma_{ii} = 3K\epsilon_{ii}$ and by the symmetry of the problem, $2\sigma_{11} + \sigma_{33} = 3K\epsilon_{33}$. Also from (*9.50a*) with $i = j = 1$, $\sigma_{11} - \sigma_{33} = -\int_0^t \frac{d\epsilon_{33}}{dt'}\phi_s(t-t')\,dt'$. Solving for σ_{33} from these two relations we obtain

$$\sigma_{33} = K\epsilon_{33} + \frac{2}{3}\int_0^t \frac{d\epsilon_{33}}{dt'}\phi_s(t-t')\,dt'$$

The relaxation function ϕ_s may be found with the help of (*9.40*). The result is

$$\phi_s = [(r_1-\lambda)e^{r_1 t} - (r_2-\lambda)e^{r_2 t}]/(r_1 - r_2)$$

where $r_{1,2} = [A\lambda - B \pm \sqrt{(A\lambda+B)^2 + 4BC\lambda}\,]/2(A+C)$. Thus finally

$$\sigma_{33} = K\dot{\epsilon}_0 t + \frac{2}{3}\int_0^t \dot{\epsilon}_0 \frac{[(r_1-\lambda)e^{r_1(t-t')} - (r_2-\lambda)e^{r_2(t-t')}]}{(r_1-r_2)} [U(t')]\,dt'$$

which upon integration gives

$$\sigma_{33} = (K\dot{\epsilon}_0 t + 2\dot{\epsilon}_0[-(r_1-\lambda)(1-e^{r_1 t})/3r_1 + (r_2-\lambda)(1-e^{r_2 t})/3r_2]/(r_1-r_2))[U(t)]$$

9.35. The "creep buckling" of a viscoelastic column may be analyzed within the linear theory through the correspondence principle. Determine the deflection $w(x_1, t)$ of a Kelvin pinned-end column by this method.

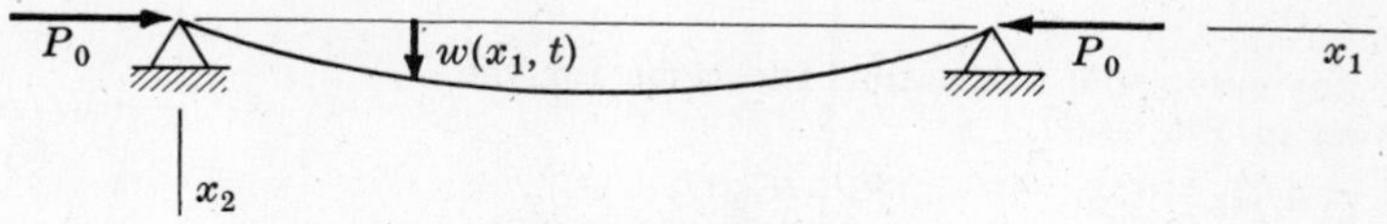

Fig. 9-30

The elastic column formula is $d^2w/dx_1^2 + P_0 w/EI = 0$, and for a Kelvin material E may be replaced by the operator $\{E + \eta\partial_t\}$ so that for the viscoelastic column $\{E+\eta\partial_t\}(d^2w/dx_1^2) + P_0 w/I = 0$. Assuming the deflection in a product form $w(x_1, t) = W(x_1)\theta(t)$, the operator leads to the differential equation

$$(E\theta + \eta\dot{\theta})(d^2W/dx_1^2) + P_0 W\theta/I = 0 \quad \text{from which} \quad \dot{\theta} + [1 + P_0 W/EI(d^2W/dx_1^2)]\theta/\tau = 0$$

where $\tau = \eta/E$. But the elastic buckling load is $P_B = -EI(d^2W/dx_1^2)/W$ and so $\dot{\theta} + (1 - P_0/P_B)\theta/\tau = 0$ which integrates easily to yield $\theta = e^{(P_0/P_B - 1)t/\tau}$. Finally then the "creep buckling" deflection is $w = We^{(P_0/P_B-1)t/\tau}$.

9.36. Formulate the steady-state vibration problem for a viscoelastic beam assuming the constitutive relations are those given by *(9.48)*.

The free vibrations of an elastic beam are governed by the equation $EI(\partial^4 w/\partial x_1^4) + \rho A(\partial^2 w/\partial t^2) = 0$. From *(9.48)* the viscoelastic operator for E is $\{9KQ/(3KP+Q)\}$, and if the deflection $w(x_1,t) = W(x_1)\theta(t)$ the resulting viscoelastic differential equation may be split into the space equation $d^4W/dx_1^4 - k^4W = 0$ and the time equation $\{3KP+Q\}(d^2\theta/dt^2) + (k^4 I/\rho A)\{9KQ\}(\theta) = 0$. The solution W_i of the space equation represents the ith mode shape, and from the time equation for $k = k_i$ the solution $\theta_i = \sum_{j=1}^{N} A_{ij}e^{\lambda_{ij}t}$ where N depends upon the degree of the operator. The total solution therefore is $w(x_1, t) = \sum_{i=1}^{\infty}\sum_{j=1}^{N} W_i(x_1)A_{ij}e^{\lambda_{ij}t}$ in which the λ_{ij} are complex.

Supplementary Problems

9.37. Determine the constitutive equation for the four parameter model shown in Fig. 9-31.
Ans. $\ddot{\sigma} + (G_1/\eta_2 + G_2/\eta_2 + G_1/\eta_1)\dot{\sigma} + (G_1G_2/\eta_1\eta_2)\sigma = G_1\ddot{\epsilon} + (G_1G_2/\eta_2)\dot{\epsilon}$

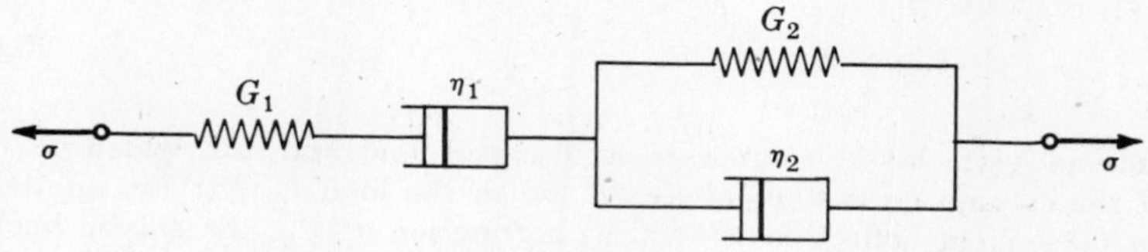

Fig. 9-31

9.38. Determine the creep response of the standard linear solid by direct integration of $\dot{\epsilon} + \epsilon/\tau_2 = \sigma_0[U(t)](G_1 + G_2)/G_1\eta_2 + \sigma_0[\delta(t)]/G_1$. (See Problem 9.7.)

9.39. Deduce the Kelvin and Maxwell stress-strain laws from the results established in Problem 9.5 for the four parameter model of that problem. (*Hint.* Let $G_3 = 0$, etc.)

9.40. Use equation (9.40) to obtain $\psi(t)$ if $\phi(t) = a(b/t)^m$ with $m < 1$. (*Hint.* Take $m = 1 - k$; then $\phi(t) = ab^m t^{k-1}$.) *Ans.* $\psi(t) = \dfrac{\sin \pi m}{am\pi}\left(\dfrac{t}{b}\right)^m$

9.41. Determine the creep and relaxation functions for the model shown in Fig. 9-32.

Ans. $\psi(t) = 1/G_2 - G_1 e^{-G_2 t/(G_1+G_2)\tau_1}/G_2(G_1 + G_2)$

$\phi(t) = G_2 + G_1 e^{-t/\tau_1}$

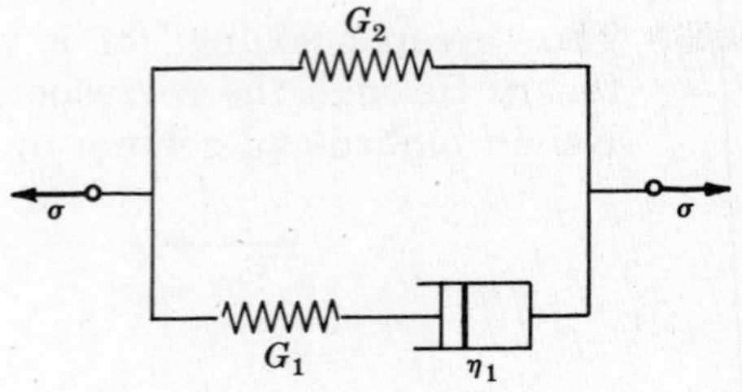

Fig. 9-32

9.42. Determine G^* for the model shown in Fig. 9-33.

$$Ans.\quad G^* = \frac{G_1(1 + \tau_2^2\omega^2) + G_2\omega^2\tau_2^2}{1 + \omega^2\tau_2^2} + i\,\frac{\omega(G_2\tau_2 + \eta_3(1 + \tau_2^2\omega^2))}{1 + \omega^2\tau_2^2}$$

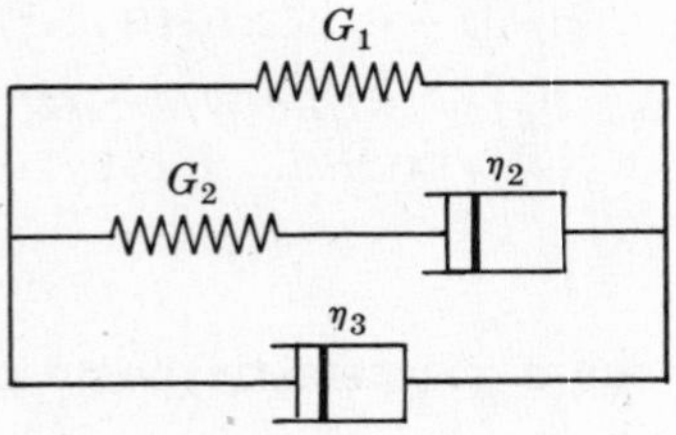

Fig. 9-33

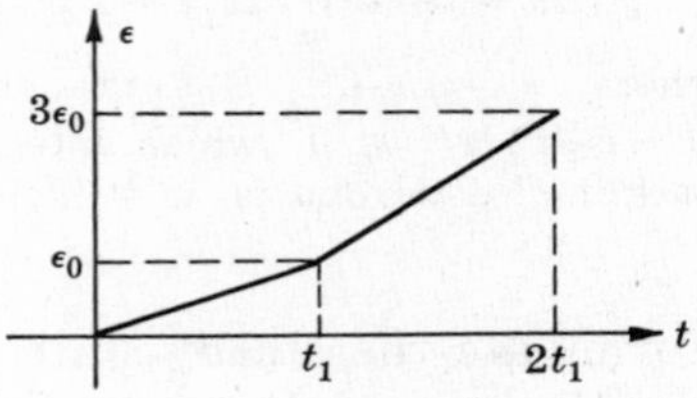

Fig. 9-34

9.43. In the model of Problem 9.42 let $G_1 = G_2 = G$ and $\eta_2 = \eta_3 = \eta$ and determine the stress history of the resulting model when it is subjected to the strain sequence shown in Fig. 9-34.

$$Ans.\quad \sigma = \frac{\epsilon_0}{t_1}(G(2t - t_1) + \eta(4 - (1 + e^{t_1/\tau})e^{-t/\tau})) \quad \text{for } t_1 < t < 2t_1$$

9.44. A viscoelastic block having the constitutive equation $\dot{\sigma} + \alpha\sigma = \beta\dot{\epsilon} + \gamma\epsilon$ where α, β, γ are constants is loaded under conditions such that $\sigma_{11} = -\sigma_0[U(t)]$, $\sigma_{22} = 0$, $\epsilon_{33} = 0$ (see Fig. 9-35). Assuming $\sigma_{ii} = 3K\epsilon_{ii}$, determine $\sigma_{33}(t)$, $\sigma_{33}(0)$ and $\sigma_{33}(\infty)$.

$$Ans.\quad \sigma_{33} = -\sigma_0\left[\frac{3\alpha K - 2\gamma}{2(3\alpha K + \gamma)\lambda} + \left(\frac{3K - 2\beta}{2(3K + \beta)} - \frac{3\alpha K - 2\gamma}{2(3\alpha K + \gamma)\lambda}\right)e^{-\lambda t}\right] \quad \text{where } \lambda = (3\alpha K + \gamma)/(3K + \beta).$$

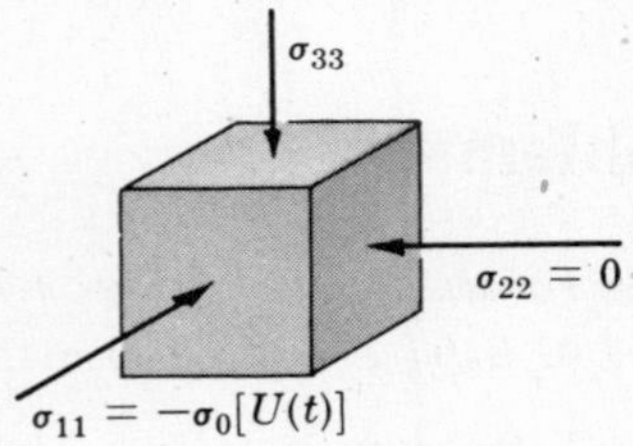

Fig. 9-35

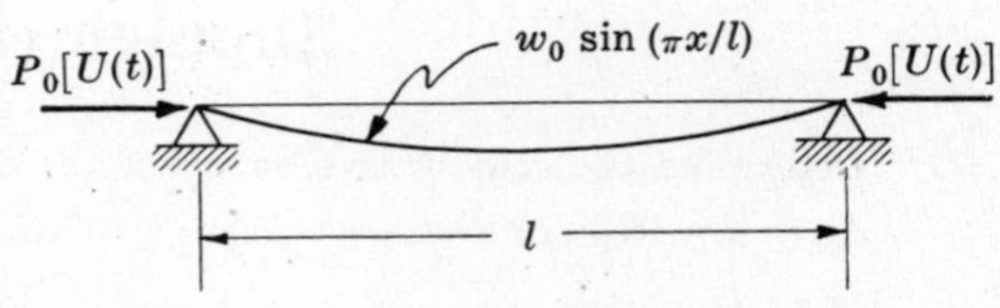

Fig. 9-36

9.45. A pinned-end viscoelastic column is a Maxwell material for which $\dot{\sigma} + \sigma/\tau = E\dot{\epsilon}$. The initial shape of the column is $w = w_0 \sin(\pi x/l)$ when the load $P_0[U(t)]$ is applied (see Fig. 9-36). Determine the subsequent deflection $w(x_1, t)$ as a function of P_B, the elastic buckling load.

Ans. $w(x_1, t) = w_0 \sin(\pi x_1/l)e^{-t/(1-P_B/P_0)\tau}$

INDEX

Absolute dynamic,
  compliance, 202
  modulus, 202
Acceleration, 111
Addition and subtraction,
  of matrices, 17
  of (Cartesian) tensors, 15
  of vectors, 2
Adiabatic process, 140
Airy stress function, 147
Almansi strain tensor, 82
Angle change, 86
Angular momentum, 128
Anisotropy, 44, 141
Antisymmetric,
  dyadic, 5
  matrix, 19
  tensor, 19
Axis of elastic symmetry, 142

Barotropic change, 161
Base vectors, 6
Basis, 6, 9
  orthonormal, 6
Bauschinger effect, 176
Beltrami-Michell equations, 144
Bernoulli equation, 164
Biharmonic equation, 148
Body forces, 45
Boundary conditions, 144, 145
Bulk,
  modulus, 143
  viscosity, 161

Caloric equation of state, 132
Cartesian,
  coordinates, 6
  tensors, 1, 12, 13
Cauchy,
  deformation tensor, 81
  strain ellipsoids, 89
  stress principle, 45
  stress quadric, 50
Cauchy-Riemann conditions, 165
Circulation, 164
  Kelvin's theorem of, 165
Clausius-Duhem inequality, 131
Column matrix, 17
Compatibility equations, 92, 114
Complex,
  modulus, 202
  potential, 166
Compliance, 201
Component, 1, 7
Compressible fluid, 165
Configuration, 77
Conformable matrices, 17
Conjugate dyadic, 4
Conservation of,
  energy, 128
  mass, 126
Constitutive equations, 132
Continuity equation, 126
Continuum concept, 44
Contraction, 16
Contravariant tensor, 11
Convective,
  derivative, 110
  rate of change, 111
Conventional stress and strain, 175
Coordinate transformation, 11
Correspondence principle, 205
Couple-stress vector, 45
Coupled heat equation, 150
Covariant tensor, 12
Creep,
  function, 200
  test, 199
Cross product, 3, 5, 16
Cubical dilatation, 90
Curvilinear coordinates, 7
Cylindrical coordinates, 8

Dashpot, 196
Decomposition,
  polar, 87
  velocity gradient, 112
Deformation, 77
  gradients, 80
  inelastic, 175
  plane, 91
  plastic, 175
  tensors, 81
  total (deformation) theory, 183
Del operator, 22
Density, 44, 126
  entropy, 130
  strain energy, 141
Derivative,
  material, 110, 114
  of tensors, 22
  of vectors, 22
Deviatoric,
  strain tensor, 91
  stress tensor, 57
Diagonal matrix, 17
Dilatation, 90
Direction cosines, 7
Displacement, 78, 83
  gradient, 80
  relative, 83
Dissipation function, 132
Dissipative stress tensor, 131
Distortion energy theory, 178

Divergence theorem (of Gauss), 23
Dot product of,
  dyads, 5
  vectors, 3
Duhamel-Neumann relations, 149
Dyadics, 1, 4
  antisymmetric, 4
  conjugate of, 4
  symmetric, 4
Dyads, 4
  nonion form, 7
Dynamic moduli, 203

Effective,
  plastic strain increment, 182
  stress, 182
Elastic,
  constants, 141
  limit, 175
  symmetry, 142
Elasticity, 140
Elastodynamics, 143
Elastoplastic problems, 183
Elastostatics, 143
Energy,
  kinetic, 129
  strain, 141
  thermal, 129
Engineering stress and strain, 175
Entropy, 130
  specific, 130
$\epsilon-\delta$ identity, 39
Equations of
  equilibrium, 48, 128
  motion, 128
  state, 130
Equivalent,
  plastic strain increment, 182
  stress, 182
Euclidean space, 11
Eulerian,
  coordinates, 78
  description, 79
  finite strain tensor, 82
  linear strain tensor, 83

Field equations,
  elastic, 143, 147
  viscoelastic, 205
Finite strain tensor, 81, 82
First law of thermodynamics, 129
Flow, 77, 110
  creeping, 169
  irrotational, 164
  plastic, 175
  potential, 175
  rule, 181
  steady, 163
Fluid,
  inviscid, 160
  perfect, 160, 164
  pressure, 160

Forces,
  body, 45
  surface, 45
Fourier heat law, 149
Fundamental metric tensor, 13

Gas,
  dynamical equation, 165
  law, 160
Gauss's theorem, 23
Generalized,
  Hooke's law, 140
  Kelvin model, 198
  Maxwell model, 198
  plane strain, 147
  plane stress, 147
Gibb's notation, 2
Gradient,
  deformation, 80
  displacement, 80
Green's
  deformation tensor, 81
  finite strain tensor, 82

Hamilton-Cayley equation, 21
Hardening,
  isotropic, 180
  kinematic, 181
  strain, 176, 183
  work, 176, 183
Harmonic functions, 166
Heat,
  conduction law, 149
  flux, 129
  radiant, 129
Hencky equations, 183
Hereditary integrals, 201
Homogeneous,
  deformation, 95
  material, 44
Hookean solid, 196
Hydrostatic pressure, 160
Hydrostatics, 163
Hyperelastic material, 149
Hypoelastic material, 149
Hysteresis, 176

Ideal,
  gas, 160
  materials, 132
Idealized stress-strain curves, 177
Idemfactor, 5
Identity matrix, 18
Incompressible flow, 126
Incremental theories, 181
Indeterminate vector product, 4
Indices, 9
Indicial notation, 8
Inelastic deformation, 175
Inertia forces, 48
Infinitesimal strain, 83
Initial conditions, 145
Inner product, 16

Integral theorems, 23
Internal energy, 129
Invariants, 21
  of rate of deformation, 114
  of strain, 89, 90
  of stress, 51
Irreversible process, 130
Irrotational flow, 113, 164
Isothermal process, 140
Isotropy, 44, 142

Jacobian, 11, 79
Johnson's apparent elastic-limit, 176

Kelvin (material) model, 197
Kinematic,
  hardening, 181
  viscosity, 163
Kinetic energy, 129
Kronecker delta, 13

Lagrangian,
  description, 79
  finite strain tensor, 82
  infinitesimal strain tensor, 83
Lamé constants, 143
Laplace,
  equation, 165
  transform, 202, 205
Left stretch tensor, 87
Levy-Mises equations, 181
Line integrals, 23
Linear,
  momentum, 127
  rotation tensor, 83
  thermoelasticity, 149
  vector operator, 8
  viscoelasticity, 196
Local rate of change, 111
Logarithmic strain, 175

Mass, 126
Material,
  coordinates, 78
  derivative, 110, 114-117
  description of motion, 79
Matrices, 17-19
Maximum,
  normal stress, 52
  shearing stress, 52, 53
Maxwell (material) model, 197
Metric tensor, 13
Mises yield condition, 178
Modulus,
  bulk, 143
  complex, 203
  loss, 203
  shear, 143
  storage, 202
Mohr's circles,
  for strain, 91
  for stress, 54-56
Moment of momentum, 128
Momentum principle, 127
Motion, 110
  steady, 112
Multiplication of,
  matrices, 18
  tensors, 15
  vectors, 3

Navier-Cauchy equations, 144
Navier-Stokes equations, 162
Navier-Stokes-Duhem equations, 162
Newtonian viscous fluid, 161, 196
Normal stress components, 47

Octahedral,
  plane, 59
  shear stress, 192
Orthogonal,
  tensor, 87
  transformation, 13
Othogonality conditions, 13, 14
Orthotropic, 142
Outer product, 15

Parallelogram law of addition, 2
Particle, 77
Path lines, 112
Perfectly plastic, 176
Permutation symbol, 16
π-plane, 179
Plane,
  deformation, 91
  elasticity, 145
  strain, 91, 145
  stress, 56-57, 145
Plastic,
  deformation, 175
  flow, 175
  potential theory, 181
  range, 176
  strain increment, 182
Point, 77
Poisson's ratio, 143
Polar,
  decomposition, 87
  equilibrium equations, 148
Position vector, 77
Post-yield behavior, 180
Potential,
  flow, 165
  plastic, 182
Prandtl-Reuss equations, 181
Pressure,
  fluid, 160
  function, 163
Principal,
  axes, 20
  strain values, 89
  stress values, 51, 58
Proper transformation, 17
Proportional limit, 177
Pure shear, 178

Quadric of,
  strain, 88, 89
  stress, 50
Quasistatic viscoelastic problem, 205

Range convention, 8
Rate of deformation tensor, 112
Rate of rotation vector, 114
Reciprocal basis, 28
Rectangular Cartesian coordinates, 6
Relative displacement, 83
Relaxation,
  function, 201
  spectrum, 201
  test, 199
Retardation,
  spectrum, 201
  time, 199
Reversible process, 130
Reynold's transport theorem, 123
Right stretch tensor, 87
Rigid displacement, 82
Rotation tensor,
  finite, 87
  infinitesimal, 83, 84
Rotation vector, 83

St. Venant's principle, 145
Scalar, 1
  field, 22
  of a dyadic, 4
  triple product, 4
Second law of thermodynamics, 130
Shear,
  modulus, 143
  strain components, 86
  stress components, 52
Slip line theory, 184
Small deformation theory, 82
Spatial coordinates, 78
Specific,
  entropy, 130
  heat, 149
Spherical,
  coordinates, 8
  tensor, 57, 91
Spin tensor, 112
Standard linear solid, 197
State of stress, 46
Stokes' condition, 161
Stokes' theorem, 23
Stokesian fluid, 161
Strain,
  deviator, 91
  ellipsoid, 88
  energy, 141
  hardening, 176
  natural, 112, 175
  plane, 91
  rate, 112
  shearing, 86
  spherical, 91
  transformation laws, 88
Stream function, 165
Stream lines, 112
Stress,
  components, 47
  conservative, 131
  deviator, 57
  effective, 182
  ellipsoid, 52
  function, 147
  invariants, 51
  Mohr's circles for, 54-56
  normal, 47
  plane, 56
  power, 130
  principle, 51
  quadric, 50
  shear, 47, 52
  spherical, 57
  symmetry, 48
  tensor, 46
  transformation laws, 49
  vector, 45
Stretch,
  ratio, 86
  tensor, 87
Summation convention, 8, 10
Superposition theorem, 145
Surface forces, 45
Symbolic notation, 2, 10
Symmetry,
  elastic, 142
  tensor, 19

Temperature, 130
Tension test, 175
Tensor,
  Cartesian, 1, 12, 13
  components of, 1
  contravariant, 11
  covariant, 12
  deformation, 81
  derivative of, 22
  fields, 22
  general, 1, 11
  metric, 12
  multiplication, 15
  powers of, 21
  rank, 1
  stretch, 87
  transformation laws, 1
Tetrahedron of stress, 47
Thermal equation of state, 132
Thermodynamic process, 130
Thermoelasticity, 133, 149
Transformation,
  laws, 13, 88
  of tensors, 1
  orthogonal, 13, 14
Tresca yield condition, 178
Triangle rule, 2
Triple vector product, 4

Two-dimensional elastostatics,
in polar form, 148
in rectangular form, 145

Uncoupled thermoelasticity, 133, 150
Uniqueness theorem, 145, 158
Unit,
dyadic, 5
relative displacements, 83
triads, 7
vector, 3

Vector,
addition, 2
displacement, 78
dual, 16
field, 22
of a dyadic, 4
position, 12, 77
potential, 126
products, 3, 4, 16
rotation, 83
traction, 46
transformation law, 14
vorticity, 113

Velocity, 110, 111
complex, 166
potential, 164
strain, 112
Viscoelastic stress analysis, 204-206
Viscoelasticity, 196
Viscous stress tensor, 160
Voigt model, 197
Vorticity,
tensor, 112
vector, 113

Work,
hardening, 176
plastic, 182

Yield,
condition, 177
curve, 179
surface, 179
Young's modulus, 143

Zero,
matrix, 17
order tensor, 1
vector, 2

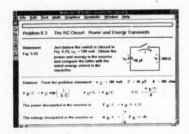